U.S. AIR FORCE YF-12A FLIGHT MANUAL

U.S. AIR FORCE

YF-12A FLIGHT MANUAL

GOVERNMENT REPRINTS PRESS
Washington, D.C.

Printed in The United States of America
Ross & Perry, Inc. Publishers
717 Second St., N.E., Suite 200
Washington, D.C. 20002
Telephone (202) 675-8300
Facsimile (202) 675-8400
info@RossPerry.com

SAN 253-8555

Government Reprints Press Edition 2001

Government Reprints Press is an Imprint of Ross & Perry, Inc.

Library of Congress Control Number: 2001093416

http://www.GPOreprints.com

ISBN 1-931641-63-3

♾ The paper used in this publication meets the requirements for permanence established by the American National Standard for Information Sciences "Permanence of Paper for Printed Library Materials" (ANSI Z39.48-1984).

UNCLASSIFIED

YF-12A UTILITY FLIGHT MANUAL (U)

YF-12A-1
COPY NO. 16
(of copies)
LAC NO. 902026
Contract F33657-70-C-0177

For official use only

WARNING

Access to this document is controlled as it pertains to a sensitive weapon system of vital importance to the defense of the United States. Distribution of this document and dissemination of information contained herein will be limited to persons determined to be included in the "Senior Crown" access list or whose "need-to-know" has been approved by the Commander, Det 51, SMAMA.

CLASSIFICATION STATEMENT

Access to this document is controlled by the Senior Crown security plan as stated in the WARNING notice above. Classified paragraphs and statements are not individually identified. Should it become necessary to identify the classification level of any information in this publication, a designated focal point at each participating location will furnish specific instructions. The following focal points shall maintain a classification guide and furnish instructions relative to program material: F12/SR-71 SPO, WPAFB, Ohio; Hq USAF (AFRDRP); Deputy IG for Insp and Safety USAF, Norton AFB, Calif; Hq AFSC (SCSZ) Wash, D.C., AFFTC, Edwards AFB, Calif; Hq SAC (DPL); 15AF (DOR), March AFB, Calif; 9th SRW, Beale AFB, Calif; Cmdr, Det 51, SMAMA, Norton AFB, Calif.

This is a GROUP 4 document. Downgraded at 3 year intervals, declassified after 12 years.

Notice - This material contains information affecting the defense of the United States within the meaning of the Espionage Laws, Title 18, U.S.C., Section 793 and 794, the transmission or revelation of which in any manner to an unauthorized person is prohibited by law.

PUBLISHED UNDER AUTHORITY OF THE SECRETARY OF THE AIR FORCE

This publication replaces YF-12A-1 dated 1 January 1968.

UNCLASSIFIED

15 NOVEMBER 1969

* Changed 15 February 1971

LIST OF EFFECTIVE PAGES

Page No.		Issue
*TITLE		02-15-71
*A		02-15-71
*B		02-15-71
*C		02-15-71
I		ORIGINAL
II		ORIGINAL
III		ORIGINAL
IV	BLANK	ORIGINAL
V		ORIGINAL
VI		ORIGINAL
SECTION I		
1-01		ORIGINAL
1-02		ORIGINAL
1-03		ORIGINAL
1-04		ORIGINAL
1-05		08-15-70
1-06		08-15-70
1-07		08-15-70
1-08		08-15-70
1-08A		11-20-70
1-08B		11-20-70
1-08C		11-20-70
1-08D	BLANK	11-20-70
1-09		ORIGINAL
1-10		ORIGINAL
1-11		ORIGINAL
1-12		ORIGINAL
1-13		ORIGINAL
1-14		01-01-71
1-14A		01-01-71
1-14B	BLANK	01-01-71
1-15		01-01-71
1-16		07-15-70
1-16A		01-01-71
1-16B		02-20-70
1-16C		02-20-70
1-16D		02-20-70
1-16E		02-20-70
1-16F	BLANK	02-20-70
1-17		ORIGINAL
1-18		ORIGINAL
1-19		ORIGINAL
1-20		ORIGINAL
1-21		ORIGINAL
1-22		ORIGINAL
1-23		ORIGINAL
1-24		ORIGINAL
1-25		ORIGINAL
1-26		ORIGINAL
1-27		ORIGINAL
1-28	BLANK	ORIGINAL
1-29		ORIGINAL
1-30		ORIGINAL
1-31		ORIGINAL
1-32		ORIGINAL
1-32A		01-01-71
1-32B		01-01-71
1-33		01-01-71
1-34		01-01-71
1-34A		01-01-71
1-34B	BLANK	01-01-71
1-35		ORIGINAL
1-36		ORIGINAL
1-37		ORIGINAL
1-38		ORIGINAL
1-39		07-15-70
1-40		02-20-70
1-40A		08-15-70
1-40B		02-20-70
1-41		01-01-71
1-42		ORIGINAL
1-43		ORIGINAL
1-44		ORIGINAL
1-45		ORIGINAL
1-46		ORIGINAL
*1-47		02-15-71
1-48		11-20-70
1-49		ORIGINAL
1-50		ORIGINAL
1-51		07-15-70
1-52		08-15-70
1-53		ORIGINAL
1-54	BLANK	ORIGINAL
1-55		ORIGINAL
1-56		ORIGINAL
1-57		ORIGINAL
1-58		ORIGINAL
1-59		ORIGINAL
1-60		ORIGINAL
1-61		ORIGINAL
1-62		ORIGINAL
1-63		ORIGINAL
1-64		ORIGINAL
1-65		ORIGINAL
1-66		ORIGINAL
1-67		ORIGINAL
1-68		ORIGINAL
1-69		ORIGINAL
1-70		ORIGINAL
1-71		ORIGINAL
1-72	BLANK	ORIGINAL
1-73		07-15-70
1-74		07-15-70
1-74A		07-15-70
1-74B	BLANK	07-15-70
1-75		ORIGINAL
1-76		ORIGINAL
1-77		ORIGINAL
1-78		ORIGINAL
1-79		ORIGINAL
1-80		ORIGINAL
1-81		ORIGINAL
1-82	BLANK	ORIGINAL
1-83		ORIGINAL
1-84		ORIGINAL
1-85		ORIGINAL
1-86		ORIGINAL
1-87		ORIGINAL
1-88		ORIGINAL
1-89		ORIGINAL
1-90		ORIGINAL
1-91		ORIGINAL
1-92		ORIGINAL
1-93		ORIGINAL
1-94		ORIGINAL
1-95		ORIGINAL
1-96		ORIGINAL
1-97		ORIGINAL
1-98		ORIGINAL
1-99		ORIGINAL
1-100		ORIGINAL
1-101		ORIGINAL
1-102		ORIGINAL
1-103		ORIGINAL
1-104		ORIGINAL
1-105		ORIGINAL
1-106		ORIGINAL
1-107		ORIGINAL
1-108		ORIGINAL
1-109		ORIGINAL
1-110		ORIGINAL
1-111		ORIGINAL
1-112		05-10-70
1-112A		05-10-70
1-112B	BLANK	05-10-70
1-113		11-20-70
1-114		ORIGINAL
1-115		ORIGINAL
1-116		ORIGINAL
1-117		ORIGINAL
1-118		ORIGINAL
1-119		ORIGINAL
1-120		ORIGINAL
1-121		ORIGINAL
1-122		ORIGINAL
1-123		ORIGINAL
1-124	BLANK	ORIGINAL
1-125		ORIGINAL
1-126	BLANK	ORIGINAL
1-127		ORIGINAL
1-128		ORIGINAL
1-129		05-10-70
1-130	BLANK	05-10-70
SECTION II		
*2-01		02-15-71
2-02		ORIGINAL
2-03		ORIGINAL
2-04		ORIGINAL
2-05		ORIGINAL
2-06		ORIGINAL
2-07		01-01-71
2-08		ORIGINAL
2-09		ORIGINAL
2-10		ORIGINAL
2-11		ORIGINAL
2-12		ORIGINAL
2-13		ORIGINAL
2-14		11-20-70
2-15		ORIGINAL
2-16		11-20-70
2-17		08-15-70
2-18		ORIGINAL
2-19		ORIGINAL
2-20		ORIGINAL
2-21		ORIGINAL
2-22		ORIGINAL
2-23		ORIGINAL
2-24		ORIGINAL
2-25		ORIGINAL
2-26		ORIGINAL
2-27		ORIGINAL
2-28		ORIGINAL
2-29		ORIGINAL
2-30		08-15-70
2-30A		08-15-70
2-30B	BLANK	08-15-70
2-31		ORIGINAL
2-32		ORIGINAL
*2-33		02-15-71
*2-34		02-15-71
*2-34A		02-15-71
*2-34B	BLANK	02-15-71
2-35		ORIGINAL
2-36		ORIGINAL
2-37		ORIGINAL
2-38		ORIGINAL
2-39		ORIGINAL
2-40		ORIGINAL
2-41		ORIGINAL
2-42		ORIGINAL
SECTION III		
*3-01		02-15-71
*3-02		02-15-71
*3-03		02-15-71
*3-04		02-15-71
3-05		07-15-70
*3-06		02-15-71
*3-06A		02-15-71
*3-06B	BLANK	02-15-71
*3-07		02-15-71
3-08		ORIGINAL
*3-09		02-15-71
3-10		ORIGINAL
*3-11		02-15-71
*3-12		02-15-71
3-13		ORIGINAL
*3-14		02-15-71
*3-15		02-15-71
*3-16		02-15-71
*3-17		02-15-71
*3-18		02-15-71
*3-19		02-15-71
*3-20		02-15-71
3-21		07-15-70
3-22		07-15-70
3-23		07-15-70
3-24		07-15-70
3-25		07-15-70
3-26		07-15-70
3-26A		07-15-70
*3-26B		02-15-71
*3-27		02-15-71
*3-28		02-15-71
*3-29		02-15-71
*3-30		02-15-71
*3-31		02-15-71
*3-32		02-15-71
3-33		ORIGINAL
3-34		ORIGINAL
3-35		01-01-71
3-36		01-01-71
*3-37		02-15-71

*** The asterisk indicates pages changed, added, or deleted by the current change. Insert latest changed and/or added pages; destroy superseded pages.**

NOTE: The portion of text affected by the change is indicated by a vertical line in the outer margins of the page. ⊣⊢ indicates deletion of text.

LIST OF EFFECTIVE PAGES

Page No.		Issue
*3-38		02-15-71
*3-38A		02-15-71
*3-38B	BLANK	02-15-71
3-39		ORIGINAL
3-40		ORIGINAL
3-41		ORIGINAL
3-42		ORIGINAL
*3-43		02-15-71
3-44		01-01-71
*3-44A		02-15-71
*3-44B		02-15-71
*3-45		02-15-71
*3-46		02-15-71
3-47		ORIGINAL
*3-48		02-15-71
3-49		ORIGINAL
*3-50		02-15-71
*3-50A		02-15-71
*3-50B	BLANK	02-15-71
*3-51		02-15-71
*3-52		02-15-71
*3-52A		02-15-71
*3-52B	BLANK	02-15-71
3-53		ORIGINAL
3-54		ORIGINAL
3-55		ORIGINAL
3-56		ORIGINAL
3-57		ORIGINAL
3-58		11-20-70
3-58A		11-20-70
3-58B		11-20-70
3-58C		11-20-70
3-58D		11-20-70
3-58E		11-20-70
3-58F		11-20-70
*3-59		02-15-71
*3-60		02-15-71
*3-61		02-15-71
*3-62		02-15-71
3-63		ORIGINAL
3-64		ORIGINAL
3-65		ORIGINAL
*3-66		02-15-71
3-67		ORIGINAL
3-68		ORIGINAL
3-69		ORIGINAL
3-70		ORIGINAL
3-71		ORIGINAL
*3-72		02-15-71
*3-72A		02-15-71
*3-72B	BLANK	02-15-71
*3-73		02-15-71
3-74		08-15-70
3-74A		08-15-70
3-74B	BLANK	08-15-70
3-75		ORIGINAL
3-76		ORIGINAL
3-77		ORIGINAL
3-78		ORIGINAL
3-79		ORIGINAL
3-80		ORIGINAL
*3-80A		02-15-71
*3-80B	BLANK	02-15-71
3-81		ORIGINAL
3-82		ORIGINAL
3-83		ORIGINAL
*3-84		02-15-71
3-85		ORIGINAL
*3-86		02-15-71
*3-86A		02-15-71
*3-86B	BLANK	02-15-71
3-87		ORIGINAL
3-88		11-20-70
*3-89		02-15-71
3-90		ORIGINAL
3-91		ORIGINAL
*3-92		02-15-71
*3-92A		02-15-71
*3-92B	BLANK	02-15-71
*3-93		02-15-71
*3-94		02-15-71
*3-94A		02-15-71
*3-94B	BLANK	02-15-71
*3-95		02-15-71
*3-96		02-15-71

SECTION V

Page No.		Issue
5-01		ORIGINAL
5-02		ORIGINAL
5-03		ORIGINAL
5-04		ORIGINAL
5-05		01-01-71
5-06		ORIGINAL
5-07		ORIGINAL
5-08		ORIGINAL
5-09		ORIGINAL
5-10		ORIGINAL
5-11		11-20-70
5-12		ORIGINAL
5-13		ORIGINAL
5-14		ORIGINAL
5-15		ORIGINAL
5-16		ORIGINAL
5-17		ORIGINAL
5-18		ORIGINAL
5-19		ORIGINAL
5-20		ORIGINAL

APPENDIX I PART I

Page No.		Issue
A-1		ORIGINAL
A-2	BLANK	ORIGINAL
A1-01		ORIGINAL
A1-02		ORIGINAL
A1-03		ORIGINAL
A1-04	BLANK	ORIGINAL
A1-05		ORIGINAL
A1-06		ORIGINAL
A1-07		ORIGINAL
A1-08		ORIGINAL
A1-09		ORIGINAL
A1-10		ORIGINAL
A1-11		ORIGINAL
A1-12		ORIGINAL
A1-13		ORIGINAL
A1-14		ORIGINAL
A1-15		ORIGINAL
A1-16		ORIGINAL
A1-17		ORIGINAL
A1-18	BLANK	ORIGINAL

APPENDIX I PART II

Page No.		Issue
A2-01		ORIGINAL
A2-02		ORIGINAL
A2-03		ORIGINAL
A2-04		ORIGINAL
A2-05		ORIGINAL
A2-06		ORIGINAL
A2-07		ORIGINAL
A2-08		ORIGINAL
A2-09		ORIGINAL
A2-10	BLANK	ORIGINAL
A2-11		ORIGINAL
A2-12		ORIGINAL
A2-13		ORIGINAL
A2-14		ORIGINAL
A2-15		ORIGINAL
A2-16		ORIGINAL
A2-17		ORIGINAL
A2-18		ORIGINAL
A2-19		ORIGINAL
A2-20		ORIGINAL
A2-21		ORIGINAL
A2-22		ORIGINAL
A2-23		ORIGINAL
A2-24		ORIGINAL
A2-25		ORIGINAL
A2-26		ORIGINAL
A2-27		ORIGINAL
A2-28		ORIGINAL

APPENDIX I PART III

Page No.		Issue
A3-01		ORIGINAL
A3-02		ORIGINAL
A3-03		ORIGINAL
A3-04		ORIGINAL
A3-05		ORIGINAL
A3-06		ORIGINAL
A3-07		ORIGINAL
A3-08		ORIGINAL
A3-09		ORIGINAL
A3-10		ORIGINAL
A3-11		ORIGINAL
A3-12		ORIGINAL
A3-13		ORIGINAL
A3-14		ORIGINAL
A3-15		ORIGINAL

APPENDIX I PART IV

Page No.		Issue
A4-01		ORIGINAL
A4-02		ORIGINAL
A4-03		ORIGINAL
A4-04		ORIGINAL
A4-05		ORIGINAL
A4-06	BLANK	ORIGINAL
A4-07		ORIGINAL
A4-08		ORIGINAL
A4-09		ORIGINAL
A4-10		ORIGINAL
A4-11		ORIGINAL
A4-12		ORIGINAL
A4-13		ORIGINAL
A4-14		ORIGINAL
A4-15		ORIGINAL
A4-16		ORIGINAL
A4-17		ORIGINAL
A4-18		ORIGINAL
A4-19		ORIGINAL
A4-20		ORIGINAL
A4-21		ORIGINAL
A4-22		ORIGINAL
A4-23		ORIGINAL
A4-24		ORIGINAL
A4-25		ORIGINAL
A4-26	BLANK	ORIGINAL

*** The asterisk indicates pages changed, added, or deleted by the current change. Insert latest changed and/or added pages; destroy superseded pages.**

NOTE: The portion of text affected by the change is indicated by a vertical line in the outer margins of the page. ‖ Indicates deletion of text.

TECHNICAL DATA CHANGE SUMMARY

TDC No.	F.S.S. or O.S. No.	Date	Subject	Status
1	- - -	02-20-70	Printed Change	Inc.
2	- - -	05-10-70	Printed Change	Inc.
3	- - -	07-15-70	Printed Change	Inc.
4	- - -	08-15-70	Printed Change	Inc.
5	O.S. 1	09-03-70	Taxiing Procedure Amendment	Inc.
6	F.S.S. 1	10-16-70	Emergency Oxygen System	Inc.
7	O.S. 2	10-23-70	Deletion of Speed Restrictions	Inc.
8	- - -	11-20-70	Printed Change	Inc.
9	- - -	01-01-71	Printed Change	Inc.
10	- - -	02-15-71	Printed Change	Inc.

Table of Contents

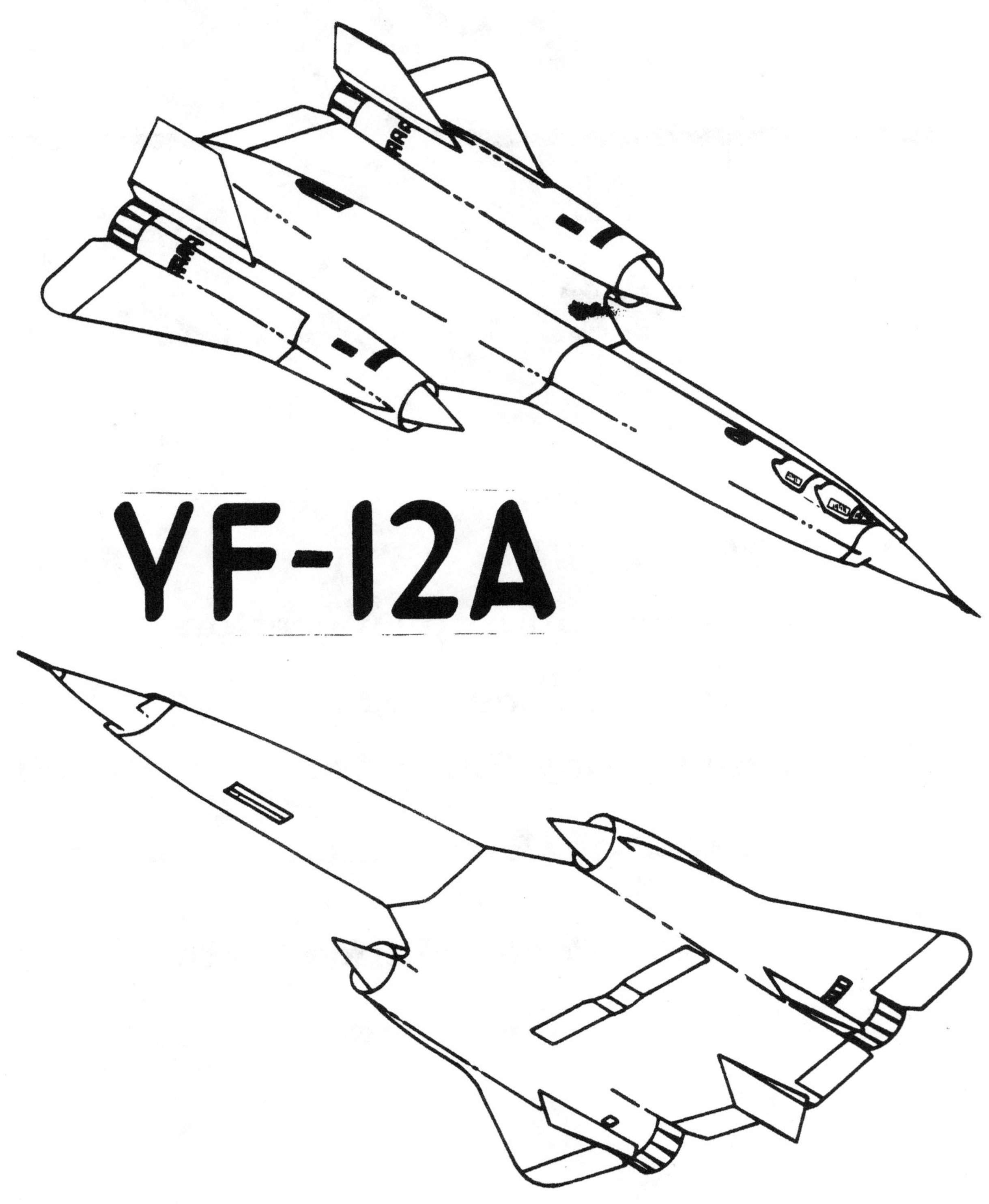
YF-12A

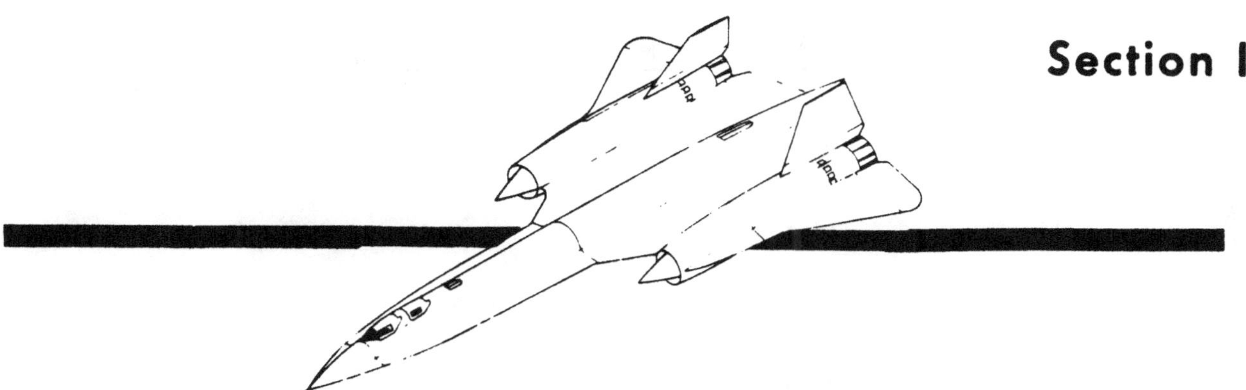

Section I

Description And Operation

TABLE OF CONTENTS

TABLE OF CONTENTS (Cont)

THE AIRCRAFT

The YF-12A is a delta-wing, two place interceptor aircraft, powered by two axial-flow turbojet engines. The aft cockpit is occupied by the fire control officer (FCO). The aircraft is designed to operate at very high altitudes and high supersonic speeds. Some notable features of the aircraft are very thin delta wings, twin canted rudders mounted on the top of the engine nacelles, a folding ventral fin attached to the lower aft fuselage, small fixed ventral fins under each nacelle, and a pronounced fuselage "chine" extending from the pilot's cockpit to the leading edge of the wing. The propulsion system uses movable spikes to vary inlet geometry. Surface controls are elevons and rudders, operated by irreversible hydraulic actuators with artificial pilot control feel. A single-point pressure refueling system is installed for both ground and inflight refueling. Dual wheels are used on the nose gear, and three wheel trucks are used on each main landing gear strut. A drag chute is provided to reduce landing roll. The aircraft is painted black to reduce unwanted internal temperatures which result from operating at high speeds.

DIMENSIONS

Length (overall)	101.7 ft
Height (to top of vertical stabilizer)	18.4 ft
Wing span	55.7 ft
Wing area (nominal)	1795 sq ft
Tread (MLG middle wheel centerlines)	16.67 ft

Gross Weight

The loaded gross weight of the aircraft including crew and full fuel is approximately 126,000 to 127,500 pounds depending on equipment installed.

ENGINE AND AFTERBURNER

Thrust is supplied by two Pratt & Whitney JT11D-20 bleed bypass turbojet engines with afterburners. (See Figure 1-1.) The engines are designed for continuous operation at compressor inlet temperatures above 400°C which are associated with high Mach number flight at high altitude. The engine has a single-rotor, nine-stage, 8.8:1 pressure ratio compressor utilizing a compressor bleed bypass cycle at high Mach numbers. When opened, bypass valves bleed air from the fourth stage of the compressor, and six ducts route it around the rear stages of the compressor, the combustion section, and the turbine. The bleed air re-enters the turbine exhaust around the front of the afterburner where it is used for increased thrust and cooling. The transition to bypass operation is scheduled by the main fuel control as a function of compressor inlet temperature (CIT) and engine speed. The transition normally occurs in a CIT range of 80° to 115°C, corresponding to a Mach number range of 1.8 to 2.0. The camber of the engine inlet air guide vanes (IGV) is automatically changed at the same time unless this function has been prevented by operation of an IGV shift lockout feature which is provided with these engines.

When on the ground or at low Mach numbers, engine speed varies with throttle movement when the throttle is between the IDLE position and a position slightly below the MILITARY stop. At higher settings, up to the position for maximum afterburner thrust,

JT11D -20 ENGINE

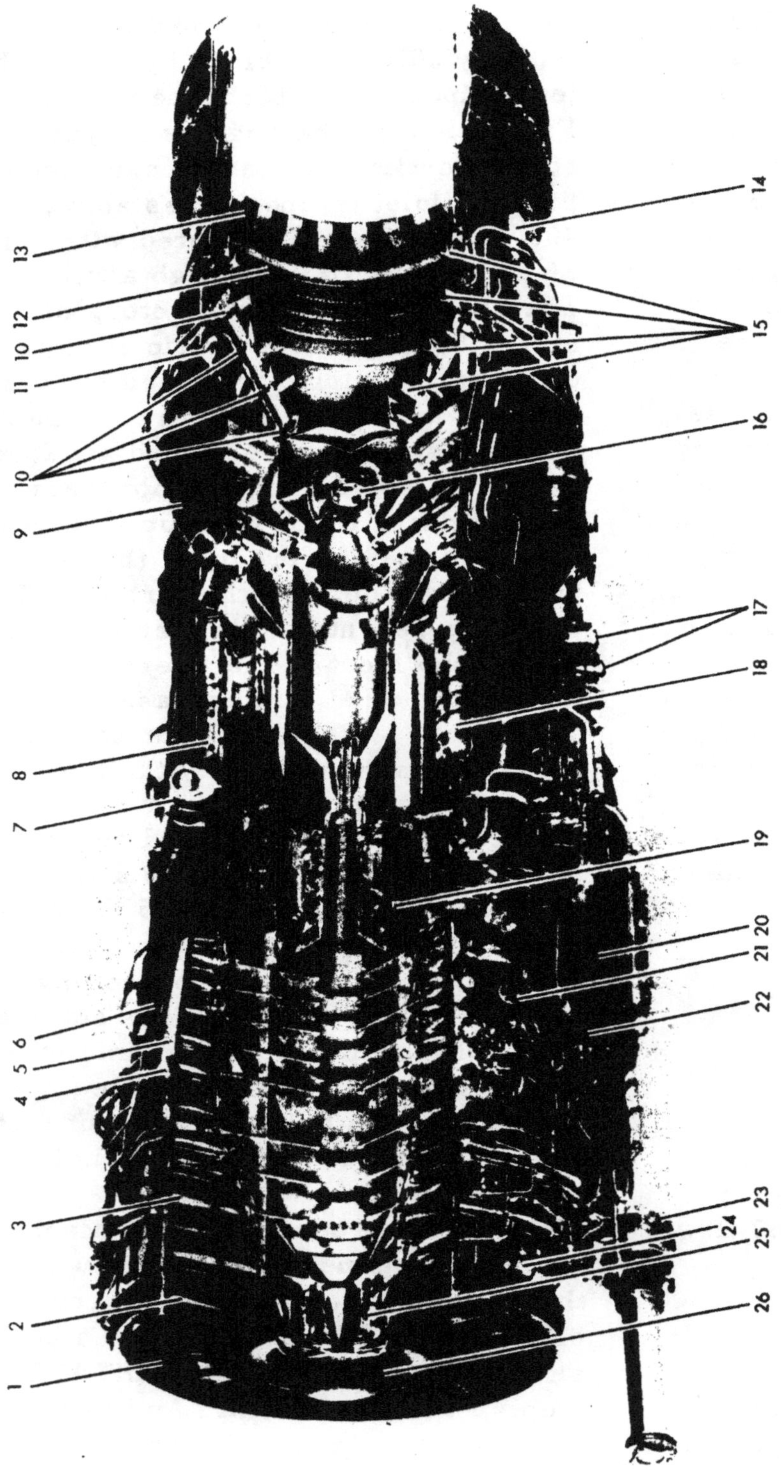

1 Inlet Case
2 Variable IGV
3 Forward Compressor Section (4 stages)
4 Internal Bleeds (24)
5 Bypass Chamber
6 External Bleeds (12)
7 Chemical Ignition Tank (TEB)
8 Main Burner Injector Probe
9 Bleed Bypass Tubes (6)
10 Afterburner Spray Bar Rings (4)
11 Aft Engine Mount Ring
12 Afterburner Liner
13 Variable Area Exhaust Nozzle
14 Exhaust Nozzle Actuators (4)
15 Flame Holders (4)
16 Turbine Section and Bearing
17 Hydraulic Filters (2)
18 Burner Can (8)
19 Aft Compressor Bearing
20 Main Gearbox
21 Main Fuel Control
22 Main Fuel Pump
23 Reduction Gear Box
24 IGV Actuators (2)
25 Front Compressor Bearing
26 Inlet Case Island Cover

F203-5(c)

Figure 1-1

the main fuel control schedules engine speed as a function of CIT and modulates the variable area exhaust nozzle to maintain approximately constant rpm. Throttle movement in the afterburning range only changes the afterburner fuel flow, nozzle position, and thrust. At high Mach number and constant inlet conditions, engine speed is essentially constant for all throttle positions down to and including IDLE. At a fixed throttle position, engine speed will vary according to main fuel control schedule when Mach number and CIT change.

The engine contains a two-stage turbine. Turbine discharge temperatures are monitored by exhaust gas temperature indication. A chemical ignition system is used to ignite the low vapor pressure fuel. A separate engine-driven hydraulic system, using fuel as hydraulic fluid, operates the exhaust nozzle, chemical ignition system dump, compressor bypass, starting bleed systems, and Inlet Guide Vanes. The main fuel pump, engine hydraulic pump and tachometer are driven by the main engine gearbox. The afterburner fuel pump is powered by an air turbine driven by compressor discharge air.

Maximum Rated Thrust

Maximum rated thrust is obtained in afterburning by placing the throttle against the quadrant forward stop. The maximum afterburning thrust rating of each engine at sea level and standard day conditions is 34,000 pounds. Typical takeoff thrust values for maximum afterburner are illustrated in Figure 1-1A at sea level pressure altitude for the engine as installed in the nacelle. This figure also shows the variation in thrust with ambient temperature and the effect of airspeed during the takeoff acceleration.

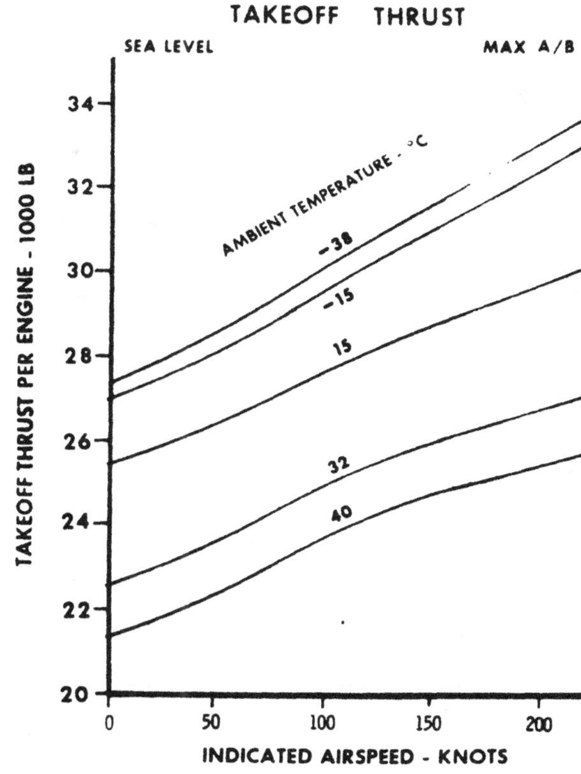

Figure 1-1A

Partial Afterburning Thrust

Afterburning fuel flow and thrust are modulated by moving the throttle between the MILITARY thrust detent and the quadrant forward stop. MINIMUM afterburning thrust is obtained with the throttle just forward of the MILITARY thrust detent on the quadrant. It provides approximately 85% of maximum afterburning thrust at sea level and approximately 55% at high altitude. Afterburner ignition is automatically actuated when the throttle is advanced past the detent, and afterburner fuel flow is terminated when the throttle is retarded below the detent. The basic engine operates at its military thrust condition during all afterburning operation.

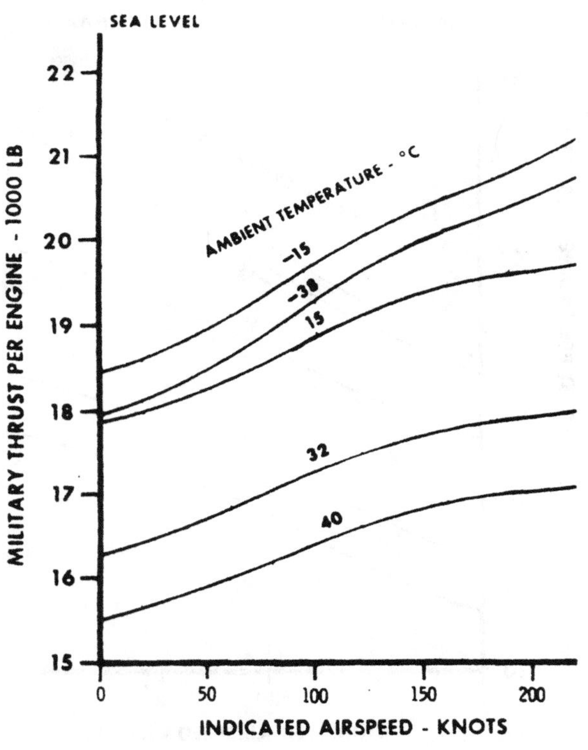

Figure 1-1B

Military Thrust

Military thrust is the maximum non-afterburning thrust and is obtained by placing the throttle against the aft side of the military thrust detent on the quadrant. At sea level static conditions, military thrust is approximately 70% of the maximum thrust rating. At high altitude, military thrust is approximately 28% of the maximum thrust available. Figure 1-1B illustrates the variation in military thrust with ambient temperature and airspeed at sea level pressure altitude.

Idle Thrust

IDLE is a throttle position for minimum thrust operation. It is not an engine rating. Minimum thrust is always obtained when the throttle is at the IDLE stop on the quadrant. With the throttle in the IDLE position, the engine operates at approximately 3975 rpm up to 32°C. At higher ambient temperatures, rpm increases at the rate of approximately 50 rpm per 1°C. This operating characteristic and the effect on idle thrust is illustrated in Figure 1-1C, at sea level pressure altitude, as a function of airspeeds typical of a landing and deceleration to a stop.

THROTTLES AND THROTTLE SETTINGS

Two throttle levers, one for each engine, are located in a quadrant on the pilot's left forward console. The right throttle is linked mechanically to the right engine main fuel control and the left throttle to the left engine afterburner fuel control. Each engine's afterburner and main fuel control are connected by a closed-loop cable. The throttle quadrant positions are labeled OFF, IDLE and AFTERBURNER. When a throttle is moved forward from OFF to IDLE, a roller drops over a hidden ledge to

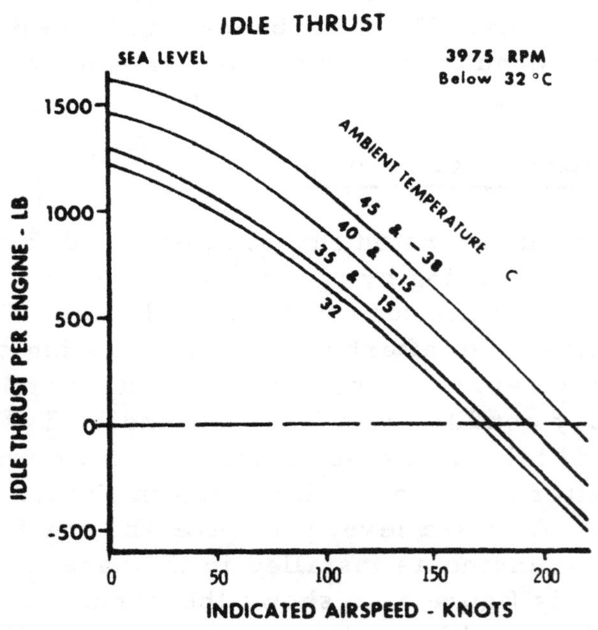

Figure 1-1C

the IDLE position. This ledge prevents inadvertent engine cutoff when the throttle is retarded to IDLE. When returning the throttle from IDLE to OFF, it must be lifted in order to clear the IDLE stop ledge.

Forward throttle movement from IDLE to the raised MILITARY detent increases the non-afterburning thrust of the engine. Additional forward movement of the throttle over the raised detent initiates afterburner fuel flow and this, in turn, actuates the chemical ignition system for afterburner ignition. The AFTERBURNER range extends from the raised detent to the quadrant forward stop.

The engine throttle knob for the right engine incorporates a pushbutton switch for radio transmission.

Off

The aft stop on the quadrant is the engine OFF throttle position. In this position, the windmill bypass valve cuts off fuel to the burner cans and bypasses it back to the aircraft system. This provides cooling for the engine oil, fuel pump and fuel hydraulic pump when the engine is windmilling.

Start

There is no distinct throttle position for starting. Starting is accomplished by moving the throttle from OFF to the IDLE position as the proper engine speed is reached. This directs fuel to the engine burners by actuation of the windmill bypass valve and actuates the chemical ignition system.

Throttle Friction Lever

The throttles are prevented from creeping by use of a friction lever located on the inboard side of the throttle quadrant. When the lever is fully aft, the throttles are free to move. Moving the lever forward increases friction to restrict throttle movement or holds the throttles in the desired position.

ENGINE FUEL SYSTEM

Engine fuel system components include the engine driven fuel pump, main fuel control, windmill bypass valve and variable area fuel nozzles in the main burner section. Refer to Figure 1-1D.

Main Fuel Pump

The engine driven main fuel pump is a two-stage unit. The first stage consists of a single centrifugal pump which acts as a boost stage. The second stage consists of two parallel gear-type pumps with discharge check valves. The parallel pump and check valve arrangement permits one pump to operate in the event the other fails. The pump discharge pressure is determined by the regulating and metering function of the main fuel control. The maximum discharge pressure is approximately 900 psia. A relief valve is provided in the second stage dischrage to prevent excessive fuel system pressure.

Main Fuel Control

The main fuel control meters main burner fuel flow and controls the bleed bypass valves, start bleed valves, inlet guide vanes, and exhaust nozzle positions. It regulates thrust by controlling EGT as a function of

ENGINE AND AFTERBURNER FUEL SYSTEM

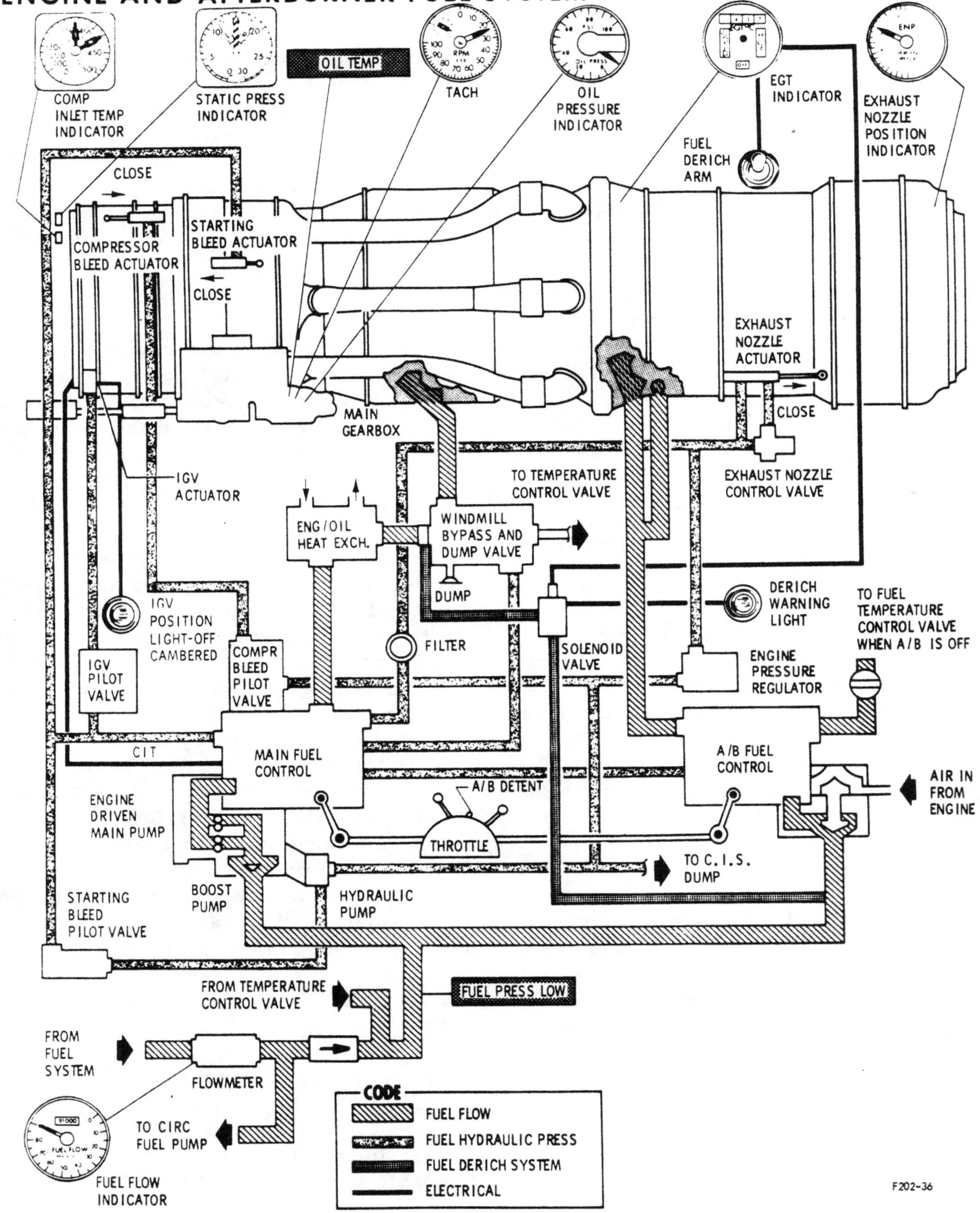

COMP INLET TEMP INDICATOR
STATIC PRESS INDICATOR
OIL TEMP
TACH
OIL PRESSURE INDICATOR
EGT INDICATOR
EXHAUST NOZZLE POSITION INDICATOR
FUEL DERICH ARM
CLOSE
COMPRESSOR BLEED ACTUATOR
STARTING BLEED ACTUATOR
CLOSE
EXHAUST NOZZLE ACTUATOR
MAIN GEARBOX
CLOSE
IGV ACTUATOR
TO TEMPERATURE CONTROL VALVE
EXHAUST NOZZLE CONTROL VALVE
ENG/OIL HEAT EXCH.
WINDMILL BYPASS AND DUMP VALVE
DUMP
IGV POSITION LIGHT-OFF CAMBERED
DERICH WARNING LIGHT
TO FUEL TEMPERATURE CONTROL VALVE WHEN A/B IS OFF
FILTER
SOLENOID VALVE
ENGINE PRESSURE REGULATOR
IGV PILOT VALVE
COMPR BLEED PILOT VALVE
CIT
MAIN FUEL CONTROL
A/B FUEL CONTROL
AIR IN FROM ENGINE
A/B DETENT
ENGINE DRIVEN MAIN PUMP
THROTTLE
TO C.I.S. DUMP
BOOST PUMP
HYDRAULIC PUMP
STARTING BLEED PILOT VALVE
FROM TEMPERATURE CONTROL VALVE
FUEL PRESS LOW
FROM FUEL SYSTEM
FLOWMETER
TO CIRC FUEL PUMP
FUEL FLOW INDICATOR

CODE
- FUEL FLOW
- FUEL HYDRAULIC PRESS
- FUEL DERICH SYSTEM
- ELECTRICAL

F202-36

Figure 1-1D

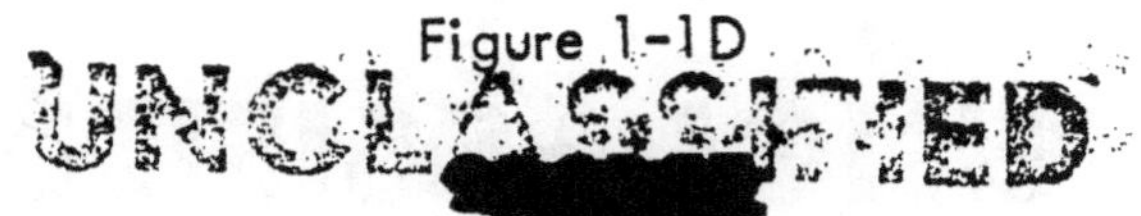

throttle position, compressor inlet air temperature, main burner pressure and engine speed. The EGT military schedule can be biased by pilot operation of a trimming control. The bypass and start bleed valve positions are controlled as a function of engine speed biased by CIT. For steady state inlet conditions at high Mach number, the control provides essentially constant engine speed at all throttle positions down to and including IDLE. On the ground and at lower Mach numbers, engine speed varies with throttle position from slightly below MILITARY down to IDLE. Afterburner operation is always accomplished at the MILITARY rated engine speed and EGT schedules. There is no emergency fuel control system.

Windmill Bypass and Dump Valve

The windmill bypass and dump valve directs fuel to the engine burners for normal operation or bypass fuel to the recirculation system for accessory, engine component and engine oil cooling during windmilling operation. The valve is actuated in response to signals from the main fuel control. The valve opens to drain the engine fuel manifold when the engine is shut down.

Fuel Nozzles

The engine has eight can-annular type combustion chambers with forty eight variable area, dual orifice fuel nozzles. The nozzles are arranged in clusters of six nozzles per burner. Each nozzle has a fixed area primary metering orifice and a variable area secondary metering orifice, discharging through a common opening. The secondary orifice opens as a function of primary orifice pressure drop.

Combustion Chamber Drain Valve

The main engine ignition system plumbing is equipped with a fuel purge or "Dribble Tee". This allows fuel from the main fuel pump interstage to flush residual ignition fluid (TEB) from the ignition probe. It prevents a condition known as "coking" from occurring which would otherwise restrict the ignition probe and prevent fuel ignition. As a result, fuel in small quantity should drain from the main burner case overboard drain fitting anytime that there is fuel pressure to the engine pump inlet, either due to fuel boost pump operation or tank pressure developed by the LN_2 fuel tank vapor inerting system. If fuel does not drain normally, either the chemical ignition system probe is plugged or the burner drain has malfunctioned. The flow is minor with normal pressure in the fuel tanks, and the leakage from the main burner case overboard drain provides a means for inspecting the main ignition probe to be sure it is functioning properly. Normal preflight boost pump operation should not cause a problem. However, it does increase the "wetted" fuel area in the burner (over that due to normal tank pressurization) and it could result in severe torching during engine start if the overboard drain was restricted.

Fuel Flow Indicators

Two fuel-flow indicators, one for each engine, are mounted on the instrument panel and display total fuel flow (engine and afterburner) plus tank return flow, if any, in pounds per hour. The dial is calibrated in 5000-pound-per-hour increments to 90,000 pph. The five-digit center window indicates the fuel flow to the nearest 100 pph. Power for the indicators is supplied from the essential ac bus.

Tachometers

Two tachometers, one for each engine, are mounted on the right side of the pilot's instrument panel. They indicate engine rpm by means of a main pointer and dial calibrated to 10,000 rpm, and a smaller dial and sub-pointer which makes one complete revolution for each 1000 rpm. Each tachometer is self-energized and operates independently of the aircraft electrical system.

Exhaust Gas Temperature Gages

Two exhaust gas temperature gages, one for each engine, are located on the right side of the instrument panel in the forward cockpit. The gages have a digital indicator that shows turbine discharge temperature from 0°C to 1198°C, a HOT (red) and COLD (yellow) condition flag, an overtemperature warning light that illuminates when EGT temperature reaches 860°C, and a power OFF warning flag. Each indicator receives power from the essential ac bus thru an EGT circuit breaker located on the left console.

AFTERBURNER FUEL SYSTEM

Afterburner fuel system components include the afterburner centrifugal fuel pump, afterburner fuel control and spray bars.

Afterburner Fuel Pump

The afterburner fuel pump is a high speed, single stage centrifugal pump. The pump is driven by an air turbine, operated by engine compressor discharge air. The compressor discharge air supply is regulated by a butterfly valve in response to the demand of the afterburner fuel control. The turbine is protected from overspeed by choking of its discharge air venturi.

Afterburner Fuel Control

The afterburner fuel control is hydromechanically operated and schedules metered fuel flow as a function of throttle position, main burner pressure and compressor inlet temperature. Fuel flow is metered on a predetermined schedule to discharge fuel from two zones of the four concentric afterburner spray bar rings simultaneously.

EXHAUST GAS TEMPERATURE TRIM SYSTEM

EGT Trim Switches

Two four-position EGT trim switches, one for each engine, are located on the left side of the instrument panel in the forward cockpit. The four positions are labeled HOLD (center), INCREASE (up), DECREASE (down), and AUTO (right). The switch is spring-loaded to the center position and must be pulled out before it can be placed in the AUTO position. Selection of the AUTO position provides automatic EGT trimming for the corresponding engine in accordance with the main fuel control automatic schedule. Selection of the HOLD position deactivates automatic trimming and maintains an existing EGT trimmer setting. Manual trimming is accomplished by holding a trim switch in the INCREASE or DECREASE position. When held in the momentary INCREASE position, a small electric motor on the engine fuel control is actuated which increases the ratio of main burner fuel flow to main burner combustion pressure and, as a result, increases the turbine discharge temperatures. Holding a switch in the momentary DECREASE position results in decreasing turbine discharge temperatures.

NOMINAL EGT SCHEDULE

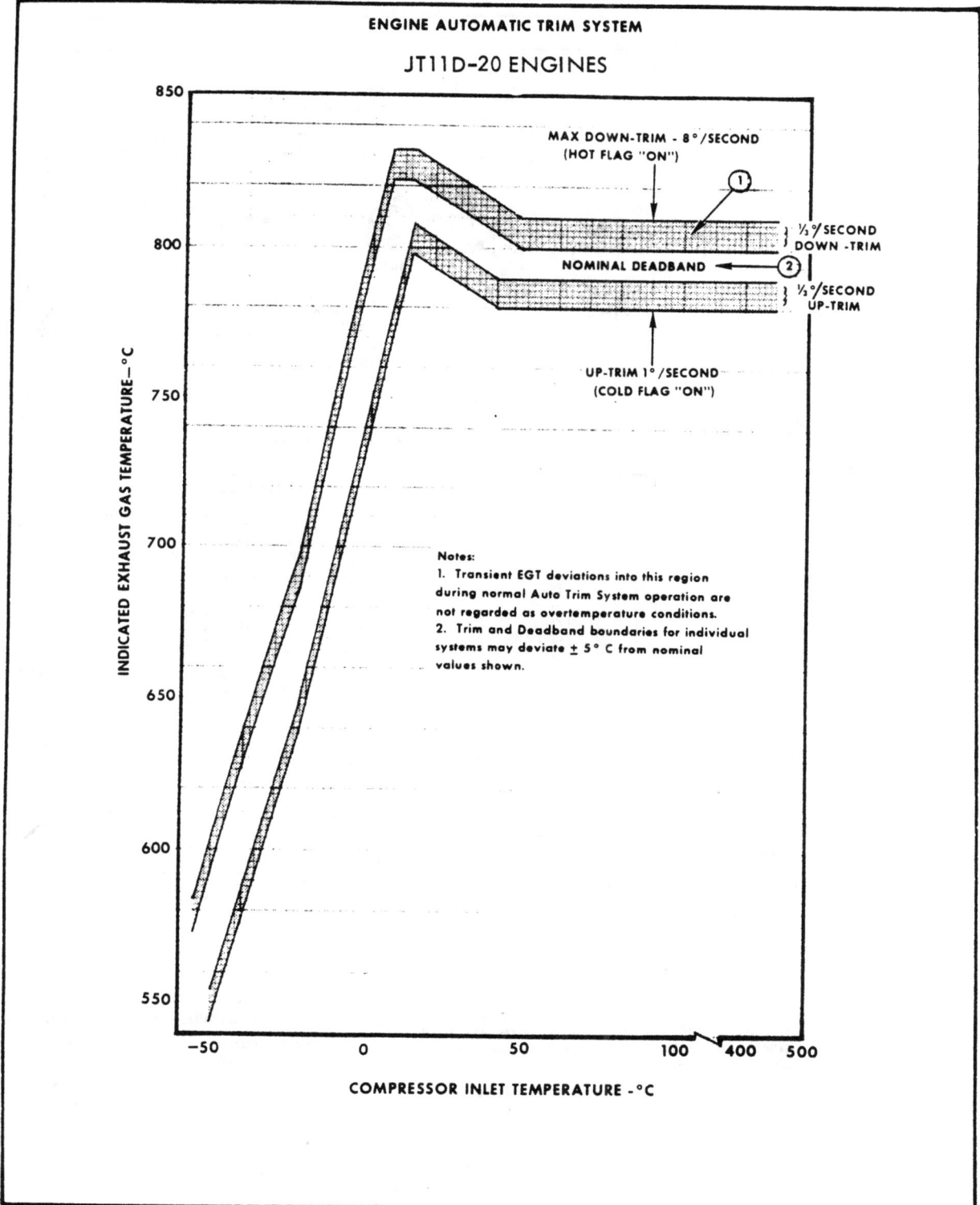

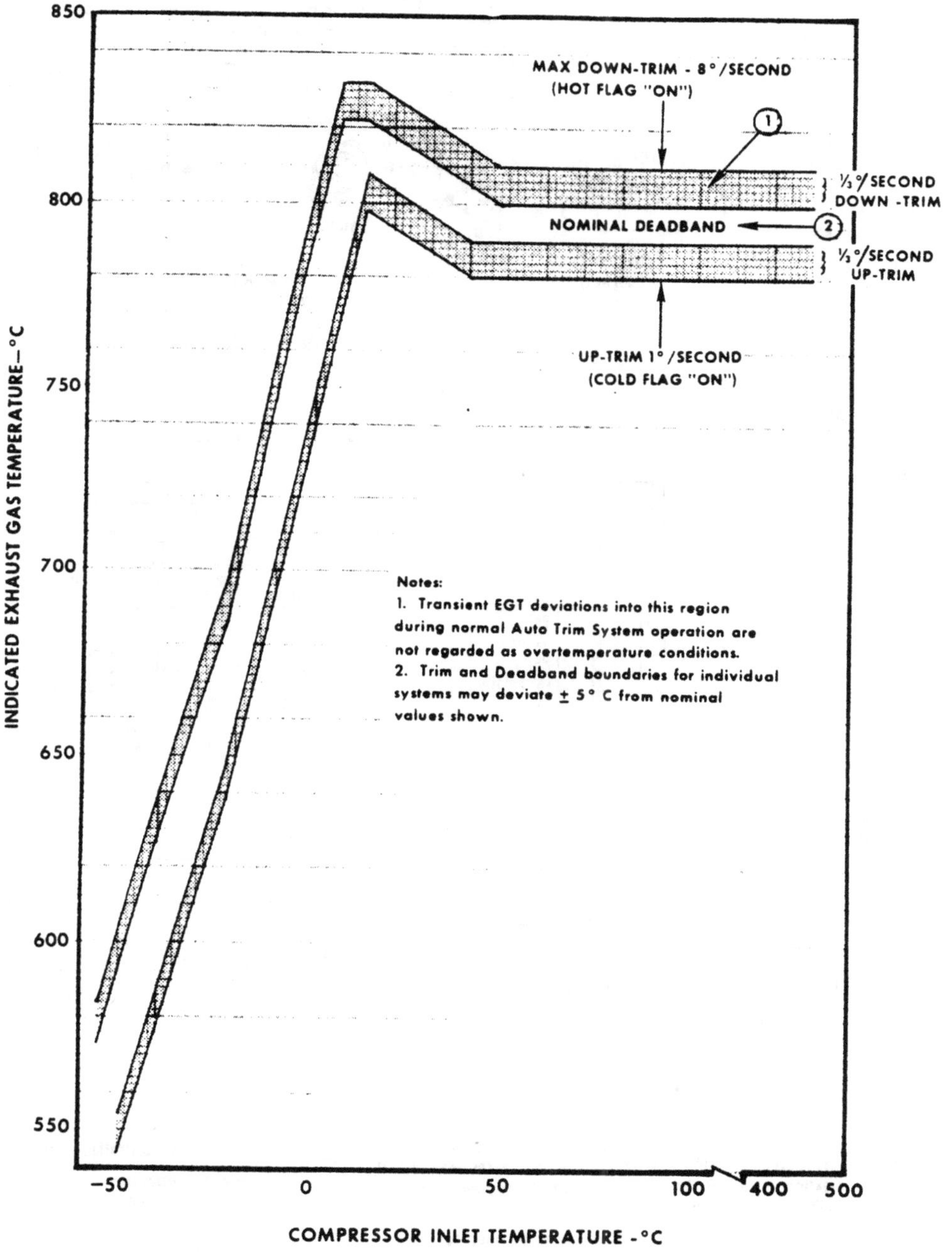

ENGINE AUTOMATIC TRIM SYSTEM

JT11D-20 ENGINES

Notes:
1. Transient EGT deviations into this region during normal Auto Trim System operation are not regarded as overtemperature conditions.
2. Trim and Deadband boundaries for individual systems may deviate ± 5° C from nominal values shown.

Figure 1-1E

EGT INDICATION AND CONTROL SYSTEM

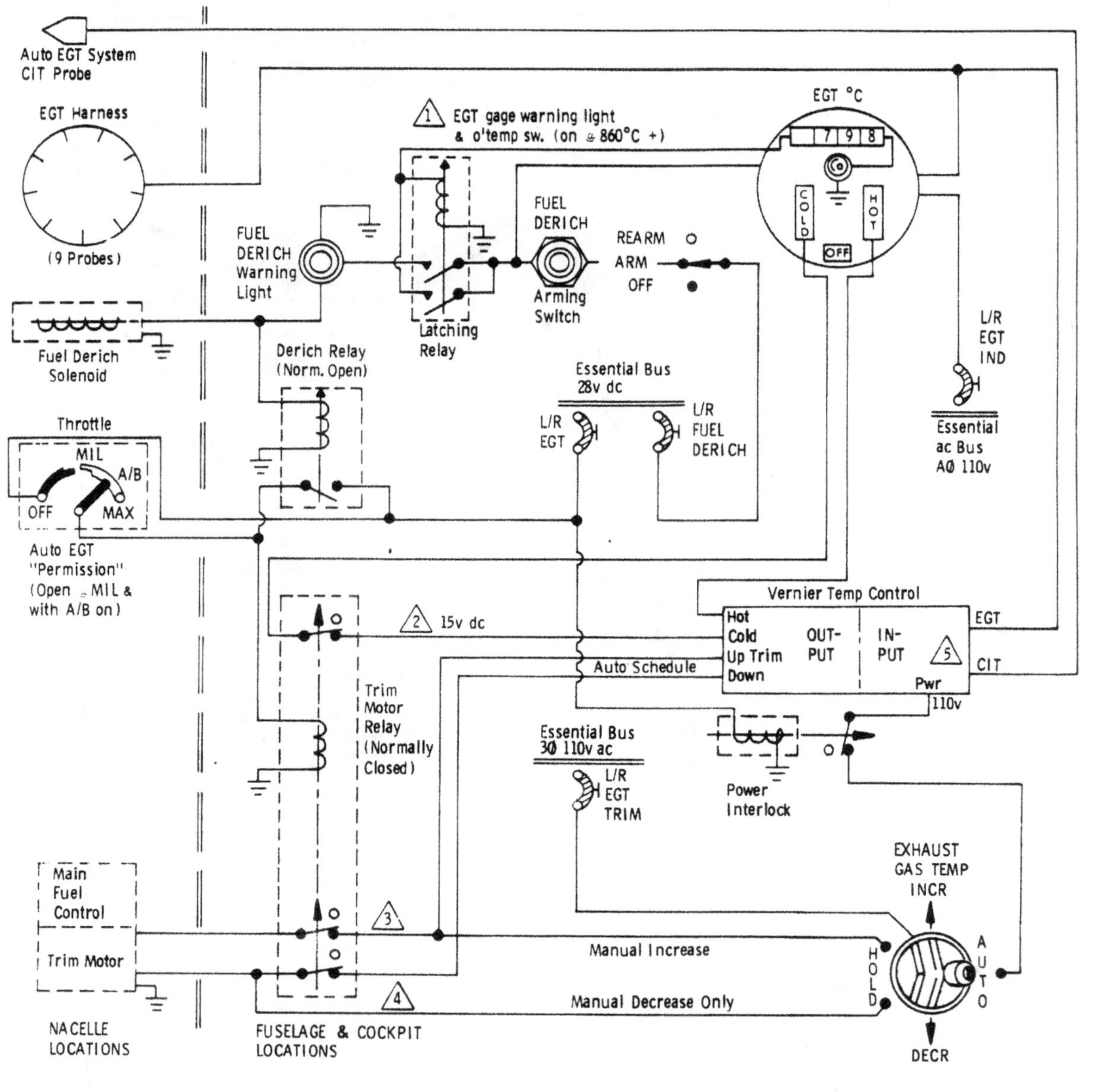

Auto EGT System
CIT Probe
EGT Harness
(9 Probes)
Fuel Derich
Solenoid
Throttle
MIL
A/B
OFF
MAX
Auto EGT
"Permission"
(Open ≥ MIL & with A/B on)
Main
Fuel
Control
Trim Motor
NACELLE
LOCATIONS
FUEL
DERICH
Warning
Light
Derich Relay
(Norm. Open)
Latching
Relay
1 EGT gage warning light & o'temp sw. (on ≥ 860°C +)
FUEL
DERICH
Arming
Switch
REARM
ARM
OFF
EGT °C
7 9 8
COLD
HOT
OFF
L/R
EGT
IND
Essential
ac Bus
AØ 110v
Essential Bus
28v dc
L/R
EGT
L/R
FUEL
DERICH
Vernier Temp Control
Hot
Cold
Up Trim
Down
OUT-
PUT
IN-
PUT
5
EGT
CIT
Pwr
110v
2 15v dc
Auto Schedule
Trim
Motor
Relay
(Normally
Closed)
Essential Bus
3Ø 110v ac
L/R
EGT
TRIM
Power
Interlock
EXHAUST
GAS TEMP
INCR
HOLD
AUTO
DECR
3
Manual Increase
4
Manual Decrease Only
FUSELAGE & COCKPIT
LOCATIONS

NOTE

1. Switch closes and light illuminates as 860°C reached. Switch stays closed and light remains on above 860°C. When EGT then decreases below 860°C, jewel light is extinguished but "latching" relay maintains power to derich solenoid until derich arming switch is cycled.

2. COLD flag does not operate when below Military power or when deriched. HOT and COLD flag operating power (15v dc) is produced within the vernier temperature control.

3. Direction of trim is controlled by ac power phase-sequencing.

4. Auto trim and manual uptrim are inoperative when below Military power or when deriched. Manual downtrim is always available if three-phase power is supplied.

5. Auto EGT vernier temperature control is powered only when AUTO is selected.

F202-19(b)

Figure 1-1F

An increase or decrease in temperature will be indicated by the respective exhaust gas temperature gage. The switches have no effect on rpm when the nozzles are modulating to provide the scheduled engine speed. Engine speed will, however, increase with increasing EGT when nozzle position is limited at the full open or closed positions. Auto trim and manual uptrim are inoperative when below Military power or when deriched. Manual downtrim is always available if three-phase power is supplied. Power for the trim motors is furnished by the essential ac bus through the L and R EGT TRIM circuit breakers.

Automatic EGT Trim

When AUTO EGT is selected and a throttle is positioned at or above the Military stop, a "permission" circuit is opened and the electric trim motor for the respective engine automatically regulates its main fuel control to provide Military power at an EGT which is within a 10°C nominal deadband. See Figure 1-1E.

If the EGT for an engine is either above or below the deadband boundaries for its auto trimmer, the system trims EGT toward its deadband. The rate at which trimming occurs is a function of the amount of temperature deviation from the deadband. In case the EGT deviates more than 10°C above the deadband, the system automatically down trims at its maximum rate (approximately 8°C per second), and a HOT flag is displayed on the EGT gage. When the EGT is more than 10°C below the deadband, the system automatically up trims at a rate of approximately 1°C per second and a COLD flag is displayed. Whenever EGT deviation is 10°C or less from its deadband, the trim system automatically operates at a slow trim rate (approximately 1/3°C per sec.) and the condition flags are retracted.

NOTE

The EGT condition flags do not operate when MANUAL EGT trim is selected. The COLD condition flag does not operate any time that the permission circuit prevents auto trim from operating when AUTO EGT is selected. See Figure 1-1F.

CAUTION

A malfunction of the Automatic EGT Trim System may exist if the EGT has a tendency to hunt or if an abnormal tendency to uptrim or downtrim is noted while AUTO EGT is selected. In this event, the corresponding engine should be operated in the MANUAL EGT trim mode only and the condition reported following flight. An EGT overtemperature may occur if continued operation in the AUTO EGT mode is attempted.

As EGT trimming is desirable only at Military or higher power settings, a switch and relay are provided that supply power to the main fuel control trim motor only if the throttle is at or above a military thrust position. This permission circuitry also prevents trimming in the event of derich actuation. Auto EGT may be used during

the engine trim check; however, in the event the AUTO EGT system is in a high EGT setting, reduce power and manually down-trim prior to resetting military power. Power for the AUTO EGT trim circuits is furnished by the essential ac bus. Power for the permission circuit is furnished by the essential dc bus through the EGT circuit breakers located on the pilot's right console.

Repeated rapid throttle movements up and down through the permission switch operating range can cause military EGT to increase sufficiently to produce derichment. This phenomenon, called "ratcheting", is caused by an inherent lag in EGT thermocouple response and can be avoided by selecting manual trim if there are flight conditions requiring repeated rapid throttle movements. A hot flag may also appear or persist at throttle positions slightly below military even though the automatic trimmer has been deactivated by throttle positioning. Since the hot flag overtemperature warning circuit is continually energized, a hot flag will appear anytime the normal EGT operating band is exceeded. This condition should be corrected by manually downtrimming or by retarding the throttle.

FUEL DERICHMENT SYSTEM

Each engine incorporates a derichment system to provide protection against severe turbine overtemperature during high altitude operation. If the EGT indication reaches or exceeds 860°C while the system is armed, the fuel/air ratio in the engine burner cans is automatically reduced, or deriched, below normal values. This is accomplished by a solenoid-operated valve and orifice which bypasses metered engine fuel from the fuel/oil cooler to the afterburner fuel pump inlet. The solenoid valve is actuated by a signal from the EGT gage when 860°C is reached. See Figures 1-1D and 1-1F. Once activated, it remains open until the system is turned off or rearmed. A fuel derich warning light is provided for each engine to indicate when the respective valve is open.

NOTE

> Inadvertent derichment at sea level will cause a thrust loss of approximately 5% if in maximum afterburning or 7% if at military power. Approximately 45% loss in thrust and 750 rpm speed suppression will occur during cruise with maximum afterburning.

Fuel Derich Arming Switch

A three-position FUEL DERICH arming toggle switch is located on the right side of the pilot's instrument panel. The positions are labeled REARM (up), ARM (center) and OFF (down). In the ARM position the derich circuit is armed and the derich solenoid valve will open automatically and remain open if the EGT indication reaches 860°C. In the OFF position the derich solenoid valve is closed and the system cannot provide derichment. The REARM position is spring-loaded and allows the pilot to rearm the fuel derich system without moving the switch to the OFF position. Power for the switch is furnished by

the essential dc bus through two FUEL DERICH circuit breakers located on the left console.

Fuel Derich Warning Light

Two fuel derich warning lights, one for each engine, are located on the right side of the center instrument panel. The respective light illuminates if EGT exceeds 860° C and will stay on until the arming switch is placed in the OFF position. The light can also be extinguished by moving the arming switch to the momentary REARM position if the EGT is below 860° C.

Fuel DerichGround Test Switch

A three-position toggle switch, labeled FUEL DERICH SYSTEM, is located aft of the left console in the pilot's cockpit. The switch is labeled LEFT (aft), RIGHT (forward), and is spring loaded to the OFF (center) position. When the switch is moved to the LEFT or RIGHT position, the digital indication on the respective exhaust gas temperature gage should slew toward 1198°C, illuminating the red jewel light in the instrument upon reaching 860°C. The respective fuel derich warning light will also illuminate and the fuel derich solenoid for the affected engine will operate if the corresponding fuel derich system is armed.

ENGINE FUEL HYDRAULIC SYSTEM

Each engine is provided with a fuel hydraulic system for actuation of the afterburner exhaust nozzle, the inlet guide vane system, and the external (start) and internal (bypass) bleed valves. Engine hydraulic system pressure is also required to dump the unused chemical ignition fluid. Pressure is supplied by a high temperature, engine driven, variable delivery, piston type pump. The pump maintains system pressures up to 1800 psi with a maximum flow of 50 gpm for transient requirements. Engine fuel is supplied to the pump from the main fuel pump boost stage. Some high pressure fuel is diverted from the hydraulic system to cool the non-afterburning recirculation line and the windmill bypass valve discharge line. This fuel is returned to the aircraft fuel system. Low pressure fuel from the hydraulic pump case is returned to the main fuel pump boost stage. Hydraulic system loop cooling is provided by the compensating fuel which is supplied from the main fuel pump.

OIL SUPPLY SYSTEM

The engine and speed-reduction gearbox are lubricated by an engine-contained "hot tank" closed system. The oil is cooled by circulation through an engine fuel-oil cooler. The oil tank is mounted on the lower right side of the engine compressor case and has a usable capacity of 4.5 US gallons. Total capacity of the tank is 6.7 US gallons. Oil is fed by gravity to the main oil pump which then forces it under pressure through a filter to the fuel-oil cooler. (The filter is equipped with a bypass in case of clogging.) From the fuel-oil cooler, the oil is distributed to the engine bearings and gears. Oil screens are installed at the lubricating jets for additional protection. Scavange pumps return the oil to the tank where it is

deaerated. A pressure-regulating valve keeps flow and pressure relatively constant during all flight conditions. Oil quantity and temperature warning lights are provided for each engine.

Main Fuel-Oil Cooler

The main fuel-oil cooler unit provides oil cooling by using the engine fuel to absorb heat. The oil temperature is controlled by fuel flow through the cooler. A bypass valve is incorporated to bypass fuel around the cooler when the fuel flow is greater than approximately 12,000 pounds per hour, the flow capacity of that cooler.

Engine Oil Pressure Gages

An oil pressure gage is provided for each engine on the right side of the pilot's instrument panel. The gages indicate output pressure of the respective engine oil pump in pounds per square inch, using electrical signals from a fuel-cooled transmitter. The dials are calibrated from 0 to 100 psi in 5-psi increments. Power for the gages is furnished by the essential ac bus through the instrument power transformer.

Engine Oil Temperature Indicator Lights

Engine oil temperature is indicated by the L and R OIL TEMP indicator lights, located on the pilot's annunciator panel. When illuminated, the lights indicate that oil temperature of the respective engine is less than 15°C or more than 282°C.

Oil Quantity Low Lights

L and R OIL QTY LOW indicator lights are located on the pilot's annunciator panel and when illuminated indicate that oil quantity in the respective engine oil tank is less than 2-1/4 gallons.

EXHAUST NOZZLE AND EJECTOR SYSTEM

The variable area, iris type, engine-afterburner nozzle is comprised of segments operated by a cam and roller mechanism and four hydraulic actuators. The actuators are operated by fuel-hydraulic system pressure. The engine afterburner nozzle is enclosed by a fixed contour, convergent-divergent ejector nozzle to which free floating trailing edge flaps are attached. In flight, the duct inlet shock trap bleed and the aft bypass doors (when open) supply secondary airflow between the engine and nacelle for cooling. During ground operation, suck-in doors in the aft nacelle area provide cooling air. Intake doors around the nacelle, just forward of the ejector, normally supply tertiary air to the ejector nozzle at subsonic speeds. The tertiary doors and trailing edge flaps open and close with varying internal nozzle pressures which are a function of Mach number and engine thrust.

Nozzle Actuation

The exhaust nozzle control and actuation system is composed of four actuators to position the exhaust nozzle, and an exhaust nozzle control which modulates pressure at the actuators in response to engine speed signals from the main fuel control. The exhaust nozzle control is mounted on the aft portion of the engine. A pressure regulator is contained in a separate unit located near the exhaust nozzle control.

Exhaust Nozzle Position Indicators

Each engine is provided with a nozzle position indicator located on the right side of the pilot's instrument panel. The indicators are labeled ENP and are marked from 0 to 10. They are an index of nozzle opening. The indicators are remotely operated by an electrical transducer located near the exhaust nozzles. The transducer is cooled by fuel and is operated by the afterburner nozzle feedback link. Power for the indicators is supplied by the essential ac bus.

ENGINE INTERNAL (BYPASS) AND EXTERNAL (START) BLEEDS

The internal (bypass) bleed control and actuation system consists of four two-position actuators to move the bleed valves, and a pilot valve within the main fuel control to establish the pressure to the actuators. The pilot valve controls the bleed valve position in response to a mechanical signal from the main fuel control. Bleed valve position is scheduled within the main fuel control as a function of engine speed and compressor inlet temperature. The starting bleed control and actuation system is similar to the bypass bleed system, except that three actuators are used and its pilot valve controls starting bleed valve position in response to pressure rise in the main fuel pump boost stage.

Bleed position schedules are given in Figure 1-1G as a function of rpm and inlet temperature.

ENGINE INLET GUIDE VANES

The JT11(K) engines incorporate a compressor inlet case which houses a variable two-position inlet guide vane (IGV) system. The guide vanes can be either in the cambered position, which is normal for cruise, or in the axial position which is normal for takeoff and acceleration to moderately high supersonic speeds. Axial positioning of the IGV parallel to the normal airflow path results in more thrust than would be developed with IGV cambered while at speeds up to approximately Mach 1.9. Actuation to the cambered position normally occurs at 80°C to 115°C during acceleration. The cambered position is mandatory when operating continuously at CIT above 125°C (approximately Mach 2.0). Shifting is normally controlled automatically by the main engine fuel control; however, the normal shift to the axial position from cambered can be prevented by selection of the LOCKOUT position of the IGV lockout switch. Refer to Figure 1-1G for IGV shift scheduling information.

IGV Lockout Switches

An inlet guide vane "lockout" system is installed on each engine. This system provides identical shift schedules for the internal bypass bleeds and inlet guide vanes of the corresponding engine. There is a positive locking feature which prevents unscheduled IGV shift to axial after the cambered position has been reached. In addition, a two-position switch has been provided for each system which allows the IGV shift feature to be locked out so that the IGV can be maintained in a cambered condition regardless of internal bypass bleed position. These lift-loc type switches are located in the

COMPRESSOR BLEED AND IGV SHIFT SCHEDULE

IGV/Bypass Bleed Functions

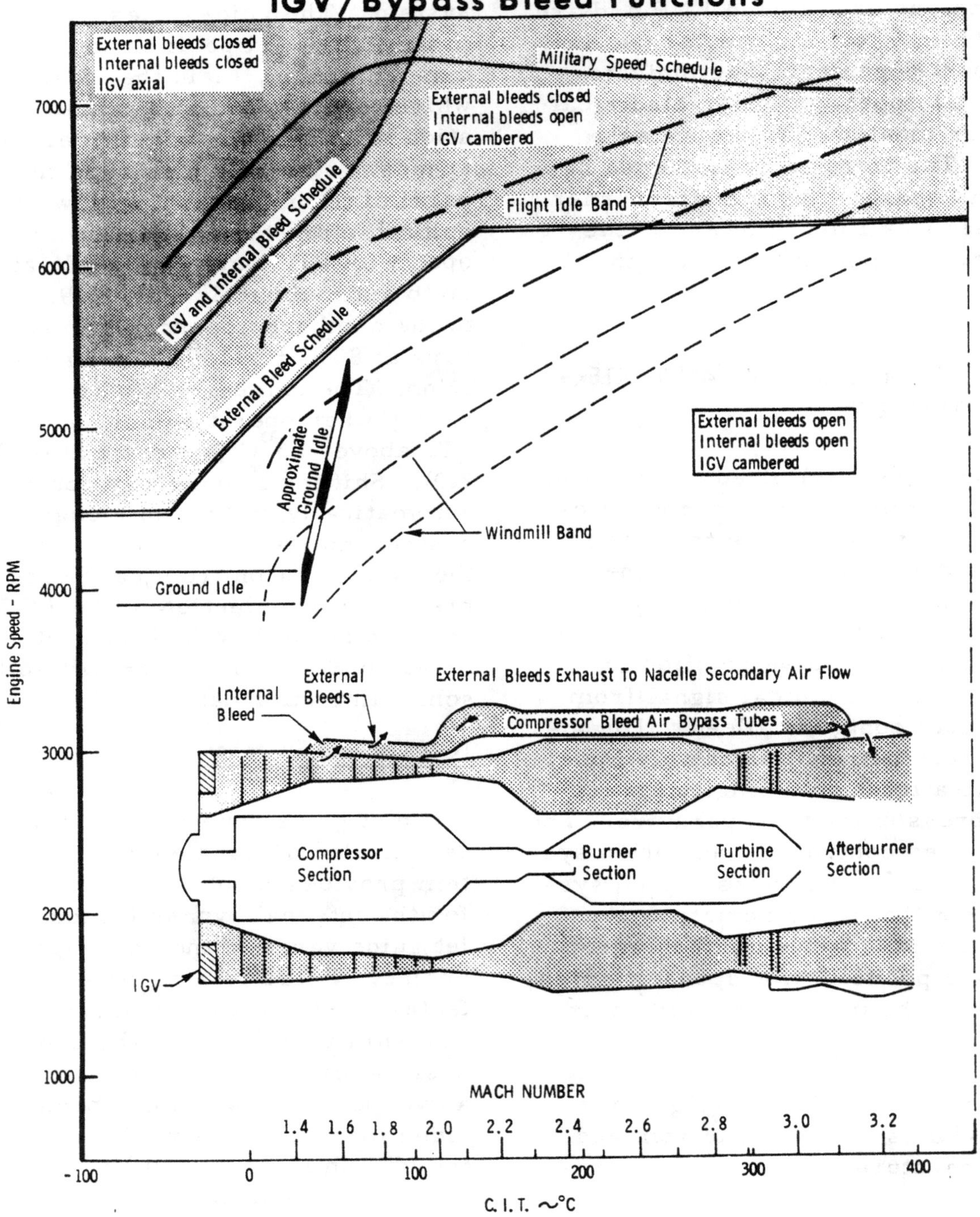

External bleeds closed
Internal bleeds closed
IGV axial

Military Speed Schedule

External bleeds closed
Internal bleeds open
IGV cambered

Flight Idle Band

IGV and Internal Bleed Schedule

External Bleed Schedule

Approximate Ground Idle

External bleeds open
Internal bleeds open
IGV cambered

Windmill Band

Ground Idle

Engine Speed - RPM

7000
6000
5000
4000
3000
2000
1000

External Bleeds

Internal Bleed

External Bleeds Exhaust To Nacelle Secondary Air Flow

Compressor Bleed Air Bypass Tubes

Compressor Section

Burner Section

Turbine Section

Afterburner Section

IGV

MACH NUMBER

1.4 1.6 1.8 2.0 2.2 2.4 2.6 2.8 3.0 3.2

-100 0 100 200 300 400

C. I. T. ~°C

F203-156(d)

Figure 1-1G

right console of the forward cockpit. The LOCKOUT position is ineffective until the guide vanes have reached their cambered position automatically. The switches cannot cause or prevent IGV shift from axial to cambered. With IGV NORM selected, IGV shift is automatic. Power for the IGV lockout solenoid circuits is supplied from the essential dc bus through the L and R INLET GUIDE VANES circuit breakers located on the pilot's right console.

Inlet Guide Vane Position Lights

Inlet guide vane (IGV) position lights are installed on the upper right instrument panel in the forward cockpit. An indicator is provided for each engine, identified by L or R adjacent to the appropriate light. The IGV lights are illuminated when the inlet guide vanes shift to the axial position as scheduled by the main fuel control. (Figure 1-1G illustrates nominal operating schedules.) The lights are extinguished when the IGV move to the cambered position. Position of the inlet guide vanes is sensed by a switch, located on the engine compressor case, which operates when the guide vanes reach or leave the cambered position. Power for the lights is furnished by the essential dc bus.

The following rules apply to IGV position and indicator light conditions.

a. The IGV lights must be off (IGV cambered) during start and at idle.

b. The IGV lights must be on (IGV axial) for takeoff.

c. The IGV lights must be off (IGV cambered) while above 150°C (approximately Mach 2.2).

The IGV schedule for other rpm and CIT conditions is shown on Figure 1-1G.

ACCESSORY DRIVE SYSTEM

An accessory drive system (ADS) is mounted forward of the engine in each nacelle. Its three major components include a constant speed drive, an accessory gearbox, and an all-attitude oil reservoir. Input power from the engine is transmitted to the ADS through a reduction gearbox on the engine and a flexible drive shaft. At the ADS, a constant speed drive unit converts the variable shaft speed to a constant rotational speed to power the ac generator. Two hydraulic pumps and a fuel circulating pump are also mounted on the ADS gearbox. The two hydraulic pumps supply power for operation of the flight control and auxiliary hydraulic systems. The fuel circulating pump supplies fuel to the aircraft heat sink system. The speeds of these pumps vary directly with engine rpm.

ADS lubrication is accomplished by an independent dry sump system with its own pump, using oil from an all-attitude reservoir. The reservoir is pressurized with nitrogen gas from the aircraft LN_2 system and supplies lubricating fluid to the accessory gearbox, the constant speed drive, and the ac generator regardless of flight attitude. (Loss of the LN_2 supply to the ADS does not affect ADS operation during flight.) The oil is cooled by circulation through a fuel-oil heat exchanger which is part of the heat sink system. Reservoir capacity is approximately 8 quarts.

EXTERNAL STARTER SYSTEM

An external starter cart is provided for ground starts. This may be either a self-contained gas engine cart or a multiple air turbine cart. The output drive gear of either cart connects to a starter gear on the main gearbox at the bottom of the engine. There are no aircraft controls for this system; it is turned on and off by the ground crew in response to signals from the pilot. Air starts are made by windmilling the engine.

CHEMICAL & CATALYTIC IGNITION SYSTEMS

Triethylborane (TEB) is used for starting ignition of main burner and afterburner fuel. Catalytic igniters attached to the afterburner flameholders maintain afterburner operation after initial ignition. Special handling procedures are required for the TEB because it will burn spontaneously upon exposure to air above -5°F. The TEB is contained in a 600 cc (1-1/4 pint) storage tank which is pressurized with nitrogen prior to flight. The nitrogen provides inerting and operating pressure to supply a metered quantity of TEB to either the main burner or afterburner section. Operation is in response to a fuel pressure signal while the engine is rotating, preferably while in the 5000 to 6000 rpm range. Actuation is automatic with throttle movement from OFF toward IDLE and from the Military position into the afterburner range. At least 16 effective TEB injections can be made with one full tank. The TEB tank is engine-mounted and is cooled by main burner fuel flow to maintain the TEB temperature within safe limits. If tank pressure exceeds a safe level, a rupture disc is provided to discharge vaporized TEB and tank nitrogen through the afterburner section. No pilot indication of TEB tank discharge is provided.

TEB Remaining Counters

A mechanical counter for each engine is located aft of each throttle which indicates the number of engine starts or afterburner lights which remain available from the chemical ignition (TEB) system. The counters are spring wound and set to 12 prior to engine start. Each time a throttle is moved forward from OFF to IDLE or from MILITARY to the AFTERBURNER range, the counter will reduce one number to indicate the number of TEB shots remaining.

Igniter Purge Button

A pushbutton labeled IGNITER PURGE is installed in the forward cockpit on the right instrument panel. When held in the on position, a solenoid-operated valve supplies fuel hydraulic system pressure to the chemical ignition system dump valve if the engine is rotating. This allows the remaining TEB to be dumped into the afterburner section, preferably while the engine is running in the 5000 to 6000 rpm range. The pushbutton must be actuated for no less than 30 seconds in order to dump a full load of TEB. At the end of the dump period it should be released and recycled in order to clean out the lines.

COMPRESSOR INLET INSTRUMENTATION

Compressor Inlet Pressure Gage

The compressor inlet pressure (CIP) gage, located on the top right side of the pilot's instrument panel, has an L needle and an R needle which indicate actual inlet static pressures at the face of the compressor, and a striped third needle that indicates pressure to be expected when the inlets are operating normally. The position of the third pointer is governed by the air data computer using pressures sensed by the pitot static system. Indications are scheduled in accordance with automatically computed values of Mach number and KEAS so that the needle shows "normal" CIP for the flight condition. A substantial difference between the "normal" hand and actual CIP indicates improper inlet operation. Higher actual pressure at normal speeds and altitudes indicates possible unstart conditions. Lower pressure indicates poor pressure recovery due to improper spike and/or bypass door settings except when at abnormal angles of attack or in yaw conditions where inlet operation is automatically biased to produce less than normal recovery. The spread between inlet CIP indications (L and R needles) should not exceep 1 psi. The difference between either needle and the striped third needle should not exceed 1 psi. The needle can be used as a guide for bypass door settings during manual operation of one or both inlets; however, it is preferable to keep the L and R needles slightly below the "normal" hand indication to maintain a margin below unstart pressures. Continued automatic or manual inlet operation at pressures below the "normal" indication can result in less aircraft range.

A CIP pushbutton test switch is located to the right of the gage. When the switch is depressed, the needles move towards zero to indicate proper operation of the gage.

Power is supplied by the essential ac bus through the CIP circuit breaker located on the pilot's left console.

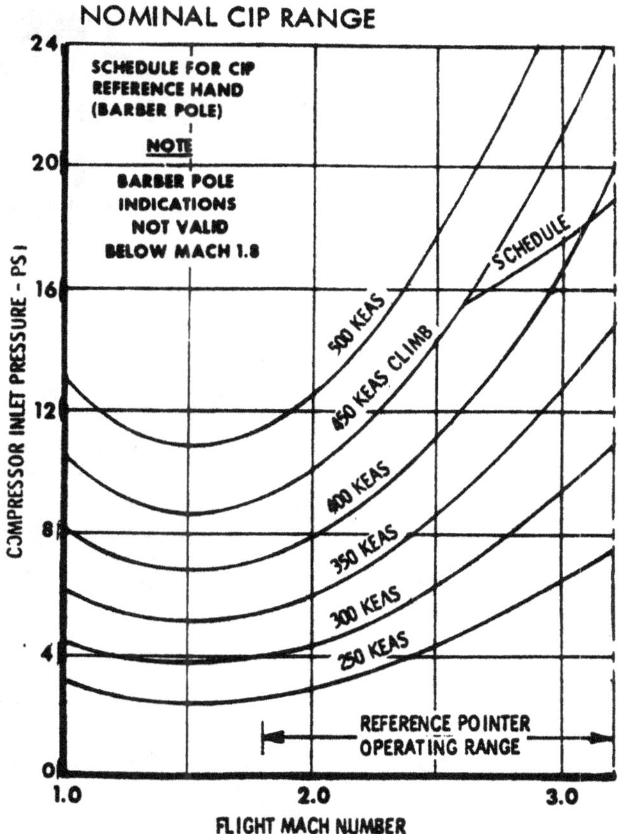

Figure 1-2

Compressor Inlet Temperature Gage

A dual-indicating compressor inlet temperature gage is mounted on the upper right side of the pilot's instrument panel. Major calibrations are marked in a range from 0° through 500°C. The needles indicate air temperature forward of the first compressor stage. Power is furnished to the gage by the essential ac bus.

CENTER INSTRUMENT PANEL - Forward Cockpit

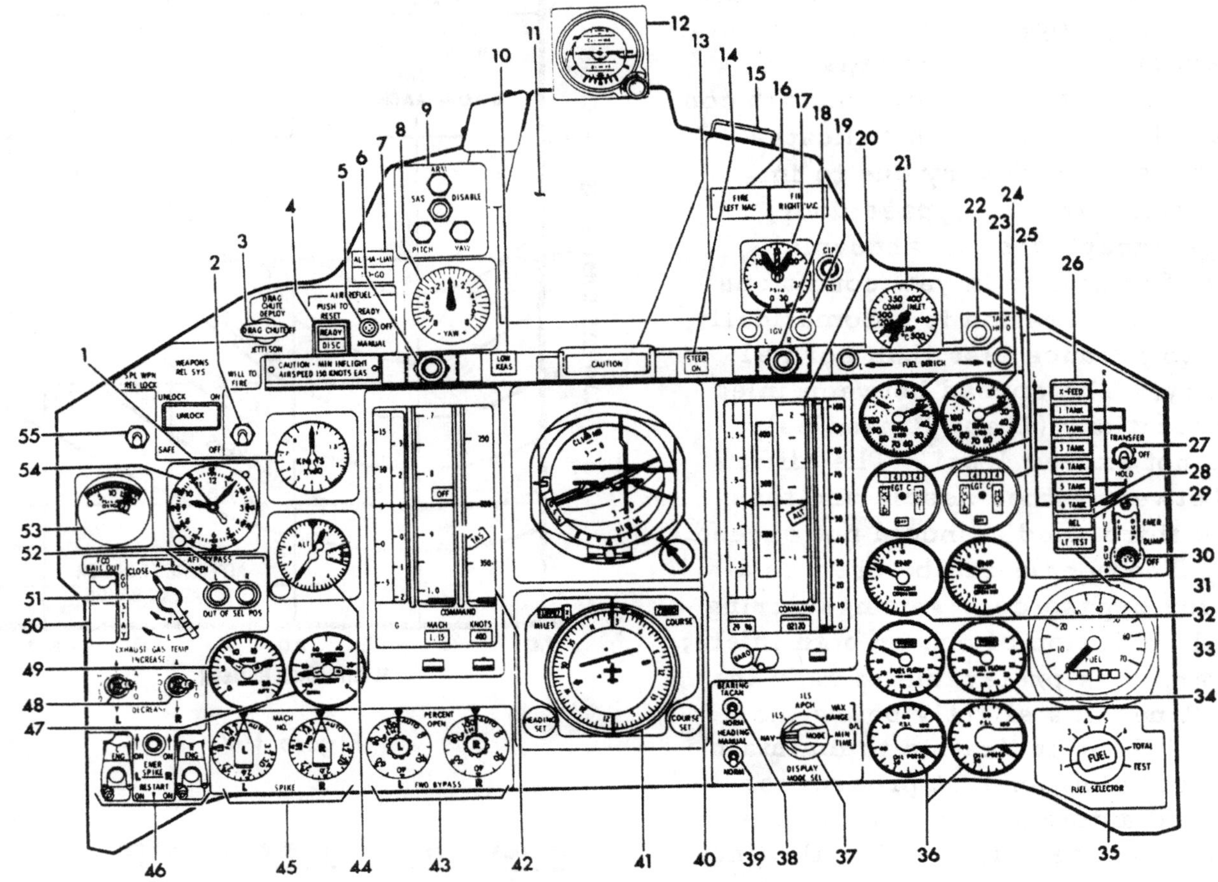

1 Standby Airspeed Indicator
2 Will-To-Fire Switch
3 Drag Chute Switch
4 IFR Reset Switch and Indicator Light
5 Air Refuel Ready Switch
6 Airspeed- Mach Indicator Test Button
7 α Limit Warning Light
8 Angle of Yaw Indicator
9 SAS Instant Recall Display Panel
10 Low KEAS Warning Light
11 Display Panel
12 Standby Attitude Indicator
13 Master Caution Light
14 Nosewheel Steering Engaged Light
15 Knife
16 Nacelle Fire Warning Lights
17 Compressor Inlet Pressure Gage
18 Altitude-Vertical Velocity Indicator Test Button
19 IGV Axial Position Lights
20 Altitude-Vertical Velocity Indicator
21 Compressor Inlet Temperature Gage
22 Tank 1 Hold Light
23 Fuel Derich Light
24 Tachometers
25 Exhaust Gas Temperature Gages
26 Fuel Crossfeed Switch
27 Fuel Transfer Switch
28 Fuel Boost Pump Switch
29 Pump Release Switch
30 Fuel Dump Switch
31 Fuel Boost Pump Light Test Switch
32 Exhaust Nozzle Position Indicators
33 Fuel Quantity Indicator
34 Fuel Flow Indicators
35 Fuel Quantity Indicator Selector Switch
36 Oil Pressure Gages
37 Display Mode Selector Switch
38 Bearing Select Switch
39 Heading Select Switch
40 Attitude-Director Indicator
41 Horizontal Situation Indicator
42 Airspeed-Mach Indicator
43 Forward Bypass Switches
44 Standby Altimeter
45 Spike Switches
46 Emergency Spike - Restart Switches
47 Forward Bypass Position Indicator
48 Exhaust Gas Temperature Trim Switches
49 Spike Position Indicator
50 FCO Bailout Switch
51 Aft Bypass Switches
52 Aft Bypass Position Lights
53 Oxygen Quantity Indicator
54 Clock
55 Special Weapon Release Lock Switch

F202-10 (g)

Figure 1-2A

INSTRUMENT SIDE PANELS - Forward Cockpit

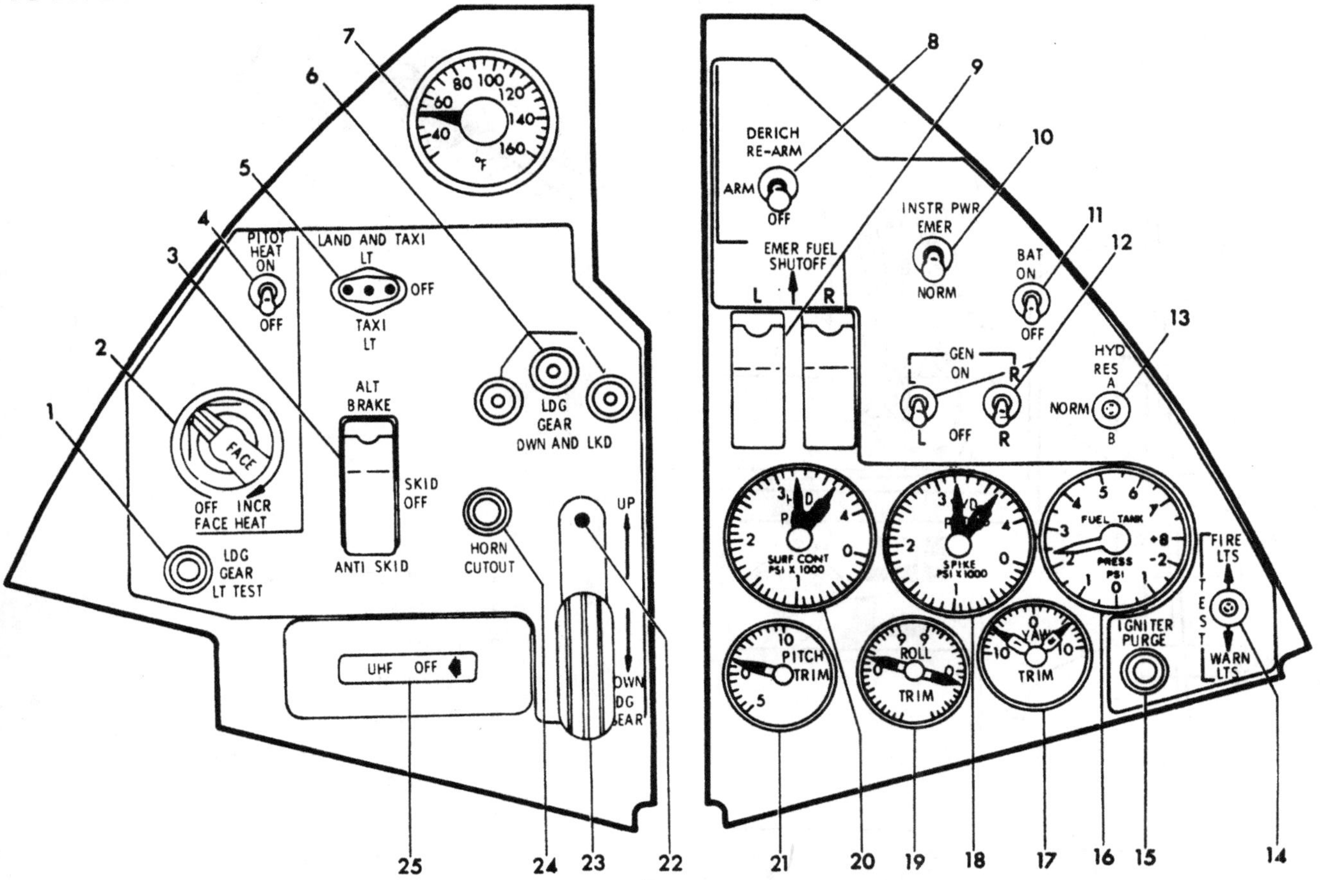

1. Landing Gear Lights Test Switch
2. Face Heat Switch
3. Brake Switch
4. Pitot Heat Switch
5. Landing and Taxi Light Switch
6. Landing Gear Indicator Lights
7. Cockpit Temperature Indicator
8. Fuel Derich Arm Switch
9. Emergency Fuel Shutoff Switches
10. Instrument Power Switch
11. Battery Switch
12. Generator Switches
13. Hydraulic Reserve Oil Switch
14. Fire and Warning Lights Test Switch
15. Igniter Purge Switch
16. Fuel Tank Pressure Indicator
17. Yaw Trim Indicator
18. L and R Hydraulic Systems Pressure Gage
19. Roll Trim Indicator
20. A and B Hydraulic Systems Pressure Gage
21. Pitch Trim Indicator
22. Landing Gear Override Button
23. Landing Gear Lever
24. Landing Gear Horn Cutout Button
25. UHF Channel Frequency Indicator Window

F202-12(d)

Figure 1-2B

LEFT CONSOLE - Forward Cockpit

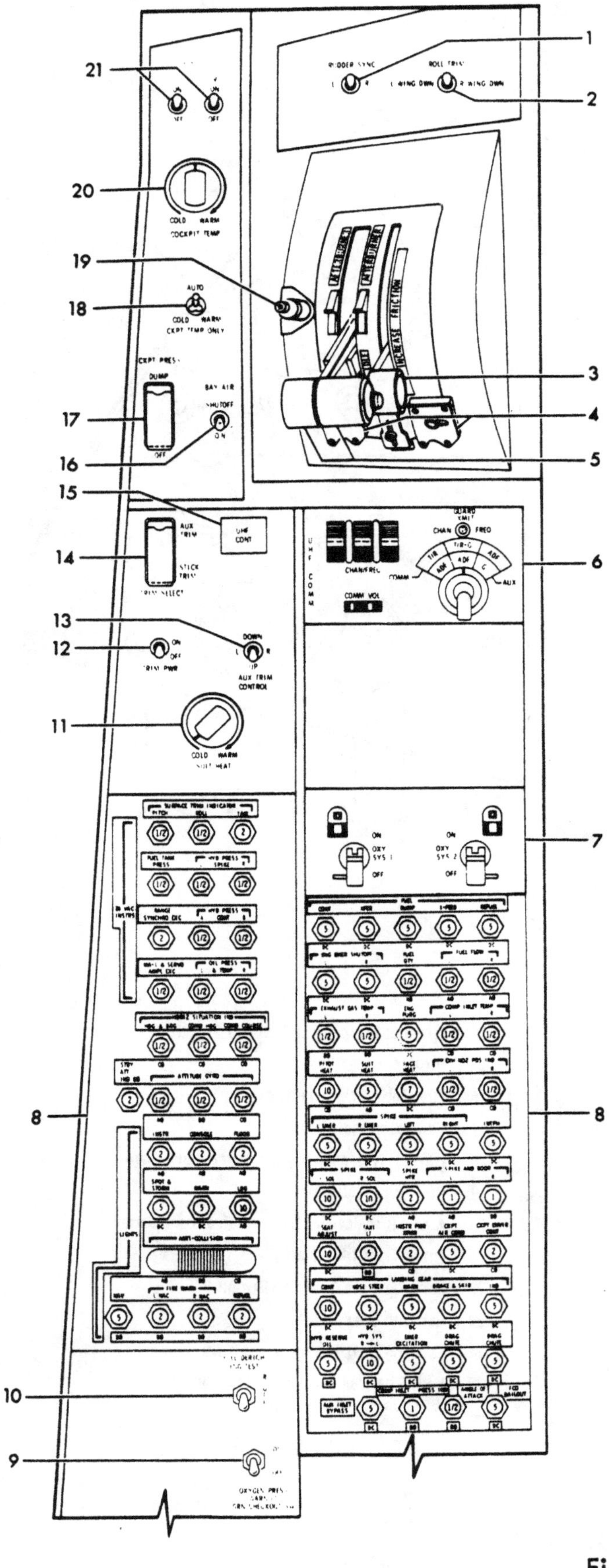

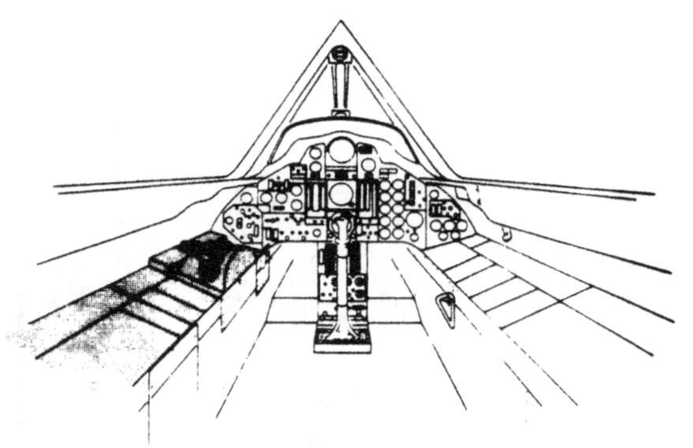

1 Rudder Synchronizer Switch
2 Roll Trim Switch
3 Throttle Friction Lever
4 TEB Counters
5 Throttles
6 UHF Command Radio Control Panel
7 Oxygen System Control Panel
8 Circuit Breaker Panels
9 Oxygen Warning Test Switch
10 Fuel Derich Test Switch
11 Suit Heat Rheostat
12 Trim Power Switch
13 Auxiliary Trim Control Switch
14 Trim Selector Switch
15 UHF Control Button
16 Bay Air Shutoff Switch
17 Cockpit Pressure Dump Switch
18 Cockpit Temperature Mode Selector Switch
19 SAS Instant Recall Switch
20 Cockpit Temperature Rheostat
21 L and R Air System Switches

F202-15(a)

Figure 1-2C

RIGHT CONSOLE - Forward Cockpit

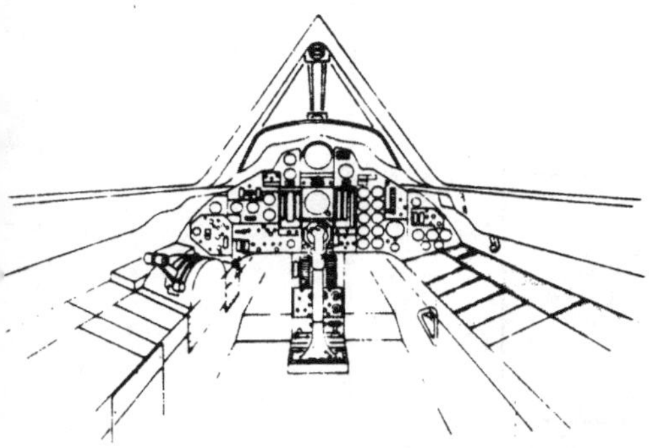

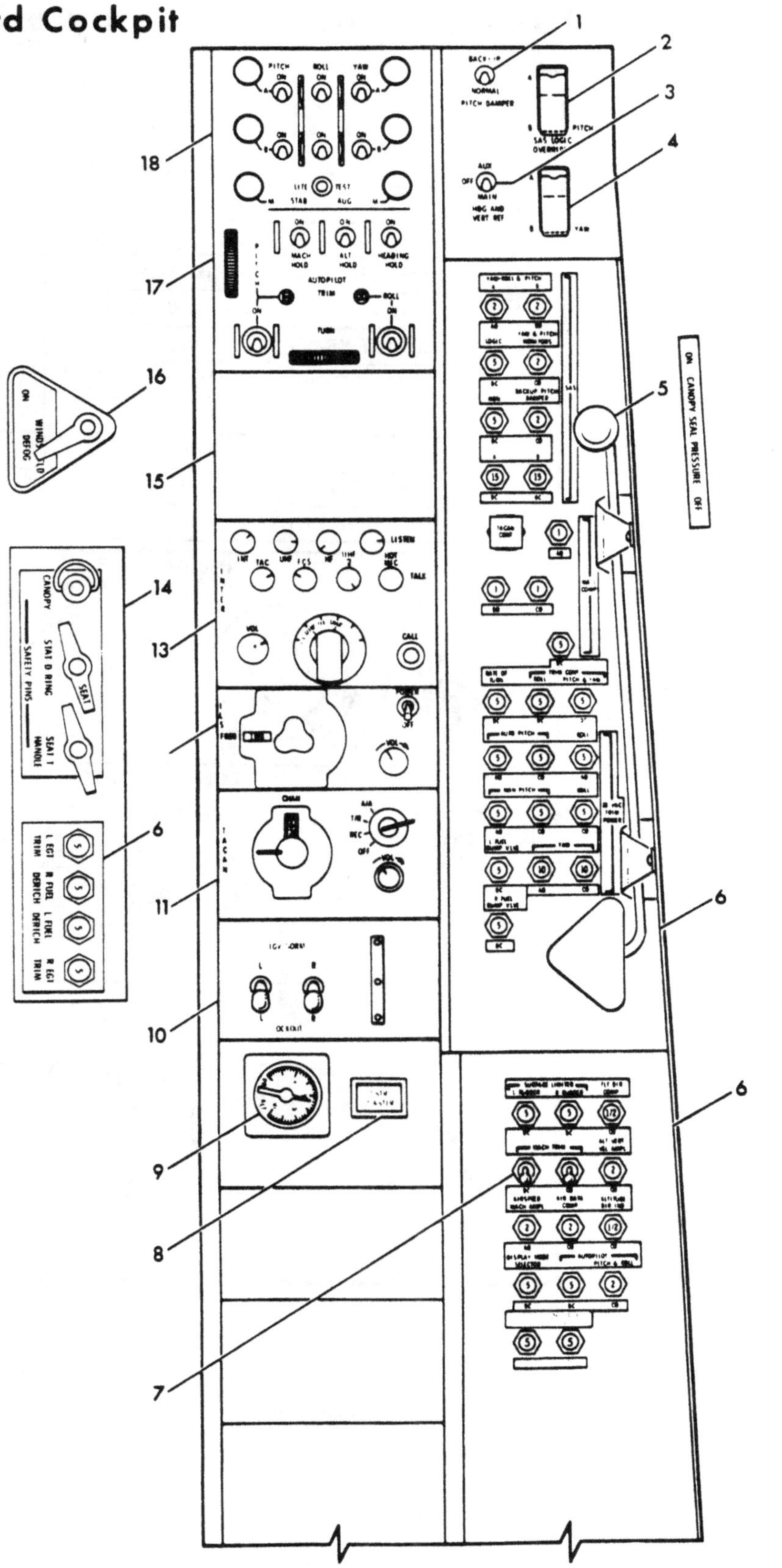

1 Backup Pitch Damper Switch
2 SAS Pitch Logic Override Switch
3 Heading and Vertical Reference Switch
4 SAS Yaw Logic Override Switch
5 Canopy Seal Selector Lever
6 Circuit Breaker Panels
7 Mach Trim System Circuit Breakers
8 Instrumentation Master Switch
9 Cabin Altimeter
10 Engine Inlet Guide Vane Lockout Switches
11 TACAN Control Panel
12 ILS Control Panel
13 Interphone System Control Panel
14 Ejection Seat Safety Pins
15 Special Instrumentation Control Panel
16 Windshield Defog Lever
17 Autopilot Controls
18 SAS Controls

F202-14(a)

Figure 1-2D

LOWER INSTRUMENT PANEL - Forward Cockpit

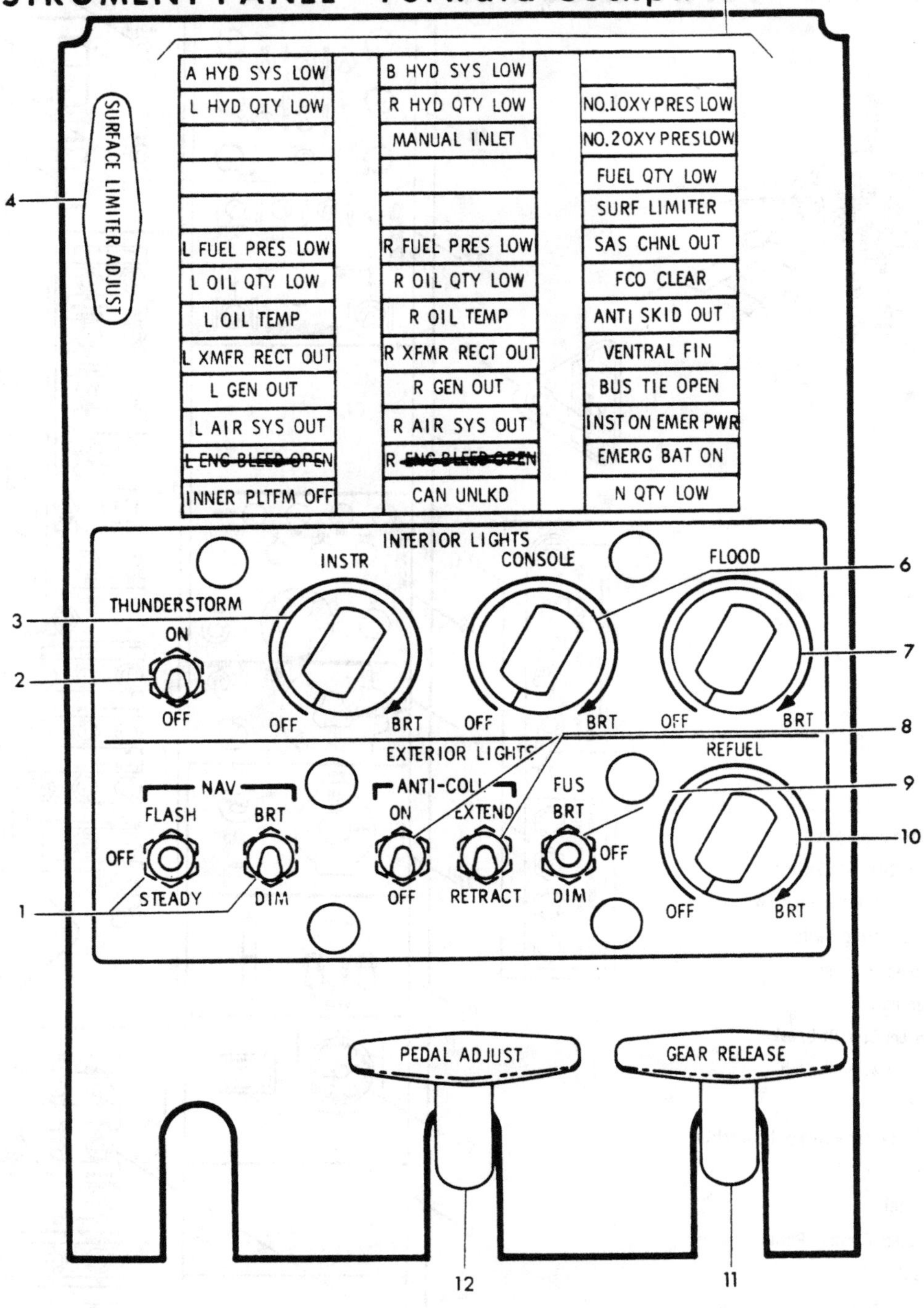

A HYD SYS LOW	B HYD SYS LOW	
L HYD QTY LOW	R HYD QTY LOW	NO.1 OXY PRES LOW
	MANUAL INLET	NO.2 OXY PRES LOW
		FUEL QTY LOW
		SURF LIMITER
L FUEL PRES LOW	R FUEL PRES LOW	SAS CHNL OUT
L OIL QTY LOW	R OIL QTY LOW	FCO CLEAR
L OIL TEMP	R OIL TEMP	ANTI SKID OUT
L XMFR RECT OUT	R XFMR RECT OUT	VENTRAL FIN
L GEN OUT	R GEN OUT	BUS TIE OPEN
L AIR SYS OUT	R AIR SYS OUT	INST ON EMER PWR
~~L ENG BLEED OPEN~~	R ~~ENG BLEED OPEN~~	EMERG BAT ON
INNER PLTFM OFF	CAN UNLKD	N QTY LOW

SURFACE LIMITER ADJUST

INTERIOR LIGHTS
INSTR
CONSOLE
FLOOD
THUNDERSTORM
ON
OFF
OFF BRT
OFF BRT
OFF BRT

EXTERIOR LIGHTS
NAV
FLASH
BRT
OFF
STEADY
DIM
ANTI-COLL
ON
EXTEND
OFF
RETRACT
FUS
BRT
OFF
DIM
REFUEL
OFF BRT

PEDAL ADJUST
GEAR RELEASE

1 Navigation Lights Switches
2 Thunderstorm Lights Switch
3 Instrument Lights Rheostat
4 Surface Limiter Handle
5 Annunciator Panel
6 Console Lights Rheostat
7 Floodlights Rheostat
8 Exterior Lights Switches
9 Fuselage Lights Switch
10 Refueling Light Rheostat
11 Landing Gear Release Handle
12 Pedal Adjust Handle

F202-13(

Figure 1-2E

AFT COCKPIT INSTRUMENT PANEL

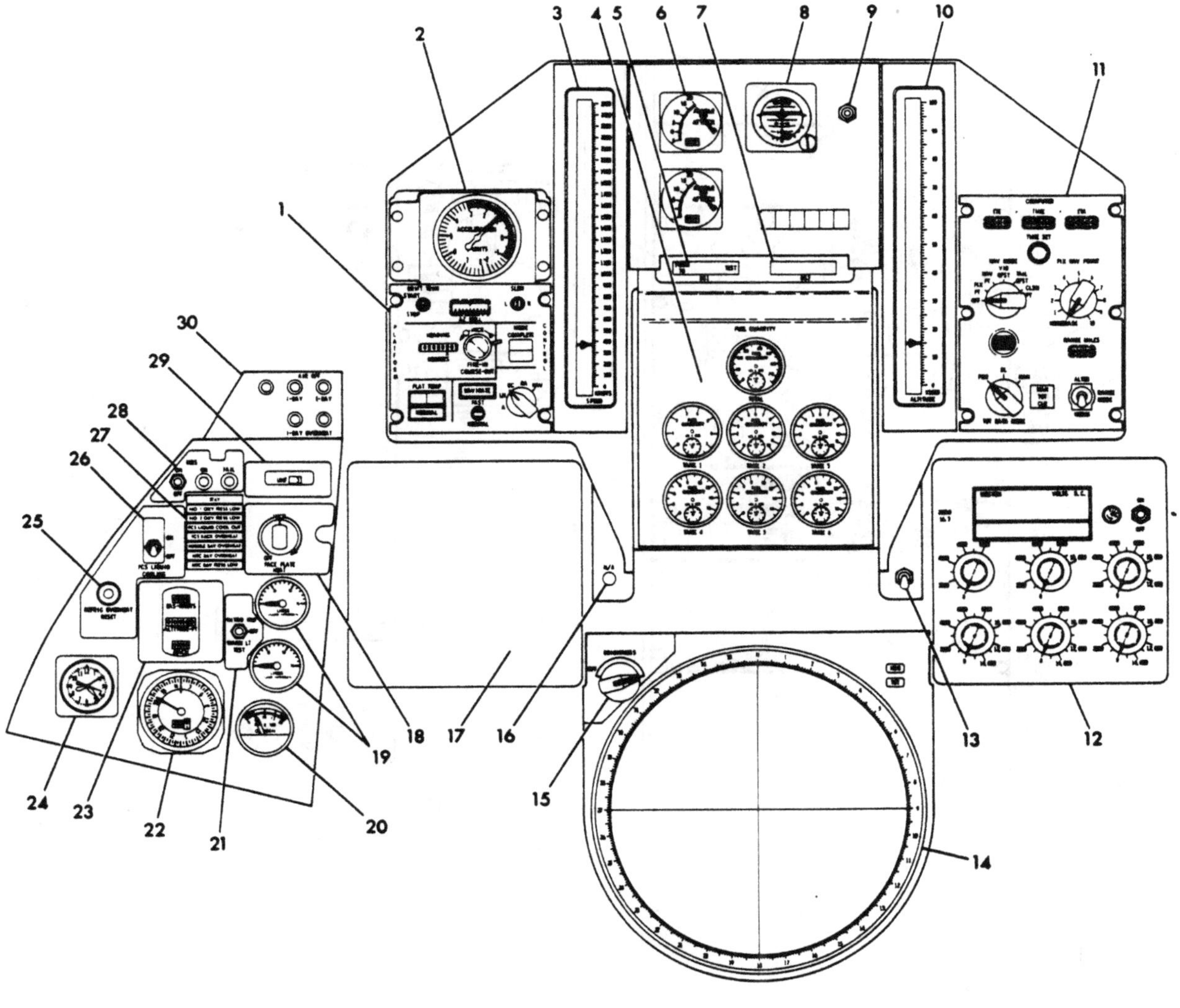

1 Alignment Control Panel
2 Accelerometer
3 True Speed Indicator
4 Fuel Quantity Indicators
5 Bailout Light
6 Angle of Attack Indicators
7 Master Caution Light
8 Attitude Indicator
9 Fast Erect Pushbutton
10 Altitude Indicator
11 Computer Display Panel
12 Center of Gravity Computer
13 Not Active
14 Horizontal Tactics Indicator (HTI)
15 HTI Brightness Rheostat
16 Not Active
17 Special Equipment Panel
18 Faceplate Heat Control
19 Liquid Nitrogen Quantity Gage
20 Oxygen Pressure Gage
21 LN_2/Warning Light Test Select Switch
22 BDHI Instrument
23 Triple Display Indicator
24 Clock
25 Refrigeration Overheat Reset Switch
26 Fire Control System Refrigeration Power Switch
27 Annunciator Panel
28 Recorder System Control Panel
29 UHF Radio Channel-Frequency Indicator
30 Air Conditioning Warning Lights Panel

Figure 1-2F

F202-37

LEFT & RIGHT CONSOLES — Aft Cockpit

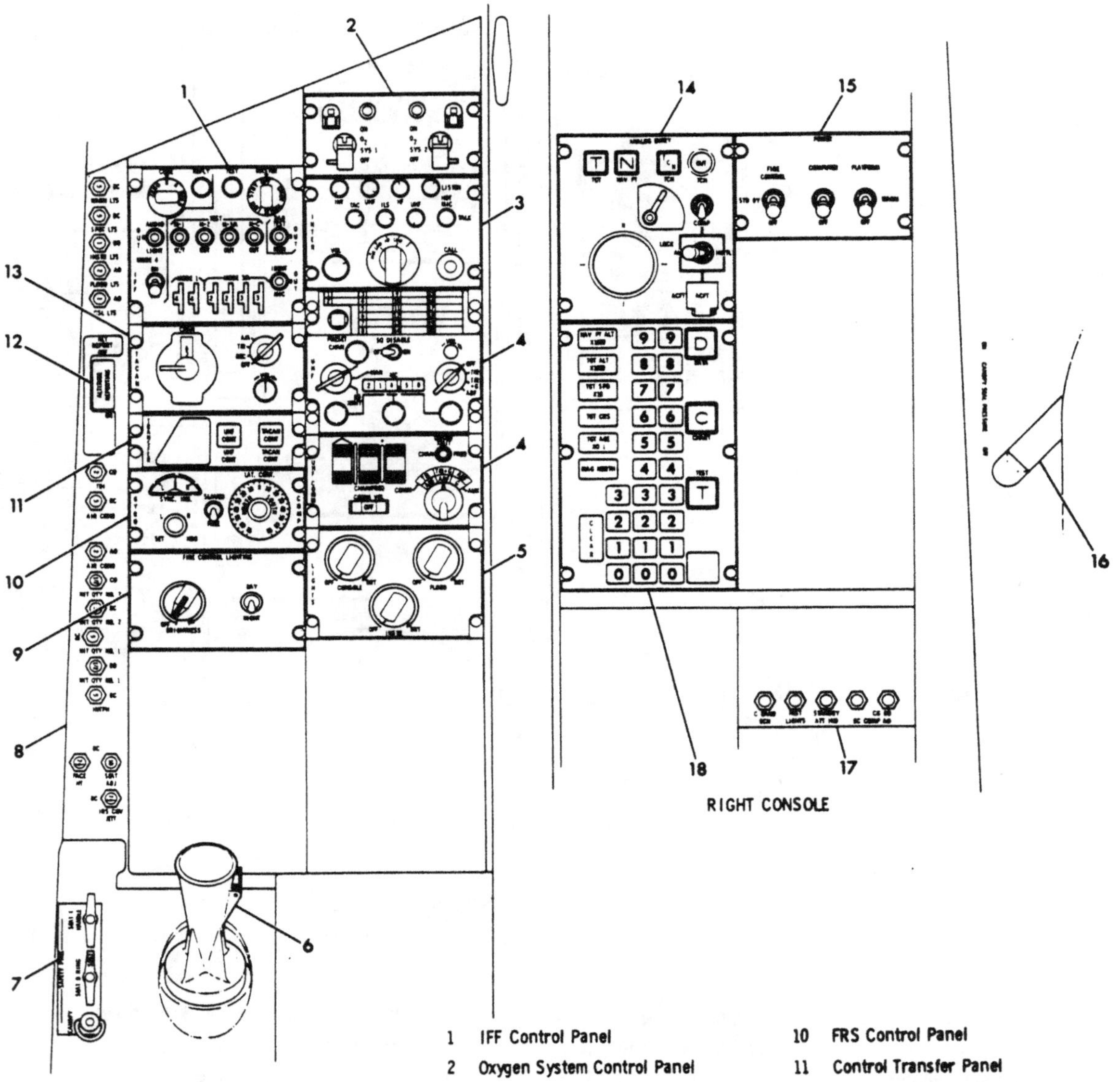

LEFT CONSOLE

RIGHT CONSOLE

1 IFF Control Panel
2 Oxygen System Control Panel
3 Interphone Control Panel
4 ARC-51 UHF Radio Control Panels
5 Cockpit Lighting Controls
6 Relief Tube
7 Seat and Canopy Safety Pins
8 Circuit Breaker Panel
9 Fire Control System Lighting Controls
10 FRS Control Panel
11 Control Transfer Panel
12 IFF Altitude Reporting Cut-Out Switch
13 TACAN System Control Panel
14 Analog Entry Panel
15 Nav System Power Control Panel
16 Canopy Seal Pressure Valve
17 Circuit Breakers
18 Computer Digital Entry Keyboard

F202-38

Figure 1-2G

AIR INLET SYSTEM

The inlet system includes the cowl structure, a moving spike to help provide optimum internal airflow characteristics, a spike porous centerbody bleed, forward and aft bypass openings, and an internal cowl shock trap bleed for internal shock wave positioning and boundary layer flow control. The air inlets for each nacelle are canted inboard and downward to align with the local airflow pattern. (See figures 1-3 and 1-4.) The forward and aft bypass openings control airflow characteristics within the inlet and mass flow to the engine. Normally, the spike and forward bypass opening is operated automatically by the air inlet control system and the aft bypass opening is scheduled manually. Overriding manual controls are provided for the spike and forward bypass. The forward bypass can be operated manually with the spike in automatic operation; however, when the spike is controlled manually the forward bypass must also be in manual control if variable bypass opening is desired. Manual operation of the spike alone while the forward bypass control is in the AUTO position will cause the forward bypass to open fully.

Inlet Control Parameters

Inlet airflow is controlled automatically at supersonic airspeed by the forward bypass and spike positioning so that the locations of shock waves ahead of the inlet and at the inlet throat produce maximum practicable ram (pressure recovery) at the engine face and supply the proper amount of air to the engine.

Manually operated controls provide incremental operation of the aft bypass for those conditions where additional bypass area is required or where a reduction in forward bypass flow is desired. The forward and aft bypass openings and the spikes for the left and right inlets are operated by the L and R hydraulic systems, respectively.

The parameters used to control automatic forward bypass and spike positioning include Mach number, angle of attack, and angle of sideslip. The inlet control also includes a load factor (g) bias to the spike and forward bypass schedules, and a shock expulsion sensor and restart system.

Load Factor Bias

The g-bias function of the air inlet control computer causes the spike to schedule forward and the forward bypass to schedule more open when load factors greater than 1.12 g or less than 0.88 g are experienced during automatic inlet scheduling. The spike biasing is approximately 2 1/2 inches forward per 1-g variation in normal acceleration. This action reduces the possibility of inlet unstarts occurring during turns or other maneuvers which affect angle of attack at the inlets. A noticeable decrease in CIP results from operation of the g-bias. The g-bias feature does not override a manually operated spike or forward bypass.

INLET SPIKES

Each spike is automatically locked in the forward position for ground operation and for flight below 30,000 feet. They are unlocked automatically above this altitude, but remain approximately full forward until Mach 1.6 is reached. During automatic operation above Mach 1.6, the spikes begin to retract to increase the inlet capture area and to de-

AIRFLOW PATTERNS

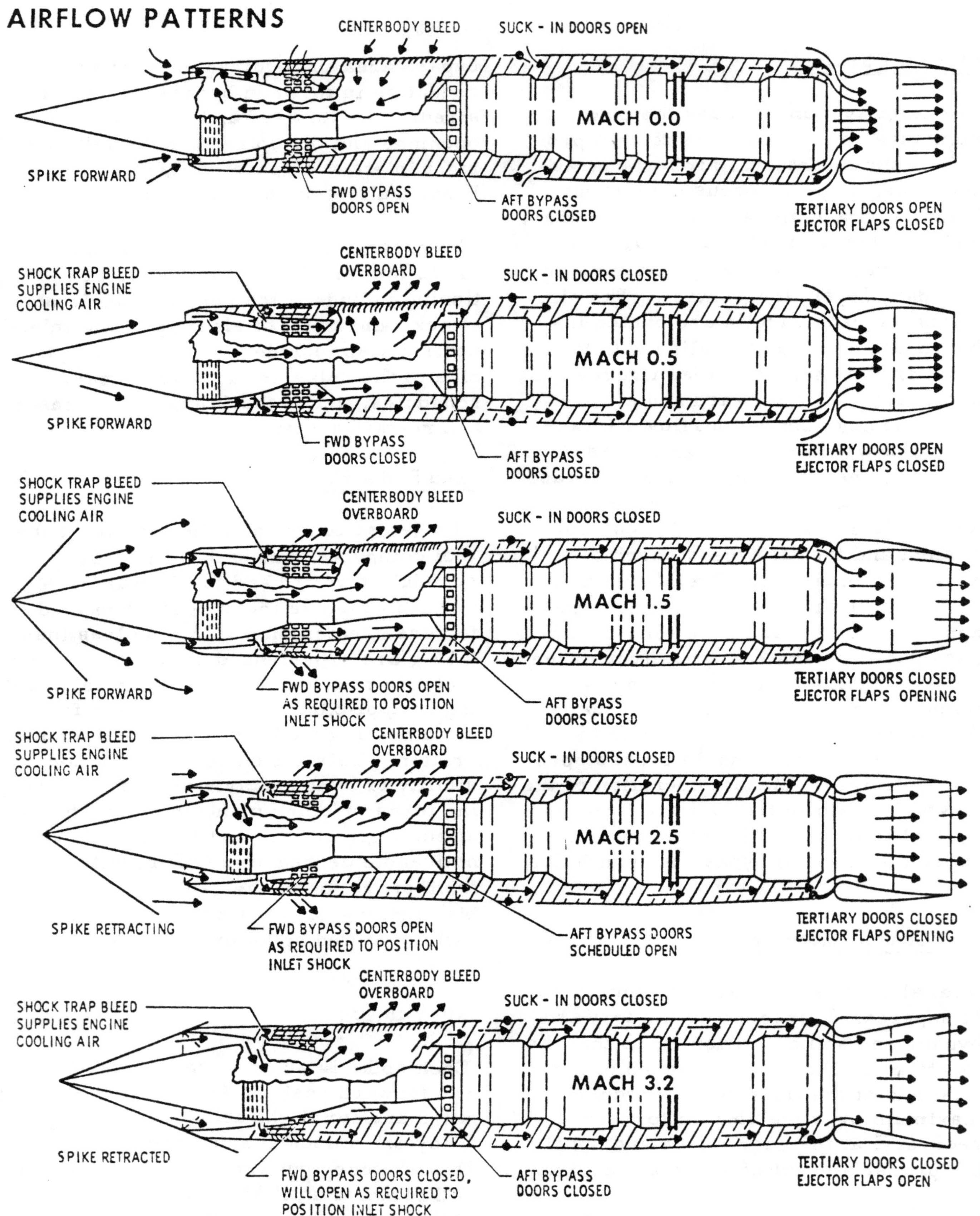

Figure 1-3

crease the inlet throat area which is the narrowest portion of the duct.

Spike position is scheduled primarily as a function of Mach number during automatic operation, with biasing for angle of attack, angle of sideslip, and normal g's. The spike moves aft approximately 26 inches during the transition from Mach 1.6 to Mach 3.2. Total and static pressures for Mach number control of spike position are sensed by the pitot-static head. Angle of attack and angle of sideslip are sensed by the attitude sensor probe located on the left side of the pitot mast. Normal g's are sensed by the inlet control normal accelerometer.

Spike position can also be set manually in accordance with a Mach schedule by use of cockpit controls unless hydraulic pressure is not available for the respective inlet. In this event, use of the appropriate emergency spike-forward switch provides one-shot pneumatic pressure to move and lock the affected inlet spike forward and open its forward bypass. A complete loss of the L and/ or R system hydraulic pressure should exist prior to using the emergency system. If system hydraulic pressure for the affected inlet has not failed completely but is low or fluctuating, it is possible for the spike to become unlocked after gas pressure released by the emergency spike switch has been depleted. (SPIKE FWD should also be selected with the inlet spike control as a precaution.)

The spike centerbody is equipped with small slots which remove spike boundary layer air from the inlet throat. This porous construction prevents flow separation. The air is ducted overboard through the nacelle louvers after passing through the spike and its supporting struts.

INLET FORWARD BYPASS

The forward bypass openings in each inlet provide overboard exhausts for that portion of the inlet air which is not required by the engine. The amount of opening is automatically scheduled by the inlet computer to control inlet diffuser pressure and position the normal (internal) shock properly. The bypass consists of a rotating basket which opens duct exhaust ports located a short distance aft of the inlet throat. The excess air exits through louvers located forward of the spike centerbody bleed louvers. When the landing gear is down, the forward bypass is held open by application of an electrical override signal from a landing gear door switch. Control of the bypass openings reverts to the manual or automatic mode of operation, as selected by the pilot, after gear retraction. In automatic operation, the forward bypass remains closed until speed is above Mach 1.4. Then it is released to modulate in accordance with the air inlet control system schedules. The inlet usually "starts" in the speed range from Mach 1.6 to Mach 1.8; that is, the normal shock is made to move from a position in front of the inlet to a position near the cowl shock trap bleed in the inlet throat area. As Mach number increases, the forward bypass opening is scheduled as required to maintain the normal shock at the throat position.

In automatic operation, the amount of forward bypass opening is controlled by the inlet computer, which senses the ratio of inlet duct static pressure (P_sD8) to a reference total pressure (P_pLM). There are four inlet duct static pressure taps which act as shock position sensors. These taps are located circumferentially on the inlet duct wall aft of the shock trap bleed. The reference

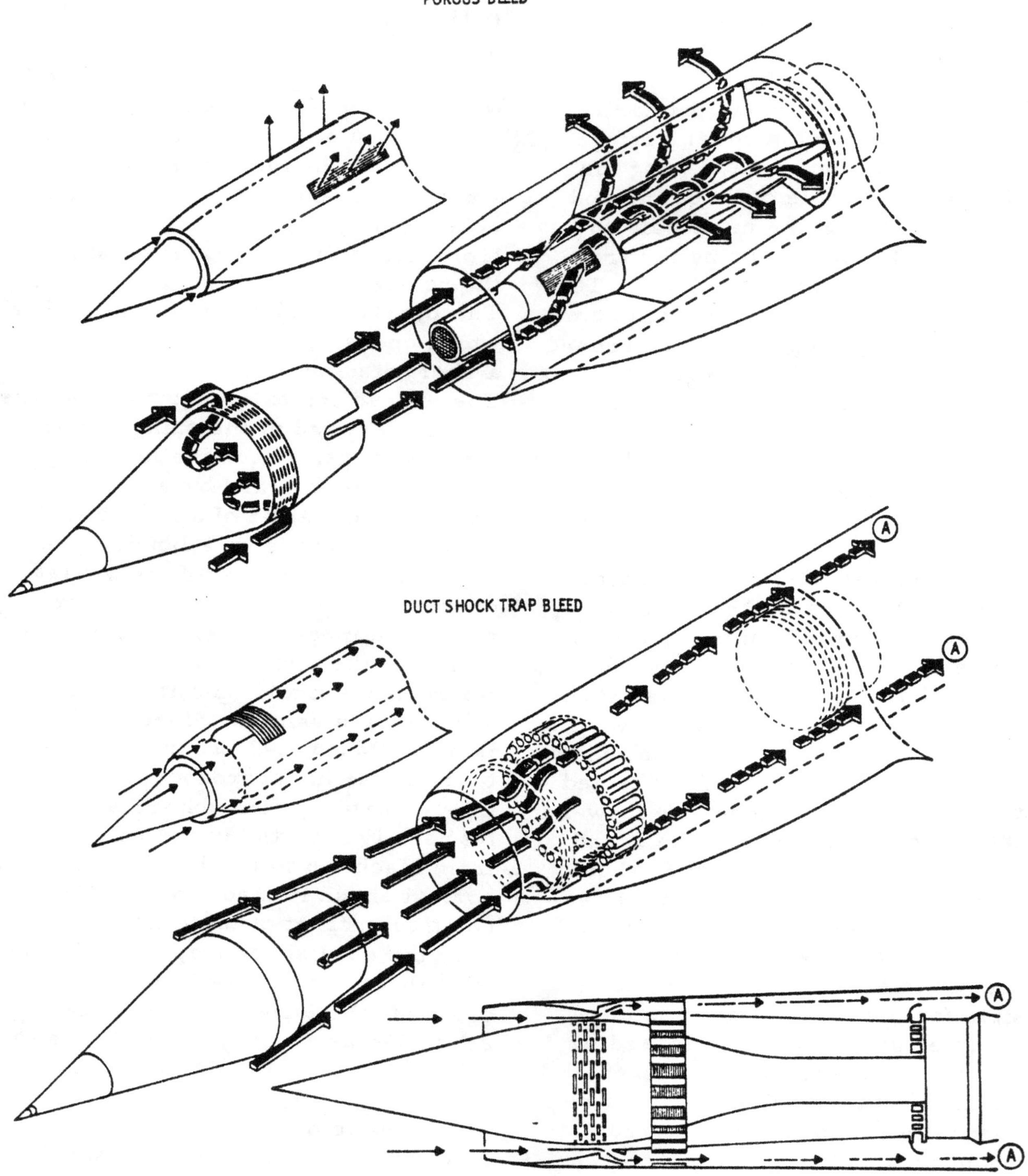

Ⓐ Duct boundary layer air is removed at the shock trap. It passes through tubes in the bypass door, becomes secondary nacelle air and exhausts through the ejector nozzle.

M203-1-4-196(b)

Boundary Layer Control

Figure 1-4 (Sheet 1 of 2)

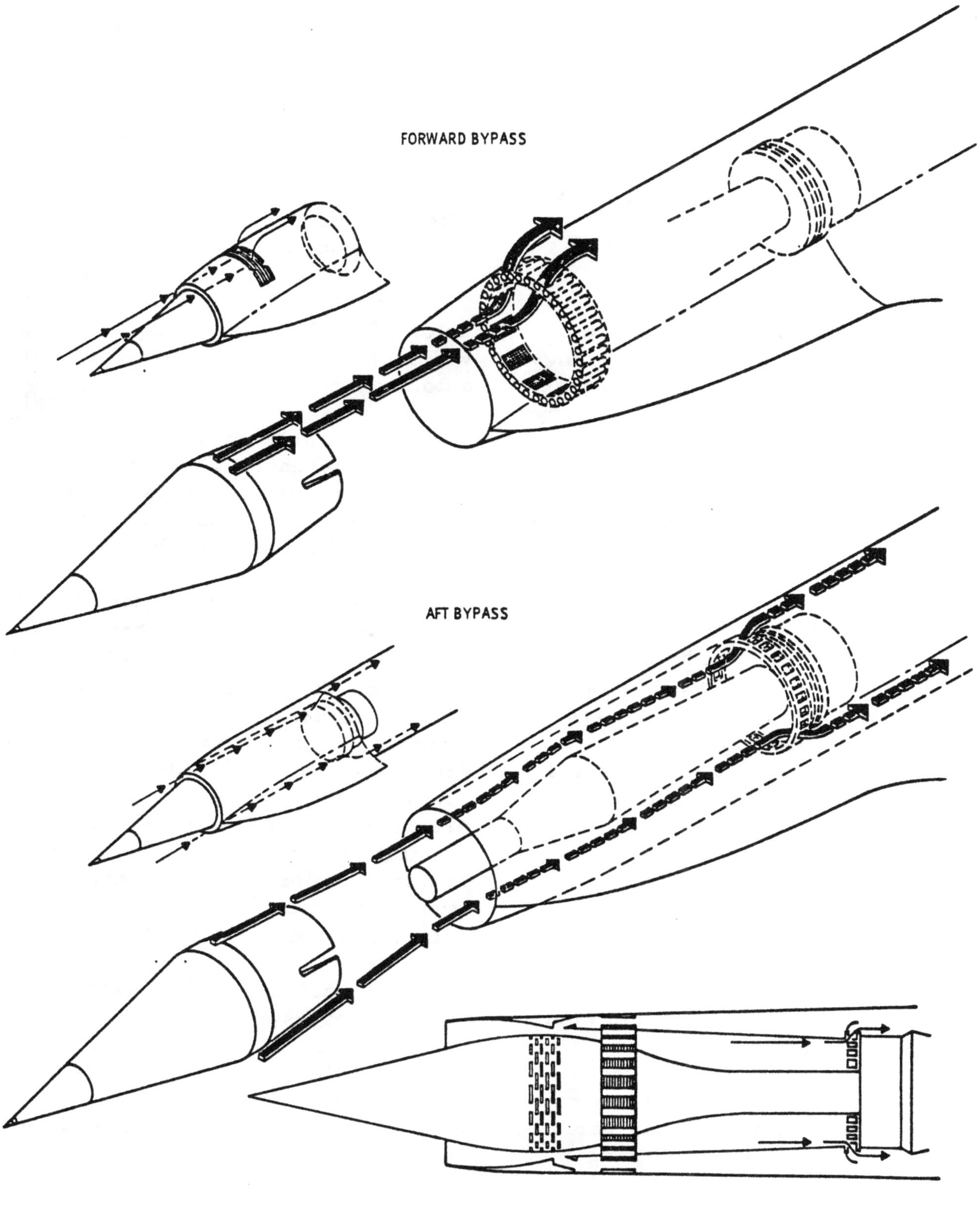

Forward and Aft Bypass Door Airflow

Figure 1-4 (Sheet 2 of 2)

total pressure is sensed by two external probes located on the outside of each nacelle. The amount of forward bypass opening can also be set manually in accordance with a Mach schedule by use of cockpit controls unless hydraulic pressure is not available. (See figure 1-5).

INLET AFT BYPASS

The air inlet aft bypass consists of a ring of adjustable, peripheral openings located just forward of the engine face. These openings allow inlet air to be bypassed around the engine. The bypassed air joins cowl shock trap bleed air and passes through the space between the outside of the engine and afterburner and the inside of the nacelle. This flow augments the exhaust gas flow in the ejector area. The aft bypass ring in each nacelle is positioned by a hydraulic actuator powered by the respective L or R hydraulic system, and controlled by a rotary switch in the cockpit. While the gear is down, the aft bypass is held closed by an electrical override signal from the nose gear downlock.

AUTOMATIC RESTART

The inlet control also includes a shock expulsion sensor (SES) and an auto restart feature which operates automatically when speeds above the minimum for inlet scheduling are reached. (Normally, SES is effective above Mach 2.0). If the inlet normal shock is expelled, the SES for that inlet overrides the auto spike and forward bypass schedules for both inlets. This "cross tie" feature keeps the other inlet from unstarting during attitude variations which might occur immediately after the original unstart occurs, reduces the asymmetric thrust which could otherwise result and minimizes undesirable sideslip angles. During the automatic restart cycle, the forward bypasses open fully and the spikes move forward as much as 15 inches. Spike retraction is started 3.75 seconds after the expulsion is sensed. Then, after the spikes return to their scheduled position, the forward bypasses are returned to automatic operation. The "cross tie" function is locked out below Mach 2.3 so that only the inlet which has unstarted will perform the automatic restart cycle.

Compressor inlet pressure (CIP) is the SES reference. Shock expulsion is sensed and the system actuated when a momentary CIP decrease of more than 23% occurs. Rapid CIP decrease is a characteristic indication of inlet unstart; however, the SES can also be actuated by compressor stalls if CIP fluctuates rapidly to more than 23% below the previously normal condition. Successive unstarts or compressor stalls may cause the SES reference pressure (CIP) to decay. The SES cannot operate if the momentary pressure drop is less than 23% of the existing reference pressure. The auto-restart function is not activated in this event, and manual restart procedures will then be necessary.

The SES feature does not override a manually operated spike or forward bypass control and, if an unstart occurs on a side which is being operated manually, neither that inlet nor the opposite inlet will respond. Also, manual operation of the air inlet restart switch for an inlet which has unstarted while in automatic operation overrides the SES automatic restart cycle for that inlet.

INLET CONTROLS, INSTRUMENTS AND INDICATOR LIGHTS

Emergency Spike - Inlet Restart Switches

Two guarded, dual-purpose, three-position emergency spike-inlet restart switches are installed on the lower left edge of the pilot's instrument panel. The three positions are EMER SPIKE (guarded, up), off (center), and RESTART (down). In order to move a switch to the EMER SPIKE position, a safety wire must be broken, the guard raised, and the switch pulled out. This causes high pressure gas to be released from a one-shot emergency pneumatic bottle to drive and lock the respective spike full forward and open the forward bypass. The guard is open at the bottom to allow the switch to be moved freely to the RESTART position without raising the guard. When the switch is moved to this position, the respective inlet spike is moved forward and the forward bypass is rotated to its fully open position by the corresponding L or R system hydraulic pressure. Power for the circuit is furnished by the essential dc bus.

CAUTION

A complete loss of the L or R system hydraulic pressure should exist prior to using the emergency system. If system hydraulic pressure for the affected inlet has not failed completely but is low or fluctuating, it is possible for the spike to become unlocked after gas pressure released by the emergency spike switch has been depleted. (SPIKE FWD should also be selected with the inlet spike control as a precaution).

Spike Switches

Two rotary control switches are provided on the instrument panel, one for each spike system. Each switch has two detent positions, AUTO and FWD indexed to the instrument panel lubber line engraved at the 12 o'clock position. The knobs are graduated counterclockwise in Mach number values from 1.4 through 3.2 to represent the manual override range of control. With the selector knob in the AUTO detent, the corresponding spike position is controlled automatically by the inlet computer. The control parameters are Mach number, load factor, and angles of attack and sideslip. The FWD position provides an override signal which moves the spike to its forward stop from any aft position, bypassing the automatic circuitry. The manual override settings schedule the spike to any desired position from full forward at the Mach 1.4 - 1.6 setting to full aft (26-inch position) at the Mach 3.2 setting. There is no automatic bias for load factor or angles of attack and sideslip during manual control. The left spike and bypass doors are powered by the L hydro system and the right spike and bypass doors are powered by the R hydro system. Power for the control circuits is furnished by the essential dc bus.

NOTE

When an emergency spike-inlet restart switch is actuated, the position selected by the respective spike control switch is overriden and that spike is driven to the full-forward position.

AIR INLET SYSTEM SCHEDULE

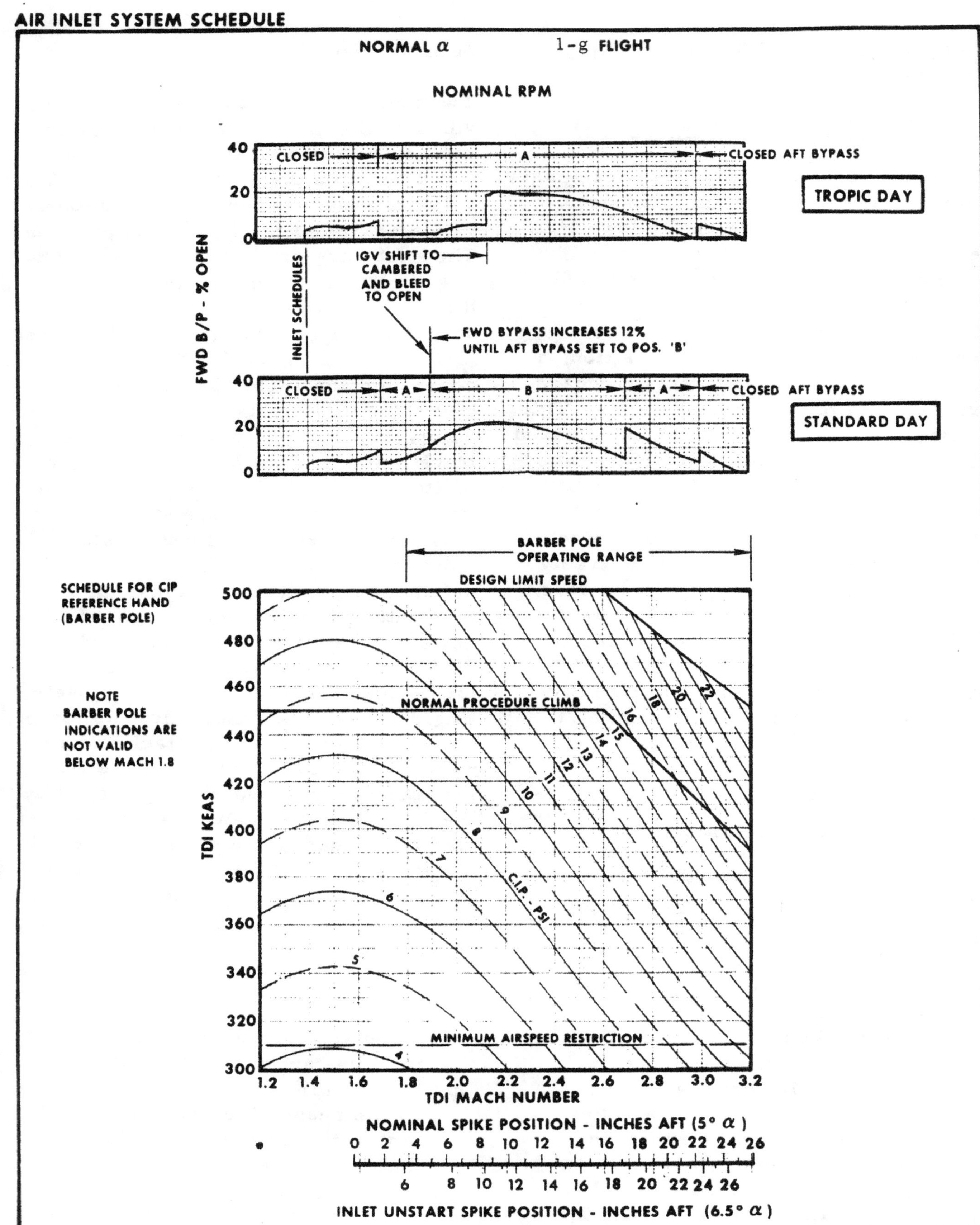

Figure 1-5

Inlet Forward Bypass Switches

Two rotary control switches are provided on the forward cockpit instrument panel, one for each forward bypass system. Each switch has two detent positions, AUTO at the full counterclockwise position, and OPEN when indexed to the instrument panel lubberline engraved at the 12 o'clock position. The rotary switch may be rotated clockwise from the OPEN position to schedule the bypass opening from any desired position between full open and full closed. Marks for setting bypass opening are engraved on the switches. In the OPEN detent, an override circuit positions the bypass fully open, bypassing the automatic circuits. The left inlet forward bypass is operated by the L system hydraulic pressure and the right by R system hydraulic pressure. Power for the circuit is furnished by the essential dc bus.

Spike Position Indicator

A dual position indicator for the left and right spikes is mounted on the left side of the forward cockpit center instrument panel to indicate the position of the spikes relative to their full forward positions. This instrument is marked in one inch increments from 0 (forward) through 26 (full aft). Power is provided by the emergency ac bus through two circuit breakers, labeled L SPIKE AND DOOR and R SPIKE AND DOOR, located on the pilot's right console.

Forward Bypass Position Indicator

A dual position indicator for the left and right forward bypass is located on the left side of the forward cockpit center instrument panel. The indicator shows the positions of the forward bypass in percent of opening from 0% (fully closed) to 100% (full open). The dial is marked in 20% increments, with additional marks at each 10% position.

A manual farther open bias schedule is applied to the forward bypass whenever the spike is in the manual mode of operation and the forward bypass is operated manually. Therefore, the indicator will show the forward bypass more open than the manual switch setting selected when operating with manual spike control. The bias schedule operates as follows:

a. Forward bypass switch in AUTO or 100 - The forward bypass will open fully.

b. Forward bypass switch selected in manual range from 100 to 0 - With the spike control setting at Mach 1.5, the bias is approximately 10% farther open. The bias increases to 25% more open when the spike setting is Mach 2.3, and decreases again to approximately 10% more open at a spike setting of Mach 2.8 or greater.

Power is provided for the indicator by the emergency ac bus through two circuit breakers labeled L SPIKE AND DOOR and R SPIKE AND DOOR, located on the right console in the forward cockpit.

NOTE

When an emergency spike-inlet restart switch is actuated to the RESTART ON position, the position selected by the respective forward bypass switch is overridden and the forward bypass for that inlet is rotated to the full open position.

Inlet Aft Bypass Switches and Indicator

The inlet aft bypass switches and
cator lights are located just abov
to the left of the spike position in
They are superimposed, four-po
rotary-type switches. The lower
(longer) handle controls the left b
and the upper (shorter) handle co
the right bypass. The switch pos
from left to right are labeled CL
A (15% open), B (50% open), and
(100%). Left and right amber lig
cated to the right of the switches
inate to indicate when an aft bypa
sition and switch setting do not c
spond. A light should illuminate
time its switch is moved, then ex
guish as the bypass reaches the
position. Approximately 5 secon
required for the aft bypass ring t
from full closed to full open. Th
bypass actuator control circuits
powered by the essential ac bus.

Manual Inlet Indicator Light

A manual inlet light, labeled MA
INLET, is located on the annunc
panel in the forward cockpit. W
illuminated, the light indicates
one or more of the four rotary
and/or forward bypass control
is not in the AUTO position, or
inlet restart switch is not in the
position. Power for the light is
nished by the essential dc bus.

FUEL SYSTEM

The aircraft fuel supply system
of six integral fuel tanks with in
necting plumbing and electrical
boost pumps for fuel feed, tran
dumping. Other provisions of
tem include nitrogen inerting,
izing, venting, single-point ref

ion. In addition
e engines, auto-
provides center-
g control at
el is also used
ockpit air, engine
l, accessory
ydraulic fluid.

ally sealed fuel
he fuselage and
re numbered
gh 6 and are in-
nd left fuel mani-
line. Submerged
ned in all tanks,
4, and 5 and
d 6. The pumps
ugh the left and
nsfer fuel
old to tank 1 for
ol. A fuel dump
ch fuel manifold.
k usage is con-
n for each pump
in an optimum
ruise. The left
plied successive-
and 6; the right
2, 5, and 6. Nor-
uencing is as

R ENGINE

Tanks 1 & 2
Tank 2
Tank 5

operated cross-
t pump switches
any engine.

Inlet Forward Bypass Switches

Two rotary control switches are provided on the forward cockpit instrument panel, one for each forward bypass system. Each switch has two detent positions, AUTO at the full counterclockwise position, and OPEN when indexed to the instrument panel lubberline engraved at the 12 o'clock position. The rotary switch may be rotated clockwise from the OPEN position to schedule the bypass opening from any desired position between full open and full closed. Marks for setting bypass opening are engraved on the switches. In the OPEN detent, an override circuit positions the bypass fully open, bypassing the automatic circuits. The left inlet forward bypass is operated by the L system hydraulic pressure and the right by R system hydraulic pressure. Power for the circuit is furnished by the essential dc bus.

Spike Position Indicator

A dual position indicator for the left and right spikes is mounted on the left side of the forward cockpit center instrument panel to indicate the position of the spikes relative to their full forward positions. This instrument is marked in one inch increments from 0 (forward) through 26 (full aft). Power is provided by the emergency ac bus through two circuit breakers, labeled L SPIKE AND DOOR and R SPIKE AND DOOR, located on the pilot's right console.

Forward Bypass Position Indicator

A dual position indicator for the left and right forward bypass is located on the left side of the forward cockpit center instrument panel. The indicator shows the positions of the forward bypass in percent of opening from 0% (fully closed) to 100% (full open). The dial is marked in 20% increments, with additional marks at each 10% position.

A manual farther open bias schedule is applied to the forward bypass whenever the spike is in the manual mode of operation and the forward bybass is operated manually. Therefore, the indicator will show the forward bypass more open than the manual switch setting selected when operating with manual spike control. The bias schedule operates as follows:

a. Forward bypass switch in AUTO or 100 - The forward bypass will open fully.

b. Forward bypass switch selected in manual range from 100 to 0 - With the spike control setting at Mach 1.5, the bias is approximately 10% farther open. The bias increases to 25% more open when the spike setting is Mach 2.3, then decreases again to approximately 10% more open at a spike setting of Mach 2.8 or greater.

Power is provided for the indicator by the emergency ac bus through two circuit breakers labeled L SPIKE AND DOOR and R SPIKE AND DOOR, located on the right console in the forward cockpit.

NOTE

> When an emergency spike-inlet restart switch is actuated to the RESTART ON position, the position selected by the respective forward bypass switch is overridden and the forward bypass for that inlet is rotated to the full open position.

Inlet Aft Bypass Switches and Indicators

The inlet aft bypass switches and indicator lights are located just above and to the left of the spike position indicator. They are superimposed, four-position rotary-type switches. The lower (longer) handle controls the left bypass and the upper (shorter) handle controls the right bypass. The switch positions from left to right are labeled CLOSE, A (15% open), B (50% open), and OPEN (100%). Left and right amber lights located to the right of the switches illuminate to indicate when an aft bypass position and switch setting do not correspond. A light should illuminate at the time its switch is moved, then extinguish as the bypass reaches the selected position. Approximately 5 seconds are required for the aft bypass ring to move from full closed to full open. The bypass actuator control circuits are powered by the essential ac bus.

Manual Inlet Indicator Light

A manual inlet light, labeled MANUAL INLET, is located on the annunciator panel in the forward cockpit. When it is illuminated, the light indicates that one or more of the four rotary switches and/or forward bypass control switches is not in the AUTO position, or that the inlet restart switch is not in the off position. Power for the light is furnished by the essential dc bus.

FUEL SYSTEM

The aircraft fuel supply system consists of six internal fuel tanks with interconnecting plumbing and electrically driven boost pumps for fuel feed, transfer, and dumping. Other provisions of the system include nitrogen inerting, pressurizing, venting, single-point refueling and

FUEL TANK CAPACITIES

Tank	Usable Fuel	
1	936 gal	6,150 lb
2	1,230 gal	8,100 lb
3	1,485 gal	9,750 lb
4	2,132 gal	14,000 lb
5	2,141 gal	14,050 lb
6	1,950 gal	12,800 lb
Total	9,874 gal	64,850 lb *

* At fuel density of 6.57 lb/gal.

REFUELING AND DEFUELING RECEPTACLE

A single-point refueling receptacle installed on top of the fuselage aft of the FCO's cockpit is used for both ground and in-flight refueling. Ground refueling is accomplished by use of an A/R probe specially modified to utilize a hand-operated locking device so that the refueling may be done without hydraulic power. Fuel from the receptacle flows through the fueling manifold to the tanks. The use of calibrated orifices for each tank allows all tanks to be filled simultaneously in approximately 12 minutes with a fueling pressure of approximately 50 psi. Dual shutoff valves in each tank shut off fuel flow when the tank is full. Defueling is accomplished through a common dump line orifice located at the extreme rear of the fuselage, using a special adapter fitting and the ground refueling probe. Refer to Section II for aerial refueling procedure.

FUEL BOOST PUMPS

Sixteen single-stage centrifugal ac-powered boost pumps are used to supply the fuel manifolds. Tanks 1 and 6, which normally feed both engines, are equipped with four pumps and tanks 2, 3, 4, and 5 have two pumps each. A single pump in each tank is capable of supplying fuel to the engine in the event of failure of the other pump. The pumps in each tank may be operated out of the normal sequence by use of the individual tank boost pump switches located on the right side of the pilot's instrument panel. Manual control of the tank pumps supplements but does not terminate automatic tank sequencing and is usually done only under abnormal conditions, such as when a tank does not feed in the proper sequence. It is necessary to actuate the pump release switch to terminate manually actuated boost pumps when a tank is empty. Normally, each pump (except pumps 1-1 and 1-2, which are protected by a common float switch) is protected by a float switch that deactivates the pump when the tank is empty. Individual circuit breakers for each pump are located in the No. 1 missile bay and are not accessible in flight.

FUEL SYSTEM CONTROLS, INSTRUMENTS AND INDICATOR LIGHTS

Fuel Boost Pump Switches and Indicator Lights

Six fuel boost pump switches, labeled 1 TANK, 2 TANK, 3 TANK, 4 TANK, 5 TANK, and 6 TANK, are installed in a vertical line on the right side of the pilot's instrument panel. These switches are self-illuminated plastic pushbuttons and light to indicate automatic sequencing and to control manual operation of the fuel boost pumps in each tank. When a set of boost pumps is actuated, either automatically or manually, the tank pushbutton lights green. When depressed, the boost pump switch will turn on

C.G. DIAGRAM
AUTOMATIC FUEL SEQUENCING

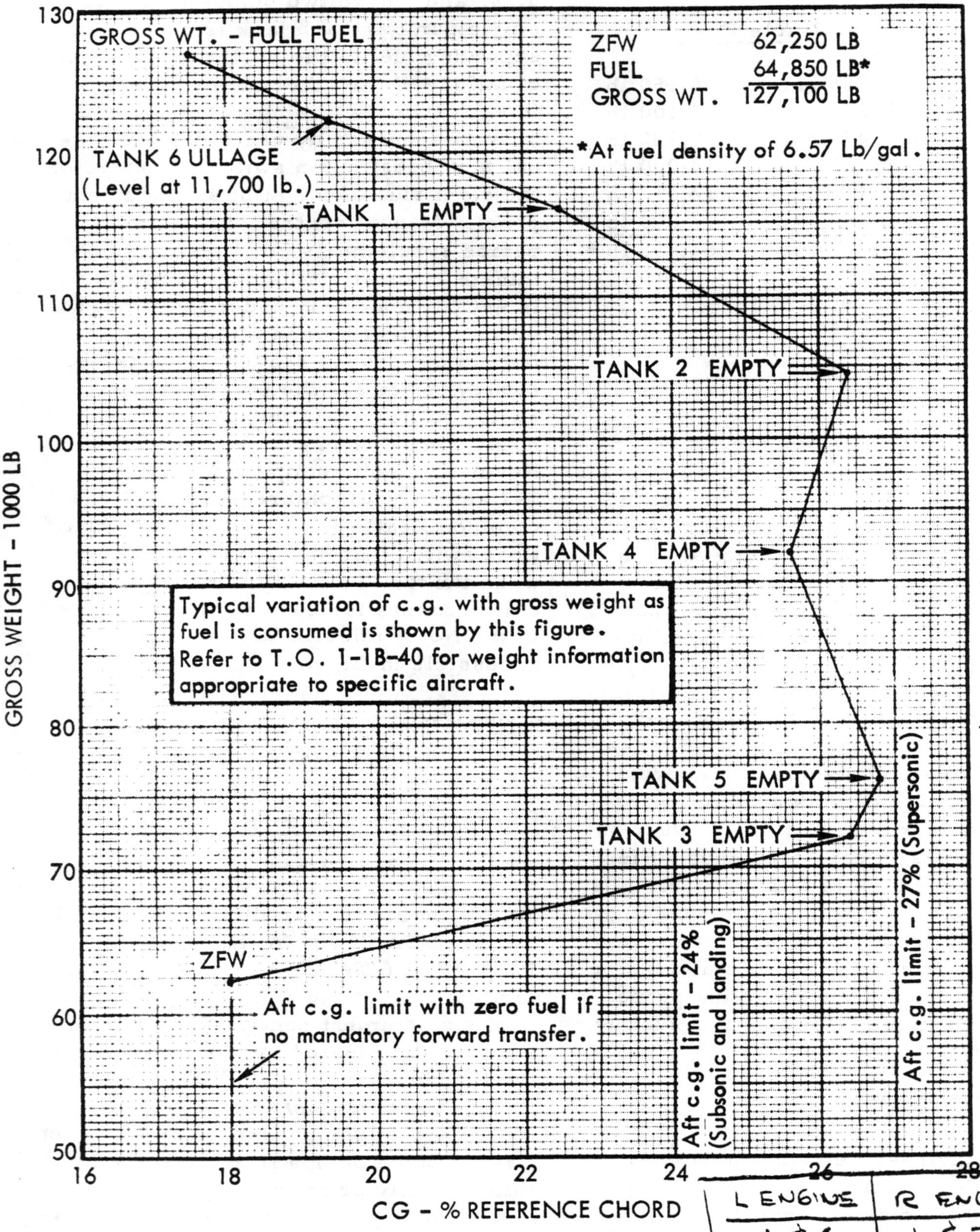

ZFW	62,250 LB
FUEL	64,850 LB*
GROSS WT.	127,100 LB

*At fuel density of 6.57 Lb/gal.

GROSS WT. - FULL FUEL

TANK 6 ULLAGE
(Level at 11,700 lb.)

TANK 1 EMPTY

TANK 2 EMPTY

TANK 4 EMPTY

Typical variation of c.g. with gross weight as fuel is consumed is shown by this figure. Refer to T.O. 1-1B-40 for weight information appropriate to specific aircraft.

TANK 5 EMPTY

TANK 3 EMPTY

ZFW

Aft c.g. limit with zero fuel if no mandatory forward transfer.

Aft c.g. limit - 24% (Subsonic and landing)

Aft c.g. limit - 27% (Supersonic)

GROSS WEIGHT - 1000 LB

130
120
110
100
90
80
70
60
50

16 18 20 22 24 26 28

CG - % REFERENCE CHORD

TANK
1 6150
2 810
3 9750
4 1400
5 1405
6 1280
64850

L ENGINE	R ENGINE
1 & 4	1 & 2
4	2
3	5

Figure 1-7

the respective boost pumps, latch, and hold electrically until released by the release switch. Power for the boost pump switch circuits and lights is furnished by the essential dc bus.

NOTE

- Manual operation supplements but does not terminate the normal automatic fuel tank sequencing.
- The switches have an electrical interlock that allows manual selection of only one tank in tank group 1, 2, and 3 and one tank in tank group 4, 5, and 6 at the same time.

Boost Pump Release Switch

An edge-lighted pushbutton-type boost pump release switch, labeled REL, is installed on the pilot's instrument panel below the fuel boost pump switches. When the switch is depressed, any boost pump switch that has been actuated manually will be released and automatic sequencing of the fuel tanks will be continued. Power for the circuit is furnished by the essential dc bus.

Crossfeed Switch

A self-illuminated, pushbutton-type crossfeed switch, labeled X-FEED, is installed at the top of the column of boost pump switches. When depressed, it illuminates green, opens a motor-operated valve in a line between the left and right fuel manifolds, allowing the right manifold to feed the left engine and vice versa. Depressing the switch a second time closes the valve, which terminates crossfeeding and causes the light to go out. Power for the circuit is furnished by the essential dc bus.

CAUTION

When the crossfeed switch is depressed, a minimum of 5 seconds delay must be observed prior to depressing the switch a second time, to allow for full travel of the crossfeed valve. Attempting to recycle the valve within 5 seconds may cause the circuit breaker in the No. 1 missile bay to open.

Fuel Shutoff Switches

A guarded fuel shutoff switch for each engine is installed on the lower right side of the pilot's right forward instrument panel. The switches are guarded in the open, or ON, position. When the switches are moved to the L or R (up) position, power from the essential ac bus closes motor-driven valves in the engine feed lines.

WARNING

When a fuel shutoff switch is moved to the L or R position, a minimum of 5 seconds delay must be observed, prior to moving the switch back to the ON position, to allow for full travel of the shutoff valve. Attempting to recycle the valve in less than 5 seconds may cause the circuit breaker in the No. 1 missile bay to open.

Fuel Transfer Switch

A three position fuel transfer switch is installed to the right of the tank boost pump switches. When TRANSFER (up) is selected, a valve in the forward end

AFT COCKPIT FUEL PANEL

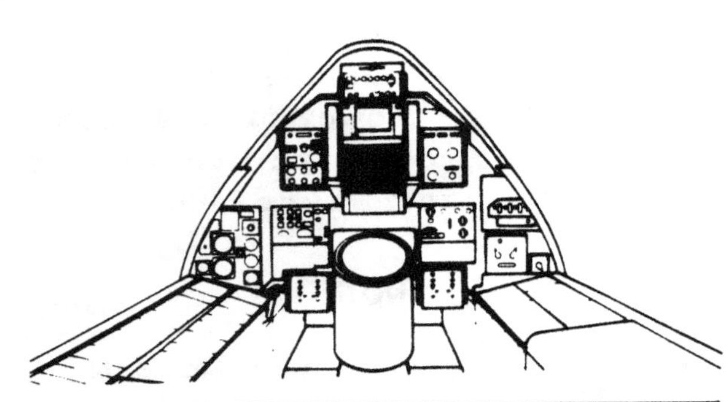

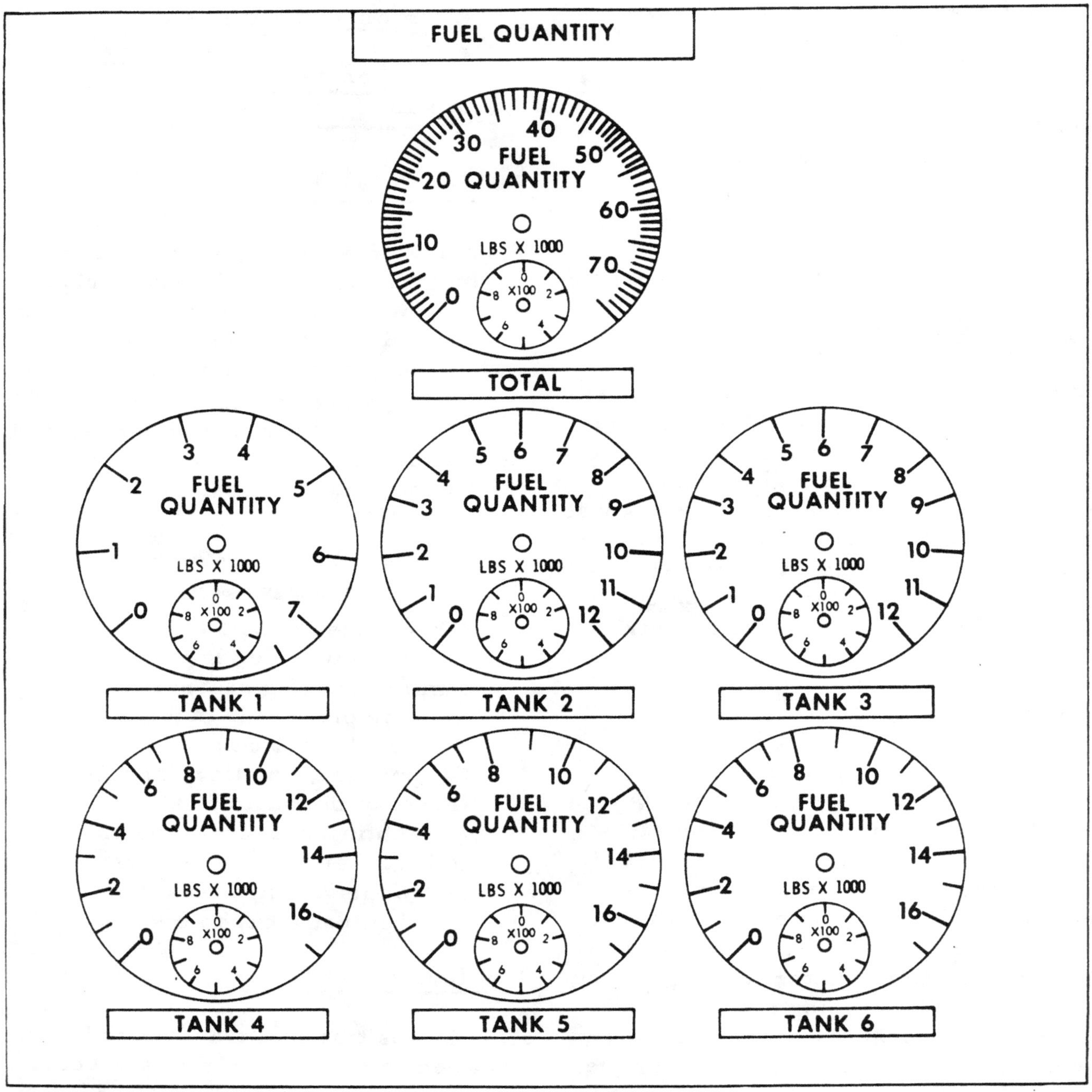

FUEL QUANTITY

FUEL QUANTITY
LBS X 1000
X100

TOTAL

FUEL QUANTITY
LBS X 1000
X100

TANK 1

TANK 2

TANK 3

TANK 4

TANK 5

TANK 6

Figure 1-7A

of the right fuel manifold is opened and and fuel is transferred into tank 1. The boost pumps in tank 1 are automatically shut off while TRANSFER is selected unless tank 6 is almost empty. Transfer is automatically terminated when tank 1 becomes full. Forward transfer is not terminated by manual selection of the tank 1 boost pumps. The pumps will resume normal operation when the OFF (center) position is selected.

Lifting the switch over a detent allows selection of the HOLD (down) position. The tank 1 pumps are de-energized in this condition, and the existing fuel quantity is held in tank 1 until:
(a) the switch is returned to the OFF or TRANSFER position, or (b) the fuel level in tank 6 decreases below approximately 4000 pounds. In this event, a float switch in tank 6 automatically overrides the "Tank 1 Hold" circuit and returns the tank 1 pumps to automatic sequencing.

Fuel Dump Switch

A three-position guarded toggle switch is installed on the right side of the pilot's instrument panel. The switch is labeled EMER (up), DUMP (center) and OFF (down). To prevent inadvertent dump, the guard is safety-wired and the wire must be broken and the guard lifted before the switch can be moved to the DUMP or EMER position. The switch must also be lifted out of the DUMP position in order to place it in the EMER position. When the switch is moved to the DUMP position, dump valves in each fuel manifold are opened and the pumps in tank 1 are inactivated. All other tanks will dump in normal usage sequence. Fuel will stop dumping when the tank 6 fuel level reaches 4000 pounds remaining. If there is any fuel in tank 1, the boost pumps will start when fuel dumping terminates unless the fuel transfer switch is in the TRANSFER position. Manual selection of tank fuel boost pumps, when the fuel dump switch is in the DUMP position, terminates dumping. In the EMER position the 4000 pound stop-dump switch in tank 6 is bypassed and dumping will continue from this tank and all others except tank 1. Power for the circuit is furnished by the essential dc bus.

Fuel Quantity Selector Switch and Quantity Indicator (Forward Cockpit)

A quantity indicator and a rotary, eight-position fuel quantity selector switch is installed on the lower right side of the pilot's instrument panel. Positions on the selector switch are marked for each of the six tanks, a TOTAL position, and a TEST position. When the selector switch is placed in the TEST position, the fuel quantity indicator needles in both cockpits should move toward zero. The switch is rotated to the position for the desired reading on the fuel quantity indicator. The indicator is calibrated in 5000-pound increments and the pointer displays a fuel quantity from 0 to 75,000 pounds. A five-digit window located on the indicator indicates fuel quantity to the nearest 100 pounds. Power for the indicator is furnished by the essential ac bus.

Aft Cockpit Fuel Panel

A fuel panel containing 7 indicators is installed on the center instrument panel in the aft cockpit. Six of the indicators display individual tank quantities and one displays total fuel quantity. The indicators have markings on the circular dial in one thousand pound increments in the following ranges:

Tank 1 indicator - 0 - 7500 pounds

Tank 2 and 3 indicators - 0 - 12,000 pounds

Tank 4, 5 and 6 indicators - 0 - 17,000 pounds

Total quantity indicator - 0 - 75,000 pounds

A small circular dial mounted at the bottom of each of the 7 indicators allows quantity readout to the nearest 100 pounds. All indicators continuously monitor respective fuel levels during ground operation and in flight. Power for the indicators is furnished by the essential ac bus.

Fuel Pressure Low Lights

Indicator lights for each engine, labeled L and R FUEL PRESS, are located on the annunciator panel in the forward cockpit. When the light is illuminated it indicates that the fuel pressure in the main fuel manifold has decreased to less than 7 (± 1/2) psi. The light extinguishes when fuel pressure rises above 10 psi. Power is furnished from the dc essential bus.

Fuel Quantity Low Light

An indicator light, labeled FUEL QTY LOW is on the pilot's annunciator panel to illuminate when fuel remaining in each of tanks 1 and 6 has dropped to 3000 pounds or less. The low-level float switches installed in tanks 1 and 6 are wired in series, and therefore the combined tank 1 and 6 quantities can actually vary from 3000 pounds to just under 6000 pounds. Power for the light is furnished by the essential dc bus.

Tank Lights Test Switch

An edge-lighted, pushbutton-type, tank lights test switch labeled LT TEST is installed below the pump release switch and is used to test the lamps in the fuel tank boost pump lights. When the switch is depressed, all the tank fuel boost pump pushbuttons will illuminate. Power for the circuit is furnished by the essential dc bus.

AIR REFUELING SYSTEM

The aircraft is equipped with an air-refueling system capable of receiving fuel at a flow rate of approximately 6000 pounds per minute from a KC-135 boom-type tanker aircraft. The system consists of a boom receptacle, receptacle doors, hydraulic valves, hydraulic actuators, a signal amplifier, control switches, and indicator lights. The refueling doors are held closed by hydraulic pressure; if hydraulic pressure is lost inadvertently, or when the engines are shut down on the ground, the doors open by spring action. This is a safety feature to permit refueling in case of partial power failure. The system requires hydraulic actuating power from the L hydraulic system to operate the doors and boom receptable. If the L hydraulic system is inoperative, the refuel system can operate from R hydraulic pressure by selecting alternate steering and brakes. Electrical power is required from the essential dc bus for operation of the controls and indicators.

Air Refuel Ready Switch

An air refuel ready switch, labeled AIR REFUEL, is installed on the upper left of the pilot's instrument panel. The switch has three positions, READY (up), OFF, (center), and MANUAL (down).

When the switch is placed in the READY position, hydraulic actuators open the refueling doors, the boom latches are armed, the receptacle lights illuminate, and the READY portion of the IFR, PUSH TO RESET light on the instrument panel illuminates. When the switch is in the MANUAL position the refuel receptacle is controlled manually by a trigger switch on the control stick grip. When in the IFR receptacle, the refueling boom will be forcibly held in place until the trigger switch is depressed.

IFR Reset Switch and Indicator Light

A square, dual-indicator, self-illuminating pushbutton, labeled IFR, PUSH TO RESET, is located on the top left of the pilot's instrument panel. The top half of the pushbutton will illuminate green and read display READY when the air refuel switch is in the READY position and the refueling receptacle is open. The READY light will extinguish when the boom is engaged. If the boom accidentally disconnects from the fueling receptacle the lower half of the switch will illuminate amber and show DISCON. In this event, the pushbutton must be depressed to reset the signal amplifier and recycle the receptacle locking mechanism. The DISCON light will extinguish when the air refuel switch is moved to OFF at termination of refueling. When the air refuel ready switch is in MANUAL the IFR reset switch and indicator light is inoperative.

Disconnect Trigger Switch

A momentary contact trigger-type switch is installed on the forward side of the control stick grip. Depressing the trigger switch will normally initiate a refueling disconnect. The trigger is also depressed to open the receptacle latches when the air refuel switch is in the MANUAL position. Releasing the trigger will close the latches.

A disconnect may be accomplished in one of the following ways:

1. Automatically.
   a. If boom envelope limits are exceeded.
   b. When tanks are full.
2. Manually.
   a. By the boom operator.
   b. By depressing the A/R DISC trigger on the control stick grip.

Pilot Director Lights (On Tanker)

Lights for receiver pilot direction are located on the bottom of the tanker fuselage, between the nose gear and the main gear. They consist of two rows of lights, the left row for elevation, and the right row for boom telescoping. The elevation lights consist of five colored panels with green strips, green trigangles, and red triangles to indicate relative position. Two illuminated letters, D and U for down and up movement, respectively, indicate elevation corrections. Background lights are located behind the panels. The colored panels are illuminated by lights controlled by boom elevation during contact. The colored panels which indicate boom telescoping are not illuminated by background lights. An illuminated white panel between each colored panel serves as a reference. The letters A for aft and F for forward are visible at the ends of the boom telescoping panel. Figures 2-7 and 2-8

show the panel illumination at various boom nozzle positions within the boom envelope. There are no lights to indicate azimuth; however, a yellow line is visible on the tanker to indicate the centerline. When contact is made, the panels automatically reflect the correction required by the pilot to maintain position.

FUEL HEAT SINK SYSTEM

Aircraft fuel is used as the cooling medium for air-conditioning systems, hydraulic fluid, and engine and accessory drive system oil. Circulated fuel is also used to cool the TEB tank and the control lines which actuate the afterburner nozzle. Engine oil is cooled by main engine fuel flowing through an oil cooler located between the main fuel control and the windmill bypass valve. This fuel is then directed to the main burner section. The other cooling is accomplished by fuel circulation through several cooling loops. If within engine consumption requirements, hot fuel returning from the accessory drive system heat exchanger, the primary and secondary air-conditioning heat exchangers, the hydraulic fluid heat exchanger, the spike heat exchanger, , and the exhaust nozzle actuators, circulates through a mixing valve and temperature limiting valve (smart valve) and returns to the main engine and afterburner fuel manifold. The quantity in excess of engine requirements is diverted to the refueling manifold. If the temperature of the mixed cooling loop and incoming engine fuel exceeds 265°F, a temperature sensitive valve starts to close and a portion of the cooling loop fuel is prevented from mixing with the incoming engine fuel. A pressure-operated valve routes the hot fuel to the refueling manifold. The valve is completely closed at 295°F and all cooling loop fuel is returned to any fuel tank that has space for it. During single-engine operation with the inoperative engine throttle OFF, actuation of the fuel crossfeed valve also allows the hot recirculated fuel from the windmilling engine to cross over and mix with the cooling loop and incoming fuel for the operating engine. If within engine consumption requirements, and if the mixed fuel temperature is below 265°F, all of the hot fuel will be burned by the operating engine and afterburner.

FUEL PRESSURIZATION AND VENT SYSTEM

The fuel pressurization system consists of 150 liters of liquid nitrogen contained in two Dewar flasks, located in the nosewheel well, associated valves, and plumbing to the fuel tanks and indicators. The flasks are equipped with automatic ac-powered heaters to supply nitrogen gas to the fuel tanks at 1.5 (± 0.25) psi. This inerts and pressurizes the ullage space above the fuel at a pressure sufficient to produce fuel flow to the engine-driven pumps in case of complete boost pump failure. The duration of the nitrogen supply depends upon the number of descents and condition of tank seals. The liquid nitrogen from the bottom of the flasks is routed through submerged heat exchangers in tanks 1 and 6 to ensure that the nitrogen has become gaseous. The nitrogen gas is then ported to the common vent line and to the top of all tanks. The venting system consists of a common vent line through all tanks, with two vent valves in each tank except tank 1. Tank 1 has only one vent valve and the open forward end of the vent line. The forward vent valves in tanks 2 through 6 are equipped to relieve tank pressure at 1.5 psi, and a float that closes the vent valve when the tank is full. The float shutoff is provided to keep fuel from entering the

vent line. The aft vent valve is similar to the forward except that it has no relief valve. The common vent line tees into two lines in tank 6 and both go through the rear bulkhead. In the tail-cone area there is a relief valve in each line with the left valve set to relieve pressure at 3.25 psi. In the event of failure of this valve, the right valve will relieve pressure at 4.0 (± .15) psi.

A suction relief line and valve connects to the common vent line in tank 1 and terminates in a bellmouth fitting in the aft end of the nosewheel well. Two valves are provided in the vent system to prevent fuel from surging forward in the vent line during aircraft deceleration. A check valve prevents fuel that is coming forward from tank 6 from going farther than the forward end of tank 5. A check valve located in tank 3 prevents fuel coming from tank 4 from going any farther than the forward end of tank 3. These float-actuated valves close the vent when fuel is moving forward in the vent line. Tank2 fuel can go forward into tank 1. Acceleration presents no problem of fuel shift between tanks.

Liquid Nitrogen Quantity Indicator

Two quantity indicators to check aircraft fuel system liquid nitrogen are installed on the FCO's forward panel. The indicators display the quantity of liquid nitrogen remaining in each of the two Dewar flasks. The indicators are marked from 0 to 75 liters, in 5-liter increments. Power for the indicators is furnished by the dc essential bus and the ac essential bus.

Nitrogen Quantity and Warning Lights Test Switch

A nitrogen quantity and warning lights test switch is installed on the FCO's left forward panel, adjacent to the liquid nitrogen quantity indicators. The switch has three positions, NITRO IND (up), OFF (center), and WARN LT (down). When the switch is moved to the NITRO IND position, the nitrogen quantity indicator needles move toward zero and the NITROGEN QTY LOW light on the pilot's annunciator panel illuminates at the 1-liter remaining indication. When the switch is moved to the WARN LT (down) position, the FCO's annunciator panel lights illuminate. Power for the switch is furnished by the essential dc bus.

Fuel Tank Pressure Indicator

A fuel tank pressure indicator is installed on the right side of the pilot's instrument panel. The gage indicates the pressure existing in fuel tank 1, and is marked from -2 to +8 in increments of 1 pound per square inch. Power for the indicator is furnished by the 26-volt instrument transformer.

ELECTRICAL SUPPLY SYSTEM

The electric power system is completely automatic. Each of the two engine-driven, 60-KVA generators is coupled to a constant-speed drive (CSD) and furnishes 115/200V, 3 phase ac power to the three ac buses (L primary, R primary, and essential) through an automatic bus transfer and protective system. (See figure 1-8.) the generators can furnish rated power at 2650 engine rpm and above. During normal operation both generators operate in parallel, the L and R primary ac buses connected by a bus tie contactor.

If a fault occurs in one generator system a logic circuit automatically disconnects the generator from its bus, and the bus transfer circuit connects its load to the good generator. The logic circuit then attempts to reset the generator line contactor three times and put it back on the bus. If the attempts to reset are unsuccessful, the generator is locked off the bus until another recycle attempt is made by moving the generator switch OFF then back to ON. The bus tie contactor will open if the system fault that caused a generator to drop off the line is a generator undervoltage, open phase, or current imbalance between the two generators. Once opened in flight, the bus tie contactor will remain open until it is reset manually on the ground. If the generators fault is cleared during the resetting attempts, the generator can be brought back on the line. However, if the generator is brought back on the bus with the bus tie contactor open, each generator will then power only its own bus until reset by the ground crew.

The 28-volt dc power is obtained from 200-amp transformer-rectifiers, one from each primary ac bus. The parallel outputs from these rectifiers supply the essential dc bus. If both generators and/or both transformer-rectifiers should fail and the battery switch is in the ON position, switching relays automatically connects the emergency battery to the essential dc bus and the inverter battery to the 600-va emergency inverter. This renders the INS system inoperative. The inverter will supply ac power to the SAS and essential flight instruments for approximately 40 minutes.

EXTERNAL POWER SUPPLY

The aircraft is equipped with an external power receptacle, located in the left chine adjacent to the No. 1 missile bay. When an external power source is connected to the aircraft with the generator switches ON, the generators are disconnected automatically from the buses and remain so until the right engine reaches IDLE rpm and the generator parallels the output of the external power cart. This causes the external power to be automatically disconnected from the ac buses. (The MD-3 ground power unit must be used for engine starting because of its paralleling feature.)

ELECTRICAL SYSTEM CONTROLS AND INDICATOR LIGHTS

Generator Switches

A generator switch for each generator system is located on the pilot's right forward panel. It is powered from the essential dc bus. The two switch positions are GENERATOR L or R (up) and OFF (down). Placing a switch to the GENERATOR L or R position will return the respective generator to normal operation if it has been removed from the bus for any reason other than complete generator failure. In the OFF

ELECTRICAL POWER DISTRIBUTION

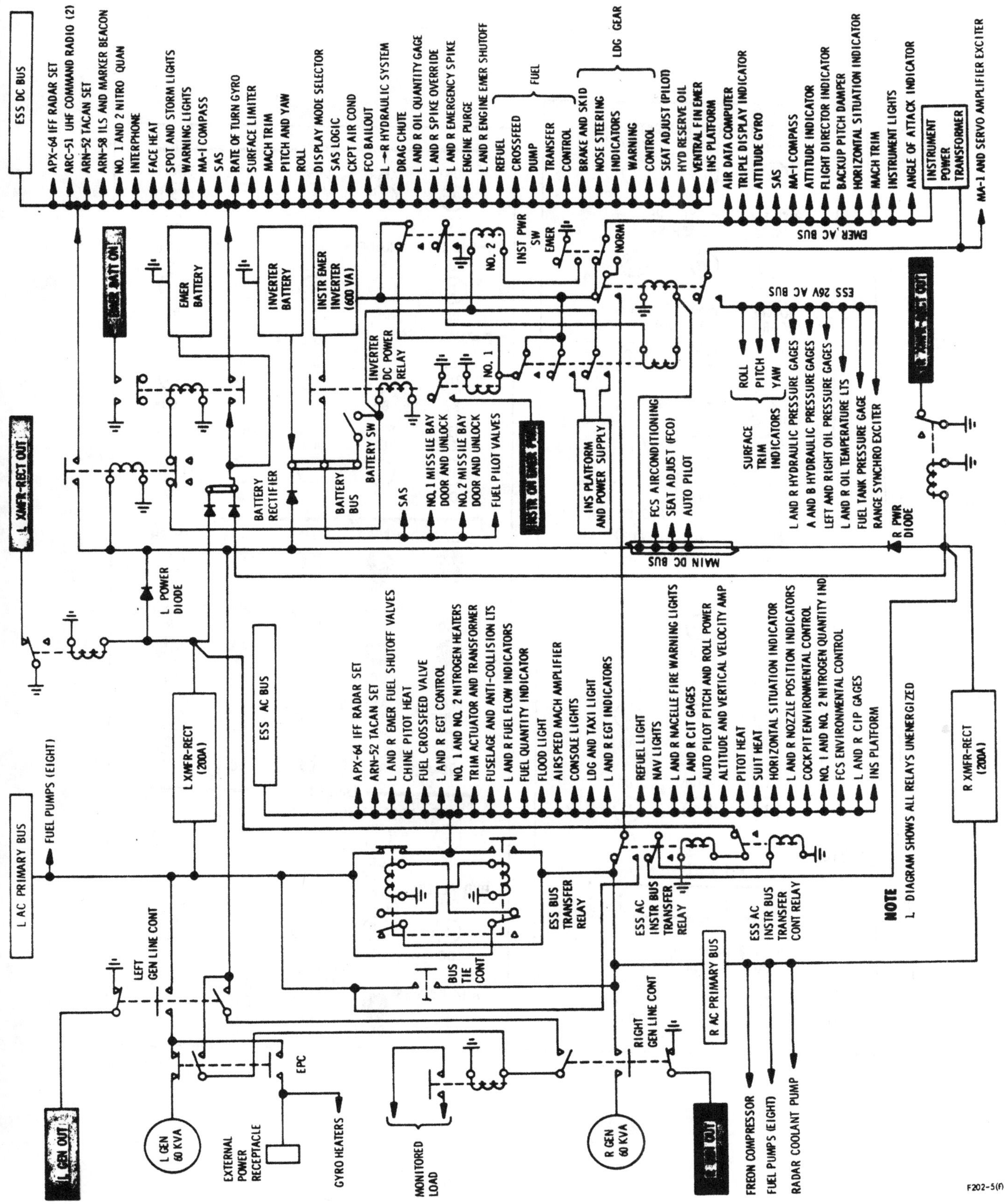

ESS DC BUS
APX-64 IFF RADAR SET
ARC-51 UHF COMMAND RADIO (2)
ARN-52 TACAN SET
ARN-58 ILS AND MARKER BEACON
NO. 1 AND 2 NITRO QUAN
INTERPHONE
FACE HEAT
SPOT AND STORM LIGHTS
WARNING LIGHTS
MA-1 COMPASS
SAS
RATE OF TURN GYRO
SURFACE LIMITER
MACH TRIM
PITCH AND YAW
ROLL
DISPLAY MODE SELECTOR
SAS LOGIC
CKPT AIR COND
FCO BAILOUT
L→R HYDRAULIC SYSTEM
DRAG CHUTE
L AND R OIL QUANTITY GAGE
L AND R SPIKE OVERRIDE
L AND R EMERGENCY SPIKE
ENGINE PURGE
L AND R ENGINE EMER SHUTOFF
FUEL
REFUEL
CROSSFEED
DUMP
TRANSFER
CONTROL
LDG GEAR
BRAKE AND SKID
NOSE STEERING
INDICATORS
WARNING
CONTROL
SEAT ADJUST (PILOT)
HYD RESERVE OIL
VENTRAL FIN EMER
INS PLATFORM

EMER AC BUS
AIR DATA COMPUTER
TRIPLE DISPLAY INDICATOR
ATTITUDE GYRO
SAS
MA-1 COMPASS
ATTITUDE INDICATOR
FLIGHT DIRECTOR INDICATOR
BACKUP PITCH DAMPER
HORIZONTAL SITUATION INDICATOR
MACH TRIM
INSTRUMENT LIGHTS
ANGLE OF ATTACK INDICATOR
INSTRUMENT POWER TRANSFORMER
MA-1 AND SERVO AMPLIFIER EXCITER

EMER BATT ON
EMER BATTERY
INVERTER BATTERY
INSTR EMER INVERTER (600 VA)
NO. 2
INST PWR SW
EMER
NORM
INVERTER DC POWER RELAY
NO. 1
BATTERY SW
BATTERY RECTIFIER
BATTERY BUS
SAS
NO. 1 MISSILE BAY DOOR AND UNLOCK
NO. 2 MISSILE BAY DOOR AND UNLOCK
FUEL PILOT VALVES
INSTR ON EMER PWR
INS PLATFORM AND POWER SUPPLY
FCS AIRCONDITIONING
SEAT ADJUST (FCO)
AUTO PILOT

ESS 26V AC BUS
ROLL
PITCH
YAW
SURFACE TRIM INDICATORS
L AND R HYDRAULIC PRESSURE GAGES
A AND B HYDRAULIC PRESSURE GAGES
LEFT AND RIGHT OIL PRESSURE GAGES
L AND R OIL TEMPERATURE LTS
FUEL TANK PRESSURE GAGE
RANGE SYNCHRO EXCITER
R XMFR-RECT OUT
R PWR DIODE
MAIN DC BUS
L XMFR-RECT OUT
L POWER DIODE

ESS AC BUS
APX-64 IFF RADAR SET
ARN-52 TACAN SET
L AND R EMER FUEL SHUTOFF VALVES
CHINE PITOT HEAT
FUEL CROSSFEED VALVE
L AND R EGT CONTROL
NO. 1 AND NO. 2 NITROGEN HEATERS
TRIM ACTUATOR AND TRANSFORMER
FUSELAGE AND ANTI-COLLISION LTS
L AND R FUEL FLOW INDICATORS
FUEL QUANTITY INDICATOR
FLOOD LIGHT
AIRSPEED MACH AMPLIFIER
CONSOLE LIGHTS
LDG AND TAXI LIGHT
L AND R EGT INDICATORS
REFUEL LIGHT
NAV LIGHTS
L AND R NACELLE FIRE WARNING LIGHTS
L AND R CIT GAGES
AUTO PILOT PITCH AND ROLL POWER
ALTITUDE AND VERTICAL VELOCITY AMP
PITOT HEAT
SUIT HEAT
HORIZONTAL SITUATION INDICATOR
L AND R NOZZLE POSITION INDICATORS
COCKPIT ENVIRONMENTAL CONTROL
NO. 1 AND NO. 2 NITROGEN QUANTITY IND
FCS ENVIRONMENTAL CONTROL
L AND R CIP GAGES
INS PLATFORM

L XMFR-RECT (200A)
R XMFR-RECT (200A)
L AC PRIMARY BUS
FUEL PUMPS (EIGHT)
ESS BUS TRANSFER RELAY
ESS AC INSTR BUS TRANSFER RELAY
ESS AC INSTR BUS TRANSFER CONT RELAY
BUS TIE CONT
LEFT GEN LINE CONT
RIGHT GEN LINE CONT
R AC PRIMARY BUS
EPC
L GEN OUT
L GEN 60 KVA
EXTERNAL POWER RECEPTACLE
GYRO HEATERS
MONITORED LOAD
R GEN 60 KVA
R GEN OUT
FREON COMPRESSOR
FUEL PUMPS (EIGHT)
RADAR COOLANT PUMP

NOTE
1. DIAGRAM SHOWS ALL RELAYS UNENERGIZED

F202-5(F)

Figure 1-8

LEFT CONSOLE CIRCUIT BREAKER PANELS - Forward Cockpit

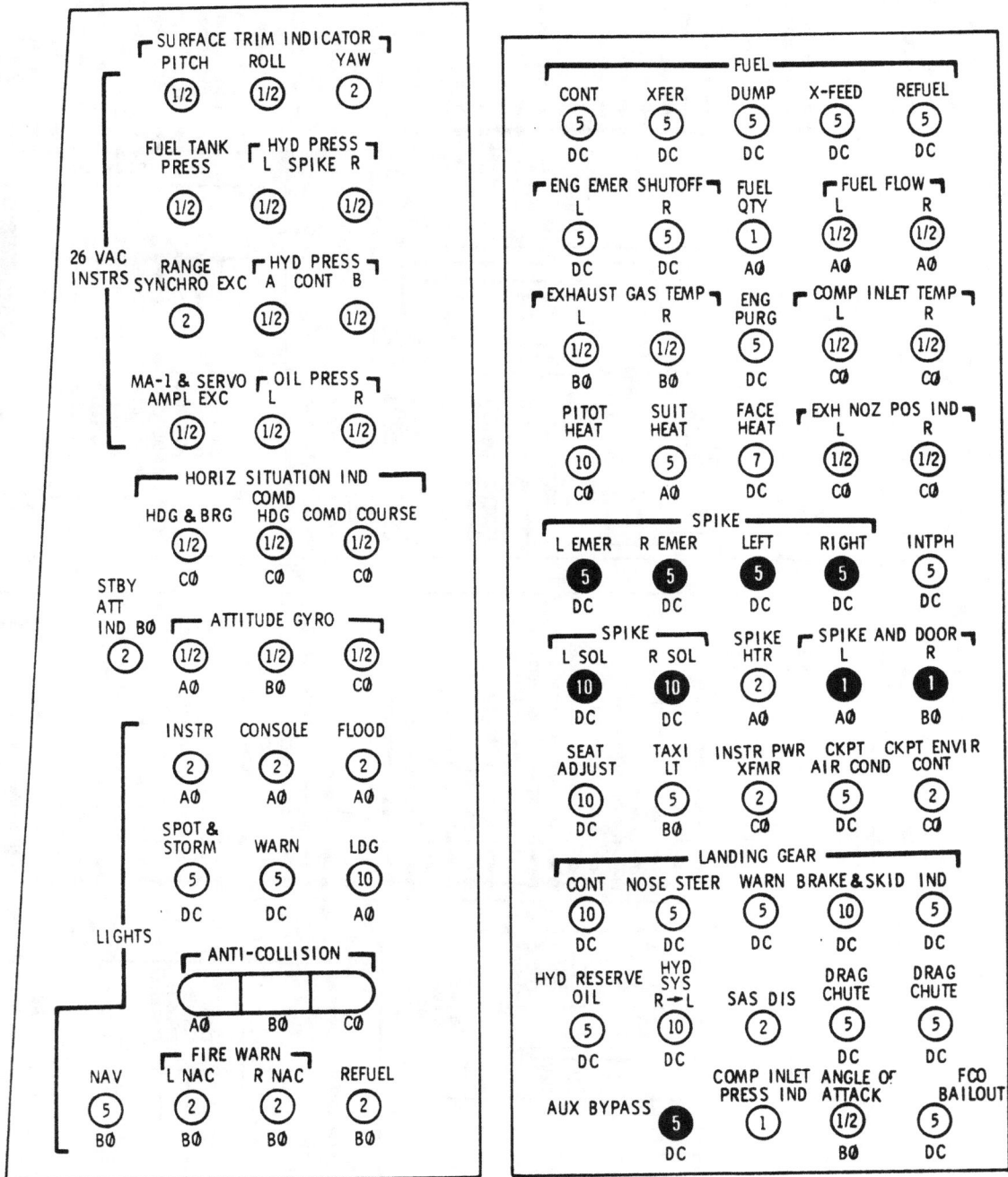

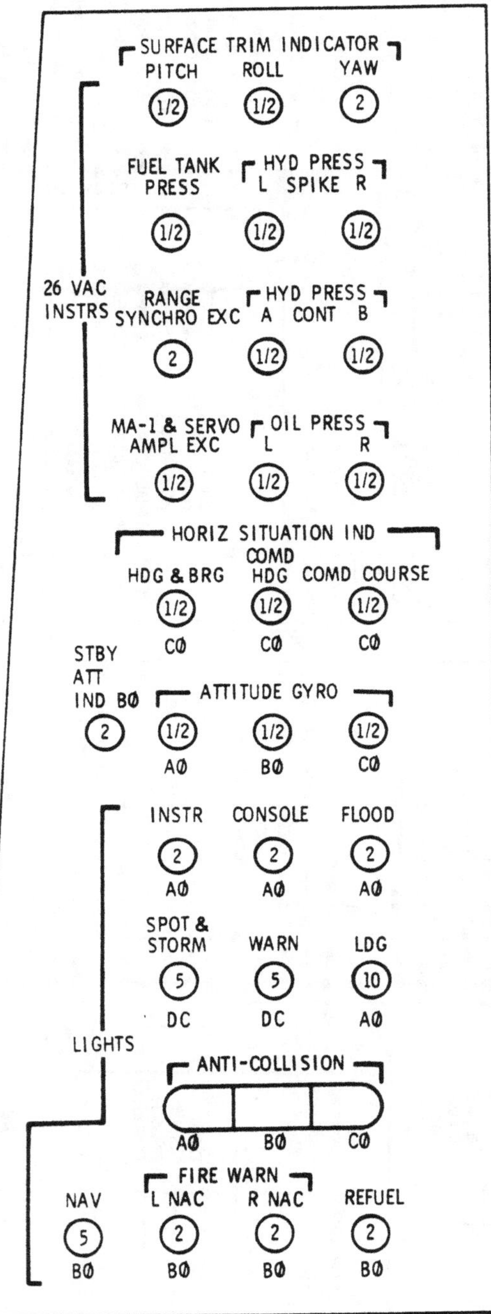

FUEL
CONT XFER DUMP X-FEED REFUEL
5 5 5 5 5
DC DC DC DC DC

ENG EMER SHUTOFF L R
FUEL QTY
FUEL FLOW L R
5 5 1 1/2 1/2
DC DC AØ AØ AØ

EXHAUST GAS TEMP L R
ENG PURG
COMP INLET TEMP L R
1/2 1/2 5 1/2 1/2
BØ BØ DC CØ CØ

PITOT HEAT
SUIT HEAT
FACE HEAT
EXH NOZ POS IND L R
10 5 7 1/2 1/2
CØ AØ DC CØ CØ

SPIKE
L EMER R EMER LEFT RIGHT
INTPH
5 5 5 5 5
DC DC DC DC DC

SPIKE L SOL R SOL
SPIKE HTR
SPIKE AND DOOR L R
10 10 2 1 1
DC DC AØ AØ BØ

SEAT ADJUST
TAXI LT
INSTR PWR XFMR
CKPT AIR COND
CKPT ENVIR CONT
10 5 2 5 2
DC BØ CØ DC CØ

LANDING GEAR
CONT NOSE STEER WARN BRAKE & SKID IND
10 5 5 10 5
DC DC DC DC DC

HYD RESERVE OIL
HYD SYS R→L
SAS DIS
DRAG CHUTE
DRAG CHUTE
5 10 2 5 5
DC DC DC DC

AUX BYPASS
COMP INLET PRESS IND
ANGLE OF ATTACK
FCO BAILOUT
5 1 1/2 5
DC BØ DC

F202-34(a)

Figure 1-8A

RIGHT CONSOLE CIRCUIT BREAKER PANELS - Forward Cockpit

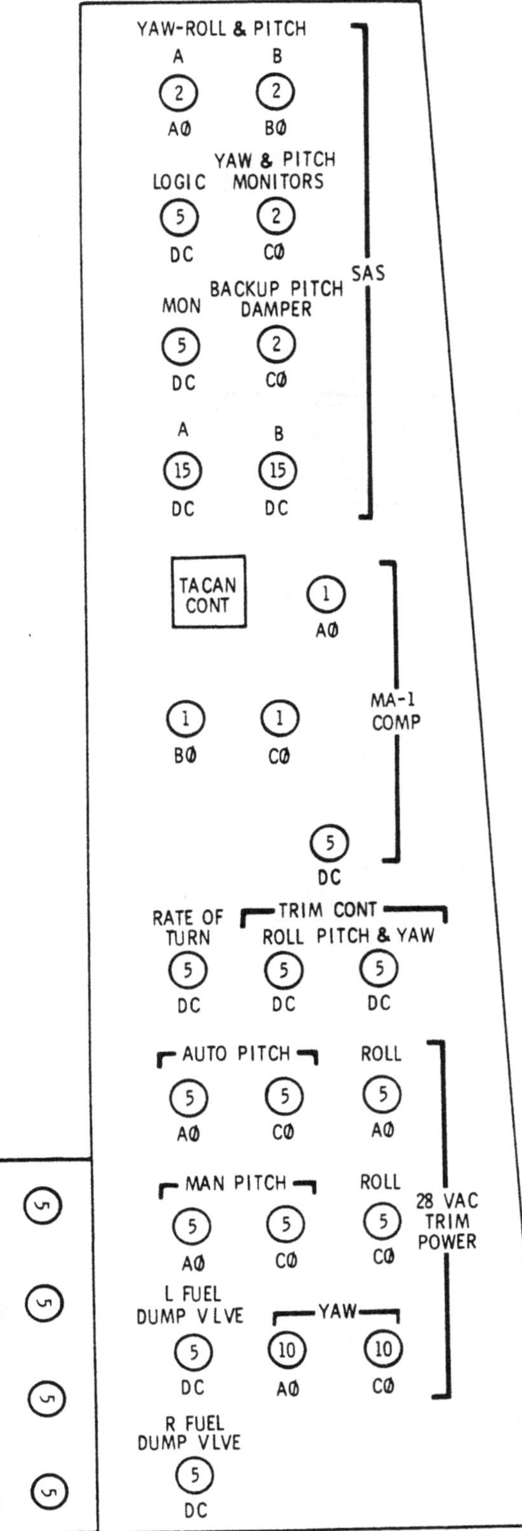

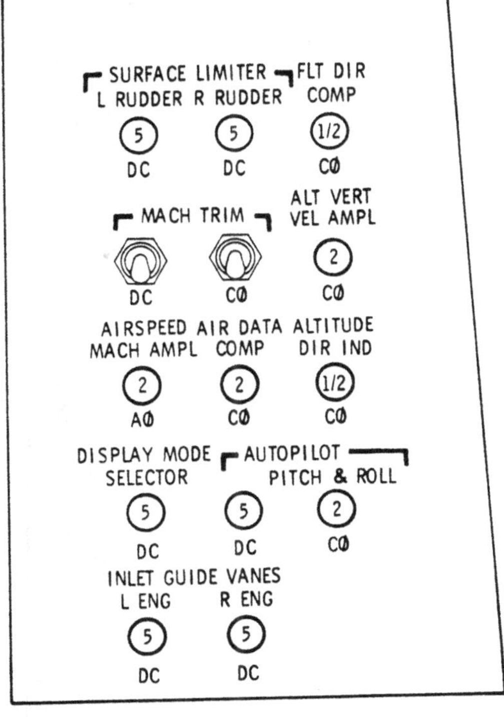

F202-17(b)

Figure 1-8B

AFT COCKPIT CIRCUIT BREAKERS

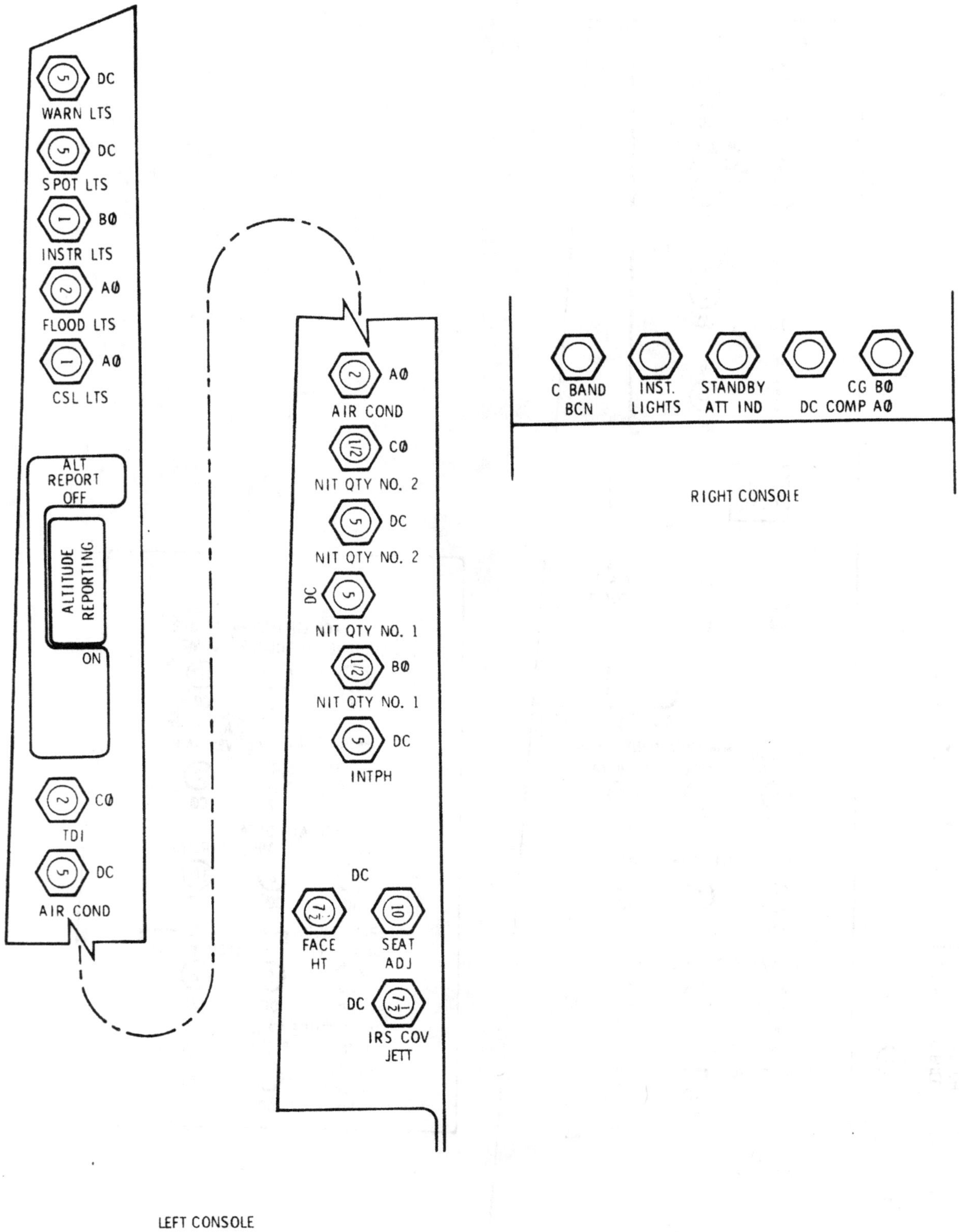

Figure 1-8C

(down) position, the generator will be removed from the ac buses and the automatic bus transfer system will supply the buses from the other generator if it is operating.

Battery Switch

A two-position battery switch is located on the pilot's right forward panel. When the switch is in the ON (up) position, the emergency battery is connected to the essential dc bus and the inverter battery is connected to the emergency 600-va inverter if neither the 200-ampere transformer-rectifier is furnishing power. If one or both transformer-rectifiers are furnishing power when the battery switch is in the ON position, the batteries will receive a trickle charge due to the slightly higher voltage of the transformer-rectifier.

Instrument Power Switch

A two-position instrument power switch is located on the pilot's right forward panel. The switch positions are NORM (down) and EMER (up). The switch should be in the NORM position for normal operation. In the event that both generators should fail and the automatic switching feature fails to connect the batteries to the essential dc bus and emergency inverter (the INST ON EMER PWR light on the pilot's annunciator panel does not illuminate) a back-up circuit can be activated by placing the switch to the EMER (up) position.

Transformer-Rectifier-Out Indicator Lights

The XFMR-RECT OUT indicator lights are located on the pilot's annunciator panel, and when illuminated indicate that the respective transformer-rectifier is not furnishing dc power to the essential dc bus.

Generator Out Indicator Lights

The GENERATOR OUT indicator lights are located on the pilot's annunciator panel. The lights illuminate when the generator is not furnishing power to the respective ac primary bus.

Bus-Tie-Open Indicator Light

The BUS TIE OPEN indicator light is located on the pilot's annunciator panel, and when illuminated indicates that the bus tie contactor connecting the L and R primary buses has opened and the generators are no longer operating in parallel.

Instruments On Emergency Power Light

The INST ON EMER PWR warning light is located on the pilot's annunciator panel, and when illuminated indicates that both generators and/or both transformer-rectifiers are inoperative and the 600 VA emergency inverter is furnishing power to the SAS and essential flight instruments.

Emergency-Battery-On Indicator Light

The EMER BATT ON warning light is located on the annunciator panel and illuminates when the battery is furnishing power to the emergency instrument inverter and essential dc bus.

CIRCUIT BREAKERS

Circuit breaker panels, located on the pilot's right and left consoles and the FCO's left console, contain push-to-reset, pullout-type circuit breakers for certain ac and dc circuits. Circuit breaker panels which are not accessible in flight, but which should be inspected before flight, are located in the No. 1 missile bay and the nosewheel well.

HYDRAULIC POWER SUPPLY SYSTEMS

Four separate hydraulic systems are installed on the aircraft (designed A and B, and L and R), each with its own pressurized reservoir and engine-driven pump. (See figures 1-9 and 1-10.) The A and B hydraulic systems provide power only for operating the flight controls; the L and R systems provide power for all other hydraulic requirements of the aircraft. The pumps for the A and L system are driven by the left engine accessory drive system, and the B and R pumps by the right engine accessory drive system. Hydraulic fluid is cooled by the fuel-oil heat exchangers, using the aircraft fuel supply as the cooling agent. Under normal operating conditions, the systems are independent of one another. The L hydraulic system provides hydraulic power to the left engine air inlet bypass doors and control, the left spike actuator and control, the landing gear (including uplocks and door cylinders), normal and anti-skid brakes, nosewheel steering and damping, air refueling door and receptacle lock, ventral fin, and for alternate missile bay door actuation if the R hydraulic system has failed. The R hydraulic system provides hydraulic power to the right engine air inlet control and bypass doors, missile bay door, right spike actuator and control, alternate ventral fin extension and retraction, and to the alternate brakes and landing gear (emergency retraction only) when the L hydraulic system has failed.

HYDRAULIC SYSTEM INSTRUMENTS AND INDICATOR LIGHTS

Two dual-indicating hydraulic gages are installed on the right side of the pilot's instrument panel. The left-hand gage indicates hydraulic pressure of the A and B (flight controls) systems, and the right-hand gage indicates hydraulic pressure in the L and R systems. The gages are calibrated in 100-psi increments, from 0 to 4000 psi. Pressure signal indications to the gages are by means of remote transmitters in the individual systems. Operating 26-volt ac power is furnished by the essential ac bus and the instrument transformer.

L and R Hydraulic Quantity Low Indicator Light

Two hydraulic quantity low warning lights are provided on the annunciator panel in the forward cockpit. The legend L HYD QUAN LOW or R QUAN LOW is illuminated when the quantity of hydraulic fluid in the corresponding reservoir decreases below approximately 1.2 gallons. Power for the lights is furnished by the essential dc bus.

A and B Hydraulic System Low Indicator Lights

Two indicator lights, one reading A HYD SYS LOW and one B HYD SYS LOW, are located on the pilot's annunciator panel to monitor the flight controls hydraulic system. When illuminated, the indicators indicate that A or B hydraulic system pressure has decreased to less than approximately 2200 psi, or quantity has dropped to 1.2 gallons remaining. The A and B hydraulic system pressure gages must be checked individually to determine which system is malfunctioning or inoperative.

HYDRAULIC OIL RESERVE SYSTEM

A reserve oil supply for the A and B hydraulic systems is contained in an 11-gallon (usable) reserve tank mounted in fuel tank 4. The reserve hydraulic oil is transferred by gravity flow and nitro-

A AND B HYDRAULIC SYSTEMS

RESERVE N_2 VALVE (TYP)
RELIEF VALVE (TYP.)
'A' RESERVOIR
RESERVE HYDRAULIC TANK
'B' RESERVOIR
RESERVOIR PRESSURE GAGE (TYP.)
FILL LEVEL VALVE
RESERVE FLUID VALVE (TYP)
CHARGE - DUMP VALVE (TYP.)
RESERVE FILL PORT
REG (TYP.)
RETURN GROUND CONNECTION (TYP.)
FILTER (TYP.)
CYLINDER PRESSURE GAGE (TYP.)
HEAT EXCHANGER
HEAT EXCHANGER
TEMPERATURE CONTROL
TEMPERATURE CONTROL
N_2 FILLER (TYP.)
N_2 CYLINDER (TYP.)
'A' HYD. PUMP
ANNUNICATOR LIGHT
'B' HYD. PUMP
A HYD SYS LOW
B HYD SYS LOW
CHECK VALVE (TYP.)
HYDRAULIC PRESS SURFACE CONTROLS
A
B
PRESSURE GROUND CONNECT (TYP.)
PRESSURE SWITCH (TYP.)
RESTRICTOR (TYP.)
PRESSURE TRANSMITTERS
ACCUMULATOR
TO FLIGHT CONTROL SERVOS
ACCUMULATOR
N_2 FILLER (TYP.)
N_2 GAGE (TYP.)
'A' SYSTEM PRESSURE
'A' SYSTEM RETURN
'B' SYSTEM PRESSURE
'B' SYSTEM RETURN
CASE FLOW LINE
N_2 PRESSURE
ELECTRICAL
F202-23(e)

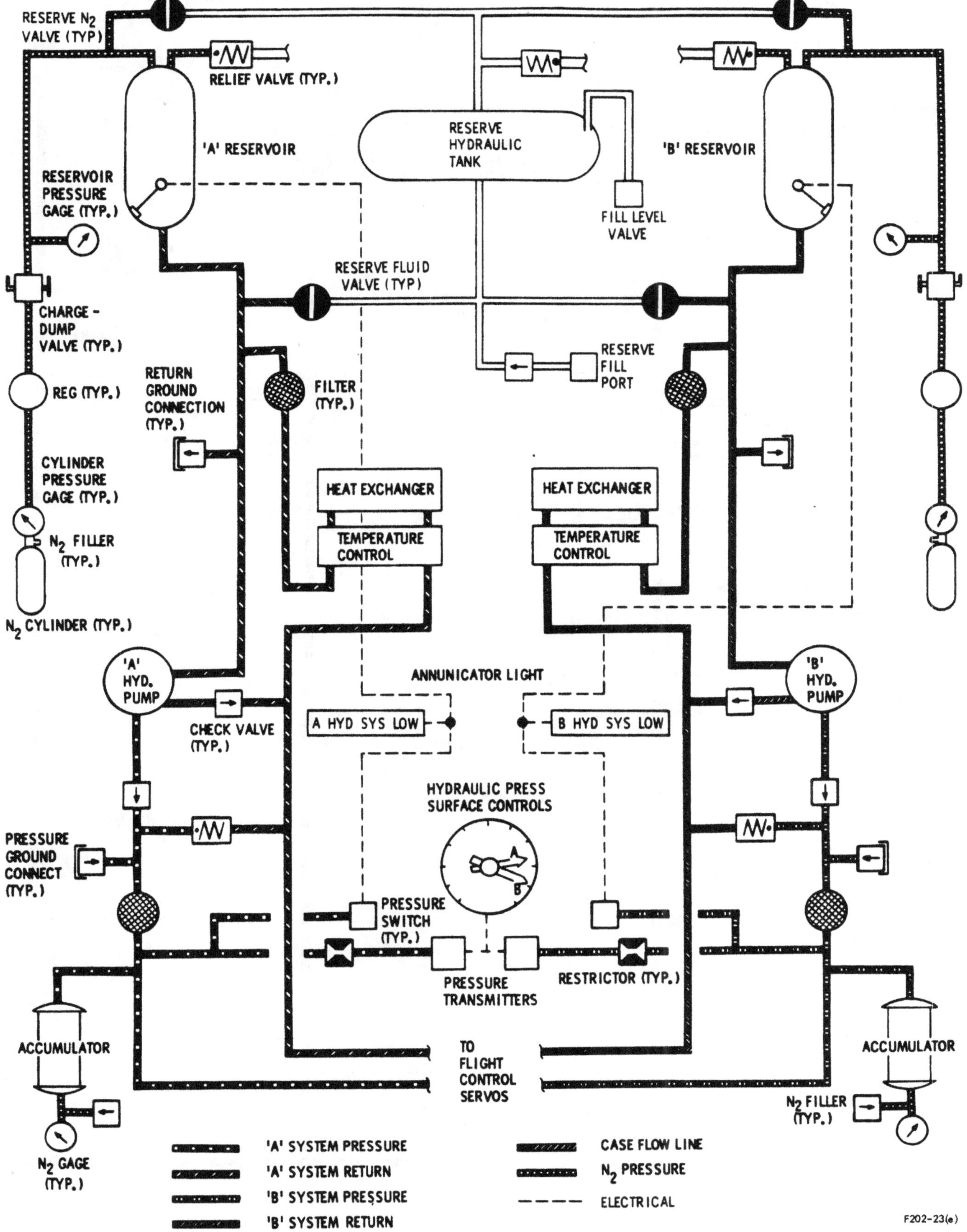

Figure 1-9

L AND R HYDRAULIC SYSTEMS

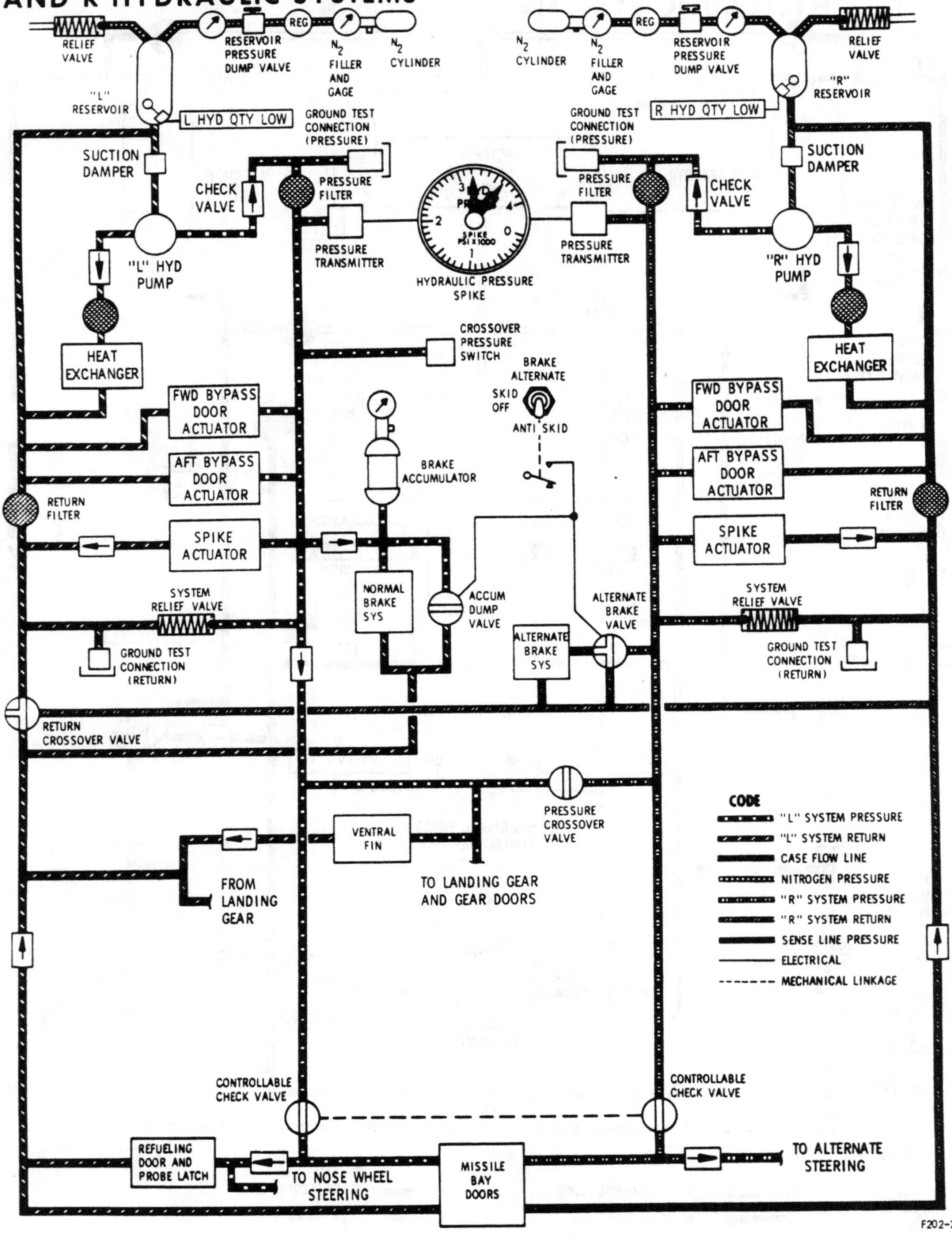

RELIEF VALVE
RESERVOIR PRESSURE DUMP VALVE
REG
N_2 FILLER AND GAGE
N_2 CYLINDER
"L" RESERVOIR
L HYD QTY LOW
GROUND TEST CONNECTION (PRESSURE)
N_2 CYLINDER
N_2 FILLER AND GAGE
RESERVOIR PRESSURE DUMP VALVE
RELIEF VALVE
"R" RESERVOIR
R HYD QTY LOW
GROUND TEST CONNECTION (PRESSURE)
SUCTION DAMPER
CHECK VALVE
PRESSURE FILTER
"L" HYD PUMP
PRESSURE TRANSMITTER
SPIKE PSI x 1000
HYDRAULIC PRESSURE SPIKE
PRESSURE FILTER
PRESSURE TRANSMITTER
CHECK VALVE
SUCTION DAMPER
"R" HYD PUMP
HEAT EXCHANGER
CROSSOVER PRESSURE SWITCH
HEAT EXCHANGER
BRAKE ALTERNATE
SKID OFF
ANTI SKID
FWD BYPASS DOOR ACTUATOR
AFT BYPASS DOOR ACTUATOR
BRAKE ACCUMULATOR
FWD BYPASS DOOR ACTUATOR
AFT BYPASS DOOR ACTUATOR
RETURN FILTER
RETURN FILTER
SPIKE ACTUATOR
SPIKE ACTUATOR
NORMAL BRAKE SYS
ACCUM DUMP VALVE
ALTERNATE BRAKE VALVE
ALTERNATE BRAKE SYS
SYSTEM RELIEF VALVE
GROUND TEST CONNECTION (RETURN)
SYSTEM RELIEF VALVE
GROUND TEST CONNECTION (RETURN)
RETURN CROSSOVER VALVE
PRESSURE CROSSOVER VALVE
VENTRAL FIN
FROM LANDING GEAR
TO LANDING GEAR AND GEAR DOORS

CODE
"L" SYSTEM PRESSURE
"L" SYSTEM RETURN
CASE FLOW LINE
NITROGEN PRESSURE
"R" SYSTEM PRESSURE
"R" SYSTEM RETURN
SENSE LINE PRESSURE
ELECTRICAL
MECHANICAL LINKAGE

CONTROLLABLE CHECK VALVE
CONTROLLABLE CHECK VALVE
REFUELING DOOR AND PROBE LATCH
TO NOSE WHEEL STEERING
MISSILE BAY DOORS
TO ALTERNATE STEERING

F202-21(e)

Figure 1-10

gen pressure through solenoid-operated shutoff valves to either the A or B hydraulic system, as selected by a cockpit switch.

Hydraulic Reserve Oil Switch

The hydraulic reserve oil switch, labeled HYD RESERVE, is mounted on the right side of the pilot's right forward panel and has three positions: A (up), NORM (center), and B (down). When in the A (up) position, solenoid-operated shutoff valves are opened to the A hydraulic system return and tank nitrogen pressurization lines. When in the B (down) position the solenoid valves to the B system are opened, and the reserve fluid supplies the B system. The reserve hydraulic fluid is supplied as needed up to approximately 3 gallons per minute. Power for the valves is furnished by the essential dc bus.

LANDING GEAR AND VENTRAL FIN SYSTEM

The tricycle-type landing gear and the main wheel well inboard doors are electrically controlled and hydrauically actuated. The main gear outboard doors and the nose gear doors are linked directly to the respective gear struts. Each three-wheel main gear retracts inboard into the fuselage and the dual-wheel nose gear retracts forward into the fuselage. The main gear is locked up by the inboard doors, the nose gear by an uplock which engages the strut. Hydraulic pressure is not on the gear when it is up and locked. Hydraulic pressure, normally from the L system, and downlocks inside the actuating cylinders hold the gear in place in the extended position. The landing gear cylinders and doors are actuated in the correct order by sequencing valves. During normal or emergency gear retraction, ventral fin extension is initiated when any of the four main gear door uplocks move to the locked position. During normal gear extension, ventral fin retraction is initiated by the nose landing gear downlock switch. Normal gear and fin operation is powered hydraulically by the L hydraulic pump on the left engine. Should pressure drop to approximately 2200 psi during retraction, the power source automatically shifts to the R hydraulic pump. In the event of an L system failure, R hydraulic pressure will not extend the gear and the landing gear must be released manually; however, the R system will provide pressure to retract the ventral fin.

A landing gear strut damper system is installed to control gear 'walking' during brake operation. The system is sensitive to less than 1G change in fore and aft acceleration. The damping is controlled through a G-monitoring valve which automatically increases or decreases the brake pressure, as required. Operating pressure for the damper system is provided by the L hydraulic system.

Landing Gear Lever

A wheel-shaped landing gear lever is installed on the lower left side of the pilot's instrument panel, just forward of the throttle quadrant. The lever has two positions, UP and DOWN. A locking mechanism is provided to prevent the gear lever from inadvertently being placed in the DOWN position. A button extending upward from the top of the lever must be pressed forward in order to release the lever lock mechanism. An override button is installed just above the gear lever and may be used to override the ground safety switch should it become necessary to raise the gear when the weight of the aircraft is on the landing gear. Once energized, the gear lever must be recycled to the DOWN position in order to bring the ground safety switch back into the circuit. A red light installed in the transparent wheel illuminates during cycling, or when the gear is in an unsafe condition. Power for the circuit is furnished by the essential dc bus.

Manual Landing Gear Release Handle

A landing gear release handle, labeled GEAR RELEASE, is installed on the center control stand. A pull of approximately 9 inches operates a cable system which releases the nose gear first, then main gear uplocks in the sequence of right door first, aft lock before forward lock. The gear will extend and lock by gravity force if no hydraulic pressure is available. If the L hydraulic system has failed but R hydraulic pressure is available, the landing gear lever must be placed in the DOWN position before pulling the GEAR RELEASE handle; otherwise, the hydraulic pressure crossover valve will operate and the R system will retract the gear. Pulling the cable also closes a switch in the nosewheel well which energizes the ventral fin retraction system. Ventral fin retraction is started automatically, using R system pressure, when the left main gear is down and locked. Gear retraction and ventral fin extension are possible in the normal way after use of the manual gear release handle, if hydraulic system pressure is again available.

Landing Gear Position Lights

Three green lights located on the pilot's left forward panel indicate the down-and-locked condition of the landing gear. The location of each light corresponds to the respective wheel in monitors. Power is from the dc essential bus.

Landing Gear Warning Light and Audible Warning

The warning light in the transparent landing gear lever handle is red when illuminated. It indicates the following:

a. Gear is cycling.

b. Gear system is not locked in selected UP or DOWN position.

c. Gear is UP and power settings are below minimum cruise.

An audible warning signal is produced in the pilot's headset when the throttles are retarded below minimum cruise setting, the landing gear is not in the down and locked position, and aircraft altitude is below 10,000 (± 500) feet. Power for the light and audible warning is furnished by the essential dc bus.

Ventral Fin Indicator Light

A VENTRAL FIN indicator light is located on the pilot's annunciator panel, and when illuminated indicates that the ventral fin is in transit to the up or down position or is not in the correct position in relation to the landing gear position. Power for the light is furnished by the essential dc bus.

Landing Gear Warning Cutout Button

The aural gear warning may be eliminated by depressing the HORN CUTOUT pushbutton switch on the pilot's left forward panel. The circuit is reactivated when the throttles are advanced past the minimum cruise setting. Power is supplied to the button circuit by the essential dc bus.

LANDING GEAR AND VENTRAL FIN SYSTEM

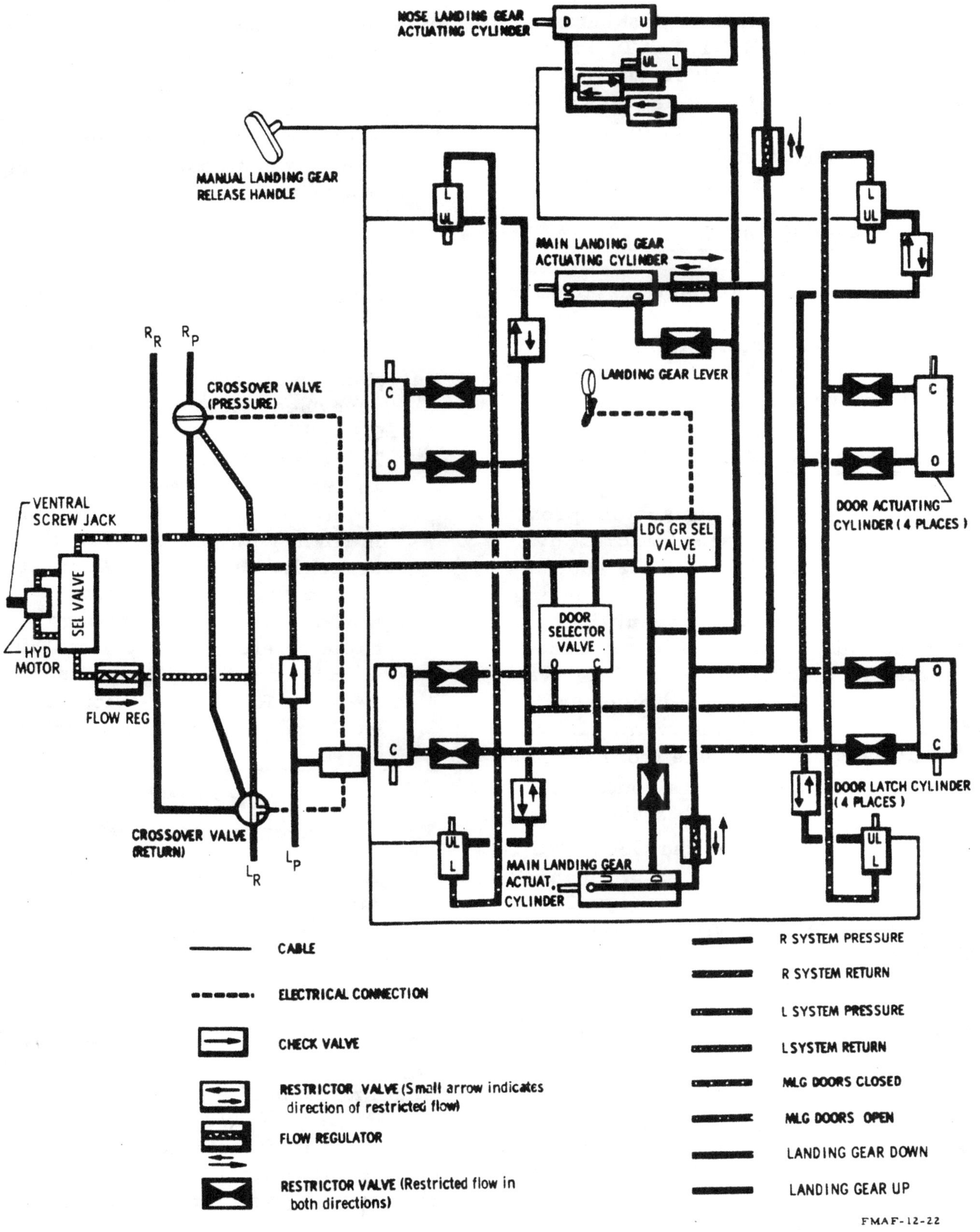

NOSE LANDING GEAR ACTUATING CYLINDER
D U
UL L
MANUAL LANDING GEAR RELEASE HANDLE
L UL
MAIN LANDING GEAR ACTUATING CYLINDER
R_R R_P
CROSSOVER VALVE (PRESSURE)
LANDING GEAR LEVER
C O
DOOR ACTUATING CYLINDER (4 PLACES)
VENTRAL SCREW JACK
SEL VALVE
LDG GR SEL VALVE
D U
HYD MOTOR
DOOR SELECTOR VALVE
O C
FLOW REG
DOOR LATCH CYLINDER (4 PLACES)
CROSSOVER VALVE (RETURN)
L_R L_P
MAIN LANDING GEAR ACTUAT. CYLINDER

Symbol	Meaning
——	CABLE
- - - -	ELECTRICAL CONNECTION
	CHECK VALVE
	RESTRICTOR VALVE (Small arrow indicates direction of restricted flow)
	FLOW REGULATOR
	RESTRICTOR VALVE (Restricted flow in both directions)
	R SYSTEM PRESSURE
	R SYSTEM RETURN
	L SYSTEM PRESSURE
	L SYSTEM RETURN
	MLG DOORS CLOSED
	MLG DOORS OPEN
	LANDING GEAR DOWN
	LANDING GEAR UP

FMAF-12-22

Figure 1-11

Landing Gear Light Test Button

A landing gear light test pushbutton, labeled LDG GEAR LT TEST, is located on the pilot's left forward panel. When this pushbutton is depressed, the three gear-down-and-locked lights and the warning light in the transparent landing gear lever knob illuminate.

Landing Gear Ground Safety Pins

Removable safety pins are installed in the landing gear assemblies to prevent inadvertent gear retraction. Warning streamers direct attention to their removal before flight.

NOSEWHEEL STEERING SYSTEM

The nosewheel steering system provides directional control when aircraft hydraulic pressure is available and aircraft weight is on any gear. Steering is accomplished by a hydraulic steer-damper unit, controlled by the rudder pedals and a cable system. The system is normally operated by power from the L hydraulic system. L hydraulic system pressure from the nose landing gear down line is routed to the steering system through a shutoff valve, which is controlled by the CSC/NWS button on the control stick grip. If the L system fails, or pressure drops to 2200 psi, nosewheel steering will be available from the R system by moving the brake switch to the ALT BRAKE position.

Nosewheel steering is engaged by depressing the CSC/NWS switch and aligning rudder pedal position with the nosewheel. A holding relay circuit keeps steering engaged when the switch is released. The switch must be depressed and released again to disengage steering.

A nosewheel steering engage light is located on the top left side of the instrument panel in the forward cockpit. The light is green when illuminated and is labeled STEER ON. The light illuminates steady when the clutch engages nosewheel steering. The light is not illuminated when nosewheel steering is disengaged.

NOTE

A hydraulically actuated clutch is located within the steering damper unit. The clutch functions to engage and disengage nosewheel steering when the CSC/NWS switch is actuated. In the event the clutch jams, and nosewheel steering does not disengage, rudder control in flight with gear down will be severely restricted. Approximately 6 degrees of rudder would be available, through cable stretch, by applying 180 pounds of force at the rudder pedals. Rudder restriction would not be noted with gear up.

WARNING

Retract the landing gear immediately to relieve restriction of rudder movement if jamming of the nosewheel steering clutch is suspected while in flight.

The nosewheel is steerable 45 degrees right and left of center position. Minimum nosewheel steering radius is approximately 54-1/2 feet. A mechanically operated centering cam automatically centers the nosewheel during retraction. Power for the system is furnished by the essential dc bus.

WARNING

The landing gear side load strength is critical. Side loads during takeoff, landing, and ground operation must be kept to a minimum.

WHEEL BRAKE SYSTEM

The aircraft is equipped with artificial feel, hydraulically operated, power brakes. Depressing the rudder pedals actuates the brake rotors on each of the six main wheels. The L hydraulic system normally furnishes brake pressure, with additional, optional, anti-skid operation. The hydraulic pressure to the brakes is approximately 950 psi. In the event of L hydraulic system failure, alternate braking is available from an independent system, using R hydraulic pressure. To operate the anti-skid system the system must be selected, the landing gear handle must be moved to the down position, and at least one main wheel or the nosewheel must be on the ground. The anti-skid brake feature is inoperative when using the alternate braking system. Also, if anti-skid braking is lost, the brake system reverts to normal operation.

Brake Switch

A three-position brake switch is located on the pilot's left forward panel. When the switch is in the SKID OFF (center) position, the brakes are operated normally, by pressure from the L hydraulic system, the anti-skid feature not being in the circuit. Moving the switch to the ANTI-SKID (down) position makes the anti-skid system operative. When the switch is in the ALT BRAKE (up) position, the brakes are powered by the R hydraulic system, through a separate system. Power for the circuit is furnished by the essential dc bus.

CAUTION

Do not switch to alternate brakes unless normal left hydraulic pressure is unavailable or normal brakes are inoperative. Pressure may be trapped in the brakes after the pedals are released, causing grabbing or locking.

NOTE

Wheel brakes cannot be applied in the air if the brake switch is in the ANTI-SKID position.

Anti-Skid Out Warning Light

Illumination of the ANTI-SKID OUT light on the pilot's annunciator panel warns that the anti-skid system is inoperative. When the aircraft is on the ground, the light will be illuminated when the cockpit switch is in the SKID OFF or ALTERNATE position. The light will be off when the switch is in the ANTI-SKID position and the anti-skid control box and wheel generators are operative. If the fail-safe within the anti-skid control box is tripped, and power from the essential dc bus is on the system, the light will illuminate. The light does not illuminate when the weight of the aircraft is off the gear.

BRAKE SYSTEM

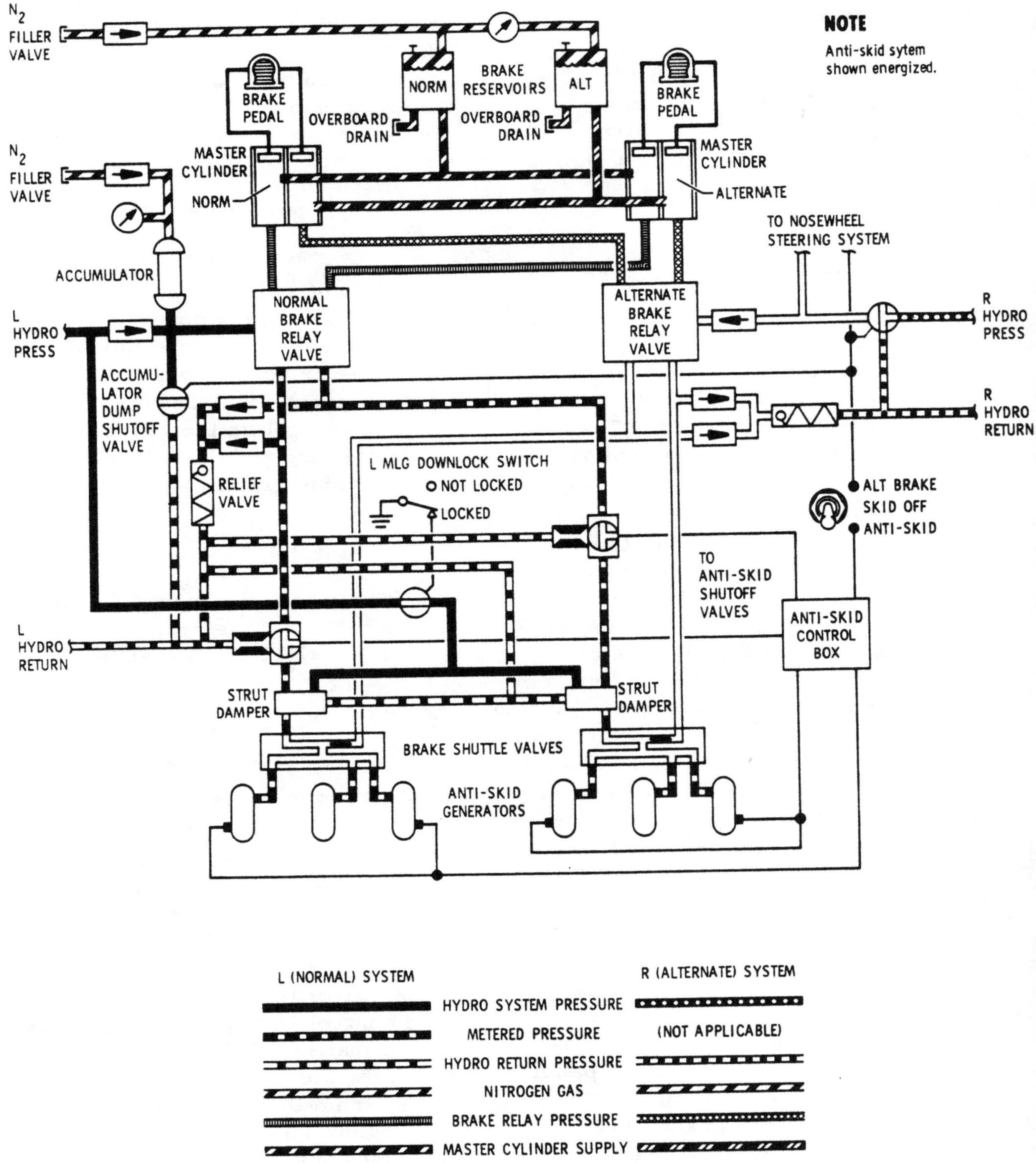

Figure 1-12

DRAG CHUTE SYSTEM

The drag chute system is provided to reduce landing roll and aborted takeoff rollout distance. A "-501" type ribbon chute of 40 foot nominal diameter is provided. The drag chute package is stowed in a compartment in the upper aft end of the fuselage. The drag chute attachment rides free in the compartment until locked to the aircraft during the initial stage of deployment. The drag chute and extraction system are packed in a deployment bag which is also the storage container.

NOTE

> The drag chute will be jettisoned if the drag chute doors should open in flight with the drag chute attachment link unlatched; with the chute latched, the chute will carry away when airspeed is over approximately 235 KEAS.

Drag Chute Switch

A three-position drag chute switch, labeled DRAG CHUTE, which controls dual drag chute actuator motors is located on the upper left side of the forward cockpit instrument panel. The switch positions are DEPLOY (up), off (center), and JETTISON (down). Deployment is accomplished by placing the switch in the DEPLOY position. Power is removed from both actuator circuits when the switch is in the off position. Placing the switch in the JETTISON position causes the chute attachment fitting to be released and this either jettisons the chute, if it has been deployed, or prevents the chute from being deployed. To prevent inadvertent movement, the drag chute switch must be pulled out slightly to clear a center detent before it can be moved from or through the off position.

Power for the drag chute actuator circuits is furnished by the essential dc bus through a circuit breaker labeled DRAG CHUTE on the pilot's left console.

FLIGHT CONTROL SYSTEM

The cockpit flight controls consist of a conventional control stick and rudder pedals. The delta wing configuration utilizes elevons instead of separate aileron and elevator control surfaces. The elevons, moving together in the same direction, function as elevators and when moving in opposite directions, function as ailerons. Each elevon consists of an inboard and outboard panel with the inboard panel located between the fuselage and the nacelle and the outboard panel outboard of the nacelle. Both panels on one side function as a single unit with the servo input of the outboard elevon connected directly to the inboard elevon surface. The dual canted rudders are full moving, one piece, pivoting surfaces with a small fixed stub at the junction of the vertical surface and the nacelle. Deflection and control of the elevons and rudders is by means of dual, full hydraulic, irreversible actuating systems.

Control surface travel limits are as follows:

	Elevons	Rudders
Pitch	* 11° Down 24° Up	- -
Roll	12° Down 12° Up	- -
Pitch plus Roll	20° Down 35° Up	- -

Yaw	20° Left 20° Right

*Relative to trim setting.

Manually operated mechanical stops are incorporated in the cockpit mechanism to limit the surface movement at high speed. Elevon travel in roll is limited to 7° up, 7° down and rudder travel is limited to 10° right, 10° left. An additional stop is installed in each rudder servo package to limit the rudder travel. These stops are electrically controlled and hydraulically operated by electrical and hydraulic systems. If no electrical power is available, the rudders will be limited to approximately 10° L and R travel. If electrical power is available to one stop, that rudder only will have the full 20° L and R travel available. The rudder cable must be stretched to obtain this travel, causing a noticeable increase in rudder pedal force.

CABLE SYSTEM

Cable systems are utilized to transfer control movements from the control stick and rudder pedals to the flight control mechanisms. The pitch and roll axis cable systems are duplicated from the pilot's cockpit to the mixing mechanism in the aft fuselage. The rudder system has two separate closed loop single cable systems, one to each rudder. Cable tension regulators and slack absorbers are incorporated in the cable systems.

ARTIFICIAL FEEL SYSTEM

The use of a full power irreversible control system for actuation of the surfaces prevents air loads and resulting "feel" from reaching the cockpit controls. Therefore, feel springs are installed in each of the pitch, roll and yaw axis mechanical control mechanisms to provide an artificial sense of control feel. The springs apply loads to the pilot controls in proportion to the degree of control deflection.

TRIM CONTROL SYSTEM

Flight control trim is accomplished by deflecting the control surfaces through the use of electrical trim actuators. The roll and pitch trim actuators are located downstream of the feel springs, so stick position remains neutral irrespective of the amount of trim. The trim actuator and feel spring location is combined in the rudder mechanism and yaw trim is reflected by rudder pedal position.

Trim Limits

Travel limits of the trim system are 3-1/2 degrees down to 6-1/2 degrees up in pitch, 4-1/2 degrees up and down (each side) in roll, and 10 degrees left to 10 degrees right in yaw. Trim position indicators are provided for each axis. Trim rates are as follows:

	Pitch	Roll	Yaw
Max.	1.5°/sec	.95°/sec Total Diff.	1.5°/sec
Min.	0.67°/sec	.47°/sec Total Diff.	0.67°/sec

Automatic pitch trim uses a separate, slow-speed motor for autopilot synchronization and Mach trim. The automatic pitch trim rate is 0.15 deg/sec maximum and 0.067 deg/sec minimum.

RUDDER PEDALS

Primary control for the rudders consists of conventional rudder pedals mechanically connected by cables, bellcranks, and pushrods to hydraulic control valves at the rudder hydraulic actuators. The rudder pedals are released for adjustment by pulling the T-handle, labeled RUDDER PEDAL ADJ, located at the bottom of the pilot's lower instrument panel. Wheel brakes are controlled conventionally by toe action on the rudder pedals; refer to Wheel Brake System, this section for additional information. Rudder pedal movement also controls nosewheel steering; refer to Nosewheel Steering System, this section.

CONTROL STICK

The control stick is mechanically connected by a torque tube, pushrods, and bellcranks to the dual cable system which operates the roll and pitch quadrants in the aft fuselage tailcone. Mechanical pushrod linkages mix the control movements and position dual hydraulic control valves which direct both A and B system hydraulic pressure to the inboard elevon actuating cylinders. Pushrods, bellcranks, and torque tubes transfer the effect of inboard elevon deflection to position the outboard dual hydraulic control valves. These valves direct both A and B system hydraulic pressure to the outboard elevon actuating cylinders. A pushrod follow-up system closes off the flow of hydraulic fluid to the actuators when the desired elevon deflection is obtained.

Control Stick Grip

Located on the control stick grip is a combination pitch and yaw trim switch (labeled TRIM), a combination nosewheel steering (NWS) and control stick command (CSC) pushbutton switch, a slide microphone switch (labeled TRANS and INPH) to select interphone or radio transmission, an autopilot disconnect (AUTOPILOT DISC) and in-flight refueling disconnect (A/R DISC) trigger switch.

Nosewheel Steering and Control Stick Command Pushbutton Switch

(Refer to the Nosewheel Steering and Autopilot System paragraphs, this section, for a description of the functions of this switch.)

Microphone Switch

The slide microphone switch is spring loaded to the center (off) position. Moving the switch to the TRANS position permits the pilot to transmit according to the setting of the AIC-18 control panel. Moving the switch to the INPH position selects interphone communications.

Autopilot Disc & A/R Disc Trigger Switch

(Refer to the Autopilot System and Air Refueling System paragraphs, this section, for a description of the function of this switch.)

Pitch and Yaw Trim Switch

Pitch and yaw trim control is provided by a spring-loaded switch installed on the control stick grip. The switch positions are center off, LEFT, RIGHT, NOSE UP and NOSE DOWN. The switch uses 28V dc from the essential dc bus to control trim motors powered by the essential ac bus through the trim actuator transformer.

CAUTION

To avert the possibility of a sticking trim switch the pilot should assure positive switch movement to the neutral position after each activation.

NOTE

The trim power switch must be in the ON position before the pitch, roll, and yaw switches will operate.

Lateral movement of the switch to the left corrects for right yaw and lateral movement to the right corrects for left yaw. Forward movement of the switch produces down-elevon operation of the trim motors and actuators, and aft movement moves the elevons up.

TRIM SYSTEM SWITCHES AND INDICATORS

Auxiliary Trim Selector Switch

A two-position, guarded, trim selector switch, located on the pilot's upper left console, is labeled AUX TRIM (up) and STICK TRIM (down). If the control stick trim switch fails, the selector switch allows the pilot to use the auxiliary trim switch for control of the pitch and yaw trim circuits. 28V dc power for the switch is furnished by the essential dc bus.

Auxiliary Trim Control Switch

A spring loaded auxiliary trim control toggle switch is located on the pilot's upper left console, adjacent to the trim power switch. The switch is usable when the auxiliary trim selector switch is in the AUX TRIM position. The AUX TRIM CONT switch positions are labeled UP, DOWN, L and R. The center position is off. The switch produces the same effects as the control stick trim switch. 28V dc power for the switch is furnished by the essential dc bus.

Trim Power Switch

The trim power switch, installed on the pilot's left console, has two positions, ON and OFF. To prevent inadvertent movement, the switch must first be pulled out before it can be moved from the ON to the OFF position. In the ON position, essential ac bus power is provided to the roll, pitch, and yaw trim actuators. The trim power circuit breaker is located in the nosewheel well and is not accessible to the pilot.

Roll Trim Switch

A three-position roll trim switch is installed just forward of the throttle quadrant. The switch positions are L WING DN (left), R WING DN (right) and center off (spring loaded). When the switch is held in the R WING DN position, the roll trim motor operates to move the right elevons up and the left elevons down. Actuation of the switch to the L WING DN position moves the right elevons down and left elevons up. Operating 28-volt ac power is furnished from the essential ac bus through the trim actuator transformer.

Rudder-Synchronization Switch

A three-position rudder synchronization switch is installed just forward of the throttle quadrant. The switch positions are L (left), R (right), and center off (spring loaded). In the L and R positions the switch provides electrical power to the right rudder trim motor

CONTROL STICK GRIP

TOP VIEW

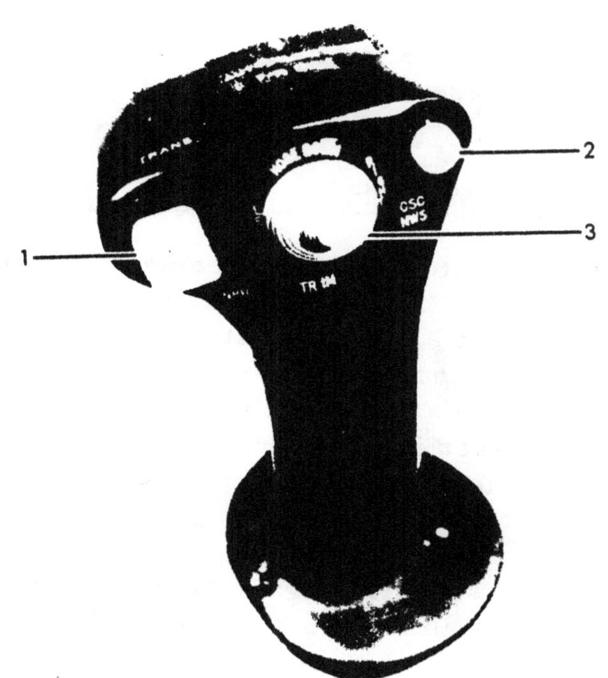

FRONT VIEW

SIDE VIEW

1 TRANSMITTER - INTERPHONE CONTROL SWITCH
2 CONTROL STICK COMMAND - NOSEWHEEL STEERING BUTTON
3 PITCH AND YAW TRIM SWITCH
4 EMERGENCY AUTOPILOT DISENGAGE SWITCH AND AIR REFUEL DISCONNECT

Figure 1-13

which moves the right rudder to agree with the position of the left. Rudder synchronization is obtained by superimposing the L and R pointers on the yaw trim gage. Operating 28-volt ac power is furnished from the essential ac bus through the trim actuator transformer.

Roll, Pitch, and Yaw Trim Indicators

Separate roll, pitch, and yaw trim indicators are installed on the instrument panel. The roll trim indicator uses double-ended pointers and displays the amount of roll trim from 0 to 9 degrees differential. The pitch trim indicator displays the amount of pitch trim from 5 degrees nose down to 10 degrees nose up, although only 8-1/2 degrees nose-up trim is available. The yaw trim indicator displays the amount of yaw trim from 10 degrees left to 10 degrees right for both rudders. Rudder synchronization is obtained by superimposing the L and R pointers on the yaw trim gage. Operating 28-volt ac power for the indicators is furnished by the essential ac bus and the instrument transformer.

Surface Limiter Release Handle

A T-handle, labeled SURFACE LIMITER RELEASE, is located on the lower left side of the center control stand just ahead of the stick. When the handle is turned 90 degrees counterclockwise and released to the forward position, the mechanical stops in the roll and yaw axis of the surface control system are activated. This action also opens an electrical switch which de-energizes a solenoid-operated valve in each rudder servo package to activate the servo package rudder stops. When the handle is pulled out and locked by 90 degrees clockwise rotation, the mechanical stops in the cockpit are released and the solenoid is energized, releasing the servo package stops. Power for the rudder limiter circuit is furnished by the dc essential bus.

Surface Limiter Warning Light

When airspeed exceeds Mach 0.5, the SURF LIMITER warning light on the pilot's annunciator panel lights and remains on until the surface limiter handle is released. If the surface limiters are on at speeds below Mach 0.5 the light will illuminate until the surface limiter handle is pulled out. Refer to Flight Control System description for control surface deflections. Power for the lights is furnished by the essential dc bus.

AUTOMATIC FLIGHT CONTROL SYSTEM

The automatic flight control system (AFCS) comprises four major subsystems, the stability augmentation, autopilot, Mach trim, and air data systems. In addition, the autopilot utilizes inputs from the inertial navigation system and the flight reference system.

The stability augmentation system provides automatic stability augmentation in the pitch, roll, and yaw axes. The autopilot provides automatic flight control in the pitch and roll axes, and the Mach trim system provides speed stability augmentation in the pitch axis. The air data system supplies airspeed, Mach, and altitude inputs to the AFCS for display and scheduling.

STABILITY AUGMENTATION SYSTEM

The three-axis stability augmentation system (SAS) is a combination of electronic and hydraulic equipment which augments the inherent stability of the aircraft. It is designed for optimum performance at the basic mission cruise speed and altitude, but also provides improved stability for takeoff, landing, and in-flight refueling. The SAS is part of the aircraft's basic control system and is normally used during all flight conditions. Because of the requirements for fail-safety and the possible consequences of complete loss of damping, redundancy is employed in the SAS. The pitch and yaw axes are each provided with four hydraulic servos, and two hydraulic servos are provided in the roll axis. Three complete, independently operating, gyro channels are provided in the yaw and pitch axes. The roll axis utilizes two completely independent gyro channels. Logic circuits compare the functioning of each pitch and yaw channel and automatically disengage a failed channel; that is, one whose channel gyro signal or servo action disagrees with other signals or actions in that system. The pilot is provided with a visual warning of a disengaged channel. In the roll axis, fail-safety is normally not as critical as in pitch or yaw; consequently, only a simplified simulated logic circuit causes disengagement of both roll channels in the event of a roll SAS failure. Reliability is provided through dual hydraulic and electrical supplies. Each active channel in each axis is powered by separate sources so that the redundant parts operate independently. Each gyro channel is operated from a separate phase of ac power (A, B, or C). In addition, each servo channel is operated from a separate hydraulic system (A or B). Consequently, only a complete loss of power (both generators and battery) or a complete loss of A and B hydraulic systems can cause loss of all channels in one axis. A single failure of one phase of ac power or a loss of one hydraulic system does not interrupt SAS operation. A backup source of ac power for the SAS is provided by an emergency battery-operated inverter, and in the event that both generators fail, SAS operation may be continued for a limited period of time.

STABILITY AUGMENTATION PITCH AXIS

Two independent servo channels (A and B) provide control through a pair of servos on each side of the aircraft. The servos are in series with pilot or autopilot control signals but do not move the control stick. A and B channels drive one servo each on the left side of the aircraft and one each on the right side. The A channel uses the A hydraulic system and the B channel uses the B hydraulic system in order to avoid loss of both channels in case of failure of either the A or the B hydraulic system. Three rate gyros, one each for the A, B, and monitor (M) systems, are located in fuel tank 3. These gyros provide signals proportional to the rate of aircraft pitch attitude change. Phasing of the gyro signals is such that an angular pitch motion produces elevon movement to oppose and restrict attitude change. The system automatically corrects for gust disturbances. Pilot inputs are also opposed; however, the elevon motion produced by the SAS aids the pilot in avoiding overcontrol and improves the handling qualities of the aircraft. The three gyro channel signals (A, B, and M) are continuously compared by electronic logic circuitry; if one channel differs from the other two, that channel is automatically disabled and the gain of the remaining channel signals is automatically increased to compensate for the loss. The three gyro signals are also added together and fed to all four servos in both servo channels. Logic circuitry is provided which will disengage both servos of a servo channel if one of the servo outputs differs from the other servo in the same channel. The operating servo channel will continue to provide the original amount of damping. Malfunctioning of the SAS is indicated to the pilot by lights on the SAS function selector panel. Gyro damping signals in the pitch axis can command a maximum elevon surface travel of 2.5 degrees up, to 6.5 degrees down. Power for the A servo channel is from phase A of the generator primary ac bus and B channel is fed from phase B. Power for gyro channels A, B, and M is taken from the A, B, and C phases, respectively. Each phase is protected by individual circuit breakers in the pilot's cockpit.

STABILITY AUGMENTATION YAW AXIS

The yaw axis of the SAS is very similar to the pitch axis, using two independent A and B channels and a monitor channel. There is one pair of hydraulic servos for each rudder, each pair mounted in a "whiffletree" arrangement. Damping signals to the rudders do not move the rudder pedals. Each A and B channel drives one servo on each side of the aircraft. The A hydraulic system is connected to the A channel and the B hydraulic system to the B channel. The rate gyro sensors for the three channels are identical to the pitch-rate gyros, except for the physical orientation to sense yawing motions. Each gyro signal is passed through a hi-pass filter circuit which allows the passage of normal short-term damping signals, but which stops the damping signals during a steady-state turn. An additional lateral accelerometer sensor is used in each channel of the yaw axis to minimize steady-state sideslip, such as caused by an engine failure, until the pilot can retrim the rudders.

NOTE

> The additional sensor provides a signal that results in a more rapid rudder response during engine failure; however, a deliberately induced sideslip will be opposed as well.

The three gyro-accelerometer channels are summed in the same manner as the pitch axis and are fed to all four yaw servos. The logic circuit is identical to that in the pitch axis and functions in a like manner. Failure of a gyro channel causes that channel to be disengaged and the gain of the two remaining channels to be increased in order to compensate for the loss. Failure of a servo channel automatically disengages each of the corresponding rudder servos. Since each servo channel in the yaw axis provides one-half of the damping action, the gain of the remaining channel is doubled when one pair of servos is disengaged. This means that damping action will remain normal during either single- or dual-channel operation. The yaw axis augmentation system can produce a maximum rudder travel of 8 degrees left to 8 degrees right. Power distribution to the yaw axis SAS is identical

to that applied to the pitch SAS. Individual circuit breakers are also installed.

STABILITY AUGMENTATION ROLL AXIS

Roll axis reliability requirements are not as severe as those for pitch and yaw; therefore, less complicated circuitry and components are used. The roll axis has two independent channels: A-channel servos position the left elevon surfaces and operate from the A hydraulic system, and B-channel servos position the right elevon surfaces and operate from the B hydraulic system. There is no monitor channel as such; there is, however, a simulated logic circuit to disengage both channels and illuminate a light on the SAS panel if one roll channel malfunctions.

Each channel can be operated individually. Although the system gain with one channel off is the same as for operating with two channels, roll control is not symmetrical, and coupling into the yaw and pitch axes is possible. With yaw and pitch SAS operating, however, undesirable aircraft motion in those axes is minimized. Maximum elevon travel in the roll axis is 2 degrees up to 2 degrees down (each side) for a total of 4 degrees differential with both roll systems operating. Power for the A channel is from the A phase of the generator ac primary bus; the B channel is powered by the B phase of the generator ac primary bus.

STABILITY AUGMENTATION SYSTEM (SAS) CONTROL PANEL

The SAS control panel on the pilot's right console contains six channel-engage switches for A and B channels of the pitch, roll, and yaw axes. The panel also contains a press-to-test switch, six indicator lights, labeled A, B, and MON in the pitch and yaw axes, and a roll channel disengage light (located between the two roll channel on-off switches). Three switches for the backup pitch damper, pitch logic override, and yaw logic override are also located on the pilot's right console, adjacent to the control panel. Individual circuit breakers are on the pilot's right console.

Channel Engage Switches

There are six toggle switches located on the SAS control panel, two each for the pitch, roll, and yaw axes. The forward switch of each pair is for A servo channel and the aft switch is for B servo channel. The switches have two positions, ON (forward) and off (aft, unlabeled). When electrical power is on the aircraft and the channel engage switches are OFF, the SAS electronics are powered but the channel servos are not engaged with the control system. Moving the switches to the ON position only engages the SAS servos.

Recycle Indicator Lights

Six recycle indicator lights are located on the SAS control panel on the pilot's right console adjacent to the pitch and yaw channel engage switches. One light is provided for each A and B servo and monitor channel in the pitch and yaw axes. The A and B lights indicate condition of the A and B servos. The MON lights indicate the condition of the A, B, and monitor pitch and yaw gyro channels. When the channel engage switches are ON and the recycle lights are not illuminated, all servo and gyro channels are functioning properly. If one or more of the gyro channels in an axis is disengaged, the corresponding axis monitor light will be ON. If an A or B light is ON, the corresponding servos are disengaged. An illuminated light may be depressed to recycle the disengaged servo and gyro (s) in a disengaged channel. If the indicator light relights when released, the channel is malfunctioning and disengaged and it is not necessary to turn the channel engage switch OFF.

NOTE

> The recycle indicator lights should be depressed firmly and released. They should not be held down. If a channel has malfunctioned, holding the light down can cause control surface transients.

The six recycle lights will be illuminated when electrical power is first turned on. After turning the channel engage switches ON the lights must be depressed individually to initiate operation, which engages the channel electronics to the servos, and the lights will go out.

Roll Channel Disengage Light

A single roll channel disengage light is located between the two roll channel switches. When illuminated it indicates that both roll channels have disengaged. Roll disengagement results when the roll servo channel outputs differ by more than an amount equivalent to 0.6 degree surface deflection. When operating on a single roll channel the light will not be illuminated and disengagement in the event of a failure is not provided. The switch must be on for the active channel and OFF for the malfunctioning channel.

Light Test Switch

A pushbutton light test switch is located in the center of the SAS control panel. Depressing the pushbutton illuminates the six recycle lights for test.

SAS Pitch Logic Override Switch

A guarded, three-position, pitch logic override switch is located on the pilot's right forward console. The switch is OFF in the center guarded position and the logic circuit is operative. Placing the switch in the A (up) position eliminates the logic circuit and selects A gyro channel operation. In the B (down) position, the logic circuit is eliminated and B gyro channel operation is selected. The proper servo channel is selected by use of the A and B channel engage switches on the function selector panel. The logic override switch must be placed in either the A or B position when the backup pitch damper is used.

NOTE

- A logic override switch is used only as part of an emergency procedure.
- Engage only one pitch channel engage switch (A or B) at a time.

SAS Yaw Logic Override Switch

A guarded, three-position yaw logic override switch is located on the pilot's right forward console, just aft of the pitch logic override switch. The switch is guarded in the OFF position. The A (up) position eliminates the logic circuit and selects A gyro channel operation. The B (down) position eliminates the logic circuit and selects B gyro channel operation. The proper servo channel is selected by use of the A and B channel engage switches. When Logic Override is selected on either A or B, the yaw axis SAS operates at only half the normal gain.

NOTE

- A logic override switch is used only as part of an emergency procedure.
- Engage only one yaw channel engage switch (A or B) at a time.

Backup Pitch Damper Switch

A backup pitch damper (BUPD) switch, located on the pilot's upper right console, adjacent to the logic override switches, has two positions, BACKUP (up) and NORMAL (down). The BACKUP position is used in case the SAS pitch channels are unusable due to gyro channel electronics malfunction or overheating of the pitch gyro package. In this position the backup gyro supplies pitch rate signals through an independent electronic channel to either the A or B servos. The pitch logic override switch must be used to select either A or B servo operation.

NOTE

The primary purpose of the BUPD is to provide an emergency system for pitch stability augmentation during refueling and landing approach. The system is optimized for use at light weight, aft center of gravity, and subsonic speeds from 0.3 to 0.8 Mach number; it is not intended as an emergency backup system during cruise.

SAS Channel Out Light

Illumination of a SAS CHANNEL OUT light, located on the pilot's annunciator panel, indicates that one SAS channel is inoperative. Power for the light is furnished by the essential dc bus.

AUTOPILOT

The autopilot portion of the AFCS relieves the pilot from continuous manual aircraft control requirements. The autopilot functions are as follows:

1. Pre-engage synchronization.
2. Attitude hold in roll and pitch.
3. Pitch and turn wheel inputs.
4. Automatic pitch trim.
5. Heading hold.
6. Mach hold.
7. Altitude hold.

The autopilot is optimized for use at the basic mission cruise speed and altitude. The altitude hold mode is optimized for use at subsonic speeds up to 40,000 feet, although reasonably good performance may be had at altitudes up to approximately 60,000 feet.

CAUTION

- Do not use the pitch autopilot below 25,000 feet at airspeeds above 300 KEAS.
- Do not use autopilot when BUPD is on.
- Do not use altitude hold mode above 60,000 feet.

The autopilot authority is limited to prevent severe maneuvers due to an

autopilot malfunction. The maximum pitch authority is 2.4 degrees up-elevon to 2.4 degrees down-elevon. The maximum roll authority is 4 degrees differential elevon. The autopilot signals are summed with SAS signals and produce control surface motion through the SAS electronics and servos.

NOTE

> Use of the autopilot when using SAS logic override is not recommended because there is no backup feature.

Autopilot control movement of the elevons is not reflected in control stick motion. Automatic pitch trim is operative when the autopilot pitch channel is engaged. The slow-speed pitch trim motor operates to correct for long period pitch trim changes and there should be no pitch transient at disengagement. Pre-engage synchronization of autopilot pitch and roll trim operates during periods when the pitch or roll channels are disengaged.

AUTOPILOT CONTROLS AND INDICATORS

The autopilot controls and indicators are on the SAS panel located on the pilot's right console. The control stick is equipped with Control Stick Command and emergency disengage switches. The circuit breakers are on the pilot's right aft console. Power is from the essential dc bus.

Heading and Vertical Reference Switch

Two systems, a MAIN and an AUX, are provided to supply heading, pitch, and roll signals. System selection is acacomplished by a three-position heading and vertical reference AUX-MAIN switch located on the pilot's right console outboard of the SAS control panel. When the switch is placed in the MAIN (down) position, the inertial navigation system supplies true heading to the HSI and pitch and roll signals to the ADI. When the switch is placed in the AUX (up) position, the MA-1 compass system supplies magnetic heading to the HSI and the MD-1 gyroscope supplies pitch and roll signals to the ADI. The OFF (center) position disengages the autopilot.

Autopilot Pitch Engage Switch

A two-position pitch engage switch is located on the inboard side of the autopilot control panel. In the ON (fwd) position, the pitch autopilot in engaged in the attitude hold mode.

NOTE

> At least one active SAS pitch channel must be engaged and bank angle must be less than 50° before the pitch autopilot can be engaged.

The switch is held in the ON position by a solenoid. The pitch channel may be disengaged by placing the switch to the OFF position, by using the disengage switch on the control stick, or by turning the Heading and Vertical Reference switch OFF.

Autopilot Pitch Trim Synchronization Indicator

The pitch trim synchronization indicator is located just aft of the pitch engage switch. The indicator shows the amount of pitch signal existing prior to the engagement. An up or down displacement of the needle indicates the direction of the transient which will occur when the pitch channel is engaged.

NOTE

> The pitch trim synchronization needle will normally be centered within one needle width. Engagement of the autopilot pitch channel with more than one needle width of misalignment is not recommended.

Autopilot Pitch Control Wheel

A serrated pitch control wheel is located just forward of the Pitch Engage Switch. The wheel is used to make pitch attitude corrections when engaged in the attitude hold mode. Forward rotation of the wheel commands nose down and aft rotation commands nose up. Pitch attitude changes 1° for 20° of wheel rotation.

Autopilot Roll Engage Switch

A two-position roll engage switch is located on the outboard side of the autopilot panel. In the ON (fwd) position, the roll autopilot is engaged in the attitude hold mode.

NOTE

> At least one SAS roll channel and one active SAS yaw channel must be engaged before the roll autopilot can be engaged. Bank angle must be less than 50°.

The switch is held in the ON position by a solenoid. Autopilot signals are supplied by either the MD-1 roll gyro or the INS, depending on the position of the Heading and Vertical Reference selector switch. The roll channel may be disengaged by placing the switch to the OFF position, by using the disengage switch on the control stick or by turning the Heading and Vertical Reference selector switch OFF.

Autopilot Roll Trim Synchronization Indicator

The roll trim synchronization indicator is located just aft of the roll engage switch. The indicator shows that a roll signal exists prior to engagement. The needle always deflects to the right and does not indicate the direction of the transient which will occur at engagement.

NOTE

> Roll engagement is not recommended if the needle is deflected to the side of the dial, indicating a hardover signal.

Autopilot Turn Control Wheel

A serrated turn control wheel is located on the autopilot panel between the pitch and roll trim indicators. It allows the pilot to make roll attitude corrections when engaged in the attitude hold mode. Right rotation of the wheel commands right roll and left rotation commands left roll. Roll attitude changes 1° for 10° of wheel rotation. The pilot can command up to 50° of bank angle in the attitude hold mode. Above 50° of bank the roll autopilot automatically disengages to prevent the steady pitch rate from bottoming the pitch servos, which would eliminate pitch damping capability.

Mach Hold Switch

A toggle-type Mach Hold switch is located on the inboard side of the autopilot panel. The Mach Hold mode is engaged when the switch is moved to the ON position provided the pitch autopilot is engaged. The switch is held in by solenoid action. The autopilot then controls the pitch attitude to maintain the same

Mach number that existed at the time of engagement. When the Mach Hold is engaged, the pitch attitude hold is discontinued and the pitch control wheel function is inoperative. Mach Hold reference signals are supplied to the autopilot from the Air Data Computer (ADC). If the Altitude Hold mode was previously engaged, it will be disengaged when Mach Hold is selected.

Altitude Hold Switch

A toggle-type altitude hold switch is located between the Mach Hold and Heading Hold switches. The Altitude Hold mode is engaged when the switch is moved to the ON position, provided the pitch autopilot is engaged. The switch is held in by solenoid action. Altitude signals are furnished by the ADC and the autopilot controls the aircraft to hold the selected altitude. If the Mach Hold mode was previously engaged, it will be disengaged when Altitude Hold is selected.

Heading Hold Switch

A toggle-type Heading Hold switch is located on the outboard side of the autopilot panel. The Heading Hold mode is engaged when the switch is moved to the ON position, provided the roll autopilot is engaged. The switch is held in by solenoid action. Heading signals from either the MA-1 compass or INS control the roll axis of the aircraft to maintain the heading existing at the time of engagement. Heading hold may be engaged while in a bank. The autopilot will roll the aircraft to a wings-level attitude and lock on the heading at time of engagement. The bank angle is limited to 35 degrees in the heading hold mode.

Control Stick Command Switch

A control stick command (CSC) pushbutton switch is located on the aft face of the control stick grip. While the pushbutton is depressed, both the roll and pitch autopilots revert to the pre-engage synchronization mode which allows attitude and heading to be changed independently of the autopilot. When the switch is released, both the roll and pitch axes engage in the attitude-hold mode, regardless of the mode that was engaged prior to depressing the CSC switch, and the autopilot holds the new attitude.

Autopilot Emergency Disengage Switch

A trigger-type switch, labeled AUTOPILOT DISC and located on the forward side of the control stick grip, will disengage the autopilot completely when depressed. The autopilot does not re-engage when the switch is released.

NORMAL OPERATION

Engagement

The autopilot is placed in normal operation as follows:

1. Check SAS engaged, recycle lights out.
2. Check pitch and roll trim pre-engag synchronization indicators aligned.
3. Pitch and roll engage switches - ON. These switches may be engaged together or separately as operation of the two is completely independent.

NOTE

To engage autopilot the bank angle must be less than 50 degrees.

4. Mach hold switch - As desired.
5. Heading hold switch - As desired.
6. Altitude hold switch - As desired.

WARNING

Do not operate manual roll or pitch trim when the autopilot is engaged.

To change attitude or heading do the following:

1. CSC switch - Depress.

After attitude and/or heading change is completed:

2. CSC switch - Release.
3. Mach hold switch - As desired.
4. Altitude hold switch - As desired.

Disengagement

To disengage autopilot do one of the following:

1. Autopilot disengage switch - Press.
2. Pitch and roll engage switches - OFF.
3. Heading and vertical reference switch - OFF.

MACH TRIM SYSTEM

In the transonic region in this aircraft the variation of elevon angle with Mach number is such that it would normally require the pilot to use nose-up trim with increasing Mach number. This characteristic is referred to as "speed instability." To compensate for this, the Mach trim system is incorporated in the aircraft control system to slowly drive the trailing edge of the elevons upward as Mach number increases, thus providing artificial stability by requiring the pilot to apply nose-down trim as Mach number increases. The system operates between Mach 0.2 and 1.5 on a schedule which varies with Mach number in the 8-1/2 degrees nose-up and 5 degrees nose-down trim limits range of the elevons. The trim change rate is 15 degrees per Mach between 0.95 and 1.30 Mach, and 5 degrees per Mach between 0.2 and 0.95 Mach and between 1.3 and 1.5 Mach. Signal input to the Mach trim system is obtained from the air data computer and the electronic components associated with the system are located in the autopilot electronic component assembly. The system is operative whether or not any SAS channel is engaged; however, the Mach trim system does not function when the pitch autopilot is engaged. The only control over the system is the circuit breaker in the forward cockpit which should be actuated to prevent undesirable Mach trim effects in the event of an air data computer malfunction. Power for Mach trim system is furnished by the essential ac bus and the essential dc bus.

PITOT-STATIC SYSTEM

The pitot-static system supplies the total and static pressures necessary to operate the basic flight instruments and air data computer system components. The pressures are sensed by an electrically heated probe mounted on the nose of the aircraft. The pitot orifice of the probe is divided inside the head to provide two separate pressure sources. It also has two circumferential sets of four static pressure ports each. The right pitot and aft set of static ports supply pressure signals to the air data computer system. The other set of pickups supply pitot and static pressures directly to the speed sensors on the ejection seats, and the standby altimeter and airspeed indicators. An offset head on the left side of the probe provides yaw and pitch pressure signals when operative. Power is furnished by the essential ac bus.

Pitot Heat Switch

The heating elements of the nose and flight recorder probes are controlled by the off-on PITOT HEAT switch, located on the pilot's left forward panel. Power to the heaters is supplied by the essential ac bus.

AIR DATA COMPUTER

The air data computer performs two functions, computation and display. The total and static pressures from the pitot-static probe are converted into electrical signals required for the pilot's airspeed-Mach indicator, altitude-vertical velocity indicator, and the automatic flight control and fire control systems. Information is also supplied to the triple display indicator located on the FCO's instrument panel. The ports which supply pressure to the air data computer are separate from those that furnish pressure to the standby flight instruments; therefore, failure of the air data computer pressure source will not leave the pilot without altitude or airspeed information. The air data computer converts pitot-static pressures into proportional rotary shaft positions which are equivalent to pressure altitude and dynamic pressure. These shaft positions are combined in a mechanical analog computer made up of cams, gears, and differentials to drive the output functions. Outputs of the air data computer and the using equipment are as follows:

OUTPUTS	USING EQUIPMENT
Pressure altitude Equivalent airspeed Mach Vertical velocity	Pilot's integrated instrument display
Pressure altitude Equivalent airspeed Mach	Triple display indicator (FCO cockpit)
Mach Dynamic pressure Mach rate Altitude rate Altitude	Autopilot
Mach	Mach trim system
Pressure altitude Mach	Fire control system

Power for the air data computer is furnished by the essential ac bus.

LOW KEAS Warning Light

A red LOW KEAS warning light is installed on the upper left side of the pilot's instrument panel. Illumination of the LOW KEAS light warns the pilot that an abnormally low airspeed condition has been reached and that appropriate corrective action must be taken in accordance with other indications.

The light flashes and a steady tone will be sounded in the headsets when airspeed is below approximately 250 KEAS. The low airspeed warning is automatically deactivated when below 0.5 Mach number to prevent nuisance warnings during takeoff and landing. The pilot should increase airspeed immediately when the LOW KEAS warning is activated while at a low airspeed condition. If the airspeed is allowed to decrease after receiving such a warning, the airplane could enter a flight region where recovery from a gust upset or high nose up attitude might not be possible.

A small hysteresis band can be expected when passing through 250 KEAS or 0.5 Mach number, as follows:

a. Decreasing airspeed - Light on at 245 to 255 KEAS while above 0.5 Mach number.

b. Increasing airspeed - Light off at 250 to 260 KEAS.

c. Decreasing Mach number - Light off with 0.50 to 0.55 Mach number.

Intensity of the light will be full bright if the instrument lights dimming rheostat is OFF, and it will be dim in all other rheostat positions. Operation of the light can be tested by placing the fire and warning lights test switch in the WARN LTS position. Power for the light is supplied from the essential dc bus through the WARN light circuit breaker in the forward cockpit.

Triple Display Indicator

A triple display indicator (TDI) is installed on the left side of the FCO instrument panel. The indicator presents digital indications of altitude in 50-foot increments, Mach number in 0.01-Mach increments, and equivalent airspeed (KEAS) in 1-knot increments. The altitude indication range is 0 to 99,950 feet. At 100,000 feet the first digit is dropped, indicating 09,950 feet at 109,950 feet pressure altitude, the maximum limit of the ADC signal to the instruments. The Mach number display capability range of the instrument is 0 to 3.99; however, the minimum indication at static conditions normally ranges from 0.11 to 0.2 Mach number and the maximum indication would be Mach 3.5 for normally functioning instruments. This range corresponds to the range of signals which the ADC is capable of providing. The instrument displays airspeed within an instrument capability of 0 to 599 KEAS; however, the minimum indication is normally 75 to 110 KEAS to correspond with the minimum ADC signal provided. The maximum signal provided by the ADC results in an airspeed indication which decreases from 599 KEAS at sea level to 523 KEAS at 66,800 feet and Mach 3.5, and then decreases further at higher altitudes to show the KEAS corresponding to Mach 3.5 and the existing pressure altitude. Power for the instrument is furnished by the essential ac bus.

INTEGRATED INSTRUMENT SYSTEM

The integrated instrument system consists of four cockpit instruments and their associated gyroscopes, computers, radio receivers, accelerometers, and sensors. The system is designed to display flight and navigation information in one cockpit location rather than on several individual instruments spread about the cockpit and whose information must be interpreted and correlated before use. These instruments, located in the center of the pilot's instrument panel, are the airspeed-mach indicator (AMI), attitude director indicator (ADI), altitude-vertical velocity indicator (AVVI), and the horizontal situation indicator (HSI).

The Tacan receiver supplies range and bearing signals to the HSI and bearing, distance, heading indicator (BDHI). The localizer and glideslope receivers supply instrument landing signals to the ADI. The FCS digital computer provides range and bearing to a target or navigation point (to the HSI) and command heading to the HSI, altitude and target altitude to the AVVI, and Mach to the AMI. The FCS digital computer uses target information provided by the radar, data link, or the FCO's manual inputs plus interceptor position in providing these inputs to the integrated instruments. Figure 1-14 shows the operational modes available for use with the integrated instruments.

Airspeed-Mach Indicator

The airspeed-mach indicator (AMI) gives a vertical presentation of speed and load factor information. The instrument contains four columns, from left to right, angle-of-attack indicator, accelerometer, Mach indicator, and equivalent airspeed indicator. In addition, the AMI has a maximum allowable Mach marker, a command Mach marker and readout window with a slewing switch, and a command airspeed marker and readout window with a slewing switch. Aircraft Mach number or EAS is shown by tape movement at the lubber line of the respective vertical display. A desired Mach number or EAS may be set in the MACH or KNOTS windows by up or down movement of the slew switches below the windows. The two-bar marker on each display will move toward the lubber line as the manually set Mach or EAS is approached. In the event of ac power failure to the drive motor of the AMI, an OFF flag will appear in the center of the Mach display column.

NOTE

> Although the OFF flag appears only in the Mach display column, it indicates that all functions of the AMI are inoperative.

An IAS flag in the EAS display column appears when a discrepancy exists between EAS indication and transmitter position. Power failure to the air data computer will also cause the OFF flag to appear.

Angle of Attack and Yaw Angle

The angle of attack display at the left side of the Airspeed Mach Indicator has been activated in the test configuration, and an indication of yaw angle is displayed separately on an instrument located above the AMI. The fire control system computer provides the signals for these displays.

INTEGRATED INSTRUMENT DISPLAY

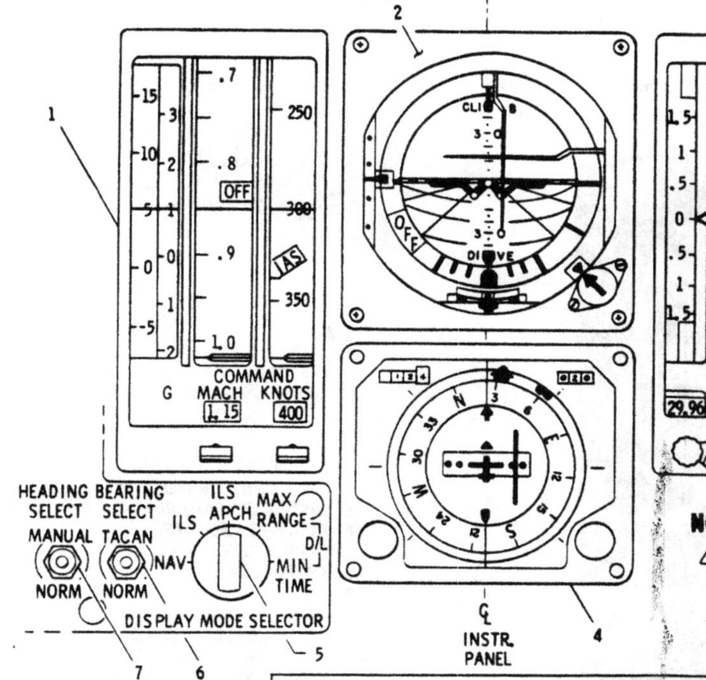

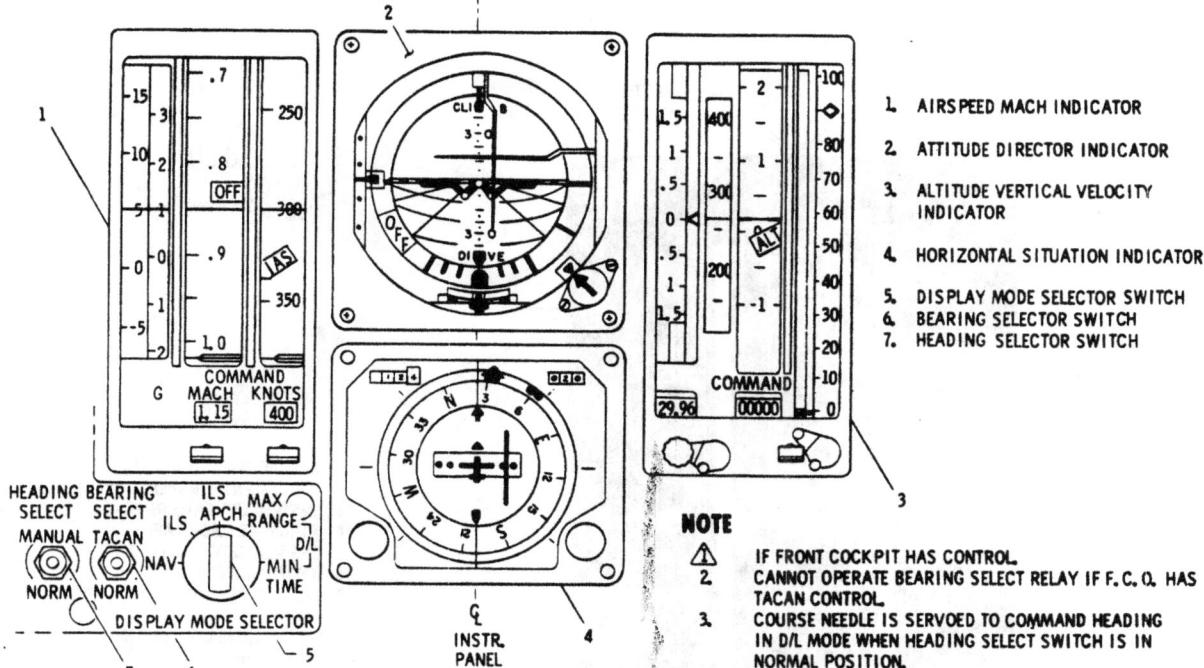

1. AIRSPEED MACH INDICATOR
2. ATTITUDE DIRECTOR INDICATOR
3. ALTITUDE VERTICAL VELOCITY INDICATOR
4. HORIZONTAL SITUATION INDICATOR
5. DISPLAY MODE SELECTOR SWITCH
6. BEARING SELECTOR SWITCH
7. HEADING SELECTOR SWITCH

NOTE

⚠1 IF FRONT COCKPIT HAS CONTROL.
2. CANNOT OPERATE BEARING SELECT RELAY IF F.C.O. HAS TACAN CONTROL.
3. COURSE NEEDLE IS SERVOED TO COMMAND HEADING IN D/L MODE WHEN HEADING SELECT SWITCH IS IN NORMAL POSITION.

INDICATOR	INDICATOR FUNCTION	BEARING SELECT SWITCH	DISPLAY MODE SELECTOR SWITCH: NAV. HEADING SEL. SW NORM	NAV. HEADING SEL. SW MAUNAL	ILS HEADING SEL. SW NORM	ILS HEADING SEL. SW MANUAL	ILS APPROACH HEADING SEL. SW NORM	ILS APPROACH HEADING SEL. SW MANUAL	D/L MAX RANGE HEADING SEL. SW NORM	D/L MAX RANGE HEADING SEL. SW MANUAL	D/L MIN TIME HEADING SEL. SW NORM	D/L MIN TIME HEADING SEL. SW MANUAL
HORIZ SITUATION INDICATOR	COMMAND HEADING	NORM TACAN	MANUALLY SET		SERVO TO LUBBER	MANUALLY SET	SERVO TO LUBBER	MANUALLY SET	COMMAND HEADING	MANUALLY SET	COMMAND HEADING	MANUALLY SET
	BEARING POINTER	NORM TACAN	TO SELECTED TACAN		SERVO TO A/P HDG. TO SELECTED TACAN		SERVO TO A/P HDG TO SELECTED TACAN		BEARING TO TARGET TO SELECTED TACAN		BEARING TO TARGET TO SELECTED TACAN	
	COURSE ARROW	NORM TACAN	MANUALLY SET FOR SELECTED TACAN		MANUALLY SET TO INBOUND LOC. COURSE		MANUALLY SET TO INBOUND LOC. COURSE		COMMAND HEADING	MANUALLY SET	COMMAND HEADING	MANUALLY SET
	COURSE DEVIATION	NORM TACAN	DEVIATION TO SELECTED TACAN		DEVIATION FROM LOCALIZER		DEVIATION FROM LOCALIZER		CENTERED		CENTERED	
	TO-FROM INDICATOR	NORM TACAN	TO - FROM SELECTED TACAN		OUT OF VIEW		OUT OF VIEW		OUT OF VIEW		OUT OF VIEW	
	DISTANCE INDICATOR	NORM TACAN	DISTANCE TO SELECTED TACAN		MASKED DIST. TO TACAN		MASKED DIST. TO TACAN		DISTANCE TO TARGET DIST. TO TACAN ⚠1		DISTANCE TO TARGET DIST. TO TACAN ⚠1	
	DISTANCE SHUTTER	NORM TACAN	TACAN VALID		TACAN INVALID TACAN VALID		TACAN INVALID TACAN VALID		COMP RANGE VALID TACAN VALID ⚠1		COMP RANGE VALID TACAN VALID ⚠1	
	"K" SHUTTER	NORM TACAN	MASKED		MASKED		MASKED		COMP RANGE VALID MASKED		COMP RANGE VALID MASKED	
	AIRPLANE HDG. REF	NORM TACAN	MAGNETIC		MAGNETIC		MAGNETIC		TRUE MAGNETIC ⚠1		TRUE MAGNETIC ⚠1	
ATTITUDE DIRECTOR INDICATOR	BANK DIR. NEEDLE	NO EFFECT	OUT OF VIEW	SELECTED HEADING	LOCALIZER COURSE	SELECTED HEADING	LOCALIZER COURSE	SELECTED HEADING	COMMAND HEADING	SELECTED HEADING	COMMAND HEADING	SELECTED HEADING
	PITCH NEEDLE		OUT OF VIEW		OUT OF VIEW		PITCH DIR G/S BEAM	OUT OF VIEW	OUT OF VIEW		OUT OF VIEW	
	G/S INDICATOR		OUT OF VIEW		POSITION RELATIVE TO GLIDESLOPE		POSITION RELATIVE TO GLIDESLOPE		OUT OF VIEW		OUT OF VIEW	
	LOCALIZER FLAG		LOCALIZER INVALID		→		→		→		→	
	WARNING FLAG		POWER OFF		→		→		→		→	
	G/S FLAG		GLIDESLOPE INVALID		→		→		→		→	
AIRSPEED MACH INDICATOR	COMMAND MACH	NO EFFECT	MANUALLY SET		NOT USED		NOT USED		COMMAND MACH		COMMAND MACH	
	COMMAND AIRSPEED		MANUALLY SET		→		→		→		→	
ALTITUDE VERTICAL VELOCITY INDICATOR	COMMAND ALTITUDE	NO EFFECT	MANUALLY SET		MANUALLY SET		MANUALLY SET		COMMAND ALTITUDE		COMMAND ALTITUDE	
	TARGET ALTITUDE		NOT USED		NOT USED		NOT USED		TARGET ALTITUDE		TARGET ALTITUDE	

Figure 1-14

AIRSPEED -MACH INDICATOR

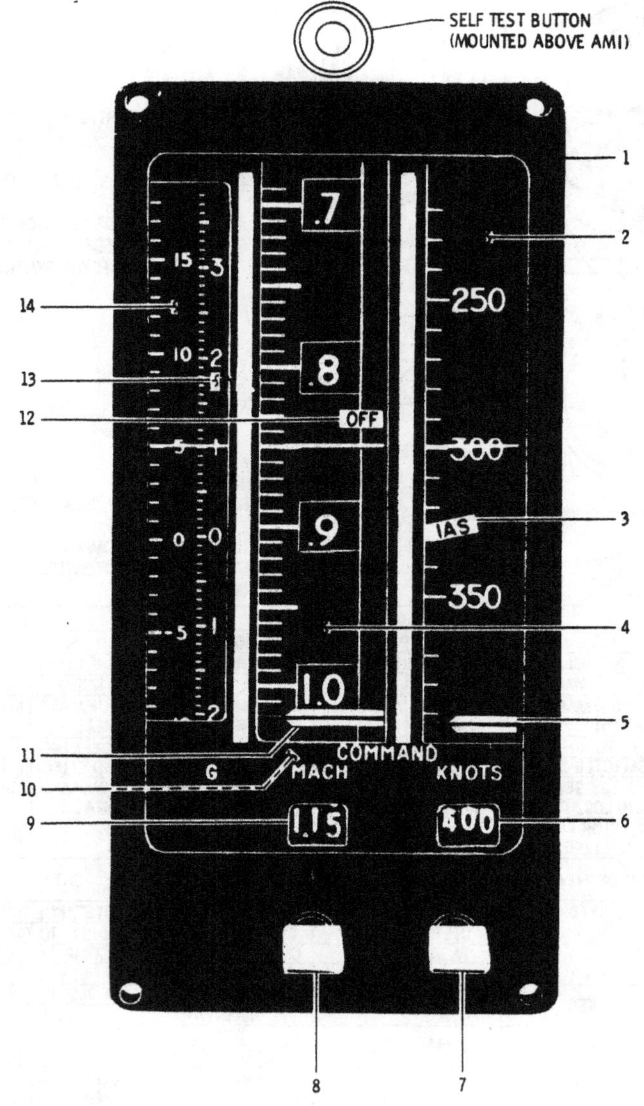

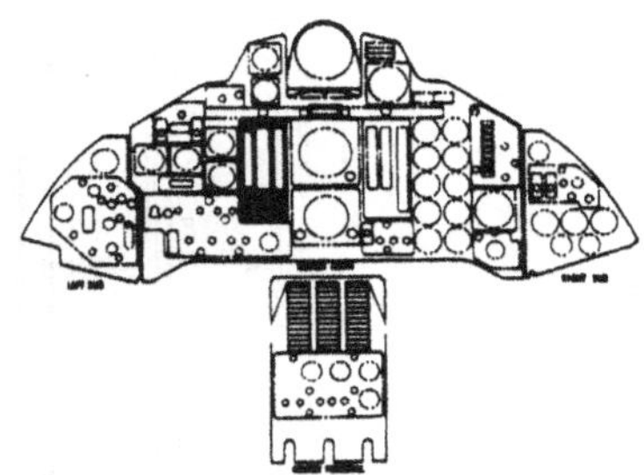

1 AIRSPEED MACH INDICATOR
2 EQUIVALENT AIRSPEED TAPE
3 INSTRUMENT FAILURE FLAG
4 AIRCRAFT MACH TAPE
5 EAS REFERENCE MARKER
6 KNOTS COUNTERS
7 EAS SLEW SWITCH
8 MACH SLEW SWITCH
9 COMMAND MACH COUNTERS
10 MAXIMUM SAFE MACH (OUT OF VIEW LOCKED AT 3.5 MACH)
11 COMMAND MACH MARKER
12 POWER OFF FLAG
13 AIRCRAFT "G" FORCE TAPE
14 ANGLE OF ATTACK

INDICATOR	INDICATOR FUNCTION	BEARING SELECT SWITCH	DISPLAY MODE SELECTOR SWITCH									
			NAV		ILS		ILS APPROACH		D/L MAX RANGE		D/L MIN TIME	
			HEADING SEL. SW.		HEADING SEL. SW.		HEADING SEL. SW.		HEADING SEL. SW.		HEADING SEL. SW.	
			NORM	MANUAL	NORM	MANUAL	NORM	MANUAL	NORM	MANUAL	NORM	MANUAL
AIRSPEED MACH INDICATOR	COMMAND MACH	NO EFFECT	MANUALLY SET		NOT USED		NOT USED		COMMAND MACH		COMMAND MACH	
	COMMAND AIRSPEED		MANUALLY SET		→							

Figure 1-15

Accelerometer

The accelerometer consists of a vertical moving scale and a fixed index line adjacent to the angle of attack indicator. The index line monitors the moving scale to indicate g load from -2 to +10 g's.

Mach Indicator

Mach number is indicated in the center of the AMI. A fixed index line monitors a vertical moving scale which is calibrated in hundreths and labeled for each tenth of a Mach from + .4 to + 3.5. A double line Command Mach marker and a command Mach readout window below the scale indicates selected or command Mach.

NOTE

> The command Mach marker will remain at the top or bottom of the indicator until the selected command Mach reading comes into view on the moving scale. At this time the marker will move with the scale toward the fixed index line. When indicated Mach equals command Mach, the command Mach marker and the index line will coincide.

Command Mach may be controlled manually by the slewing switch under the Mach readout window. The command system is capable of selecting speeds from Mach .4 to 3.5.

Equivalent Airspeed Indicator

The equivalent airspeed indicator displays equivalent airspeed in knots. A fixed index line monitors a vertical moving scale which is graduated in 10 knot markers with figure numbers at each 50 knot level from 50 through 1000 knots. At speeds below 50 knots the scale will continue to indicate 50. A double line command airspeed marker and a knots readout window below the indicator show manually selected command airspeed.

NOTE

> The command airspeed marker will remain at the top or bottom of the indicator until the manually selected airspeed reading comes into view on the moving scale. At this time the marker will move with the scale toward the fixed index line. When the indicated airspeed equals selected command airspeed, the command marker and the index line will coincide.

Command airspeed is controlled manually by the command airspeed slewing switch under the knots readout window.

NOTE

> The command airspeed slewing switch has a center detent position. Moving the switch to the right, when centered, locks the switch and causes digital equivalent airspeed values to be displayed at the command airspeed readout window.

Airspeed-Mach Indicator Test Switch

An airspeed-Mach indicator test button is installed above the airspeed-Mach indicator. When the button is depressed, the Mach tape should move to 2.44 Mach and the EAS tape move to 800 knots. Power for the switch is furnished by the essential dc bus.

ANGLE OF ATTACK INDICATOR

In the test configuration, two angle of attack indicators are mounted near the center of the instrument panel in the rear cockpit. Each dial displays positive and negative angles of attack on the wing in one-degree increments through a range of -5° to +22°. The instruments respond to synchro signals from the angle of attack transducer which is part of the pitot-static system. The transducer functions in response to differential pressures sensed by the pitch probe static pressure orifices and total pressure sensed by the pitot orifice on the nose of the airspeed probe. Power for the indicator and transducer is provided from the emergency ac bus through the ANGLE OF ATTACK circuit breaker located in the forward cockpit. An OFF flag is displayed on the face of the indicator if power to the instrument is interrupted.

Angle of Attack Warning

The aircraft is provided with an amber warning light in the forward cockpit and an aural signal which sounds a steady tone in the pilot's and FCO's headsets if high angles of attack are attained. The frequency of the signal is the same as the intermittent tone used to warn of an unsafe landing gear position. The tone sounds and the light illuminates when α exceeds 8° when at high supersonic speeds, or when α exceeds 14° when at low supersonic speeds or if subsonic. The transition point is 1.4 Mach number.

The α signal to the warning light and aural warning is disconnected by a relay associated with a scissor switch when aircraft weight is on the nose gear.

Angle of Attack Indicator Operating Characteristics

The angle of attack indicator normally provides wing angle of attack while in flight. The angle of the fuselage reference line is 1.2° greater because of the negative angle of incidence of the wing.

The angle of attack transducer requires pitot pressure to balance pitch probe differential pressure. If pitot pressure is not sufficient, as during stationary or low speed ground operation, or in the event that pitot pressure is lost, the transducer tends to generate a high angle of attack signal which results in full scale positive deflection of the angle of attack indicator. The transducer also drifts toward a high angle of attack condition if electrical power is lost. The angle of attack instrument indication should not reflect the drift, as its synchro signals are provided from the same bus; however the APW α potentiometer follows the drift to high α, the high angle of attack warning system switch should close, and an aural warning should result while in flight. The warning can be stopped by pulling the LDG GR WARN essential dc bus circuit breaker in the forward cockpit.

NOTE

> If the LDG GR WARN circuit breaker is pulled, power to the entire Landing Gear Warning System is interrupted.

ATTITUDE-DIRECTOR INDICATOR (ADI)

The attitude-director indicator combines an attitude indicator, turn and slip indicator, and a flight director and

ATTITUDE DIRECTOR INDICATOR

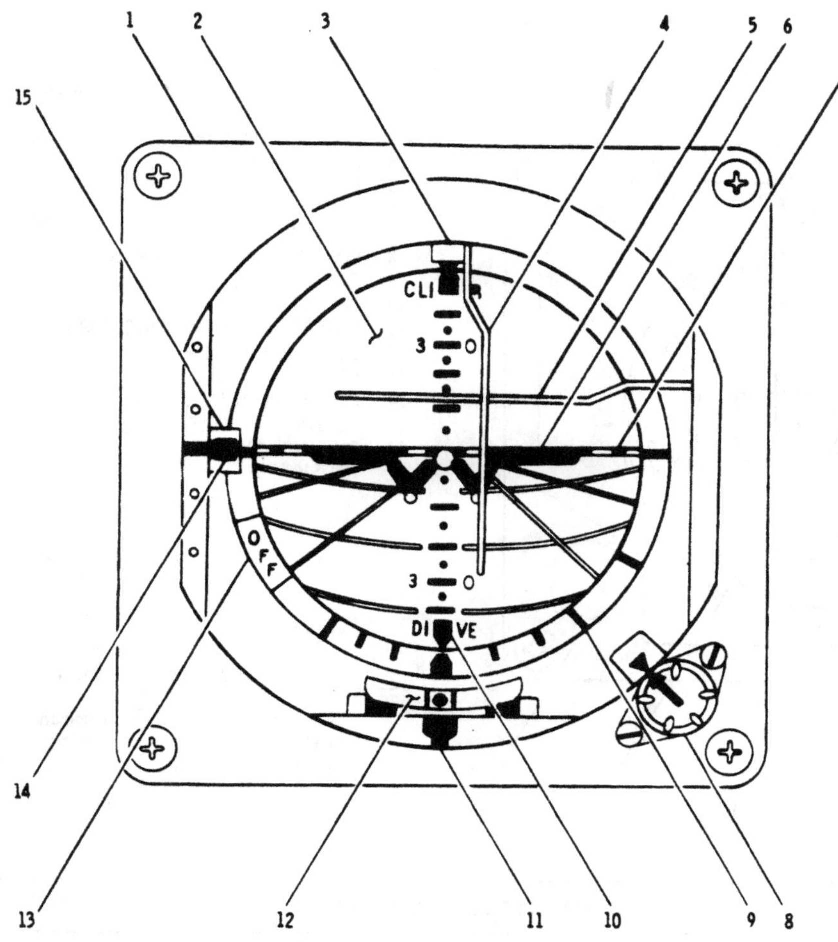

1. ATTITUDE DIRECTOR INDICATOR
2. ATTITUDE SPHERE
3. COURSE WARNING (LOCALIZER) FLAG
4. BANK DIRECTOR NEEDLE
5. PITCH NEEDLE
6. MINIATURE AIRCRAFT
7. HORIZON BAR
8. PITCH TRIM KNOB
9. BANK SCALE
10. BANK POINTER
11. TURN NEEDLE
12. SLIP INDICATOR
13. WARNING FLAG
14. GLIDESLOPE INDICATOR
15. GLIDESLOPE WARNING FLAG

NOTE

CANNOT OPERATE BEARING SELECT RELAY IF F.C.O. HAS TACAN CONTROL.

INDICATOR	INDICATOR FUNCTION	BEARING SELECT SWITCH	DISPLAY MODE SELECTOR SWITCH									
			NAV		ILS		ILS APPROACH		D/L MAX RANGE		D/L MIN TIME	
			HEADING SEL SW		HEADING SEL SW		HEADING SEL SW		HEADING SEL SW		HEADING SEL SW	
			NORM	MANUAL	NORM	MANUAL	NORM	MANUAL	NORM	MANUAL	NORM	MANUAL
ATTITUDE DIRECTOR INDICATOR	BANK DIR NEEDLE	NO EFFECT	OUT OF VIEW	SELECTED HEADING	LOCALIZER COURSE	SELECTED HEADING	LOCALIZER COURSE	SELECTED HEADING	COMMAND HEADING	SELECTED HEADING	COMMAND HEADING	SELECTED HEADING
	PITCH NEEDLE		OUT OF VIEW		OUT OF VIEW		PITCH DIR G/S BEAM	OUT OF VIEW	OUT OF VIEW		OUT OF VIEW	
	G/S INDICATOR		OUT OF VIEW		POSITION RELATIVE TO GLIDESLOPE		POSITION RELATIVE TO GLIDESLOPE		OUT OF VIEW		OUT OF VIEW	
	LOCALIZER FLAG		LOCALIZER INVALID →									
	WARNING FLAG		POWER OFF →									
	G/S FLAG		GLIDESLOPE INVALID →									

NAF12-2-8-2

Figure 1-16

HORIZONTAL SITUATION INDICATOR

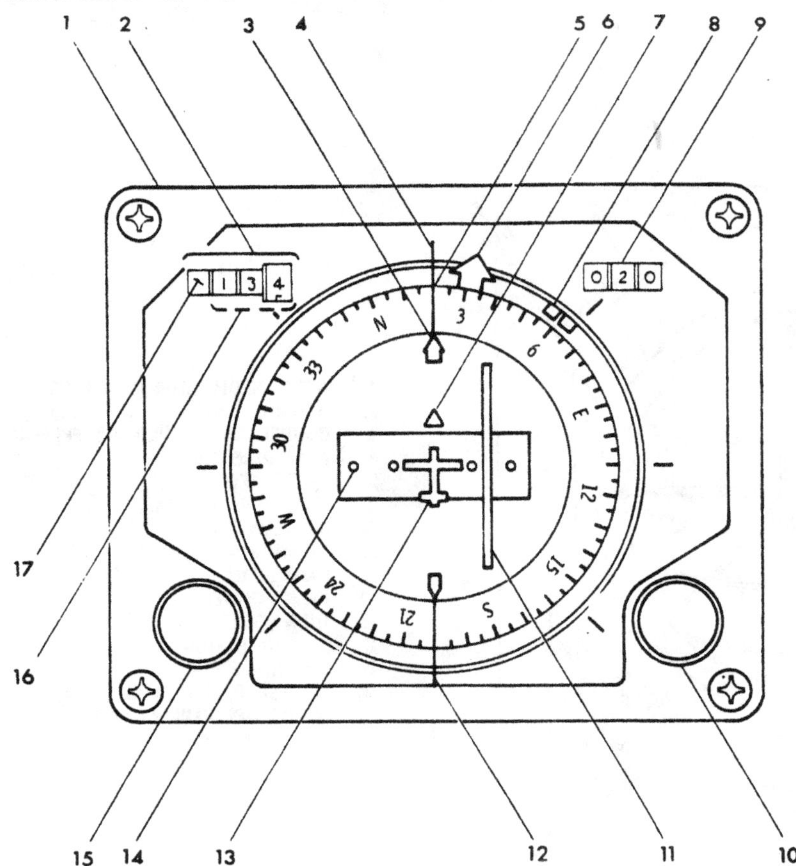

1 HORIZONTAL SITUATION INDICATOR
2 DISTANCE INDICATOR
3 COURSE ARROW (HEAD)
4 UPPER LUBBER LINE
5 AIRCRAFT HEADING REF.
6 BEARING POINTER
7 TO-FROM INDICATOR
8 COMMAND HEADING MARKER
9 COURSE SELECTOR WINDOW
10 COURSE SET KNOB
11 COURSE DEVIATION INDICATOR
12 LOWER LUBBER LINE
13 AIRCRAFT SYMBOL
14 COURSE DEVIATION DOTS
15 HEADING SET KNOB
16 DISTANCE SHUTTER
17 "K" SHUTTER

NOTE

△1 Only if forward cockpit has control

Course needle is servoed to command heading in D/L mode when heading select switch is normal.

INDICATOR	INDICATOR FUNCTION	BEARING SELECT SWITCH	DISPLAY MODE SELECTOR SWITCH									
			NAV		ILS		ILS APPROACH		D/L MAX RANGE		D/L MIN TIME	
			HEADING SEL SW		HEADING SEL SW		HEADING SEL SW		HEADING SEL SW		HEADING SEL SW	
			NORM	MANUAL	NORM	MANUAL	NORM	MANUAL	NORM	MANUAL	NORM	MANUAL
HORIZ SITUATION INDICATOR	COMMAND HEADING	NORM	Manually Set		Servo To Lubber	Manually Set	Servo To Lubber	Manually Set	Command Heading	Manually Set	Command Heading	Manually Set
		TACAN										
	BEARING POINTER	NORM	To Selected TACAN		Servo To Autopilot Heading				Bearing To Target			
		TACAN			To Selected TACAN △1							
	COURSE ARROW	NORM	Manually Set For Selected TACAN		Manually Set To Inbound Localizer Course				Command Heading	Manually Set	Command Heading	Manually Set
		TACAN										
	COURSE DEVIATION	NORM	Deviation To Selected TACAN		Deviation From Localizer				Centered			
		TACAN										
	TO-FROM INDICATOR	NORM	To-From Selected TACAN		Out Of View							
		TACAN										
	DISTANCE INDICATOR	NORM	Distance To Selected TACAN		Masked				Distance To Target			
		TACAN			Distance To TACAN △1							
	DISTANCE SHUTTER	NORM	TACAN Valid		TACAN Invalid				Comp Range Valid			
		TACAN			TACAN valid △1							
	"K" SHUTTER	NORM	Masked						Comp Range Valid			
		TACAN							Masked			
	AIRPLANE HDG REF	NORM	Magnetic						True			
		TACAN							Magnetic △1			

F202-X 35

Figure 1-17

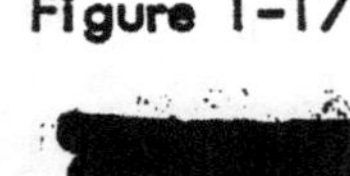

ILS glide slope presentation. Unrestricted motion of the attitude sphere allows presentation of pitch and roll through 360°. The sphere moves behind a small aircraft fixed at the center of the instrument. A pitch trim knob allows manual positioning of the sphere with relation to the small aircraft. Pitch angle is marked on the sphere by small dots for 5° increments, small lines for each 10°, number markers for each 30°, and large dots which indicate the poles. Bank angle is shown at the top circumference of the instrument. 10° graduations are provided for angles to 30° and 30° graduations for angles up to 90° of bank. The turn and slip indicator is mounted at the bottom of the ADI and is centered with the vertical axis. A deflection of one needle width indicates a four minute 360° standard turn. The rate of turn transmitter receives power from the essential dc bus. Bank and pitch steering needles are superimposed on the indicator to provide command information. The bank steering needle shows amount and direction of bank required to position the aircraft on a desired heading or course. The bank needle will center (1) when on course, (2) when applied bank is correct for return to course, and (3) when bank is correct for rollout on course. The maximum indicated command bank angles are dependent upon the setting of the Display Mode Selector Switch located on the pilot's instrument panel just below the AMI. These angles, set within the Flight Director Computer are as follows:

NAV	35°
ILS	35°
ILS APCH	15°
DL MAX RANGE	35°
DL MIN TIME	60°

The pitch steering needle indicates amount and direction of pitch angle required to position the aircraft on the ILS glide slope. The needle centers (1) when on glide slope with proper pitch angle, (2) when pitch angle is correct for return to glide slope, and (3) when pitch angle is correct for leveling out on glide slope. The steering needles indicate the corrective action required, not direction or displacement from desired course or glide slope. The two needles are automatically stowed when not in use, and warning flags appear to show that the glide slope or localizer signals are invalid. The glide slope pointer at the left center of the instrument shows actual position with reference to the glide slope. A power failure warning flag appears in the lower left hand portion of the ADI when ac power to the instrument is interrupted. This includes power to the MD-1 gyroscope, the platform electronics unit, or the digital computer depending upon the position of the HDG and VERT REF selector switch located on the pilot's right console. Power for the ADI is furnished by the essential ac bus.

HORIZONTAL SITUATION INDICATOR (HSI)

The horizontal situation indicator integrates and visually presents information from Tacan, Data Link messages, the Inertial Navigation System, the MA-1 Compass, the Digital Computer and the ILS Receiver. Power for the HSI is furnished by the essential ac bus. Individual components of the HSI are described below.

Rotary Compass Card

The compass card is a rotating azimuth ring which is read at a stationery lubber line located at the 12 o'clock position. The card reflects aircraft true or magnetic heading, depending on the position of the Display Mode Selector Switch. During normal operation the Display Mode Selector Switch is placed in one of the D/L (MAX RANGE or MIN TIME) positions. With these settings, the compass card displays true heading derived from the Inertial Navigation System. An exception occurs when the Bearing Select Switch is in the TACAN position, in which case the card reads magnetic heading. Magnetic heading from the auxiliary (MA-1 compass) system is presented by the compass card when the Display Mode Selector Switch is placed in the NAV, ILS or ILS APCH positions.

Command Heading Marker

The heading marker is a rectangular marker located just outside of the rotating compass card. The command heading marker may be manually set to any heading when the Heading Select Switch is in the MANUAL position. The MANUAL position would be used only in conjunction with data link equipment. When the Heading Select Switch is in the NORM position, the command heading marker is slaved to the HSI lubber line in either ILS mode. When the Display Mode Selector Switch is in the NAV position, the command heading marker may be manually set regardless of the position of the Heading Select Switch.

Bearing Pointer

The bearing pointer is a heavy arrowhead located on the outside of the rotating compass card. Depending on the positions of the Bearing Select and Display Mode Selector Switches, the bearing pointer will display the relative bearing of a Tacan station or target to the heading of the aircraft. When the Display Mode Selector Switch is in the ILS or ILS APCH position, and the Bearing Select Switch is in the NORM position, the bearing pointer is slewed to the lubber line and will not show bearing to the target. The pointer will show bearing to a Tacan station with the Display Mode Selector Switch in any position and the Bearing Select Switch in the TACAN position. It is not necessary to move the Bearing Select Switch to the TACAN position when the Display Mode Selector Switch is in the NAV position.

Course Arrow and Course Deviation Bar

The course arrow and course deviation bar are located inside (toward center of instrument) the rotating compass card. When the Display Mode Selector Switch is in the NAV or ILS modes, the course arrow may be manually set with the course set knob (lower right hand corner of the HSI) to the desired course. The course deviation bar then indicates angular deviation from the selected course. When the Display Mode Selector Switch is in either DL mode the course is slaved to the command heading. The course deviation bar shows angular deviation to a Tacan station in the NAV mode and deviation to a localizer station in the ILS mode.

NOTE

In the ILS or NAV mode, maximum deflection of the needle represents 5° or greater angular deviation from desired course.

To-From Arrow

The to-from arrow is located above or below the miniature aircraft at the center of the HSI. The arrow shows Tacan ambiguity when the Display Mode Selector Switch is set to the NAV position. The to-from arrow is inoperative and hidden from view in all other mode positions.

Distance Shutter

The distance shutter is located in the upper left hand corner of the HSI and covers the numerals in the distance digital readout window. When the Display Mode Selector Switch is in the NAV or ILS modes, shutter movement to the open position indicates valid Tacan reception. When the switch is in the DL mode, shutter movement to the open position indicates valid DL range reception.

Distance Readout Window

The distance readout window behind the distance shutter is a three digit counter which displays slant range in nautical miles to a Tacan station or target.

K Shutter

A K shutter covers a 4th digit of the distance readout window. When the shutter moves to the open position, valid DL range information at a distance greater than 999 miles is being received. The window displays a fixed numeral 1. The shutter remains closed unless the Display Mode Selector Switch is positioned to one of the DL modes.

Display Mode Selector Switch

The Display Mode Selector Switch located directly below the Altitude-Vertical Velocity Indicator has 5 positions; NAV, ILS, ILS APCH, MAX RANGE and MIN TIME. In the NAV, ILS or ILS APCH positions, the MA-1 Compass System furnishes magnetic heading to the HSI. In the MAX RANGE or MIN TIME positions, the HSI displays true heading from the Inertial Navigation System. The NAV position selects the Tacan and bearing inputs, the ILS and ILS APCH positions select the ILS inputs to the HSI and ADI. The ILS APCH position also activates the Pitch Director Indicator on the ADI.

Heading Select Switch

A Heading Select Switch installed below the Altitude-Vertical Velocity Indicator has two positions: NORM (down) and MANUAL (up). In the NORM position, the command heading as selected by the Display Mode Selector Switch is available in the DL modes to the HSI, and command bank angles are displayed by the bank steering needle of the ADI. In the MANUAL position, the command heading may be manually set on the HSI. The bank steering needle (ADI) then indicates bank steering angle in reference to the selected heading.

Bearing Select Switch

A Bearing Select Switch installed below the Altitude-Vertical Velocity Indicator has two positions: NORM (down) and TACAN (up). In the NORM position, bearing and distance information displayed by the HSI are deter-

mined by the Display Mode Selector Switch position. In the TACAN position, the bearing and distance inputs are derived directly from the Tacan receiver.

Heading and Vertical Reference Selector Switch

Two systems, a MAIN and an AUXILIARY are provided to supply heading, pitch and roll signals. System selection is accomplished by a three-position Heading and Vertical Reference Aux-Main switch located on the pilot's right console outboard of the SAS control panel. When the switch is placed in the MAIN (down) position, the Inertial Navigation System makes available true heading to the HSI and pitch and roll signals to the ADI, depending on the positions of the Display Mode and Bearing Selector Switches. When the switch is placed in the AUX (up) position, the MA-1 compass system supplies magnetic heading to the HSI and BDHI and the MD-1 gyroscope supplies pitch and roll signals to the ADI. The OFF (center) position disengages the autopilot.

ALTITUDE-VERTICAL VELOCITY INDICATOR (AVVI)

The altitude-vertical velocity indicator displays a vertical presentation of altitude information and contains 5 columns. From left to right the columns indicate the following: vertical velocity, vernier altitude, sensitive altitude, command altitude and gross altitude. In addition, the AVVI has a barometric pressure readout window and set knob, a target altitude marker, two command altitude markers with a single readout window and a command altitude slewing switch. The command altitude slewing switch has a side detent which allows sensitive altitude values to be displayed digitally. Barometric pressure is set in the lower left hand window. An OFF flag appears when the instrument power supply is off or to indicate instrument failure.

Vertical Velocity Indicator

The vertical velocity indicator indicates vertical velocity on a fixed scale calibrated from 0 to ± 1500 feet per minute. All vertical velocities between ± 1500 fpm and ± 40,000 fpm are displayed on a moving scale and are viewed through readout windows located at the top and bottom of the indicator. The moving scale shows thousand-fpm increments up to 10,000 fpm, five thousand fpm increments from 10,000 to 30,000 fpm, and a terminal indication of 40,000 fpm. A moving pointer monitors vertical velocity and moves to the appropriate window as ± 1500 fpm is exceeded. The vertical velocity tape is actuated by an output from the Air Data Computer.

Vernier and Sensitive (Pressure) Altitude Indicators

Aircraft true pressure altitude is shown by movement of two tapes which are actuated by the Air Data Computer. The right altitude tape is read at the lubber line and is calibrated from -1000 to +100,000 feet. The command altitude window may be used to set any desired altitude by moving the slew switch beneath the window. The left altitude tape is a vernier altitude scale also read at the lubber line and is calibrated from 0 to 1000 feet in 50-foot increments. This tape is geared to and moves with the sensitive altitude tape.

ALTITUDE - VERTICAL VELOCITY INDICATOR

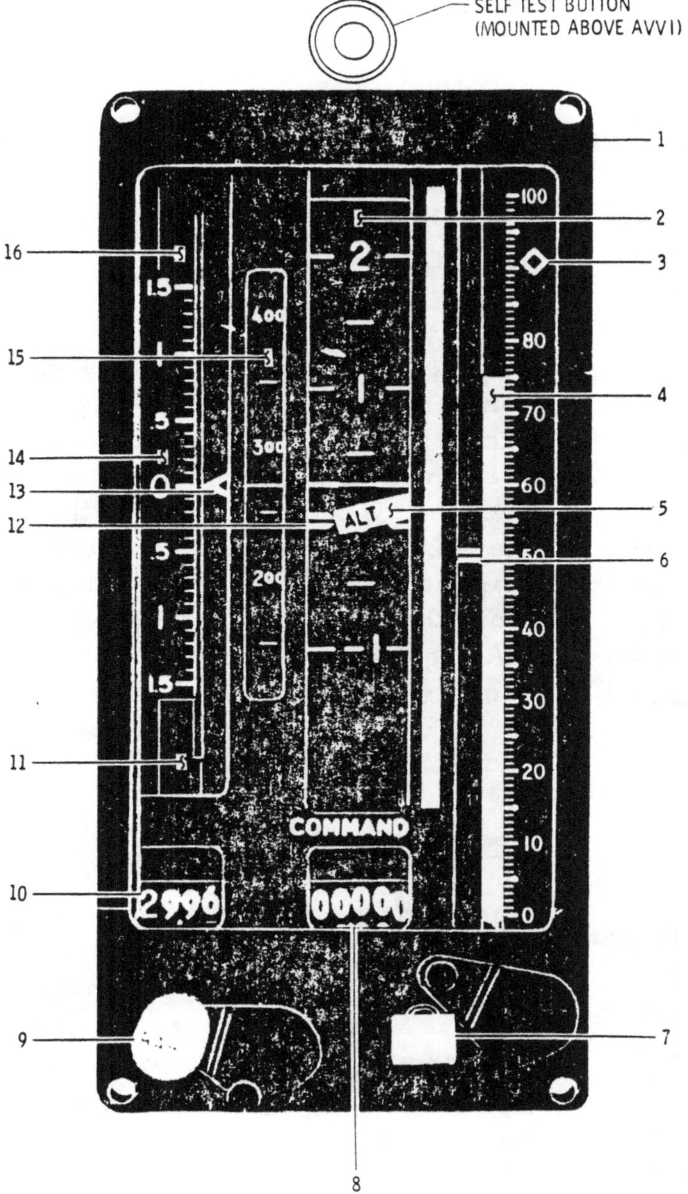

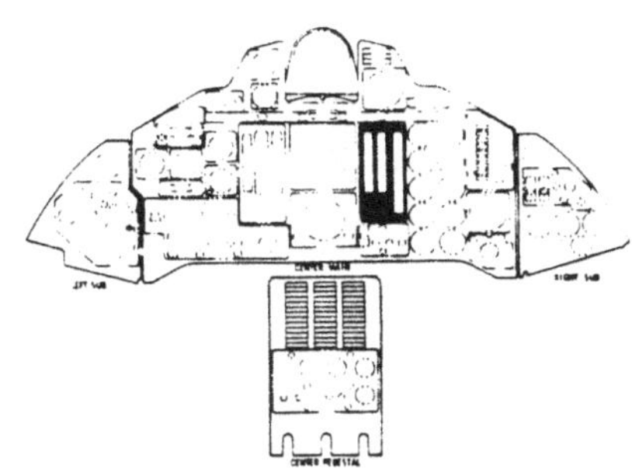

1. ALTITUDE VERTICAL VELOCITY INDICATOR
2. FINE ALTITUDE TAPE
3. TARGET ALTITUDE INDICATOR
4. COARSE ALTITUDE TAPE
5. INSTRUMENT FAILURE FLAG
6. COARSE ALTITUDE COMMAND MARKER
7. ALTITUDE SLEW SWITCH
8. COMMAND ALTITUDE COUNTERS
9. BAROMETRIC PRESSURE CONTROL KNOB
10. BAROMETRIC PRESSURE INDICATOR
11. VERTICAL SPEED CLIMB TAPE (COVERED)
12. FINE ALTITUDE COMMAND MARKER
13. VERTICAL VELOCITY MARKER TAPE
14. VERTICAL VELOCITY SCALE
15. VERNIER ALTITUDE TAPE
16. VERTICAL SPEED DIVE TAPE (COVERED)

INDICATOR	INDICATOR	BEARING SELECT SWITCH	DISPLAY MODE SELECTOR SWITCH									
			NAV		ILS		ILS APPROACH		D/L MAX RANGE		D/L MIN TIME	
			HEADING SEL. SW.		HEADING SEL. SW.		HEADING SEL. SW.		HEADING SEL. SW.		HEADING SEL. SW.	
			NORM	MANUAL	NORM	MANUAL	NORM	MANUAL	NORM	MANUAL	NORM	MANUAL
ALTUTUDE VERTICAL VELOCITY INDICATOR	COMMAND ALTITUDE	NO EFFECT	MANUALLY SET		MANUALLY SET		MANUALLY SET		COMMAND ALTITUDE		COMMAND ALTITUDE	
	TARGET ALTITUDE		NOT USED		NOT USED		NOT USED		TARGET ALTITUDE		TARGET ALTITUDE	

Figure 1-18

Command and Gross Altitude Indicators

The command altitude tape has a two bar marker imposed on it and will stop at the commanded altitude. The gross altitude indicator contains a fixed vertical scale calibrated from 0 to 100, each increment representing 1000 feet of altitude. In the data link mode target altitude is indicated by a diamond shaped marker over the gross altitude scale.

Altitude-Vertical Velocity Indicator Test Switch

A pushbutton type altitude-vertical velocity indicator test switch is installed above the altitude-vertical velocity indicator. When the switch is depressed, the gross altitude tape should indicate 74,000 feet. Power for the switch is furnished by the essential dc bus.

BEARING, DISTANCE, HEADING INDICATOR (BDHI)

The bearing, heading, distance indicator installed on the FCO's left instrument panel displays magnetic heading derived from the MA-1 compass system and bearing and distance information derived from the Tacan receiver. Power for the BDHI is furnished by the essential ac bus.

MA-1 COMPASS SYSTEM

The MA-1 compass system utilizes a gyro-stabilized compass which is designed for use at all latitudes. The system consists of a magnetic flux valve, MA-1 gyro and amplifier, control panel and the rotating compass card of the HSI and BDHI. The system may be operated either as a magnetically slaved compass or as a free directional gyro with correction for the effect of earth's rotation. The system provides secondary reference signals for directional control of the autopilot and heading information for the HSI and BDHI. Power for the system is furnished by the essential dc bus.

Magnetically Slaved System

Operation as a magnetically slaved compass may be used in any locality except near the magnetic poles or in areas where severe magnetic distortion occurs. When operated in this mode, the flux valve defines the direction of magnetic north. The electrical signals from the flux valve are supplied to the MA-1 gyro and amplifier. These components in turn furnish secondary heading information to the autopilot, HSI, and BDHI.

Free Directional Gyro

Operation as a free directional gyro may be used in any latitude, but is especially useful where the magnetic field is weak or distorted. Distortion occurs at any latitude when in close proximity to large masses of iron. Above approximately 60° latitude, the declination of the earth's magnetic lines of force creates errors in a magnetically slaved system, and the free mode should be used. In this mode, the heading information is provided by the orientation of the MA-1 gyro and is in turn transmitted to the autopilot and heading indicators.

Manual Fast Slaving

Before Takeoff

The normal slaving rate of the system is about 2° per minute. When the compass system is energized before takeoff, the gyro may be as much as 180° from the proper heading. About 1-1/2 hours would be required to slave to the correct heading at normal slaving rates. Manual fast slaving is provided by actuating the set heading switch, which increases the rate to 720° per minute. This corrects a 180° error in 15 seconds.

In-Flight

Normally, if the compass is properly slaved before takeoff, no in-flight manual fast slaving is required unless free directional gyro operation is selected. When operating as a free gyro, the desired heading can be established by using the set heading switch.

NOTE

> The autopilot must be turned OFF during manual fast slaving when MA-1 is being used as a heading reference.

MA-1 Compass Control Panels

A control panel is installed on the right console in the pilot's cockpit and the left console in the FCO's cockpit. The panels contain a slaved-free switch, latitude control dial, set heading switch and a synchronization indicator.

Slaved-Free Switch

The two-position SLAVED-FREE switch allows selection of either a magnetic heading reference or a free gyro. In the FREE mode, the heading is preset and is retained by the gyro. In the SLAVED mode the sytem is oriented to the earth's magnetic field.

Latitude Control Dial

The latitude control (Lat Cont) dial is used only when operating in the FREE mode to provide electrical compensation for drift caused by the rotation of the earth. The dial is turned to set the existing latitude at the index marker on the panel. When set properly, the gyro spin axis stays fixed in relation to the earth and provides a stable directional reference. Since drift varies with latitude, the Lat Cont dial must be adjusted for each 5° change in latitude. When operating as a magnetically slaved system, the latitude control is disconnected and may be left at any setting.

Set Heading Switch

The set heading switch (Set Hdg) allows synchronization of the stabilizing gyro with the proper magnetic orientation, and is used initially in both the FREE and SLAVED modes of operation. The synchronization normally takes place at a rate of 2° per minute. Actuation of the switch to the L or R positions increases the rate to 720° per minute. The positions L and R refer to the direction of travel of the HSI and BDHI and assist in establishing correct synchronization in the shortest time.

Synchronization Indicator

The synchronization indicator (Sync Ind) operates only when the slaved-free switch is in the SLAVED mode and provides a check on the degree of synchronization between the gyro and the magnetic reference (flux valve).

Operation of the MA-1 Compass System

1. Obtain control of set by pressing the COMP CONT (transfer) button and check button illumination.
2. Slaved-Free switch - As desired
3. Latitude Control Dial - Set to proper latitude when operating on free gyro.
4. Set Heading switch - Fast slave compass card of HSI and BDHI to proper heading.
5. Synchronization Indicator - Center needle when operating as a magnetically slaved system.

STANDBY ATTITUDE INDICATOR

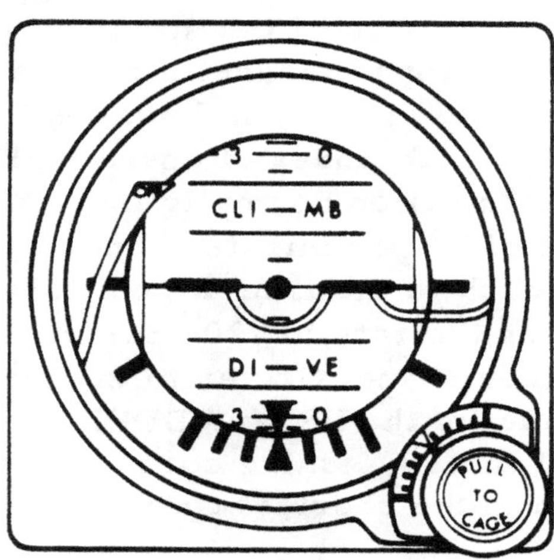

Figure 1-18A

STANDBY INSTRUMENTS

Standby Airspeed Indicator

The standby airspeed indicator, located adjacent to the Airspeed-Mach Indicator on the pilot's instrument panel, is a direct reading, pressure sensitive instrument plumbed to the ship pitot static system. It does not require electrical power for operation. The indicator displays knots indicated airspeed (KIAS). This results in a difference from the indication provided by the equivalent airspeed column of the Airspeed-Mach Indicator.

Standby Altimeter

A standby altimeter is installed below the standby airspeed indicator. It is a direct reading pressure instrument plumbed to the ship pitot-static system and does not require electrical power for operation.

Standby Attitude Indicator

A self-contained standby attitude indicator is located at the top of the center instrument panel in the pilot's cockpit. The indicator contains a cylinder that is detailed to represent sky and earth areas, and the cylinder rotates to indicate pitch angles of 80° in climb, 90° in dive and roll angles of 360°. The cylinder is inscribed with an artificial horizon and 5° graduations for pitch angle. The bank angle scale is marked on the periphery of the instrument case in 10° increments up to 30° and then 30° increments up to 90°.

A cutout is provided which disables the self erection feature whenever pitch or roll angle exceeds 12 ± 1/2°. This puts

the indicator into a free drift mode when this angle is exceeded and prevents the indicator from erecting to a false vertical.

The indicator has a built-in 7° pitch attitude bias. With this bias, the gyro horizon will center with reference to the case index marks when the aircraft pitch attitude is 7° nose-up (normally 5.8° angle of attack). When the aircraft is at a 0° pitch attitude, such as when on the ground, the gyro will erect to a 7° dive indication.

A combination pitch trim and caging knob is provided on the lower right corner of the indicator. The knob can be turned to position the miniature airplane through a range from 5° up to 5° down. The knob is also used for caging the gyro to the case. This is accomplished by pulling the knob out to its fully extended position. Releasing the knob erects the gyro to indicate 7° nose-down and 0° roll within five minutes if the aircraft is level. Pulling the knob out and turning it CW locks the cylinder in the caged position. (Use of this position is not recommended except for shipping the instrument.) No separate fast erect switch is provided.

The indicator receives power from the emergency ac bus through the STBY ATT IND circuit breaker located on the left console in the pilot's cockpit. An OFF flag becomes visible in the face of the indicator if power is interrupted or while the cage knob is extended. The indicator will display useful pitch and roll information for nine minutes after power loss.

Attitude Indicator Operating Characteristics

The standby attitude indicator has a maximum free drift rate of 0.9 degrees per minute. In level flight, any tendency to drift is continuously corrected to the effective vertical axis by the automatic erection system at a rate of from two to three degrees per minute. The indicator contains an acceleration sensor which will disable the erection system at bank angles greater than twelve degrees to prevent accumulation of substantial bank errors during turns. However, in this free gyro operating mode, the random combination of errors due to gyro drift, earth's rate and earth's profile can accumulate at a maximum rate of approximately 1.6 degrees per minute for the duration of the turn.

In wings level flight, the normal aircraft acceleration and deceleration rates are not sufficient to cause the acceleration sensor to disable the erection system. Therefore, under these flight conditions, the indicator will erect to the apparent vertical. During climbout and descent, pitch errors up to approximately four and eight degrees, respectively, can be expected. If a turn is initiated immediately after climbout or descent, the free gyro errors may add to the previous acceleration induced error. For example, the indicator error could reach four degrees in roll after a normal supersonic descent and 90 degree subsonic turn (with more than twelve degree bank).

If the bank or pitch error does not exceed approximately twelve degrees, the indicator will automatically correct the above errors at the normal erection

rate when the aircraft is in level flight and at constant velocity. However, at this time, the indicator can be aligned to the true vertical more rapidly by momentarily operating the manual caging knob. Use of this control should be limited to level flight conditions, as the gyro is erected to the case reference when caged manually.

NOTE

> If a bank or pitch error of more than approximately twelve degrees is accumulated during a turn (or any maneuver) while in the free drift mode, the cut-out feature prevents automatic correction on roll-out from the turn. Manual erection by caging will be necessary when level flight is resumed.

Standby Compass

A standby magnetic compass installed on the pilot's instrument panel is furnished for navigational purposes in the event of electrical failure or the failure of navigational equipment.

Clocks

Elapsed time clocks are installed on the pilot's and FCO's instrument panels. They contain an elapsed time mechanism that is started by pushing in on the winding knob.

COMMUNICATIONS AND ASSOCIATED ELECTRONIC EQUIPMENT

Control Transfer Buttons

Square push-type buttons labeled COMP CONT, UHF CONT, and TACAN CONT installed in each cockpit allow either the pilot or the FCO to assume control of the MA-1 compass, UHF, and TACAN control panels. The buttons are located adjacent to each control panel in the pilot's cockpit and on a separate panel adjacent to the UHF panel in the FCO's cockpit. Verification of control is made by checking transfer button illumination. When aircraft dc power is interrupted, such as by aircraft shutdown, control of each set automatically reverts to the pilot's cockpit.

Microphone Switches

Press-to-talk switches for the pilot are located on the control stick and the right engine throttle handle. The FCO's press-to-talk switches are located in the right and left foot rest.

INTERPHONE SYSTEM, AN/AIC-18

The interphone system provides a means of crew intercommunication plus air to air and air to ground communication. The system provides the crew with the following capabilities:

1. Either press-to-talk or "hot mic" communication between pilot and FCO.

2. UHF and HF radio transmission and reception for pilot and FCO.

3. Emergency communication between pilot and FCO, regardless of control settings.
4. Communication with the ground crew by means of an external receptacle.
5. Landing gear and low airspeed audio warning to the pilot
6. Monitoring of audio signals from the navigation radios.

Interphone Control Panel

The pilot's interphone control panel is located on the right console in the forward cockpit. The FCO's panel is located on his left console. The panels do not contain an ON-OFF switch, as the equipment is on whenever the essential dc bus is energized. Each panel contains a Volume Control, monitoring buttons, rotary selector knob and call button.

Volume Control

The volume control adjusts the listening level for all monitored incoming signals ot the respective headset.

Monitoring Switches

The six monitoring switches are labeled INT, UHF, HF, TACAN, ILS and UHF 2. The switches can be used to monitor the equipment individually or all at the same time regardless of the selector knob setting. To monitor a desired audio source, the switch is pulled out and rotated clockwise to increase the volume. The ILS switch in the FCO's cockpit is inoperative.

Hot Microphone and Listen Switch

Hot microphone operation allows the crewmembers to communicate with each other without depressing a microphone switch. Two-way hot microphone operation is accomplished by each crewmember pulling out his HOT MIC and LISTEN switch and setting his rotary selector switch to any position except INT.

Rotary Selector Knob

The rotary selector knob, labeled INT, UHF 1, HF, UHF 2, 1 and 2, enables the crewmember to transmit and receive on the system selected.

INT position - For cockpit inter-communication using the microphone buttons.

UHF 1 position - For voice transmission and reception using the ARC-51 UHF command radio.

HF position - For voice transmission and reception using the 618 T HF long range radio.

UHF 2 position - For voice transmission and reception using the UHF #2 radio.

No. 1 and No. 2 positions - These positions are inoperative.

Call Button

Pressing the CALL button permits the user to interrupt and override any signals being received by the other crew member. CALL volume is preset and cannot be adjusted in flight.

UHF COMMAND RADIO, AN/ARC-51

The UHF command radio provides two-way communications on any of 1750 different frequencies extending from 225.0 through 399.9 megacycles. Any of these frequencies may be selected manually; however, the radio is preset on the ground to the 20 frequencies most commonly used. In addition to the main receiver, the set utilizes a second guard receiver which can be preset in a frequency range between 238.0 and 248.0 megacycles, but which is normally pretuned to 243.0 megacycles. A Frequency-Channel indicator is installed on the instrument panel in each cockpit. Power for the set is furnished by the essential dc bus.

AN/ARC-51 Control Panels

A control panel is installed on the left console in each cockpit. Each panel contains frequency slew switches, mode and function selector switches, and a volume control.

Frequency Slew Switches

Three rocker-type frequency slew switches are used in conjunction with the mode selector switch to obtain either a preset channel or a manually tuned frequency on the frequency-channel indicator. When the channel selector switch is placed in the CHAN position and the left frequency selector switch is depressed, a preset channel number will appear in the frequency-channel indicator window. The switch must be held down until the indicator slews to the desired number. A frequency number appears when the mode selector switch is moved to the FREQ position. The desired frequency may be manually selected by depressing and holding the left frequency slew switch to slew the first two digits, the middle switch for the third digit, and the right switch for the digit after the decimal point. When the upper portion of a slew switch is depressed, the frequency numbers will increase, and conversely, when the lower portion of the switch is held down, the frequency numbers will decrease.

Mode Selector Switch

The mode selector switch has three positions; CHAN (left), GUARD XMIT (middle), and FREQ (right). The CHAN position is used when selecting a preset channel, the FREQ position permits manual frequency selection, and the GUARD XMIT position allows transmission on the pretuned guard frequency.

Function Selector Switch

The function switch has two bands: The upper band is labeled COMM and has three positions; T/R (transmit - receive), T/R + G (Transmit - receive + guard) and ADF (inoperative). In the T/R position with the mode selector switch in CHAN or FREQ, both the receiver and transmitter are tuned to the preset or manually selected frequency while the guard portion of the receiver remains inoperative. When the switch is in the T/R + G position, the radio will receive signals simultaneously from the main and guard channels of the receiver. The lower band labeled AUX is inoperative.

Volume Control

The set is turned ON and then audio level increased by rotating the volume (VOL) control wheel to the right.

Channel-Frequency Indicators

The channel-frequency indicator windows located on the pilot's and FCO's instrument panels display the numerals of a preset frequency, a manually selected frequency, or the word GUARD depending on the position of the channel selector switch. An OFF flag appears in the windows when the equipment is not operating.

ADF Antenna

The ADF antenna is flush mounted on the aircraft fuselage bottom centerline just below the pilot's cockpit. The main transmit and receive antenna is under the forward left chine and its radiation pattern is omnidirectional below the horizontal plane.

Operation of UHF Command Radio

1. Interphone rotary selector knob - UHF
2. Obtain control of set by pressing UHF CONT (transfer) button. Check button illuminated.
3. Function switch - As desired
4. Channel selector switch - As desired
5. To select a frequency other than one of the preset channels:
   a. Mode selector switch - FREQ
   b. Frequency slew switches - Press to obtain desired frequency. A numerical indication of the selected frequency will be shown by frequency-channel indicator on the instrument panel.
6. To transmit and receive on guard channel:
   a. Mode selector switch - GUARD - XMIT
   b. Function switch - T/R + G

UHF COMMAND RADIO AN/ARC-51A (FCO's Cockpit Only)

An additional UHF command radio is installed in the FCO's cockpit. It is similar to the ARC-51 provided in the forward cockpit except that control can not be transferred to the pilot and it does not have a direction finding capability. The radio provides two-way communications on any of 3500 different frequencies extending from 225.0 through 399.9 megacycles. Any of these frequencies may be selected manually; however, the radio is preset on the ground to the 20 frequencies most commonly used. In addition to the main receiver, the set utilizes a second guard receiver which is preset to 243.0 megacycles. Power for the set is furnished by the essential dc bus.

AN/ARC-51A Control Panel

A control panel is installed on the left console in the FCO's cockpit. The panel contains 3 manual frequency selector switches, mode and function selector switches, a squelch disable switch and a volume control.

Manual Frequency Selector Switches

Three manual frequency selector switches are used in conjunction with the mode selector switch to obtain either a preset channel or a manually tuned frequency on the frequency-chan-

nel indicator. When the mode selector switch is placed in the PRESET CHAN position and the selector switch is rotated, a preset channel number will appear in the frequency-channel indicator window. A frequency number appears when the mode selector switch is moved to the MAN position. The desired frequency may be manually selected by rotating the left switch to slew the first two digits, the middle switch for the third digit, and the right switch for the digit after the decimal point.

Mode Selector Switch

The mode selector switch has three positions: PRESET CHAN, MAN, and GD XMIT. The PRESET CHAN position is used when selecting a preset channel, the MAN position permits manual frequency selection, and the GD XMIT position allows transmission on the pretuned guard frequency.

Function Selector Switch

The function selector switch has four positions: OFF, T/R, T/R + G, and ADF. In the T/R position with the mode selector switch in the PRESET CHAN or MAN position, both the receiver and transmitter are tuned to the preset or manually selected frequency while the guard portion of the receiver remains inoperative. When the switch is in the T/R + G position, the radio will receive signals simultaneously from the main and guard channels of the receiver. The lower band labeled ADF is inoperative.

Squelch Disable Switch

When placed in the ON position, the receiver squelch circuit is disabled.

Volume Control

The set is turned ON and then audio level increased by rotating the volume (VOL) control wheel to the right.

UHF Antenna

The antenna is flush mounted on the aircraft fuselage in the chine just outside of the left forward missile bay. The antenna radiation pattern is omnidirectional below the horizontal plane.

Operation of UHF Command Radio

1. Function switch - As desired.
2. Channel selector switch - As desired.
3. Volume control - ON and set.

To select a frequency other than one of the preset channels:

4. Mode selector switch - MAN.
5. Manual frequency switches - Rotate to obtain desired frequency. A numerical indication of the selected frequency will be shown by frequency-channel indicator.

To transmit and receive on guard channel:

6. Mode selector switch - GD XMIT.
7. Function selector switch - T/R + G.

618-T HF RADIO EQUIPMENT

NOTE

The HF radio is not installed when in the test configuration.

The HF radio is a long range airborne Single Side Band (SSB) voice communications transceiver which transmits and receives in the 2 to 30 megacycle range. The transceiver can be tuned in one kilocycle steps. The primary operating mode is SSB, using either the upper or lower side of the modulated signal. The equipment can also transmit and receive AM signals. The equipment consists of the transceiver with an automatic antenna tuner which is mounted in the left chine bay. The Pitot Boom is the antenna. AC electrical power for the equipment is furnished from the essential ac bus. Control circuit power is supplied from the dc essential bus.

HF Control Panel

The control panel for the HF equipment is located on the left console and contains the following:

Service Selector Switch

This switch turns the equipment on or off and selects the desired operating mode. In the USB (upper side band) position, only the upper side band signal is transmitted or received. This is the sum of the voice signal and the radio frequency (rf) signal. In the LSB (lower side band) position, only the lower side band signal is transmitted or received. This signal is the difference of the voice signal and the rf signal. In the AM position the signal is amplitude modulated and both side bands and the original rf signal are transmitted and received.

Frequency Selector Switches

The first switch selects the proper megacycle point as indicated by the digits in the first two windows. It will indicate from 02 to 29. The frequency will increase as the knob is rotated clockwise and decrease as the knob is rotated counterclockwise. The 100 kc knob selects the proper one hundred kilocycle point and indicates from 0 to 9 in the third window. The 10 kc knob selects the proper ten kilocycle point and indicates from 0 to 9 in the fourth window. The one kc knob selects the proper one kilocycle point and indicates from 0 to 9 in the right window.

RF Sensitivity Knob

This knob is used to adjust the audio level in the headphones.

Normal Operation

1. Interphone rotary selector knob - HF
2. Service selector switch - Set to desired mode. This will turn the equipment on. For normal communication this may be USB, LSB or AM.
3. Frequency selector knobs - Set to desired operating frequency. The muting of sound in the headphones will indicate the equipment is setting to the new frequency.

NOTE

The Service Selector Switch may have been moved from the OFF position to an operating mode with the desired operating frequency already set up. In this case, rotate the ten kilocycle select knob one digit off frequency and then back to the operating frequency. This will allow the equipment to return to the desired frequency.

When background sound is again heard in the headphone:

4. XMTR - Button - Press. Wait for the equipment to tune - a 1000 cps tone will be heard until tuning is complete.

When the equipment is tuned (no 1000 cps tone):

5. RF SENS Knob - Adjust so that background noise in headphones is barely audible.

Emergency Operation

If an overload exists in the output of the power supply, a protective circuit turns off the equipment. Restore to operation as follows:

1. Service selector switch - OFF, then back to desired operating mode.

NOTE

When the antenna coupler is required to complete several consecutive tuning cycles, a thermal relay will de-energize the equipment. Restore to operation as follows:

1. Service selector switch - OFF. After two minutes the thermal relay will cool.

2. Service selector switch - Set to desired operating mode.

INSTRUMENT LANDING SYSTEM, AN/ARN-58

Glideslope, localizer and marker beacon receivers are provided for instrument landings and supply vertical and horizontal guidance during an ILS approach. The localizer receiver supplies signals to the bank director needle of the Attitude Director Indicator and to the course deviation bar of the Horizontal Situation Indicator. The glideslope receiver supplies signals to the glideslope deviation pointer and pitch needle of the Attitude Director Indicator. The bank director and glideslope needles are automatically stowed when not in use and warning flags appear to indicate that the glideslope or localizer signals are invalid. The localizer and glideslope receivers operate on 20 fixed, separate frequency channels. Localizer frequencies range from 108.1 to 111.9 megacycles and glideslope frequencies from 329.3 to 335 megacycles. The proper glideslope frequency is automatically tuned when the localizer frequency is selected. The marker beacon receiver operates on a fixed frequency of 75 megacycles and provides an audio signal to the pilot through the interphone during station passage when ILS is selected. No marker beacon light is installed at the present time. Power for the receivers is furnished by the essential dc bus. See Integrated Instrument System, Section I for further information.

ILS Control Panel

A control panel located on the right console in the pilot's cockpit contains a power switch, volume control and a frequency selector knob. The localizer and glideslope frequencies are selected by rotating the knob until the desired localizer frequency appears in the window.

Operation of ILS System

1. Interphone ILS monitor switch - Pull and set volume
2. Display mode selector switch - ILS or ILS APCH
3. Power switch - ON

Allow approximately 90 seconds for warmup.

4. Frequency selector knob - Set to desired frequency
5. Volume - Adjust as desired

RT-728A/APX-64(V) IFF SYSTEM

The APX-64(V) IFF system enables the aircraft to identify itself when challenged by proper signals from appropriate radar interrogation equipment at land bases, aboard ships, or in other aircraft. The system also includes altitude reporting, tracking and identification, and emergency reporting features. A Mode 4 function, which would provide a secure (encrypted) IFF capability, is not operational.

The IFF transponder control panel is located on the left console in the aft cockpit. The controls and indicators for Modes 1, 2, 3A, and C are located on the right side of the panel. A separate Altitude Reporting switch is provided which causes the information pulses of the altitude reporting function to be cut out when desired.

IFF CONTROL PANEL

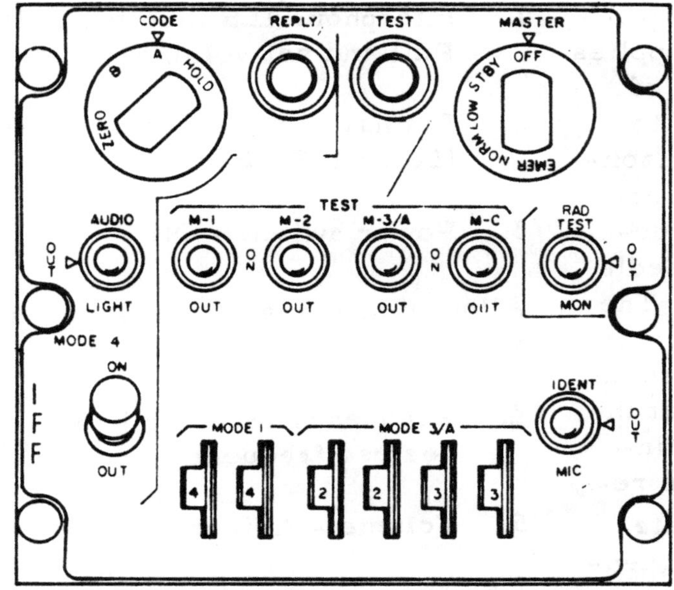

*1. Mode 4 ON-OUT Switch
*2. Mode 4 Indication Switch
*3. Mode 4 Code Select Switch
*4. Mode 4 Reply Light
*5. Test Indicator Light
6. Mode Control Switches
7. Master Selector Switch
8. RAD-OUT-MON Switch
9. Identification of Position Switch
10. Code Selector Switches
* Inactive

Figure 1-19

NOTE

The following descriptions apply to specific controls and indications for IFF system modes 1, 2, 3A and C. The Mode 4 controls illustrated by Figure 1-19 are not operational.

Master Selector Switch

A rotary type five-position master selector switch is used to control the IFF transponder in all modes. The switch positions and functions are:

OFF - The IFF system is de-energized. The switch must be pulled out before it can be placed in the OFF position.

STBY - The system is energized and maintained in a standby state. Transmission is inhibited.

LOW - Only strong (local) interrogations can be recognized and answered. This position should not be used except at the request of the ground controller.

NORM - The system is operating with a full range of recognition and reply capabilities on the modes selected.

EMER - The switch must be pulled out before it can be moved to the EMER position. The EMER position is used for transmitting special coded signals on Modes 1, 2 and 3A. Mode C transmits a normal reply when interrogated.

Mode 1, 2, 3A, and C Control Switches

Four three-position toggle switches are provided for selecting and controlling operation of the desired modes 1, 2, 3A, and C. Switch positions are labeled TEST (up), ON (center), and OUT (down). The ON position places the corresponding mode in operation. The modes are inoperative when the OUT position is selected. The TEST position is inactive. The switches are spring loaded to return from TEST to the ON position. A separate Altitude Reporting switch is provided which can be used to cut out the information pulses from a mode C response when mode C ON has been selected.

RAD-OUT MON Switch

The three-position RAD-OUT-MON switch is provided for testing and monitoring the IFF system. The spring-loaded momentary RAD position is used in conjunction with ground equipment to test the transponder. The OUT and MON positions, which would otherwise be used in conjunction with the mode switches for self test, are inactive as the self test feature is inoperative.

Mode 1 and 3A Code Selector Switches

Six thumb-wheel type switches and adjacent windows are provided for selecting and indicating the codes for Mode 1 and Mode 3A. The two switches on the left side are used for selecting Mode 1 codes and the other four switches are used for Mode 3A selection. The left Mode 1 switch is numbered 0 through 7, the other is numbered 0 through 3 on each half of the drum. The Mode 3A switches are numbered 0 through 7. Spring-loaded centering detents prevent ambiguity of selection.

Identification-of-Position Switch

The identification-of-position (I/P) switch controls transmission of I/P pulse groups. The switch has three positions, MIC, OUT, and a spring-loaded IDENT position. The I/P timer is energized for thirty seconds when the switch is momentarily held in the IDENT position, and I/P replies will be made if a Mode 1, 2 or 3A interrogation is recognized within this thirty second period. Transmission of the I/P pulse group is withheld when the I/P switch is in the OUT position. The MIC position is inactive.

Inactive Controls and Indicators

The controls and indicators which would be associated with a Mode 4 system if it were installed are inactive. The Test indicator light is also inactive.

Mode C Altitude Reporting Cut-Out Switch

A guarded two-position ALT REPORT switch is provided on the left console circuit breaker panel in the aft cockpit. When in its normal (guarded) ON position, the Mode C altitude reporting capability of the IFF system is functional if Mode C ON has been selected and the IFF Master switch is in the NORM position. The altitude information pulses are suppressed when the guard is raised and the OFF position is selected.

Normal Operation, Modes 1, 2, 3A, and C

1. Master Selector switch - STBY.

   Three minutes are required for warmup. Response to interrogation is inhibited while in the STBY position.

2. Master Selector switch - NORM.

   The equipment is usually operated with NORM selected. The LOW position should only be used at the request of the ground controller.

3. Mode 1 and 3A Code Selector switches - As required.

NOTE

The Mode 2 Code is preset on the ground.

4. Mode switches - Set as required.

5. Altitude Reporting switch - ON.

To transmit identification of position:

1. Identification switch - IDENT.

To suppress the Mode C altitude reporting capability:

1. Altitude Reporting - OFF (guard raised).

To turn IFF off:

1. Master selector switch - OFF.

Emergency Operation, Modes 1, 2, and 3A

1. Master selector switch - EMER.

NOTE

With the master selector switch in the EMER position, Mode C transmits a normal reply when interrogated.

RT-864/ARN-52V TACAN SYSTEM

The TACAN System provides continuous indications of bearing and slant distance to selected surface beacons or to other aircraft equipped with the proper ARN-52 equipment. Bearing and distance information are displayed by the Horizontal Situation Indicator in the pilot's cockpit and by the BDHI on the FCO's instrument panel.

The system is capable of operating on any one of 126 channels and has a range of about 300 nautical miles. The transmitting frequency range is from 1026 to 1150 MHz, providing 126 channels with 1 MHz separation between each. Receiver frequencies are divided as follows:

For ground-to-air (T/R) mode: 962 to 1024 MHz and 1151 to 1213 MHz.

For air-to-air (A/A) mode: 1025 to 1087 MHz and 1088 to 1150 MHz.

A pair of channels with 63 MHz spacing is required for air-to-air operation. In addition, since the air-to-air (A/A) mode can operate on or near the same frequencies as the IFF system, IFF

interference can cause unreliable TACAN operation on channels 1 through 8 and these channels should not be used.

Power for control and operation of the TACAN equipment is furnished from the essential ac and dc buses.

NOTE

- During multiple refuelings, a separation of at least three channels between channels in use should be maintained to prevent A/A TACAN operating from one aircraft from interfering with the TACAN of other aircraft operating in the same area and on the same channel.
- If break-lock is being experienced during tanker rendezvous, request the tanker to temporarily suppress IFF operation.
- When TACAN is being used for refueling rendezvous, TACAN lock may be broken if either or both aircraft are in a turn.
- Do not use channels 1 through 8 for operation in the A/A mode, as interference from IFF signals may cause unreliable TACAN operation.

AN/ARC-52 Control Panels

A Tacan control panel is installed on the right console in the forward cockpit and on the left console in the aft cockpit. Each panel contains a channel selector switch, mode selector switch, and a volume control.

Channel Selector Switch

A channel selector is used to select any one of 126 available channels. Channel selection is accomplished by setting the desired channel number in the digital readout window, using the concentric knobs. The outer knob selects the first two digits and the inner knob selects the third digit.

Volume Control Knob

Audio level of the TACAN station identification signals is increased by rotating the volume control clockwise.

Mode Selector Switch

The mode selector switch has four positions, with functions as follows:

OFF - Set is de-energized.

REC - Set is energized and presents bearing and course information on the HSI and BDHI.

T/R - Same as the REC position and also presents range in nautical miles to a Tacan station.

A/A - Same as the REC position and also presents range in nautical miles to another aircraft.

Operation of the TACAN System

1. Interphone TACAN monitor switch-Pull and set volume control.
2. Obtain control of TACAN set by pressing TACAN CONT (transfer) button. Check button illuminated.
3. Display mode selector switch - NAV.

4. Heading select switch - NORM.

5. Bearing select switch - NORM.

6. TACAN mode selector switch - REC.

WARNING

Improperly adjusted or malfunctioning ground or airborne TACAN equipment may lock onto a false bearing. Such an error will probably be ± 40 degrees of correct bearing, but may be a multiple of 40 degrees. To guard against such an error, verify correct TACAN operation by the use of other navigation equipment. After takeoff, cross-check TACAN with ground radar, VOR, or airborne radar.

7. Channel selector switch - Tune desired channel.

8. Verify station identification.

9. Observe bearing pointer on HSI and BDHI; to-from indicator on HSI.

10. Function sel switch - T/R or A/A.

11. Observe range to station or aircraft on HSI and BDHI.

WINDSHIELD

The pilot's windshield is composed of two glass assemblies secured and sealed in a V-shaped titanium frame. The glass surfaces are coated with low reflective magnesium fluoride. A plastic, V-shaped air duct runs along the lower edge of the windshield. Hot bleed air for defogging is supplied through this duct when selected by a switch on the side panel below the right console in the pilot's cockpit. The air is directed to the windshield through a series of holes on the upper surface of the duct. Holes are also provided at the aft ends of the duct to direct air toward the canopy glass. A vision splitter is located in the Vee of the windshield. It may be pulled down to eliminate reflections between the windshield glasses during night flying and refueling.

CANOPIES

The forward and aft cockpit canopies are essentially identical in construction and operation. Each canopy consists of a reinforced titanium frame containing two high-temperature-resistant glass windows and is hinged at the aft end. A canopy counterbalance system (nitrogen-powered) aids normal opening and closing of the canopy. The canopies can be opened without the aid of the counterbalance system, but upwards of 112 pounds lifting force is required to do so. Small recesses on the outside of each side of the canopies provide lifting points so the canopies can be opened from outside the aircraft. A latch on top of the nitrogen counterbalance cylinder engages when the canopy if fully open, and holds the canopy in that position until the latch is re-

leased by pressing the latch release lever. The two canopies are independent in operation, except for the external jettison feature.

WARNING

- Canopies should be opened or closed only when the aircraft is completely stopped. Hold canopy securely when opening to prevent adverse wind forces from shearing the canopy hinge pins.
- Maximum taxi speed with the canopy open is 40 knots. Gusts and severe wind conditions should be included as part of the 40 knot limit speed.

Canopy Latching Mechanism

Each canopy is latched closed by two hooks in each canopy sill. The canopy is latched and unlatched by operating a handle at the forward right side of each cockpit. Moving the handle rotates a transverse torque tube behind the seat to simultaneously position all four hooks. With the canopy closed, forward movement of the handle latches and locks the canopy, while aft movement releases the hooks and unlocks the canopy. During emergency egress the canopies are unlatched by gas pressure from the canopy unlatch thruster behind the seat, fired when seat ejection or canopy jettison is initiated.

Canopy Counterbalance System

Normal opening and closing of the canopy is assisted by gaseous nitrogen pressure from the canopy counterbalance cylinder at the left rear of the seat. Without counter-balance assistance, the canopy is difficult to raise, and can drop with sufficient force to injure personnel. A counterbalance system pressure gage is located above the nitrogen cylinder, and pointer should indicate in the green if system pressure is normal.

WARNING

To prevent possible injury, always verify that counterbalance pressure indication is normal prior to pushing the canopy latch release.

CAUTION

Do not rotate the canopy upward beyond its normal full-open position or canopy shear pins may be severed.

Canopy Seal System

An inflatable seal is installed along the edge of each canopy frame. The seal is inflated by engine bleed pressure to provide an airtight seal between the canopy and the canopy sills and windshield. An individual pressure regulator and control valve is provided in each cockpit. The seal selector lever is located in the right forward corner of each cockpit, and controls pressure to the seals for inflation, or vents the seals to ambient pressure for deflation. Seal pressure is regulated to approximately 20 psi.

CAUTION

Do not inflate seals unless the canopy is latched closed, or damage to seals may occur.

Canopy Unlocked Warning Light

When the CPY UNLKD light on the annunciator panel is illuminated, it indicates that one or both of the canopies is not latched down and/or properly sealed. The light will go out when both canopies are positioned on the sill, the canopy release handles are in the latched position (full forward), and the canopy seals are fully inflated. The light receives power from the essential dc bus.

Canopy Jettison Systems

The canopy may be jettisoned by independent action or as part of the seat ejection sequence. In either case, the canopy latches are opened by gas pressure from a canopy unlatch thruster on the aft bulkhead of the respective cockpit. The unlatch thruster may be fired by pulling the seat D-ring or the CANOPY JETTISON T-handle on the left console.

When the unlatch thruster fires, it rotates the canopy hooks to the unlatch position and releases the canopy; then, the thruster charge is ported to the canopy-removal thruster and seal-hose cutter. The seal-hose cutter severs the canopy seal hose, and the canopy-removal thruster forces the canopy up and aft, shearing the hinge attach points and thrusting the canopy clear of the aircraft.

The canopy can also be removed in flight by manually unlatching it and pushing it up into the airstream. If the cockpit is pressurized when the canopy latches are released, the canopy will be blown upward into the airstream by cockpit pressure.

WARNING

Do not enter or leave the cockpit unless ground safety pins are installed in the seat ejection D-ring, the catapult jettison T-handle, and the canopy jettison T-handle.

Canopy External Latch Controls

Individual external latch release fittings are flush-mounted on the left side of the fuselage, just below the canopy hinge points. Fittings accept a 1/2-inch-square bar extension to permit opening the canopies from outside the aircraft. Canopy must be raised manually after releasing the latch hooks externally.

Canopy External Jettison System

Both canopies may be jettisoned on the ground actuation of the external jettison system. An external jettison handle is located under an access panel on top of the left chine, at aft end of forward cockpit. When the external jettison handle is pulled, both canopies are jettisoned in sequence, the forward cockpit first, the aft cockpit 1 second later. A long jettison cable requires the man pulling the handle to be well clear of the fuselage during jettison.

REAR VIEW PERISCOPE

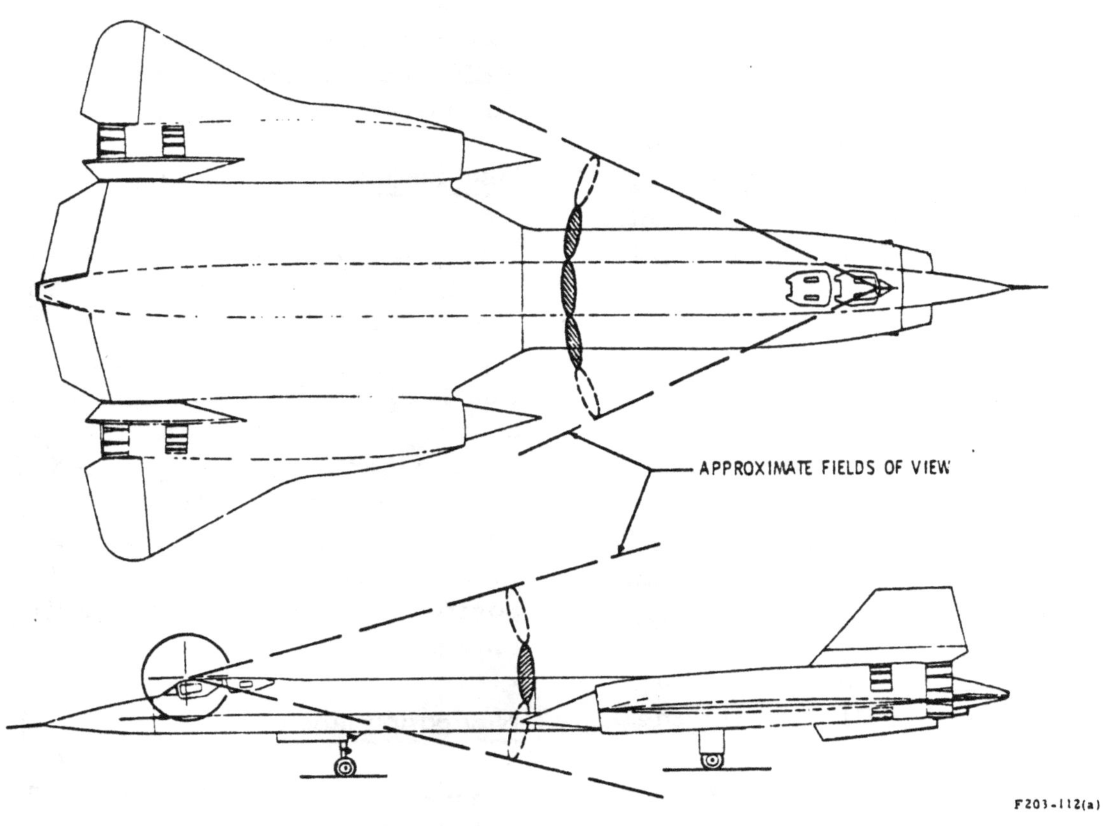

Figure 1-20

REAR VIEW PERISCOPE

A manually extended rear-view periscope is mounted in the top of the forward cockpit canopy. Normally, the periscope is locked in a fully retracted position. It is moved by using a white nylon pad, mounted on the aft side of the viewing tube, as a handle. Pushing the handle to the left unlocks the tube, allowing the periscope to be extended. Then, pushing the tube upward to a spring-detented position makes the rear view available. Cockpit pressure tends to assist extension, and resists retraction. The diameter of the instantaneous cone of view is approximately 10 degrees; however, head movement extends the viewing cone to approximately 30 degrees total angle. (See figure 1-20.) When extended, the periscope can be rotated horizontally to move the center of the viewing arc up to 10 degrees from the aft centerline. The lens system provides a 2 to 1 reduction ratio.

LIGHTING EQUIPMENT

EXTERIOR LIGHTING

Landing and Taxi Lights

Two lights are mounted on the nose gear strut. One light, rated at 1000 watts is the landing light; the other is the taxi light, rated at 450 watts, Both lights receive power from the essential ac bus.

Landing and Taxi Lights Switch

A luminous, three-dot switch for control of landing and taxi lights is located on the pilot's left forward panel. The switch has three positions, LAND & TAXI LT (up), TAXI LT (down), and OFF (center). Moving the switch to the LAND & TAXI LT position illuminates both lights; moving the switch to the TAXI LT position illuminates the taxi light only.

Anti-Collision and Fuselage Lights

Two retractable lights are located near the top and bottom midpoints of the fuselage. They are flush with the fuselage contour when retracted and show a white light from above and below when turned on in the retracted position. The lights extend approximately 2 inches and, when turned on in this position, the illuminated red lights and reflectors rotate at 45 rpm to give an effect of 90 flashes per minute. The lights and rotating and retracting mechanism are powered from the essential ac bus.

Anti-Collision Lights Switches

A pair of two-position switches, located on the pilot's lower instrument panel and placarded ANTI-COLL, control the anti-collision lights. The left switch is labeled ON (up) and FUS LT (down); the other is labeled EXTEND (up) and RETRACT (down). The switches are moved to the ON and EXTEND positions, respectively, to turn on and extend the red lights.

Fuselage Lights Switch

A fuselage lights switch, labeled FUS and located adjacent to the anti-collision light switches, has three positions, BRT (up), OFF (center), and DIM (down). The BRT or DIM positions are used to control illumination and intensity of the lights when they are operated in the flush (retracted) position as fuselage lights. The anti-collision light switch must be moved to the FUS LT (down) position before the fuselage lights switch may be operated.

Navigation Lights

Lights for navigation purposes are installed on the left and right wing tips and on the top and bottom of the aft fuselage. Power for the lights is furnished by the essential ac bus.

Navigation Lights Switches

The navigation lights are controlled by a selector switch and a dimming switch located on the pilot's lower instrument panel. The lights are energized when the selector switch is moved from OFF to STEADY or FLASH. In the STEADY position, all the lights are continuously illuminated. In the FLASH position, all the lights are energized intermittently through a flasher unit at a rate of approximately 85 flashes per minute. Intensity of the lights is controlled by the position (BRIGHT or DIM) of the dimming switch.

Refuel Light Rheostat

A refuel light rheostat installed on the pilot's lower instrument panel may be rotated from OFF to BRT as desired to vary the intensity of a refueling light located in the refueling receptacle.

INTERIOR LIGHTING

The cockpit lighting system includes instrument lights, console panel lights, console floodlights, thunderstorm lights, utility spotlights, circuit breakers and controls. The instruments and consoles are illuminated by integral and edge lighting. Utility spotlights are mounted on the left console in both cockpits. These spotlights are detachable and may be moved about the cockpit. White light may be focused by rotating the lens. Thunderstorm lights are located on the left and right sides of the pilot's canopy and direct light forward onto the instrument panel to avoid blindness caused by lightning.

Cockpit Lighting Switches

A rheostat on the aft end of each spotlight is used to vary light intensity. A pushbutton switch on each light may be used to bypass the rheostat and obtain maximum spotlight brilliance instantaneously. The instrument panel lights, console panel lights and floodlights are controlled by three rheostats located on the pilot's lower instrument panel and the FCO's left console. These rheostats, labeled INSTR, CONSOLE, and FLOOD may be rotated clockwise from OFF to BRT as desired to vary the intensity of the associated lights. The thunderstorm lights are controlled by a two-position switch, on the pilot's lower instrument panel, labeled ON and OFF.

ENVIRONMENTAL CONTROL SYSTEM

The environmental control system consists of air-conditioning and liquid cooling subsystems.

AIR-CONDITIONING SUBSYSTEM

Two complete and independent air-cycle refrigeration systems are provided which use ninth-stage compressor bleed air from each engine to pressurize and cool the pilot and FCO cockpits and ventilated flying suits, electronics bay, and missile bays. (See figure 1-21.) Each system is provided with ram air and fuel-air heat exchangers for primary bleed air cooling. The air temperature of each system is further lowered by fuel-cooled, air-cycle refrigerators before it is used. Cooling fuel is obtained from the engine supply manifold and is circulated by pumps driven by the accessory drive gearboxes. After the fuel is heated it is returned to temperature sensitive valves and then either used by the engines, or diverted to tank 6. A hot-air bypass manifold, connecting the left and right systems upstream of the refrigerators, provides pressure-regulated air for cockpit heating, windshield defogging, and ASG-18 FCS air-cooled rack temperature control.

ENVIRONMENTAL CONTROL SYSTEM
(Air cycle subsystem)

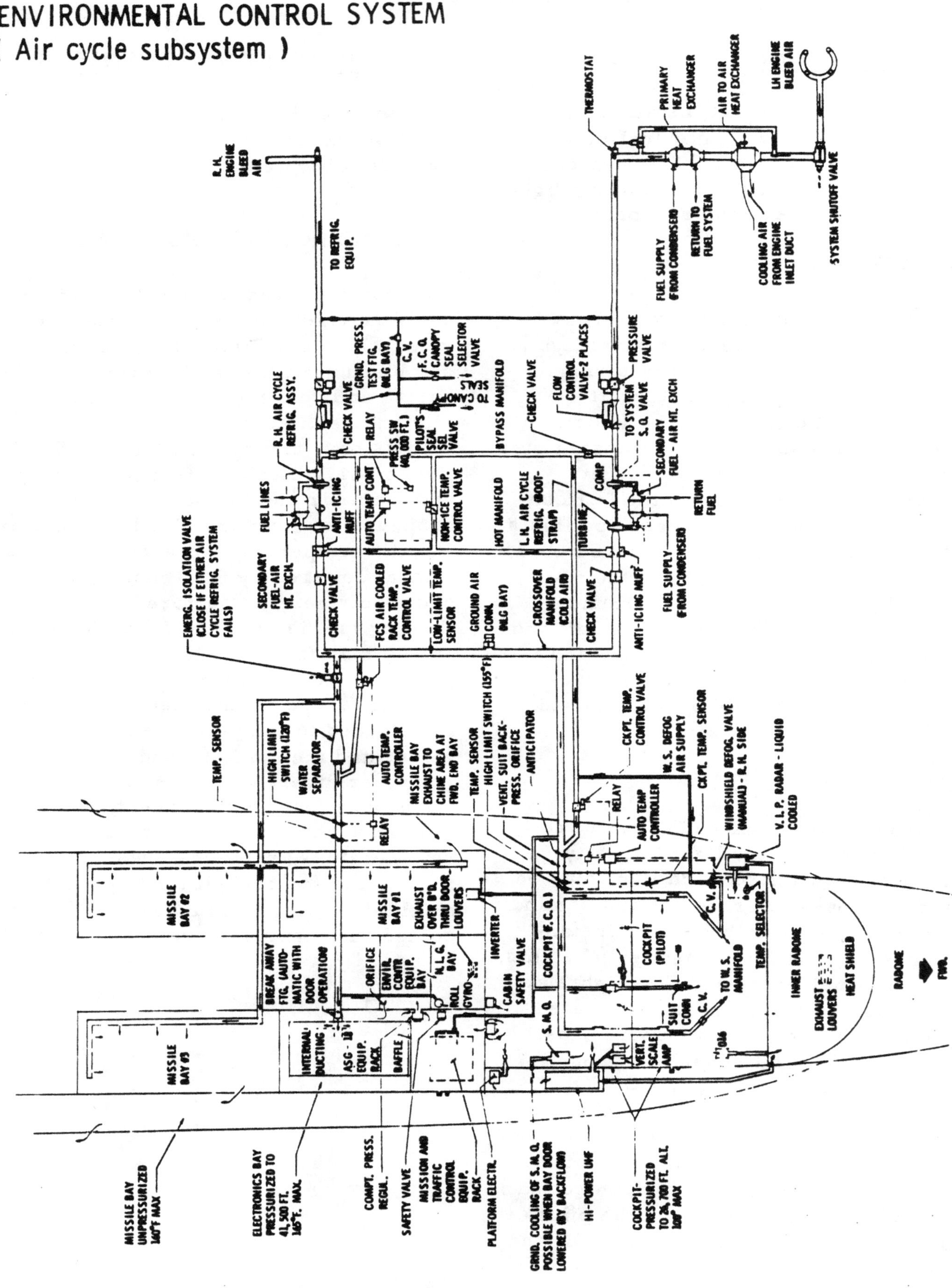

R.H. ENGINE BLEED AIR
TO REFRIG. EQUIP.
THERMOSTAT
PRIMARY HEAT EXCHANGER
AIR TO AIR HEAT EXCHANGER
LH ENGINE BLEED AIR
SYSTEM SHUTOFF VALVE
FUEL SUPPLY (FROM CONDENSER)
RETURN TO FUEL SYSTEM
COOLING AIR FROM ENGINE INLET DUCT
PRESSURE VALVE
FLOW CONTROL VALVE-2 PLACES
TO SYSTEM S.O. VALVE
SECONDARY FUEL - AIR HT. EXCH
CHECK VALVE
BYPASS MANIFOLD
GRND. PRESS. TEST FTG. (NLG BAY)
C.V.
F.C.O. CANOPY SEAL SELECTOR VALVE
TO CANOPY SEALS
PILOT'S SEAL SEL VALVE
PRESS SW (40,000 FT.)
RELAY
R.H. AIR CYCLE REFRIG. ASSY.
FUEL LINES
ANTI-ICING MUFF
AUTO TEMP CONT
NON-ICE TEMP. CONTROL VALVE
HOT MANIFOLD
L.H. AIR CYCLE REFRIG. (BOOT-STRAP)
COMP
TURBINE
RETURN FUEL
SECONDARY FUEL-AIR HT. EXCH.
CHECK VALVE
FCS AIR COOLED RACK TEMP. CONTROL VALVE
LOW-LIMIT TEMP. SENSOR
GROUND AIR CONN. (NLG BAY)
CROSSOVER MANIFOLD (COLD AIR)
CHECK VALVE
ANTI-ICING MUFF
FUEL SUPPLY (FROM CONDENSER)
EMERG. ISOLATION VALVE (CLOSE IF EITHER AIR CYCLE REFRIG. SYSTEM FAILS)
TEMP. SENSOR
HIGH LIMIT SWITCH (120°F)
WATER SEPARATOR
RELAY
AUTO TEMP. CONTROLLER
MISSILE BAY EXHAUST TO CHINE AREA AT FWD. END BAY
TEMP. SENSOR
HIGH LIMIT SWITCH (155°F)
VENT. SUIT BACK-PRESS. ORIFICE
ANTICIPATOR
CKPT. TEMP. CONTROL VALVE
RELAY
W.S. DEFOG AIR SUPPLY
CKPT. TEMP. SENSOR
AUTO TEMP CONTROLLER
WINDSHIELD DEFOG. VALVE (MANUAL) - R.H. SIDE
V.L.P. RADAR - LIQUID COOLED
MISSILE BAY #2
MISSILE BAY #1
EXHAUST OVER B'D. THRU DOOR LOUVERS
INVERTER
CABIN SAFETY VALVE
COCKPIT (F.C.O.)
COCKPIT (PILOT)
C.V.
TO W.S. MANIFOLD
TEMP. SELECTOR
INNER RADOME
EXHAUST LOUVERS
HEAT SHIELD
RADOME
FWD.
BREAK AWAY FTG. (AUTO-MATIC WITH DOOR OPERATION)
ORIFICE
ENVIR. CONTR EQUIP. BAY
N.L.G. BAY
ROLL GYRO
S.M.Q.
SUIT CONN
C.V.
VERT. SCALE AMP
MISSILE BAY #3
INTERNAL DUCTING
ASG-18 EQUIP. RACK
BAFFLE
MISSILE BAY UNPRESSURIZED 140°F MAX
ELECTRONICS BAY PRESSURIZED TO 41,500 FT. 165°F. MAX
COMPT. PRESS. REGUL.
SAFETY VALVE
MISSION AND TRAFFIC CONTROL EQUIP. RACK
PLATFORM ELECTR.
GRND. COOLING OF S.M.Q. POSSIBLE WHEN BAY DOOR LOWERED (BY BACKFLOW)
HI-POWER UHF
COCKPIT-PRESSURIZED TO 26,700 FT. ALT. 100° MAX

Figure 1-21

ENVIRONMENTAL CONTROL SYSTEM (Liquid cooling subsytem)

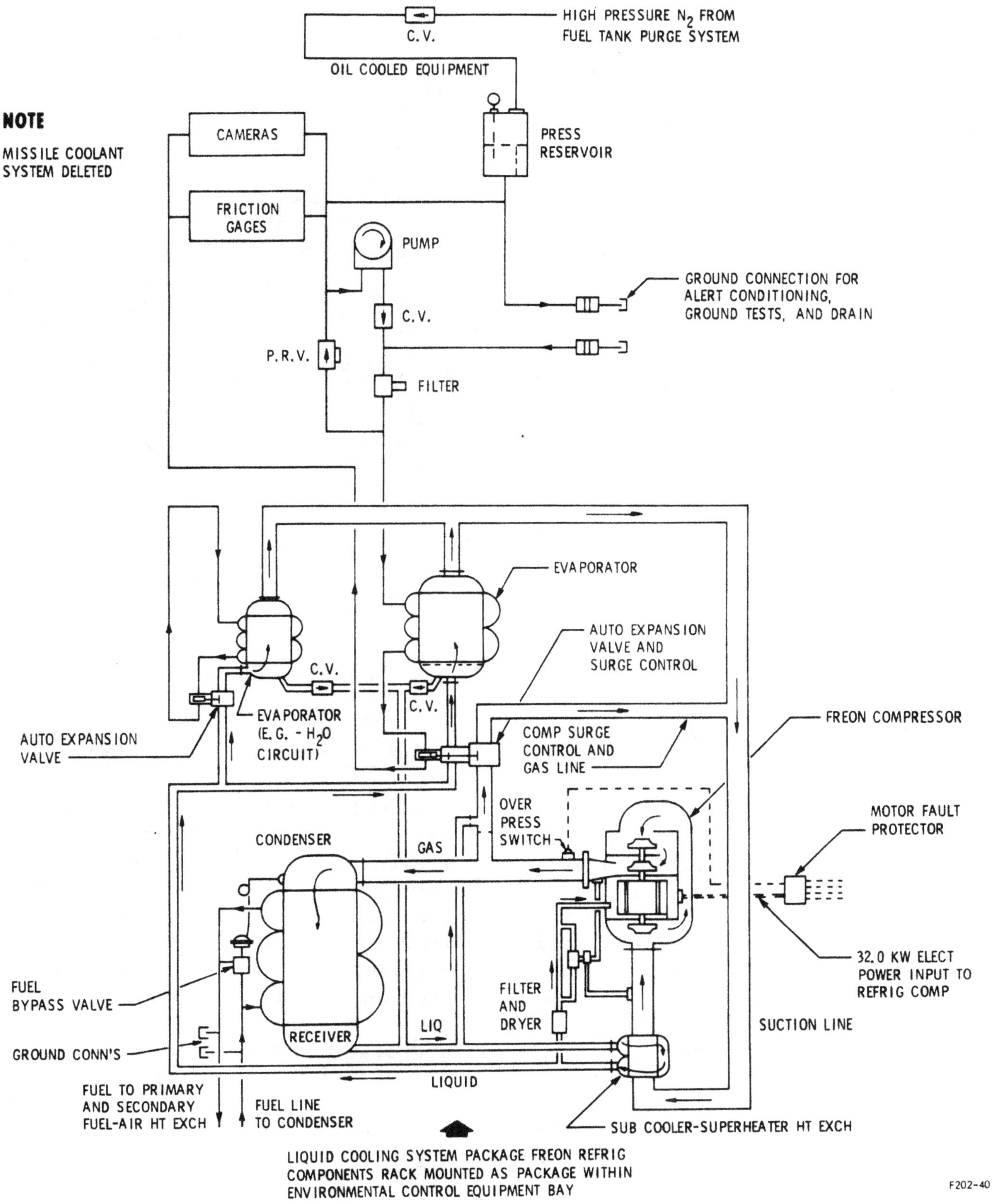

HIGH PRESSURE N_2 FROM FUEL TANK PURGE SYSTEM

C.V.

OIL COOLED EQUIPMENT

NOTE

MISSILE COOLANT SYSTEM DELETED

CAMERAS

FRICTION GAGES

PRESS RESERVOIR

PUMP

GROUND CONNECTION FOR ALERT CONDITIONING, GROUND TESTS, AND DRAIN

C.V.

P.R.V.

FILTER

EVAPORATOR

AUTO EXPANSION VALVE AND SURGE CONTROL

C.V.

C.V.

EVAPORATOR (E.G. - H_2O CIRCUIT)

AUTO EXPANSION VALVE

COMP SURGE CONTROL AND GAS LINE

FREON COMPRESSOR

OVER PRESS SWITCH

MOTOR FAULT PROTECTOR

CONDENSER

GAS

32.0 KW ELECT POWER INPUT TO REFRIG COMP

FILTER AND DRYER

FUEL BYPASS VALVE

SUCTION LINE

LIQ

RECEIVER

GROUND CONN'S

LIQUID

FUEL TO PRIMARY AND SECONDARY FUEL-AIR HT EXCH

FUEL LINE TO CONDENSER

SUB COOLER-SUPERHEATER HT EXCH

LIQUID COOLING SYSTEM PACKAGE FREON REFRIG COMPONENTS RACK MOUNTED AS PACKAGE WITHIN ENVIRONMENTAL CONTROL EQUIPMENT BAY

F202-40

Figure 1-22

LIQUID COOLING SUBSYSTEM

The liquid cooling subsystem is independent of the air-conditioning units. It consists of a Freon vapor-cycle refrigeration package and two liquid coolant circulation systems. (See figure 1-22.) The refrigeration unit is mounted between the two air-cycle refrigerators in the air-conditioning bay, and includes a motor-driven compressor, a condenser, a subcooler-superheater heat exchanger, associated valves and plumbing, and two evaporators. One evaporator, rated at 1-1/2 tons capacity, was used to cool an ethylene glycol and water mixture which was circulated through the missile components. This system is no longer active. A larger evaporator of 11-ton capacity rating cools hydraulic oil used in the SP 302 coolant circulation system (special test instrumentation installation). The Freon vapor cycle starts at the compressor. Freon gas is compressed and sent through a fuel-cooled condenser and a subcooler-superheater heat exchanger where it is liquified. Liquified Freon then passes through expansion valves and enters distributor tubes in the bottom of each evaporator. Heat absorbed from the liquid coolants circulating through the evaporators causes the Freon to boil and change to a gas. The Freon gas is drawn from the top of the evaporators by a manifold and passed through the subcooler-superheater heat exchanger, into the compressor where the cycle repeats. The expansion valve at the base of the larger evaporator incorporates a surge control port so that compressed Freon gas can be bypassed around the condenser and evaporators during periods when cooling requirements are low. This arrangement insures satisfactory compressor operating speed. During times when the surge control is in operation, liquid Freon from the heat exchanger is introduced into the expansion and surge valve and mixed with gaseous Freon to prevent an overheat condition.

The vapor cycle refrigeration unit receives 28 VDC and 200 VAC, three-phase, 400-cycle power from the right primary ac bus. Both generators (or ground power) must be operating in order to turn on the refrigeration unit. After the unit is operating, the L engine generator may be taken off the line and the unit will continue to operate; however, if the R engine generator is inoperative, the unit will shut off automatically.

NOTE

> If the bus tie contactor has opened in flight, the liquid cooling system cannot be restarted even if both generators resume operation.

SP 302 Coolant Circulation System

A closed-loop liquid coolant system circulates hydraulic oil from a separate pressurized reservoir to a transfer pump, then through a double walled chamber which houses two 16 mm instrumentation camera. From there it circulates the oil through a skin friction gage jacket. The oil returns from this equipment to the larger Freon evaporator unit where it is cooled before it is recirculated again. The SP 302 coolant transfer pump receives three phase, 400 cycle power from the right primary ac bus.

SYSTEM CONTROLS AND INDICATORS

L and R Air System Switches

The L (left) and R (right) air system switches, on the pilot's forward left console, have two positions, ON (up) and OFF (down). When a switch is placed in the ON position, the normally open system shutoff valve is deenergized and that system's air is supplied to the cockpits, fire control electronics, and missile bays when the respective engine is running. In the OFF position, the shutoff valves are closed, shutting off the air. The circuits are powered by the essential dc bus.

L and R Air System Out Lights

L AIR SYS OUT and R AIR SYS OUT warning lights are installed on the pilot's annunciator panel. Illumination of a light indicates that a shutoff valve has closed and the respective air system has become inoperative. Power for the lights is furnished by the essential dc bus.

Temperature Control Selector and Override Switch

An air temperature selector and override switch, labeled CKPT TEMP OVRD, is installed on the pilot's upper left console. The switch has four positions, AUTO (up), COLD (down left), WARM (down right), and MAN HOLD (center). The switch is spring-loaded to MAN HOLD from the COLD and WARM manual control position. The switch will normally be in the AUTO position; however, in case of a malfunction in the automatic feature of the system, the pilot can manually control the system by moving the switch to either the momentary COLD or WARM position. The essential dc bus powers the temperature control system.

Temperature Indicator

A temperature indicator, located on the pilot's canopy only, allows the pilot to monitor the air temperature in his cockpit. Power for the indicator is furnished by the essential dc bus.

NOTE

> If the cockpit temperature approaches 140°F it will exceed the ability of the suit to keep the pilot comfortable.

Cockpit Temperature Control Knob

An air temperature control rheostat, labeled CKPT TEMP, is installed on the pilot's upper left console. An arrow on the panel adjacent to the knob indicates the direction of rotation necessary to increase cockpit temperature. Electrical power for the cockpit temperature control circuit is furnished by the essential dc bus.

Cockpit Depressurization (Dump) Switch

A guarded depressurization switch, labeled CKPT PRESS DUMP, is install-

ed on the pilot's upper left console. When the switch is moved to the up position, a safety valve opens to depressurize both cockpits. When moved to the down (guard down) position the safety valve will close and the cockpits will repressurize.

WARNING

Depressurization and repressurization occur at an extremely rapid rate.

Bay Air Shutoff Switch

A bay air shutoff switch, located on the pilot's left-hand forward console and labeled BAY AIR, has two positions, SHUTOFF (up) and ON (down). The switch is normally in the ON position during flight; however, in the event of an environmental air system failing, switching to the SHUTOFF position will divert all bay cooling air to the cockpits and ventilated flying suits. A failure of one air system will be indicated when the pilot notices a decrease in suit flow. Power for the switch is furnished by the essential dc bus.

Refrigeration Overheat Reset Switch

A momentary contact, spring-loaded switch, labeled REFRIG OVERHEAT RESET, is installed on the FCO left-hand forward panel adjacent to the FCS LIQUID COOLING switch. The switch is operated whenever the FCS LIQUID COOL OUT light on the FCO annunciator panel lights. If the system has become inoperative due to an overheat or overpressure condition, depressing the switch will re-energize the liquid cooling Freon compressor circuit after a 3-second delay. Power for the switch is furnished by the essential dc bus.

FCS Liquid Cooling Switch

A guarded FCS LIQUID COOLING switch, installed on the FCO left-hand forward panel and on the pilot's right console, has two positions, ON and OFF (guarded). When the switch is placed in the ON position the Freon compressor motor is energized. Power for the switch is furnished by the essential dc bus.

NOTE

During ground operation the FCS liquid cooling switch should only be actuated while the engines are near idle rpm. This practice minimizes transient loads which occur during Freon package start and shutdown.

CAUTION

To prevent excessive loads from being applied to the CSD, ensure that FCO allows 10 seconds to elapse between in-flight Freon package turnon and FCS radar final power amplifier turnon.

Air-Conditioning Warning Lights

Five air-conditioning warning lights, labeled FCS LIQUID COOL OUT, FCS RACK OVERHEAT, MISSILE BAY OVERHEAT, MTC BAY OVERHEAT, and MTC BAY PRESS LOW, are installed on the FCO annunciator panel.

An illuminated FCS LIQUID COOL OUT light indicates that the liquid cooling compressor motor is not operating. If

the light comes on, check the FCS liquid cooling switch to confirm that it is in the ON position. Then, operate the refrigeration overheat reset switch in an attempt to re-energize the circuit. After this, the liquid cooling system should begin to operate and extinguish the light. If a generator fails and the

FCS LIQUID COOL OUT light illuminates, an automatic lockout feature in the electrical system prevents a restart of the liquid cooling system.

If the FCS RACK OVERHEAT light illuminates it is an indication that the air surrounding the FCS rack is too hot. Check that the bay air shutoff switch is in the ON position. If the bay air shutoff switch is ON, turn off the FCS equipment and descend to a lower altitude as soon as practicable.

MISSILE BAY OVERHEAT light illumination indicates that the air in at least one of the missile bays is too hot. Check that the bay air shutoff switch is in the ON position. If the light remains illuminated, descend to a lower altitude as soon as practicable.

MTC BAY OVERHEAT light illumination indicates an overheat condition in the MTC bay. Turn off all FCS equipment and all non-essential radio and navigation equipment. Descend to a lower altitude as soon as practicable.

MTC BAY PRESS LOW light illumination indicates that the MTC bay has become unpressurized. Descend to a lower altitude as soon as practicable.

COCKPIT PRESSURIZATION

At high altitude, the pressurization systems maintain a constant 26,000-foot altitude in the cockpits. Normal cockpit pressurization can be maintained by either system if bay air is shut off.

TYPICAL COCKPIT PRESSURIZATION SCHEDULE

Aircraft Alt	Cockpit Alt
10,000 feet	8,000 feet
20,000 feet	16,000 feet
30,000 feet	24,000 feet
35,000 feet and higher	26,000 feet

DEFOG SYSTEM

The defog system delivers hot air from both left and right air systems through check valves to defog the windshields and canopies.

Defog Lever

A defog lever, located below the right console in the pilot's cockpit, provides incremental control as required.

CAUTION

The lever should always be OFF except when defogging is necessary as the hot defogging air will increase cockpit temperature and can damage the windshield.

COCKPIT PRESSURIZATION SCHEDULE

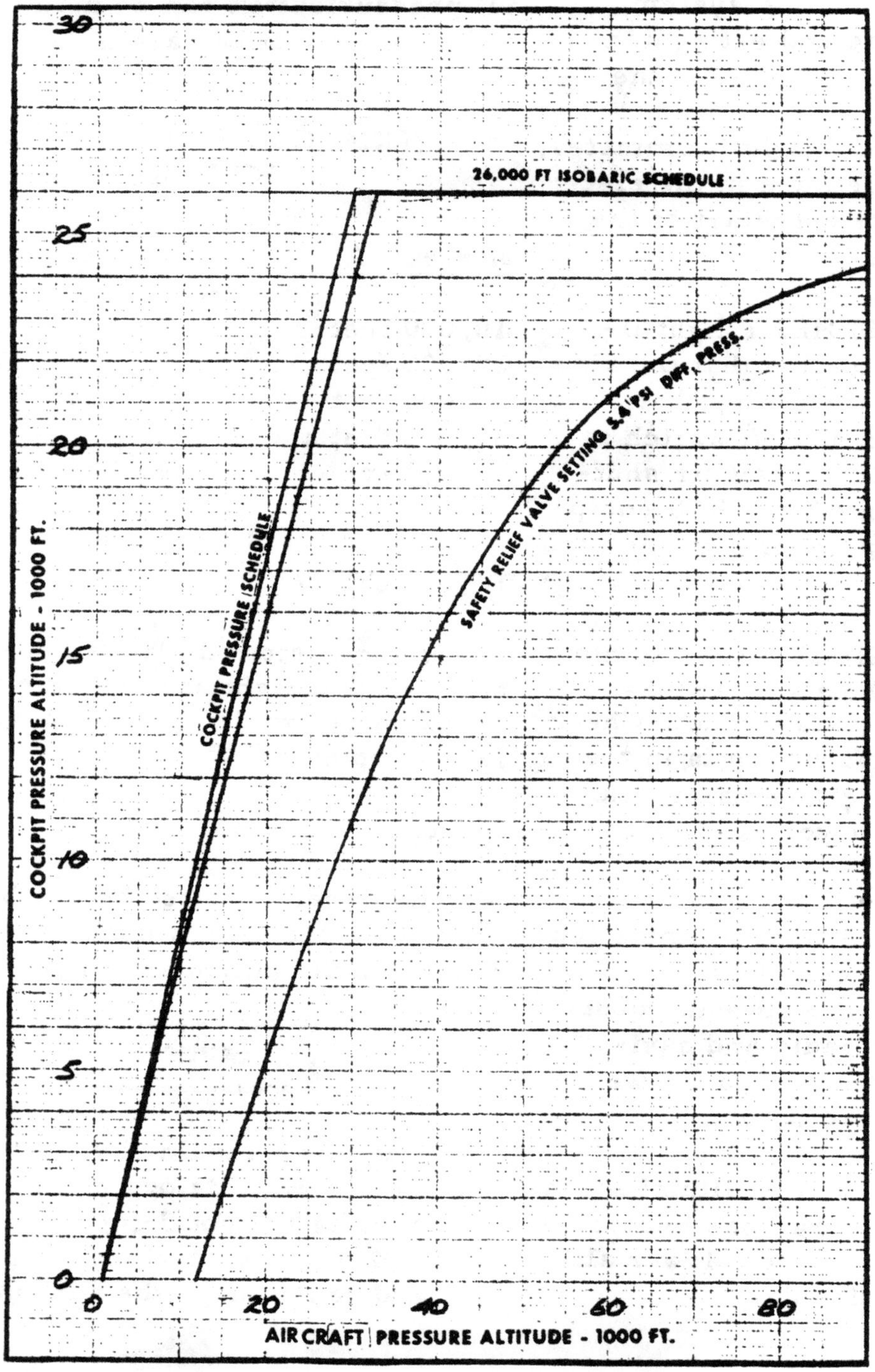

26,000 FT ISOBARIC SCHEDULE

COCKPIT PRESSURE SCHEDULE

SAFETY RELIEF VALVE SETTING 5.4 PSI DIFF. PRESS.

COCKPIT PRESSURE ALTITUDE - 1000 FT.

30
25
20
15
10
5
0

0 20 40 60 80

AIRCRAFT PRESSURE ALTITUDE - 1000 FT.

Figure 1-23

LIFE SUPPORT SYSTEMS

AIRCRAFT OXYGEN SYSTEM

The aircraft is equipped with two independent oxygen systems. Each crewmember is supplied with a dual, high pressure gaseous oxygen system. The systems are supplied by four 875 cubic inch, 1800 psi oxygen cylinders; two for each cockpit. Oxygen is consumed simultaneously from the two cylinders servicing the respective cockpit. In the event that one cylinder should fail, the other cylinder will continue to supply oxygen but the duration will be shortened. As oxygen leaves the cylinders, the pressure is reduced to 75 psi. ON-OFF levers for the two systems are located on the oxygen control panels installed on the left consoles. Oxygen cylinder pressure is displayed on dual indicating gages located on the instrument panels. The No. 1 OXY PRESS LOW or the No. 2 OXY PRESS LOW lights on the annunciator panel in the respective cockpit will illuminate when oxygen pressure decreases below approximately 50 psi.

EMERGENCY OXYGEN SYSTEM

Each crewmember is supplied with two independent emergency oxygen systems. Each system is supplied by a 45-cubic-inch capacity cylinder, pressurized at approximately 2000 psi, and located in the survival kit container. Emergency oxygen is supplied automatically to the helmet by each oxygen system in the event of ejection. Both oxygen systems can be activated manually by pulling the "green apple". When actuated, check valves prevent oxygen flow into the aircraft system. Emergency oxygen system pressure is slightly lower than the aircraft system pressure to prevent emergency oxygen flowing if aircraft system pressure is still available. Pressure in the emergency system containers is indicated by gages on the forward right side of the survival kit (figure 1-26) container.

Emergency Oxygen System Actuation

The emergency oxygen system may be actuated either manually by pulling the conventional green apple, or automatically by the upward motion of the seat during ejection. Once the emergency oxygen system is actuated, it cannot be shut off. The emergency oxygen system should be actuated if the aircraft is not delivering the desired amount of oxygen, or hypoxia or noxious fumes are suspected.

WARNING

Attempt to positively ascertain that the ship oxygen supplies are not available or that they have been lost before pulling the green apple to activate the crewmember's emergency oxygen supply. If the ship systems have not actually been lost and the green apple has been pulled, there is the very remote possibility that the total emergency system might not be available if all ship systems are subsequently lost.

FULL-PRESSURE SUIT

A full-pressure, model J suit is provided which is capable of providing the crewmember with a safe environment exclusive of pressure conditions in the cockpit. The suit consists of four

OXYGEN SYSTEM SCHEMATIC

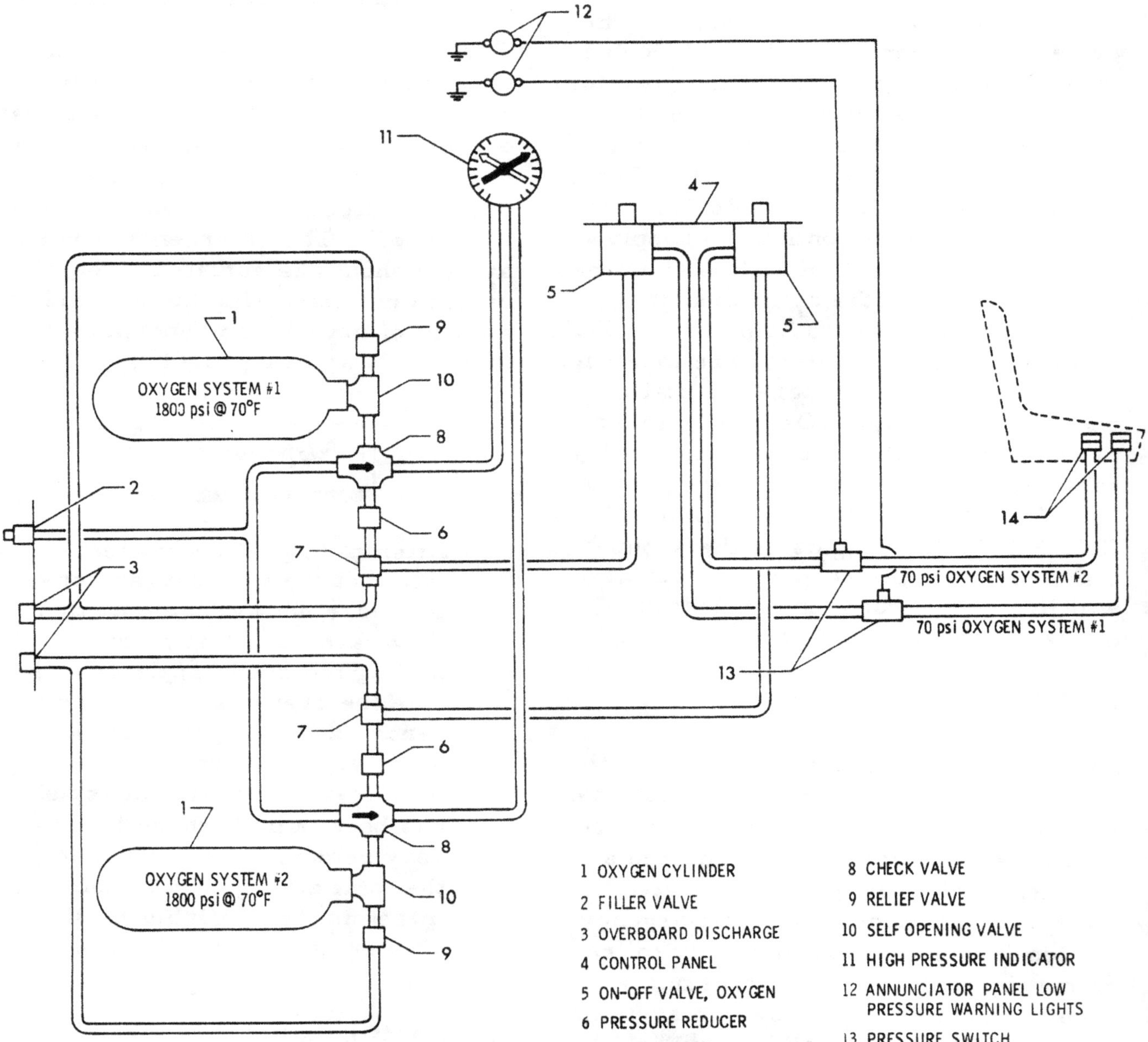

1 OXYGEN CYLINDER
2 FILLER VALVE
3 OVERBOARD DISCHARGE
4 CONTROL PANEL
5 ON-OFF VALVE, OXYGEN
6 PRESSURE REDUCER
7 RELIEF VALVE
8 CHECK VALVE
9 RELIEF VALVE
10 SELF OPENING VALVE
11 HIGH PRESSURE INDICATOR
12 ANNUNCIATOR PANEL LOW PRESSURE WARNING LIGHTS
13 PRESSURE SWITCH
14 SLIDE VALVE COUPLINGS FOR HOSES TO PRESSURE REGULATORS IN HELMET

NOTE

Each cockpit has a complete system as shown above

Figure 1-23A

F202-16(a)

layers; ventilation garment, bladder, link net, and heat-reflective outer garment. The ventilation garment layer allows vent air to circulate between the crewmember's cotton underwear and the bladder layer. The bladder provides an air-tight seal to hold pressurized air in the suit. The link net is a mesh which holds the suit in conformance with the crewmember's body. The outer garment is heat-reflecting aluminized cloth that provides some protection from a high temperature environment. The suit has oxygen hoses in the back, and the parachute harness is part of a torso harness worn over the pressure suit and under the coveralls. A vertical entry zipper is located in the back.

Pressure Suit Vent Air

Air for suit ventilation comes directly from the air-conditioning cold air manifold. In cruise, the temperature of this air may be as cold as -30°F when suit heat is OFF, or warmed to any desirable temperature by use of the suit heat control. Vent air and exhaled breathing air is exhausted through the suit controller valve.

Suit Heat Rheostat

A suit heat rheostat, located on the left console in the forward cockpit, controls the input to an electric heater for temperature regulation of the pilot's and FCO's pressure suits. Individual comfort adjustment can be accomplished by varying suit airflow thru the ventilated air valve on each suit.

Suit Controller Valve

Air pressure to the suit is regulated by dual suit controller valves, located on the front of the suit just above the waist. Each controller valve contains a sensor which programs airflow to keep internal suit pressure at 3.5 psi in the event of cockpit depressurization. A press-to-test button for each oxygen system (SYS 1 and SYS 2) is installed on both controller valves, allowing the crewmember to check suit inflation. (The press-to-test button for system 1 has a screw-in feature to permit the crewmember to partially inflate the suit for individual comfort.)

HELMET

The helmet head area for all helmets is divided into two separate sections by a rubber face seal. The front area between the face plate and the face seal receives oxygen from either the aircraft or emergency oxygen system through regulators built into the helmet. Oxygen flows across the face plate from the inhalation valves inside the helmet and accomplishes some faceplate defogging before it is inhaled. The rear area receives ventilating air. The face seal is not positive; however, the pressure of the oxygen in the front area is slightly higher than the rear area and prevents air from leaking forward. If the oxygen supply to the helmet is interrupted or exhausted, an anti-suffocation valve in the helmet senses the drop in pressure and allows cockpit air to enter the helmet, thus preventing suffocation.

The helmet is distinguished by a "Baylor" bar, which is a transverse bar encircling half of the helmet. It is pivoted at each side attachment point. When raised, it lifts the faceplate and closes the helmet oxygen regulator valves. When lowered, it opens the helmet oxygen regulator valves and then closes and seals the faceplate. A latch below the face opening must be engaged by the bar and a lock on the bar rotated 90 degrees to the right to hold the faceplate in the sealed position. A microphone adjustment knob is provided on the front of the helmet below the face opening. When turned, the fore-and-aft position of the helmet microphone is adjusted.

CAUTION

Do not force the microphone adjustment knob after full travel of the microphone support has been reached, as the helmet shell may be damaged.

The faceplate is raised and lowered by the Baylor bar. With the bar raised, the faceplate is held in the up position by a light spring, uplock plunger, and cam arrangement on the both sides of the helmet. A transparent dark visor is provided as part of the helmet. It is also pivoted at each side connection, and lowers from a stowed position on top of the helmet to cover the faceplate (but not the bar) when its position control at either pivot is rotated down toward the front. A friction clutch holds the visor in the selected open or closed position. An external cranking knob is provided at the rear of the helmet for headband adjustment. External helmet connections include two oxygen supply hoses and a combination electrical lead to provide power for faceplate heat and microphone and earphone connections.

Faceplate Heat Switches

A rheostat-type faceplate heat switch is installed in each cockpit on the left side of the instrument panel. Markings show the position for power off and for low, medium, and high heat, increasing in the clockwise direction. Crewmember faceplate defogging is accomplished by the combination of oxygen flow and individual adjustment of the face heat control.

Power for the individual switches is supplied from the essential dc bus through FACE and FACE HTR circuit breakers in the forward and aft cockpits, respectively.

BOOTS

The sock or boot liner fastens onto the suit at the thigh by means of a zipper. The boots are made of white leather for heat reflection, and fit snugly over the socks. A spur that fastens to the seat foot retention cable is strapped onto each heel.

GLOVES

Leather gloves fasten onto the suit at the wrist rings. The inner liner of the glove is similar to the suit inner liner and will retain pressure. The outer glove has a leather palm with high-temperature-resistant material over the back hand area.

Pressure Suit Preflight Check

1. After donning the suit, check inflation, breathing, and ventilation air while in the personal equipment room prior to entering the aircraft.
2. After entering the aircraft:
   a. Check that the aircraft oxygen system is ON, then plug in oxygen leads to front of seat.

WARNING

One portable oxygen connection must be connected and that system turned ON before the second portable oxygen connection is disconnected.

b. Connect ventilation air hose from aircraft to suit.

This can be accomplished at a later period of preflight if external ventilation air is available. The aircraft will not deliver ventilation air until an engine is running.

c. Fasten lap belt and shoulder harness, making sure that the parachute timer arming key is placed on the lap belt prior to securing the belt.

d. Hook up spurs.

The spurs may be hooked from a sitting position. Or, crewmember may prefer to step down on the ball prior to sitting down in the seat.

e. Connect radio leads to helmet.

f. Check suit inflation and breathing for both systems, using system ON-OFF switches on oxygen panel.

It is not necessary to adjust the helmet during an inflation check. The pulley arrangement on the tiedown strap allows adequate mechanical advantage for helmet adjustment after inflation.

g. Check accessibility of emergency oxygen actuator (green apple).

OXYGEN MASK AND REGULATOR

A suitable oxygen mask assembly may be used in lieu of the pressure suit for flights at low and intermediate altitudes when permitted by appropriate regulations. The mask assembly consists of an MBU-5/P oxygen mask, a specially designed oxygen regulator, anti-suffocation valve, and two oxygen personal leads with connectors for both aircraft and emergency oxygen systems. In the event that the regulator should malfunction or the oxygen supply is exhausted, the anti-suffocation valve installed between the regulator and the mask will sense a drop in oxygen pressure and permit ambient air to enter the mask to prevent crewmember suffocation.

EMERGENCY ESCAPE SYSTEM

The aircraft emergency escape system comprises an SR-1 stabilized ejection seat, canopy jettison system, and the operating controls and indicators in both cockpits.

EJECTION SEAT

The SR-1 seat is usable from zero speed and altitude to the extremities of the flight envelope. The ejection seat is a rocket-propelled, upward ejecting unit, which uses a drogue chute to stabilize seat descent until man-seat separation occurs and the personnel (main) chute is deployed at approximately 15,000 feet. The ejection seat system is connected to the canopy system to provide canopy jettison prior to initiation of the normal seat ejection sequence.

STABILIZED EJECTION SEAT

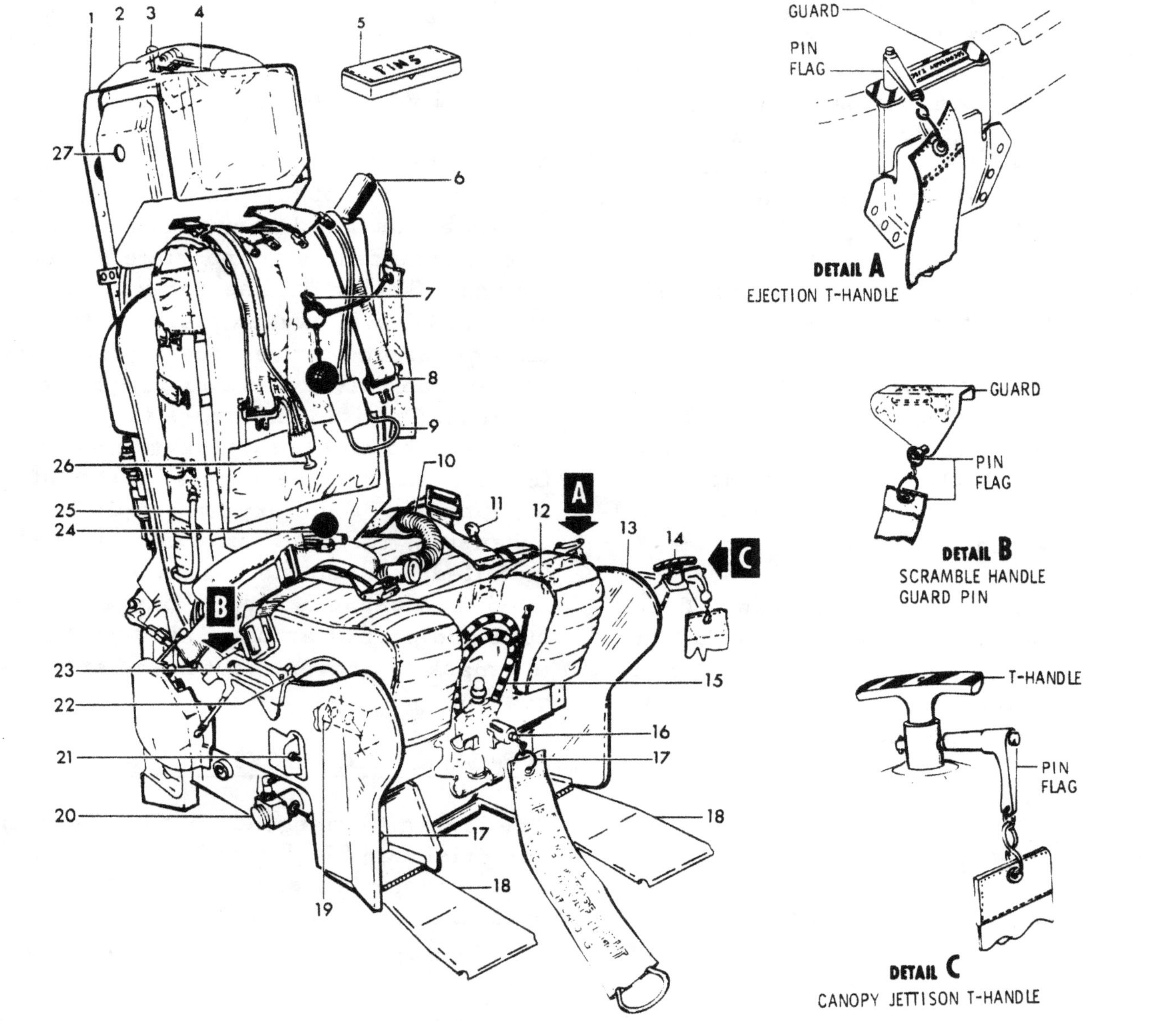

DETAIL A
EJECTION T-HANDLE

DETAIL B
SCRAMBLE HANDLE GUARD PIN

DETAIL C
CANOPY JETTISON T-HANDLE

1 AIRCRAFT GUIDE RAIL
2 DROGUE PARACHUTE
3 DROGUE PARACHUTE GUN
4 DROGUE PARACHUTE PIN
5 PIN STOWAGE - FWD COCKPIT ONLY
6 PARACHUTE GUN SAFETY CAP
7 PARACHUTE GUN SAFETY PIN
8 SHOULDER HARNESS
9 MANUAL PARACHUTE DEPLOY RING
10 SUIT VENT AIR HOSE
11 INERTIA REEL LOCK
12 EJECTION TEE HANDLE
13 LEG GUARD
14 CANOPY JETTISON HANDLE
15 EJECTION D RING
16 SAFETY PIN
17 FOOT RETENTION CABLES
18 FOOT RAMPS
19 EMERG OXYGEN PRESSURE INDICATOR
20 FOOT RETENTION CABLE CUTTER (2)
21 SEAT VERTICAL ADJUSTMENT SWITCH
22 SURVIVAL KIT RELEASE
23 SCRAMBLE HANDLE
24 EMERGENCY OXYGEN HANDLE (GREEN APPLE)
25 MAIN PARACHUTE ARMING CABLE
26 RADIO BEACON CONTROL
27 LEAK DETECTOR, ANEROID ACTUATED INITIATOR

F202-29

Figure 1-24

NOTE

Automatic jettison of the canopy occurs only during the primary ejection sequence, initiated by pulling the seat D-ring. If it becomes necessary to use the secondary ejection system (initiated by pulling the seat T-handle), the canopy must be jettisoned by separate action prior to pulling the T-handle.

The parachute, emergency oxygen supply, and the survival kit are normally attached to the pressure suit integrated harness and installed in the seat before the crewmember enters the cockpit. Hookup to the crewmember is by means of five quick-disconnect attachments.

Seat Spacer and Headrest

On modified seats, a seat spacer and an extended headrest cushion assembly may be installed to provide adequate fit and support of the crewmember in the seat during ejection. The extended headrest cushion and the standard cushion assembies are interchangeable, but the extended cushion must be used whenever the seat spacer is installed. The spacer consists of a 1.5-inch-thick balsa wood core, covered with a fiber glass frame.

Seat Vertical Adjustment

An electric actuator on the lower end of the seat catapult moves the seat up or down as controlled by a 3-position switch on the right side of the seat bucket. Seat movement is in the same direction as switch movement. Maximum vertical seat travel is 9 inches for the forward cockpit seat, and 6.75 inches for the aft cockpit seat. Seat operating power is from the essential dc bus through the SEAT ADJ circuit breaker in the respective cockpit.

Emergency Faceplate Heat

A method of providing faceplate heat after ejection is provided on modified ejection seats. The source of electric power for this heat is provided by an automatically activated silver-zinc battery, mounted on the left side of the seat. The battery is hose-connected to the D-ring ballistic line, and a power cable is connected (by pull-to-release plug) to the helmet. During ejection, a squib in the battery is fired to rupture diaphragms at both ends of the battery, and force electrolyte into the battery cells; providing a freshly energized battery for faceplate heat after ejection. The battery cable plug is pulled loose from the helmet during man-seat separation, as faceplate heating is no longer necessary at this time.

Inertia Reel

An inertia reel in the seat headrest maintains a constant pressure on the shoulder harness straps to keep them taut and permit normal restricted movement of the crewmember. The inertia reel may be manually locked to retain the straps in a fixed position by placing the shoulder harness manual control knob (on left side of seat) to the LOCK position. The reel will release and resume normal tension on the shoulder straps when the knob is returned to the UNLOCK position. When the inertia reel control knob is in the normal (UNLOCK) position, the reel will automatically lock with an instantaneous forward load of approximately 2 g to 3 g, and will remain locked until manually released when the control knob is moved to LOCK and back to UNLOCK.

Early in the ejection sequence, an initiator is fired to roll up and lock the shoulder straps in a "tight" condition. Later in the ejection sequence, the shoulder straps are automatically severed as strap cutters are fired. If the cutters do not sever the straps they will pull free of the inertia reel during man-seat separation. The strap cutters may be fired manually (while seat is in the airplane) by pulling the manual release (scramble) handle on the right side of the seat.

D-Ring Assembly

A D-ring is located on the forward part of the seat, between the crewmembers knees, and is pulled to initiate ejection. After being pulled, the D-ring serves as a handhold for protection of the arms during ejection, and a compression spring within the D-ring assembly acts as a shock absorber. During man-seat separation, the D-ring cable is automatically severed to release the D-ring from the seat. (The D-ring cable cutter may be fired manually as well, by pulling on the scramble handle.) A safety pin hole is provided at the base of the D-ring assembly so it may be pinned to prevent accidental firing of the primary ejection system on the ground.

T-Handle Assembly

A backup ejection control, a T-handle, is available in the event of D-ring malfunction (D-ring initiator not firing). The T-handle is located to the left of the seat, under a cover shield which must be pushed away to gain access to the handle. A safety pin hole is provided in the cover shield and T-handle so it can be "pinned" to prevent accidental firing on the ground. Pulling the T-handle initiates only the seat ejection sequence; the canopy must be jettisoned by separate action prior to pulling the T-handle.

On modified aircraft a safety pin has been installed in the ejection seat T-handle system that prevents accidental actuation of the secondary ejection system in flight. The D-ring must be pulled before the T-handle can be actuated. Pulling the D-ring automatically removes the safety pin and arms the T-handle system (secondary ejection sequence).

Scramble Handle

The manual release (scramble) handle provides a means of quick release from the seat for bailout with ejection seat inoperative and for over-the-side egress on the ground. This handle is on the right side of the seat, covered by a guard and pinned to prevent accidental actuation. A button at the forward end of the scramble handle must be depressed to permit release of the scramble handle. When pulled, the scramble handle mechanically rotates the seat torque tube to release the lap belt (both sides) and the parachute

arming cable, and fires the no-delay initiator to release the crewmember from the seat through the following actions:

a. Foot retraction cables are cut.

b. Shoulder harness straps are severed.

c. Inertia reel is released (so shoulder straps pull free of reel if cutters fail).

d. D-ring cable is cut.

e. Main (personnel) chute lanyard is released from the seat.

NOTE

The release cartridge fires rather loudly as the scramble handle is pulled, and could cause some concern unless expected.

If the crewmember must use the personnel chute after an over-the-side egress, the chute must be manually deployed, using the shoulder D-ring.

Foot Retention System

The crewmember connects his shoe spurs to the seat by attaching them to foot retention fittings at the rear of the footrest. The fittings are cable-connected to reel assemblies, which maintain the cables under slight spring tension during normal operations to permit unrestricted foot movement. When the ejection sequence is initiated, the foot spur cables are automatically reeled in to retract and hold the crewmembers feet in the footrests. The spur cables are severed automatically during man-seat separation, but may be severed while in the airplane by pulling on the scramble handle.

Parachute Beacon

Modified parachute assemblies have a battery powered beacon radio (URT-27) installed. The beacon radio has a minimum operating life of 15 hours, and transmits an automatic signal on a frequency of 243.0 megacycles when turned on. During chute deployment the beacon is automatically turned on when a plastic plug is pulled from between the beacon control switch contacts by a lanyard attached to one of the chute risers. The beacon radio is provided with a 22-inch telescoping antenna and a flexible removable antenna.

Some parachute assemblies have a "push-pull" control added to the radio beacon so the crewmember can prevent automatic arming and transmission during ejection. The push-pull control is attached to the chute harness at the right front shoulder, and pulling the knob out prior to ejection will disable the beacon. If ejection is initiated with the control knob pulled (beacon disabled), the beacon cannot then be turned on until reaching the ground.

Seat Catapult

During the ejection sequence, the catapult gas charge is pressure-fired to initiate seat ejection. The catapult gas charge has a duration of 0.15 seconds, sufficient to raise the seat above the canopy sills, at which point the seat rocket motor is automatically ignited to complete the ejection. The seat

rocket motor provides sufficient thrust and duration (0.5 second) to provide a seat elevation (relative to aircraft flight path) of approximately 300 feet.

When ejection is initiated by the D-ring, the catapult fires through a 0.3-second-delay initiator to provide time for complete canopy jettison prior to initiation of seat ejection. When the T-handle is used to eject, the catapult is fired immediately, as no canopy ballistics are fired when the seat T-handle is pulled. (The canopy must be jettisoned by separate action prior to pulling the T-handle.)

Ejection Seat Drogue Parachute

A 6.5-foot-diameter drogue chute (stowed in the headrest) is deployed by a ballistic-type drogue gun, which fires 0.2 second after the seat catapult is fired. The drogue (stabilizing) chute is connected to the seat at four points by a "bridle" and 10 feet of webbing. Ten seconds after drogue chute deployment (to permit seat deceleration), the lower two bridle lines are automatically severed to stablize the seat, and the seat continues its descent in an upright position. The drogue chute controls rate of descent and attitude until man-seat separation is initiated at approximately 15,000 feet.

Man-Seat Separation

Dual aneroid-actuated initiators start the man-seat separation at a seat altitude of 15,000 feet (or less if ejection occurs at a lower altitude). The aneroid initiators are located on the upper left and right sides of the seat. The face of each unit contains a small dial marked in feet of altitude, or colored red and green. The dial serves as a leak indicator for the aneroid bellows; and will "peg out" (or indicate in red band) if the bellows leak. When the man-seat separation sequence is initiated, the following events occur:

a. Foot retention cables are severed.

b. Drogue chute upper risers are severed (to release drogue chute).

c. Lap belt is released.

d. Shoulder harness is severed and inertia reel is unlocked.

e. D-ring cable is severed.

f. Rotary actuator (butt-snapper) is fired.

When the rotary actuator is fired, the crewmember is forcibly separated from the seat as seat webbing is rapidly retracted by reel rotation. The seat webbing is secured to the front of the seat and passes under the seat pack up to the rotary actuator in the headrest. Rapid tightening (retraction) of the seat webbing as the actuator fires "snaps" the man from the seat with a sling-shot action. The main chute is automatically deployed after man-seat separation by a lanyard.

Main Parachute

The 35-foot-diameter main chute is normally deployed automatically after man-seat separation, but may be deployed manually. Automatic deployment is initiated by a drogue gun in the upper left corner of the pack. The drogue gun is connected to the seat by a lanyard, and is fired as man and seat separate. Accidental deployment is prevented by a lanyard housing, disconnected from the seat at separation. The

STABILIZED SEAT EJECTION TRAJECTORIES

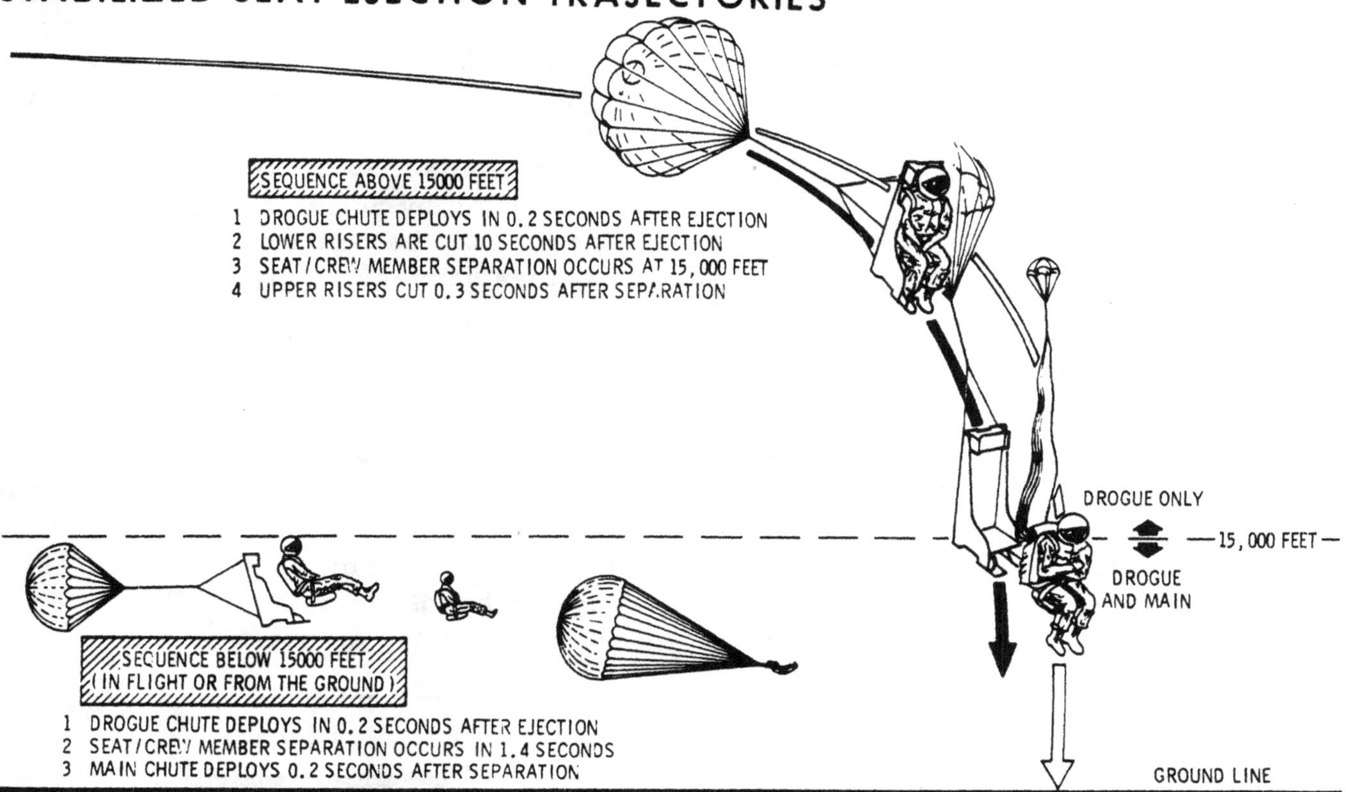

SEQUENCE ABOVE 15000 FEET
1 DROGUE CHUTE DEPLOYS IN 0.2 SECONDS AFTER EJECTION
2 LOWER RISERS ARE CUT 10 SECONDS AFTER EJECTION
3 SEAT/CREW MEMBER SEPARATION OCCURS AT 15,000 FEET
4 UPPER RISERS CUT 0.3 SECONDS AFTER SEPARATION
DROGUE ONLY
15,000 FEET
DROGUE AND MAIN
SEQUENCE BELOW 15000 FEET (IN FLIGHT OR FROM THE GROUND)
1 DROGUE CHUTE DEPLOYS IN 0.2 SECONDS AFTER EJECTION
2 SEAT/CREW MEMBER SEPARATION OCCURS IN 1.4 SECONDS
3 MAIN CHUTE DEPLOYS 0.2 SECONDS AFTER SEPARATION
GROUND LINE

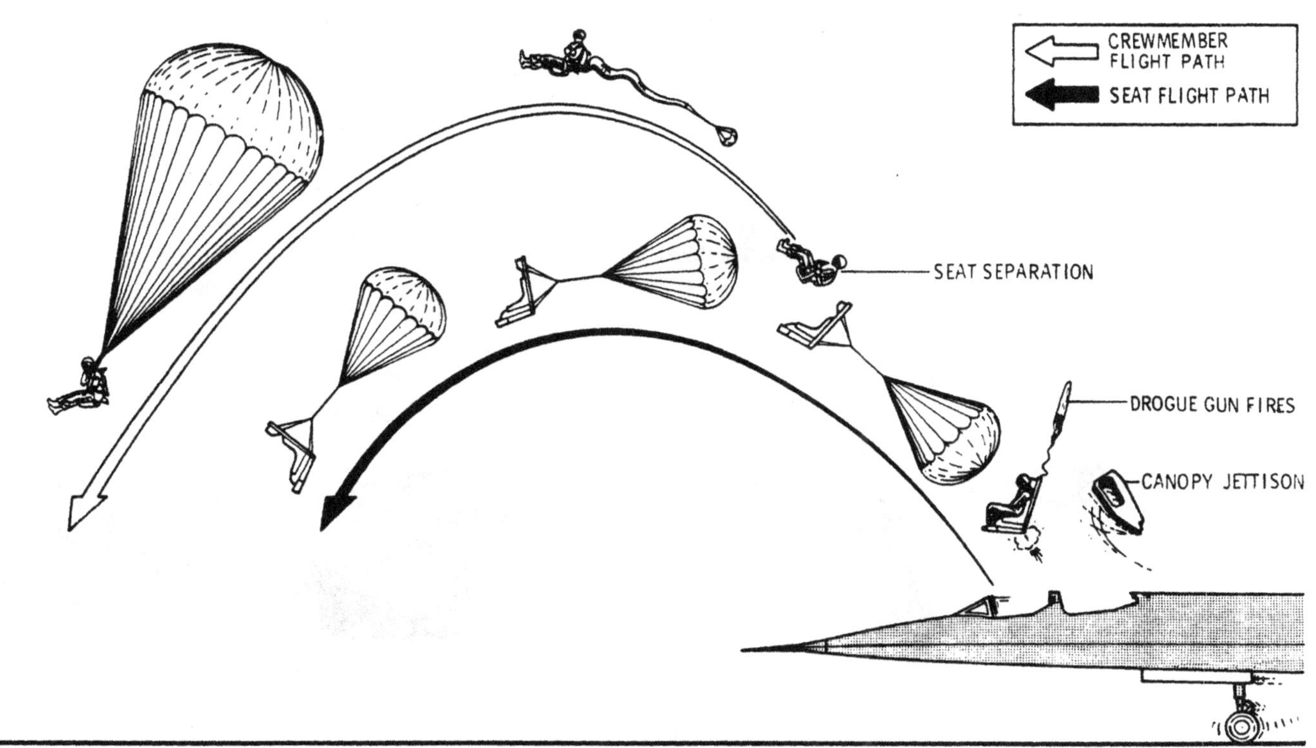

Figure 1-25

F203-96(b)

chute may be manually deployed by pulling the shoulder D-ring (on upper left of shoulder harness). The drogue gun is not fired when the chute is manually deployed.

NOTE

> The main chute must be deployed manually if the crewmember has used the scramble handle for an "over-the-side" exit from the aircraft, as the main chute lanyard is released from the seat when the scramble handle is pulled.

Survival Kit

The survival kit is a hard-shell-type seat pack, containing an emergency oxygen supply and survival equipment. (The seat cushion may contain a 130-foot nylon tape for descending from a tree.) The emergency oxygen outlets (quick disconnects) are on the aft right corner of the kit, and two small oxygen quantity gages are at the forward right corner. (See figure 1-26). Emergency oxygen is automatically turned on during ejection by a lanyard attached to the seat rail and may be manually turned on at any time by pulling the green apple. The kit may be released from the crewmember harness by pulling the kit release handle at the right side of the kit. The kit release handle should be pulled while descending in the main parachute except when a tree landing is anticipated. When the release handle is pulled, the kit will fall away to the end of a 25-foot lanyard. The lanyard is disconnected from the harness if the release handle is pulled while seated normally in the aircraft.

WARNING

> Pulling the kit release handle while in the aircraft will disconnect both normal and emergency oxygen supplies.

SURVIVAL KIT

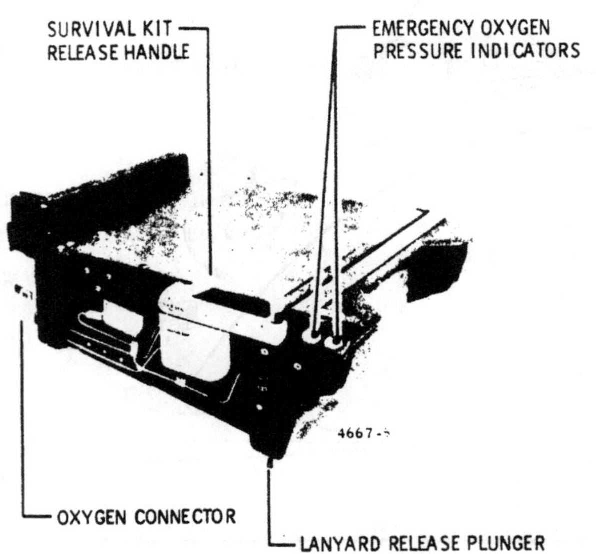

Figure 1-26

PRIMARY EJECTION SEQUENCE

The D-ring is pulled to initiate the primary ejection sequence, which includes automatic canopy jettison. Pulling the D-ring fires an initiator on the front apron of the seat. Gas pressure from the D-ring initiator is reported directly to the foot retractor, the shoulder harness reel, and the canopy unlatch thruster. Pulling the D-ring also arms the secondary ejection sequence.

When the canopy unlatch thruster reaches full throw, gas pressure is ported to the canopy seal cutter and canopy removal thruster. Another initiator (pin-fired as the canopy is raised) ports pressure through the jettison valve to a 0.3-second delay initiator, which allows time for complete canopy jettison prior to firing the catapult initiator. Gas pressure from the delay initiator provides pressure to fire the catapult initiator and "arm" the 0.2 and 1.4-second delay initiators in the drogue chute. The seat rocket motor is ignited as the catapult raises the seat above the canopy sills. The 0.2-second-delay initiator fires the drogue gun to deploy the drogue chute. The 1.4-second-delay initiator fires to arm the drogue chute lower riser cutters and the dual aneroid actuators. The lower riser cutters are armed through a 10-second delay, which provides time for seat deceleration before severing the lower riser lines. When the lower risers are severed, the seat is stabilized in an upright position by the drogue chute. The dual aneroid actuators initiate the man-seat separation sequence at an altitude of 15,000 feet (or less if ejection occurs at lower altitude). At completion of man-seat separation, the main chute is automatically deployed.

SECONDARY EJECTION SEQUENCE

In the event of primary ejection sequence malfunction, the ejection seat T-handle is pulled to initiate the secondary ejection sequence. The secondary sequence does not include canopy jettison.

NOTE

> The secondary ejection sequence can not be initiated before the D-ring is pulled.

When the T-handle is pulled, the catapult initiator fires immediately, and gas is ported to the two delay initiators in the drogue chute. Action downstream of the 1.4-second-delay initiator in the drogue chute is identical to that described under primary ejection sequence

EGRESS (COORDINATION) SYSTEM

An egress light system installed in the aircraft permits bailout coordination between the pilot and FCO in the event that interphone communication is impossible. With this system the pilot has the capability to order bailout and check compliance by a light signal. Power for the system is furnished by the essential dc bus.

Egress Lights and Switches

The pilot's instrument panel contains a guarded toggle switch, labeled FCO BAILOUT, with three positions; STAY (down), off (center), and GO (up). The pilot's annunciator panel contains an FCO CLEAR light. The FCO's instrument panel contains two lights, labeled STAY and BAILOUT. When the pilot actuates the FCO BAILOUT switch to

ANNUNCIATOR PANELS

A HYD SYS LOW	B HYD SYS LOW	
L HYD QTY LOW	R HYD QTY LOW	NO. 1 OXY PRESS LOW
	MANUAL INLET	NO. 2 OXY PRESS LOW
		FUEL QTY LOW
		SURF LIMITER
L FUEL PRESS LOW	R FUEL PRESS LOW	SAS CHNL OUT
L OIL QTY LOW	R OIL QTY LOW	FCO CLEAR
L OIL TEMP	R OIL TEMP	ANTI-SKID OUT
L XMFR RECT OUT	R XMFR RECT OUT	VENTRAL FIN
L GEN OUT	R GEN OUT	BUS TIE OPEN
L AIR SYS OUT	R AIR SYS OUT	INST ON EMERG PWR
		EMERG BATT ON
INER PLTFM OFF	CAN UNLKD	N QTY LOW

FORWARD COCKPIT

STAY
NO. 1 OXY PRESS LOW
NO. 2 OXY PRESS LOW
FCS LIQUID COOL OUT
FCS RACK OVERHEAT
MISSILE BAY OVERHEAT
MTC BAY OVERHEAT
MTC BAY PRESS LOW

AFT COCKPIT

Figure 1-27

STAY or GO position, either the STAY or BAILOUT light on the FCO instrument panel will illuminate. The FCO CLEAR light on the pilot's annunciator panel is wired directly to the FCO ejection seat tracks and illuminates when the FCO seat is ejected.

EMERGENCY EQUIPMENT

MASTER WARNING SYSTEM

An annunciator panel is mounted on the pilot's lower instrument panel. The panel contains individual warning lights that indicate malfunctions or failure of equipment and systems. The master warning system does not include the fire warning and landing gear unsafe lights. Illumination of any individual annunciator light also illuminates the red master caution light on the upper portion of the instrument panel. Once illuminated, the master caution light will be extinguished (reset) when depressed so that another malfunction will illuminate the light. The individual annunciator panel light will remain illuminated until the malfunction is corrected. Warning lights are automatically dimmed when the instrument panel lights are on. Power to the lights is furnished by the essential dc bus.

NACELLE FIRE WARNING SYSTEM

A fire warning system detects and indicates the presence of a fire in the engine nacelles. A hot spot anywhere along the length of the detection circuit will cause the light of that particular nacelle to illuminate. The lights are located on the pilot's instrument panel above the respective column of instruments pertaining to each engine.

Nacelle Fire Warning Lights

Left and right nacelle FIRE warning lights are located on the top right side of the pilot's instrument panel. These lights illuminate when nacelle temperature at the turbine or at the afterburner exceeds 1200° (±50°) F. Power for the lights is furnished by the dc essential bus.

Fire-Warning Lights Test Switch

A fire-warning lights test switch, installed on the right side of the pilot's instrument panel, has three positions, FIRE LT (up), center (spring loaded) off, and WARN LT (down). When the switch is moved to the FIRE LT position, the nacelle fire warning circuit is activated and the L and R nacelle fire warning lights should illuminate. When the switch is placed in the WARN LT position, the upper and lower annunciator panel, the master caution and the LOW KEAS lights illuminate.

Section II

Normal Procedures

TABLE OF CONTENTS

CREW COORDINATION

Crew coordination is paramount to mission success and safety of flight during all phases of operation. Communication between crewmembers should be continuous when accomplishing checklists. Verbal coordination between crewmembers will be required prior to any of the following actions:

a. A crewmember going off interphone.

b. A crewmember going off aircraft oxygen system or opening faceplate.

c. FCO making a change to the programmed mission or steering reference points, or changing navigational system mode.

d. A crewmember changing the attitude reference.

e. A crewmember checking the warning system.

PREPARATION FOR FLIGHT

FLIGHT RESTRICTIONS

Refer to Section V for operating restrictions and limitations.

FLIGHT PLANNING

Refer to Appendix I.

TAKEOFF AND LANDING DATA

Refer to Appendix I.

WEIGHT AND BALANCE

For detailed loading information, refer to Handbook of Weight and Balance data. Before each flight, check takeoff and anticipated landing gross weights and weight-and-balance clearance (Form 365F).

AIRCRAFT STATUS

Refer to AF Form 781 for engineering, servicing, and equipment status.

EXTERIOR INSPECTION

Because it is not practical for the flight crew to perform an exterior inspection while wearing pressure suits, the exterior inspection should be accomplished by other qualified personnel.

PREFLIGHT CHECK

ENTRANCE

Ladder platform stands which overhang the chine are used for entering the cockpits.

a. Check external power connected and on.

PERSONAL EQUIPMENT HOOKUP - Stabilized Seat

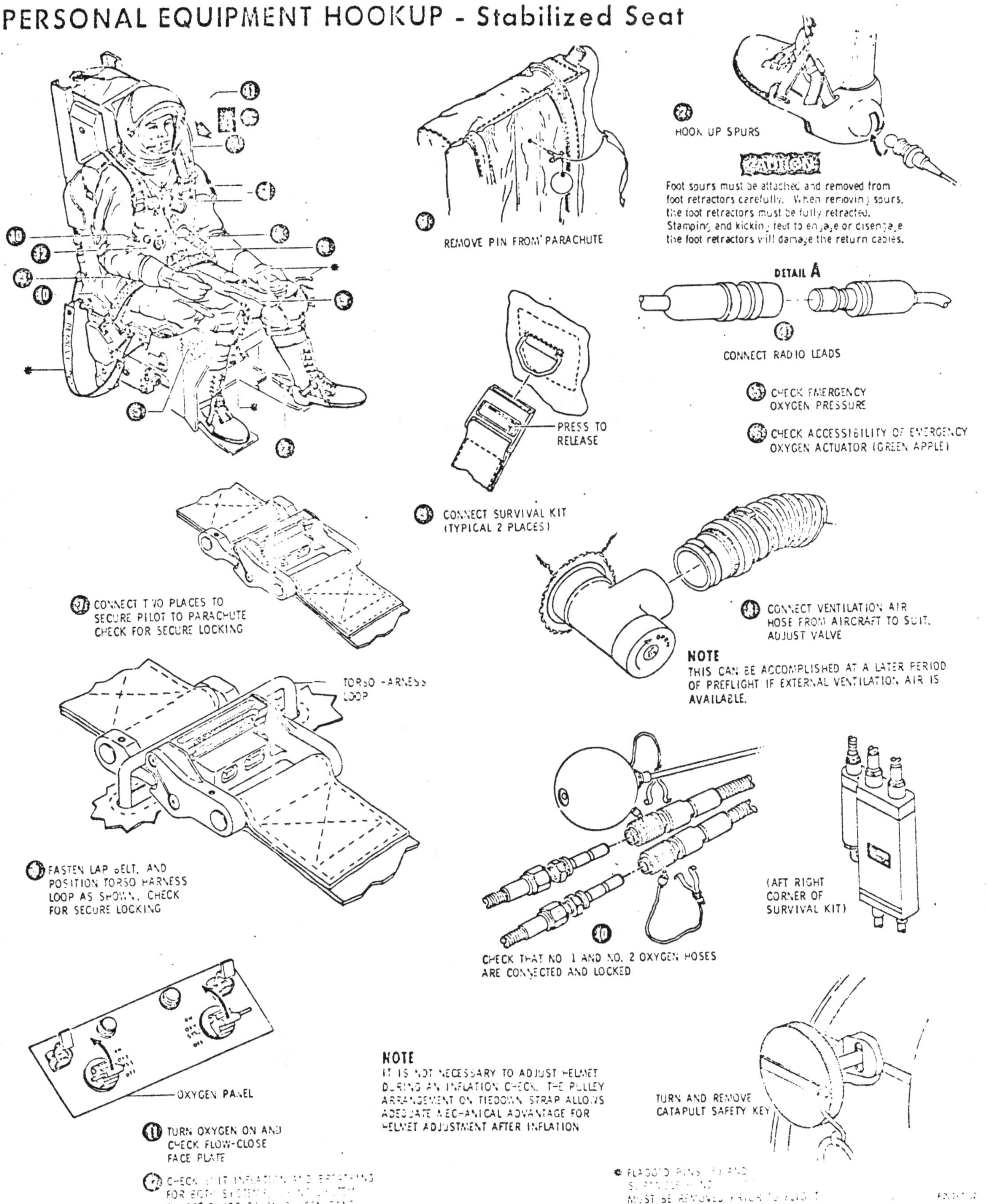
REMOVE PIN FROM PARACHUTE
HOOK UP SPURS
CAUTION
Foot spurs must be attached and removed from foot retractors carefully. When removing spurs, the foot retractors must be fully retracted. Stamping and kicking feet to engage or disengage the foot retractors will damage the return cables.
DETAIL A
CONNECT RADIO LEADS
CHECK EMERGENCY OXYGEN PRESSURE
CHECK ACCESSIBILITY OF EMERGENCY OXYGEN ACTUATOR (GREEN APPLE)
PRESS TO RELEASE
CONNECT SURVIVAL KIT (TYPICAL 2 PLACES)
CONNECT TWO PLACES TO SECURE PILOT TO PARACHUTE CHECK FOR SECURE LOCKING
CONNECT VENTILATION AIR HOSE FROM AIRCRAFT TO SUIT. ADJUST VALVE
NOTE
THIS CAN BE ACCOMPLISHED AT A LATER PERIOD OF PREFLIGHT IF EXTERNAL VENTILATION AIR IS AVAILABLE.
TORSO HARNESS LOOP
FASTEN LAP BELT, AND POSITION TORSO HARNESS LOOP AS SHOWN. CHECK FOR SECURE LOCKING
(AFT RIGHT CORNER OF SURVIVAL KIT)
CHECK THAT NO 1 AND NO. 2 OXYGEN HOSES ARE CONNECTED AND LOCKED
OXYGEN PANEL
TURN OXYGEN ON AND CHECK FLOW-CLOSE FACE PLATE
NOTE
IT IS NOT NECESSARY TO ADJUST HELMET DURING AN INFLATION CHECK. THE PULLEY ARRANGEMENT ON TIEDOWN STRAP ALLOWS ADEQUATE MECHANICAL ADVANTAGE FOR HELMET ADJUSTMENT AFTER INFLATION
TURN AND REMOVE CATAPULT SAFETY KEY
ON-OFF SWITCHES ON OXYGEN PANEL
MUST BE REMOVED PRIOR TO FLIGHT

Figure 2-1 (Sheet 1 of 2)

PERSONAL EQUIPMENT HOOKUP - Shirt Sleeve Flight

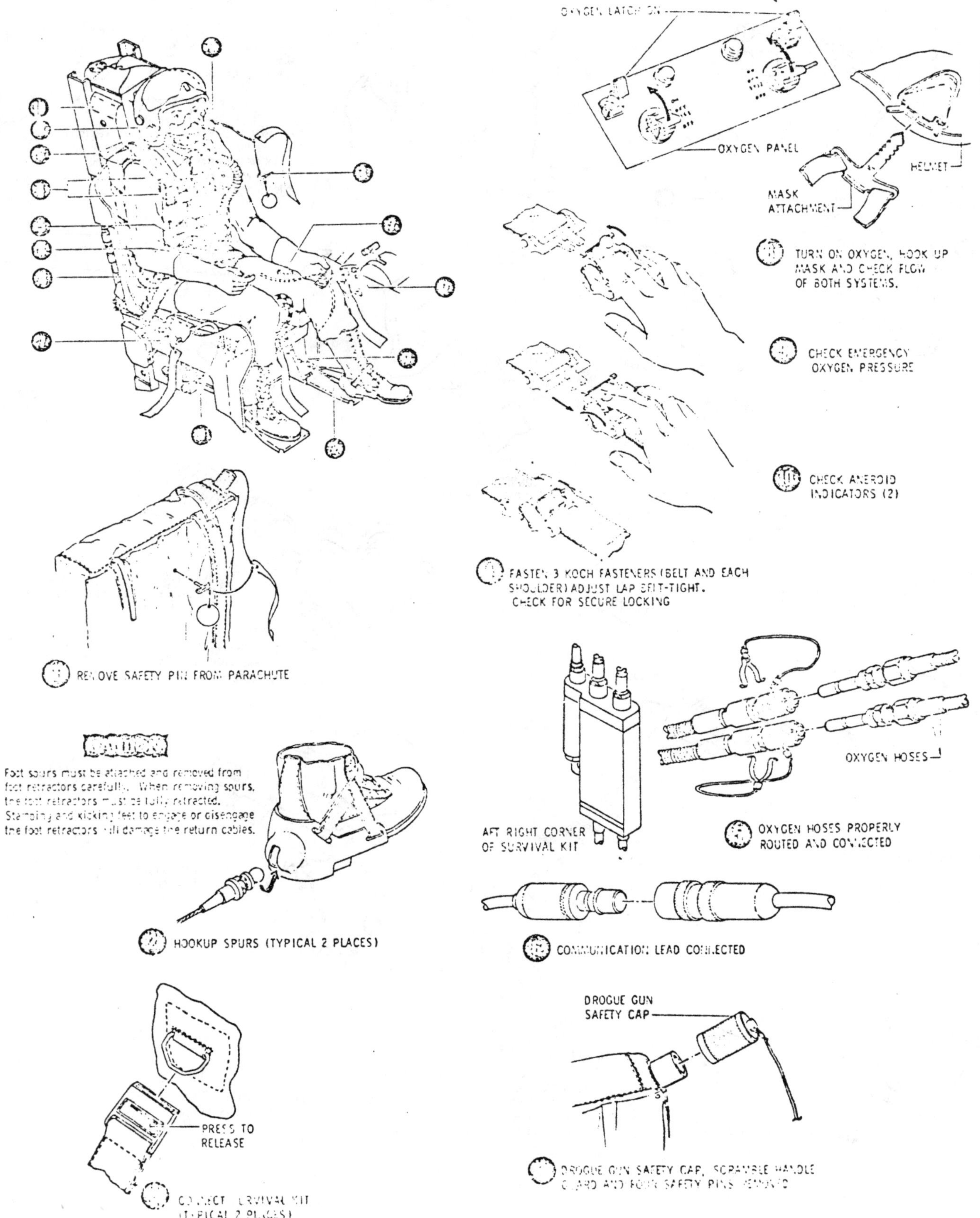
OXYGEN LATCH ON
OXYGEN PANEL
HELMET
MASK ATTACHMENT
TURN ON OXYGEN, HOOK UP MASK AND CHECK FLOW OF BOTH SYSTEMS.
CHECK EMERGENCY OXYGEN PRESSURE
CHECK ANEROID INDICATORS (2)
FASTEN 3 KOCH FASTENERS (BELT AND EACH SHOULDER) ADJUST LAP BELT-TIGHT. CHECK FOR SECURE LOCKING
REMOVE SAFETY PIN FROM PARACHUTE
CAUTION
Foot spurs must be attached and removed from foot retractors carefully. When removing spurs, the foot retractors must be fully retracted. Stamping and kicking feet to engage or disengage the foot retractors will damage the return cables.
HOOKUP SPURS (TYPICAL 2 PLACES)
AFT RIGHT CORNER OF SURVIVAL KIT
OXYGEN HOSES
OXYGEN HOSES PROPERLY ROUTED AND CONNECTED
COMMUNICATION LEAD CONNECTED
DROGUE GUN SAFETY CAP
DROGUE GUN SAFETY CAP, SCRAMBLE HANDLE GUARD AND FOUR SAFETY PINS REMOVED
PRESS TO RELEASE
CONNECT SURVIVAL KIT (TYPICAL 2 PLACES)

Figure 2-1 (Sheet 2 of 2)

BEFORE ENTERING COCKPIT

> Steps without special notation apply to the forward cockpit and steps with a circle around the number apply to the aft cockpit only. All items marked with a ▲ must be checked in both the pilot and FCO cockpits.

▲ 1. Ejection seat and canopy safety pins - Installed.

▲ 2. All circuit breakers - In.

INTERIOR CHECK

1. Accomplish and check personal equipment hookup. (See figure 2-1.) Hookup will be performed by personal equipment personnel.

Left Console - Forward Cockpit

1. Oxygen pressure warning light ground checkout switch - OFF.
2. Fuel derich ground test switch - OFF.
3. Oxygen system switches - As required.
4. Interphone control panel - As desired.
5. UHF radio - Set.
6. Suit heat rheostat - COLD.
7. Trim power switch - ON.
8. Aux trim control switch - Centered.
9. Aux trim selector switch - STICK TRIM, guard down.
10. Cockpit pressure dump switch - OFF, guard down.
11. Bay air shutoff switch - ON.
12. Cockpit temperature override switch - AUTO.
13. Cockpit temperature rheostat - As desired.
14. L and R air system switches - ON.
15. Throttles - OFF.
16. Throttle friction lever - As desired.
17. TEB counters - Checked. (Both counters should read 12.

Instrument Panel - Forward Cockpit

1. Landing gear lever - DOWN.
2. Landing gear light test button - Press.
3. Faceplate heat rheostat - As desired.
4. Pitot heat switch - OFF.
5. Landing and taxi lights switch - OFF.
6. Brake switch - ANTI-SKID.
7. Emergency spike and restart switches - OFF (guards down).

8. Spike and forward bypass controls - AUTO.

9. EGT trim switches - AUTO

10. Aft bypass switches - CLOSE.

11. Aft bypass position lights - Checked. (Press to test)

12. Oxygen quantity indicator - Check both systems full.

13. Clock - Set.

14. Standby altimeter - Set to field elevation.

15. Standby airspeed indicator - Checked.

16. Special weapon release lock switch-SAFE.

17. Will-to-fire switch - OFF.

18. Drag chute switch - JETTISON.

19. Air refuel ready switch - OFF.

20. Compressor inlet temperature gage - Check needles together and ambient temperature indicated.

21. Standby attitude indicator - Erecting.

    Pull the cage knob to erect the instrument, then release the knob. The instrument will erect and then seek a 7° nose-down and 0° roll position (if the aircraft is level) within five minutes.

    NOTE

    A jitter of ± 1/2° in the pitch attitude is acceptable. This jitter may occur when at any pitch attitude.

22. Compressor inlet pressure (CIP) gage - Checked.

23. Airspeed - Mach indicator test button - Press. Mach tape should indicate 2.44 Mach and knots tape should indicate 800 KEAS.

24. Vertical altimeter check.

    a. Test button - Press. Both altitude tapes should indicate 74,000 feet.

    b. Barometric knob - Set as desired.

25. Surface limiter control handle - Pulled.

    Check SURF LIMITER warning light off.

26. Manual gear release handle - Stowed.

27. Interior and exterior lights panel- As desired.

28. Engine instruments - Checked.

29. Fuel transfer switch - OFF.

30. Fuel panel - Checked.

    a. Fuel system panel lights test- Checked.

       Check all pump lights, tank empty lights, crossfeed light and pump release light on.

    b. Crossfeed switch - Press ON.

    c. Boost pump switches - Checked.

       Press pump switches 1 thru 6 ON in sequence.

d. Pump release switch - Checked.

Check tank 6 pump switch released when pump release pressed.

e. Crossfeed switch - OFF.

f. Takeoff tank boost pump lights - ON. (Tanks 1, 2 and 4.)

31. Fuel dump switch - OFF (guard down).

32. Fuel quantity indicating system - Checked.

a. TEST - Check that quantity indication decreases and confirm with FCO that aft cockpit fuel indicator needles travel toward zero.

b. Individual tank quantities - Check.

c. Compare individual tank readings with total (within 1200 lbs).

d. Compute and verify c.g. if fuel load is less than maximum tank capacity.

33. Fuel derich switch - ARM.

34. Emergency fuel shutoff switches - Fuel on (guards down).

35. Instrument power switch - NORM.

36. Battery switch - ON.

37. Generator switches - OFF.

The generator circuitry is such that the switches may either be off or on; however, it is more advantageous at this point for the switches to be off.

38. Hydraulic reserve oil switch - NORM.

39. Igniter purge switch - Off.

40. Fire and warning lights test switch - TEST, then Off.

Right Console - Forward Cockpit

1. SAS channel switches - OFF.

If all SAS panel lights do not illuminate:

2. SAS lights test switch - Press (all SAS panel lights should illuminate).

3. Autopilot - OFF.

4. Backup pitch damper switch - NORMAL.

5. Pitch logic override switch - OFF, guard down.

6. Heading vertical reference selector switch - As desired.

7. Yaw logic override switch - OFF, guard down.

8. Canopy seal pressurization lever - OFF.

9. TACAN - OFF; set channel.

10. ILS - OFF.

11. MA-1 compass control switch - SLAVED.

Check latitude set, fast slave to proper heading and center synchronization needle.

12. Cabin altimeter - Checked.
13. Defog lever - OFF.

Aft Cockpit Check

1. All circuit breakers - In.
2. Missile glycol heater - OFF.
3. Camera pod cooling - OFF.
4. I-bay cooling - OFF.
5. Interior lighting rheostats - As necessary.
6. Fire control lighting - DAY.
7. MA-1 compass control switch - SLAVED.
8. UHF radio(s) - Both set.
9. Interphone panel - As desired.
10. TACAN - OFF.
11. IFF - OFF.
12. Oxygen system switches - As required.
13. HF radio - OFF.
14. SIF panel - As desired.
15. FCS liquid cooling switch - OFF.
16. Faceplate heat rheostat - OFF.
17. FCS instruments - OFF.
18. FCS power switch - OFF.
19. Computer power switch - OFF.
20. Platform power switch - OFF.
21. MRS power switch - ON.
22. INS/AUX switch - INS.
23. INS - As desired.

CAUTION

Do not turn on the FCS liquid cooling system switch during INS platform alignment.

24. Computer - As desired.
25. I-bay cooling - RAM.

BEFORE STARTING ENGINES

▲ 1. Interphone - Check.

▲ 2. BAIL OUT light test - Coordinate light test with pilot.

▲ 3. Nitrogen quantity and indicating system - Check. Coordinate with FCO.

▲ 4. MA-1 compass control transfer button - Press. Check button illumination.

▲ 5. TACAN control transfer button - Press. Check illumination.

▲ 6. TDI and integrated instruments - Check.

DANGER AREAS

WARNING

THE ENGINE TURBINE SECTION AND NACELLE INTAKE AND EXHAUST AREAS CAN BE DANGEROUS. KEEP CLEAR.
ENGINE NOISE CAN DAMAGE HEARING PERMANENTLY. DURING ENGINE RUNUP, USE EAR PLUGS AND MUFFS WITHIN 400 FEET DURING AFTERBURNER OPERATION AND WITHIN 200 FEET DURING MILITARY POWER OPERATION.

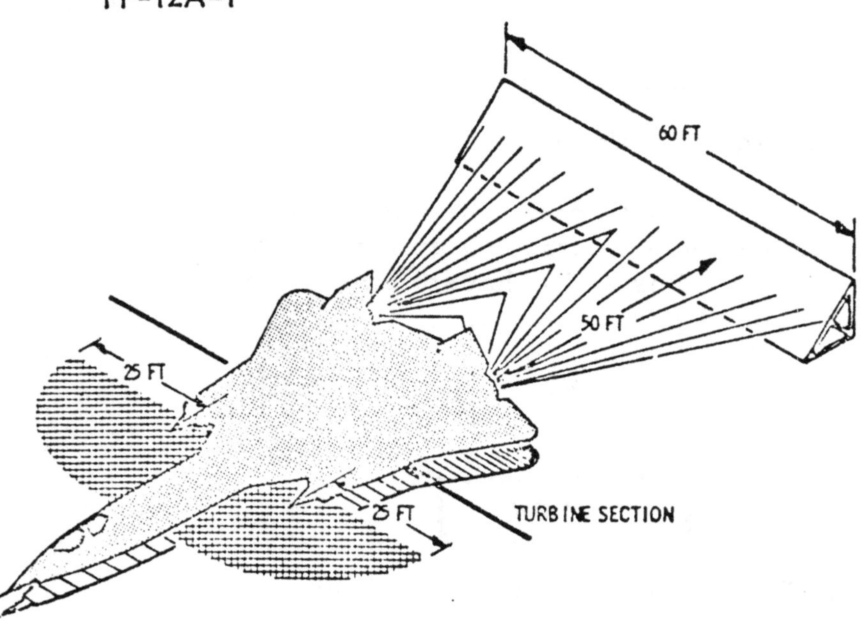

ESTIMATED JET WAKE DIAGRAMS

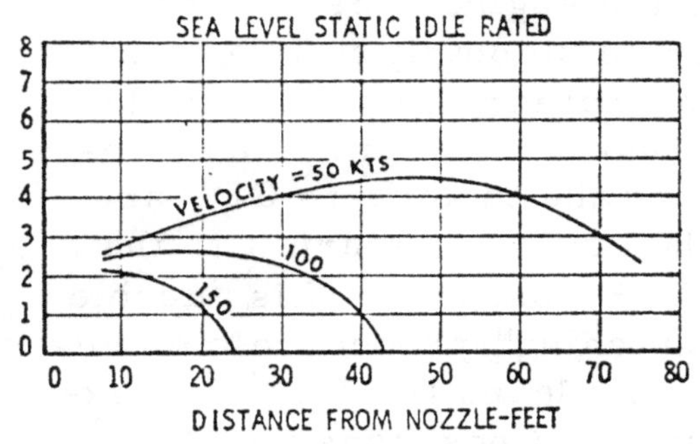

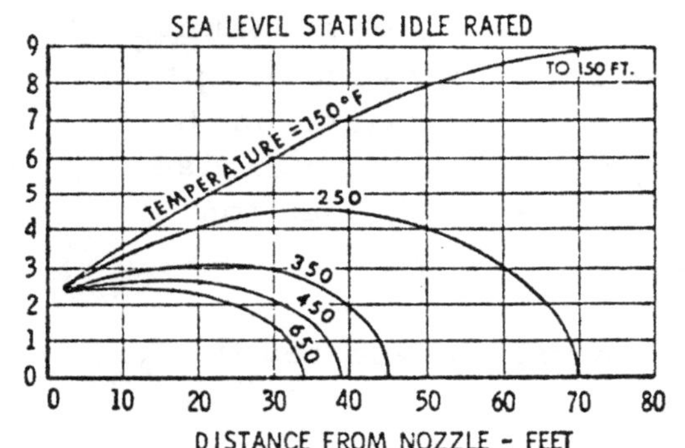

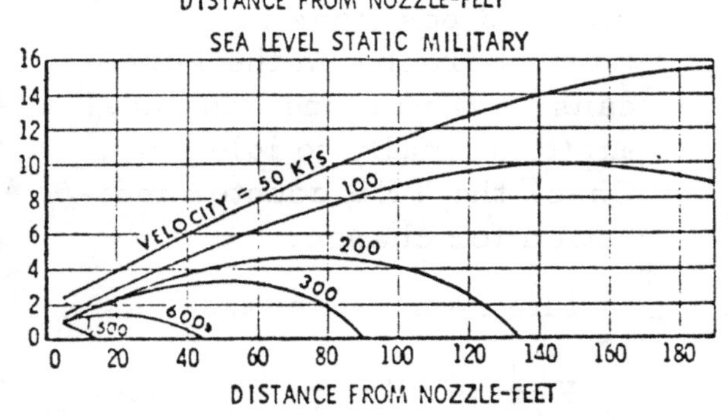

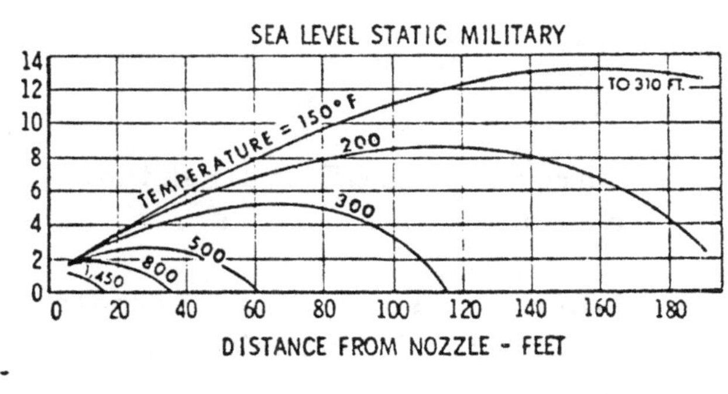

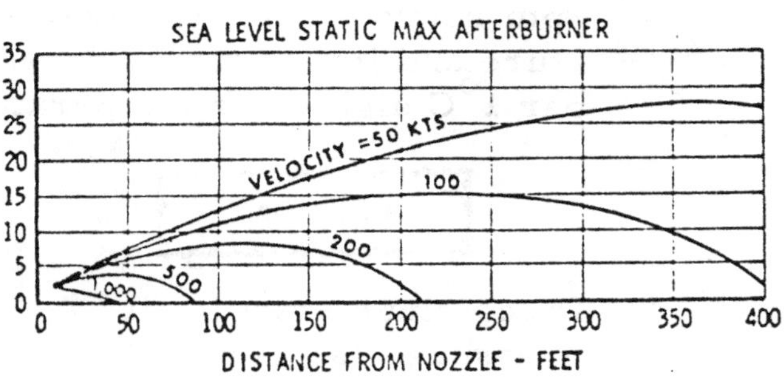

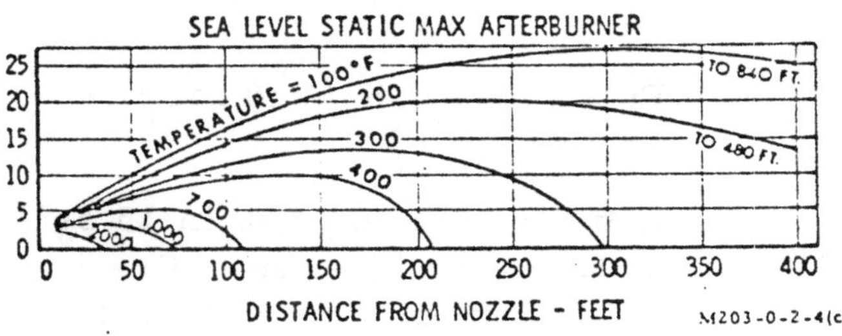

M203-0-2-4(c)

Engine Run Danger Areas

Figure 2-2

STARTING ENGINES

1. Ready to start - Confirmed.

   Confirm with crew chief that wheels are chocked, starting unit connected and intakes and exhausts are clear. Ground personnel using interphone equipment shall be in position to observe exhaust nozzle and nacelle inspection panels during start.

   CAUTION

   - There is no parking brake and brakes are inoperable until hydraulic pressure is available.
   - Do not move control stick until at least 1500 psi hydraulic pressure can be maintained on the A or B system gages; otherwise, a control system inspection will be necessary.

Either engine may be started first.

2. First engine - Started.

   a. Pilot - Signal for engine rotation.

   b. Throttle - IDLE at first indication of rpm increase.

      When necessary, an alternate technique of advancing the throttle at 1000 rpm may be used for some engines.

   c. Fuel flow - Checked for increase.

   d. Ignition - Verify within 15 seconds. If no ignition indicated by an rpm increase and a rise in EGT within 15 seconds, move throttle to off and continue cranking engine for 30 seconds at 1000 rpm.

      CAUTION

      In case of a false start, use engine clearing procedure, this section.

   e. Ground starting unit - Signal for disconnect at 3200 rpm.

   f. EGT - Check during acceleration, if 565°C is reached or exceeded, move throttle to off.

      NOTE

      If the engine does not accelerate smoothly to idle rpm, but appears to "hang" in the 2600 to 2800 rpm range, retard the throttle to OFF and then quickly return it to the IDLE position. This "double clutching" procedure momentarily leans the fuel/air mixture and positions the flame front correctly in the burner cans so the engine can accelerate normally to idle rpm. Check the TEB counter indication for change.

   g. Idle rpm - Checked.

      Engine idle speed is 3600 ± 50 rpm below 32°C air temperature and increases 50 rpm per 1°C higher air temperature.

      CAUTION

      Discontinue start if oil pressure rise is not observed by the time IDLE rpm is obtained.

h. Engine and hydraulic instruments - Check normal indications.

(1) Fuel flow - Approximately 4500 pounds per hour.

(2) EGT - 350° to 565°C.

(3) Oil pressure gage - Checked.

Minimum oil pressure 35 psi.

(4) Hydraulic system pressures - Checked.

i. Forward bypass door - Confirmed open.

3. Start second engine - Started.

Use same start procedure as for first engine.

4. TEB - Checked.

5. Right generator switch - ON, light off.

Check R GEN OUT light extinguished to confirm that generator is engaged.

6. Left generator switch - ON, light off.

Check L GEN OUT light extinguished to confirm that generator is engaged.

7. External power and ground air - Disconnected.

Signal ground crew to disconnect.

8. BUS TIE light - Check off.

9. Spike and door position indicators - Checked.

a. Spikes - 0 in. aft.

b. Forward bypass door - 100% open.

CLEARING ENGINE

Cool the engine and remove trapped fuel and vapor as follows:

1. Throttle - OFF.

CAUTION

Allow a minimum of 1 minute for fuel drainage and coast down before motoring engine.

2. Start cart - Engage and motor engine for at least 30 seconds and until EGT is below 150°C.

Signal ground crew to motor engine at 1000 rpm. Crew chief will advise pilot when engine is clear and ready for start.

CAUTION

Do not signal for cranking the engine with the fuel shut off switch in the fuel OFF position except in case of emergency. Damage to the engine may result with the engine fuel-hydraulic system off.

BEFORE TAXIING

(1) HF radio - ON.

WARNING

Do not transmit on ground until safe to do so.

2. IFF - STBY.

▲ 3. UHF and Nav radios - ON.

▲ 4. Ejection seat and canopy pins - Removed and stowed.

▲ 5. Canopy - Close and lock.

▲ 6. Canopy seal pressurization lever - ON.

Check CAN UNLKD light off.

A cockpit will not pressurize unless seal is inflated.

CAUTION

The canopies should be opened or closed only when the aircraft is completely stopped. Maximum taxi speed with the canopy open is approximately 40 knots. Gusts or severe wind condition should be considered as a portion of the 40-knot limit taxi speed.

7. FCS liquid cooling switch - ON at idle rpm after engines started and exercised.

Turning on Freon package at idle rpm minimizes effects of current transients when compressor motor starts.

8. FCS instruments - ON.

9. FCS power switch - STDBY.

10. INS - As briefed.

11. SAS and autopilot - Checked.

a. SAS channel engage switches - ON.

b. SAS recycle lights - Press; lights should go out.

NOTE

The SAS monitor recycle lights must be pressed off before A or B recycle lights can be pressed off.

c. Cycle elevons in Pitch and Roll and check for abnormal control surface oscillation or vibration.

d. Autopilot pitch and roll engage switches - ON.

e. Mach hold switch - ON.

f. Altitude hold switch - ON; Mach hold should disengage.

g. Heading hold switch - ON.

h. CSC switch - Depress; altitude hold and heading hold should disengage.

i. Stick trigger - Pull; pitch and roll switches should disengage.

j. SAS roll and yaw channel engage switches - OFF.

12. Pitch A and B control transfer valve function - Checked and pitch SAS channels OFF.

With pitch SAS channel engage switches ON:

a. Pitch SAS A channel - Press recycle light off.

b. Pitch autopilot engage switch - ON.

c. Pitch autopilot trim wheel - Rotate at least 30 degrees of wheel rotation. After 10 seconds, note that SAS pitch channel light does not illuminate, and that slow-speed trim motor is functioning properly as indicated by the pitch trim indicator.

d. Pitch SAS A channel engage switch - OFF.

e. Repeat steps (a), (b), (c) using pitch SAS channel B.

f. Pitch SAS B channel engage switch - OFF.

NOTE

- The autopilot pitch engage switch should automatically move to OFF.
- Illumination of either pitch SAS channel recycle light during this test indicates failure of a flight control transfer valve to operate properly, and the flight should be aborted.

13. Trim - Checked.

Check pitch, roll and yaw trim and set to zero. Confirm that direction of movement corresponds with indication in the following sequence: nose up, nose down, left roll, right roll, nose left and nose right. Check RH rudder synchronizer.

14. Flight control system - Checked.

Individually check each axis in both directions and confirm correct deflection of control surfaces in the following sequence; nose up, nose down, left roll, right roll, nose left and nose right.

15. Shotgun cartridge - Checked.

Confirm left and right cartridges engaged.

16. A/R system - Checked.

a. Air refuel switch - READY.

Check ready light ON, confirm doors open and light on, toggles unlatched.

b. Set refuel switch MANUAL, confirm door open, receptacle light ON, toggles latched.

c. Actuate stick trigger, confirm toggle latch movement.

d. Set refuel switch OFF, light OUT, confirm door closed.

17. Fuel derich - Checked and armed.

a. Set both engines 400 rpm above idle rpm.

b. Derich test switch - Left.

When EGT indication exceeds 860°C:

c. Verify EGT gage light on and fuel derich light on.

d. Note engine speed decreases 50 to 400 rpm.

e. Recycle fuel derich switch to ARM.

Verify engine rpm returns to 400 rpm above the idle rpm and EGT indication is normal.

f. Repeat b, c, d, e for right derich check.

g. Throttles idle.

A18. Flight instruments and navigation equipment - Checked and set.

NOTE

The radar must be OFF or in STDBY if the ILS is to be used. Do not turn radar ON while ground personnel are within 250 feet in line with radiation pattern.

19. Exterior lights - ON.

a. Fus and tail lights switch to BRT and STEADY.

b. Rotating beacon switch in ANTI-COLLISION position.

c. Landing lights - Set.

20. Periscope - Checked and set.

21. Nosewheel steering - Engaged.

Check STEER ON engage light illuminated.

22. Brakes - Checked.

Check brake pedals for evidence of L & R system pressure; however, normal feel does not necessarily indicate braking action.

23. Panels secured, missile bay doors closed, gear pins removed - Secured and removed.

Crewchief confirms all panels and doors secured. Crewchief disconnects interphone and displays landing gear down lock pins.

24. Taxi clearance - Obtained.

Obtain taxi clearance and altimeter setting from control tower.

25. Chocks - Removed.

Pilot signals crew chief to remove chocks.

TAXIING

Observe crewchief for clearance signal to taxi. See figure 2-3 for minimum turning radius and taxi clearance.

CAUTION

All taxiing and turns should be accomplished at low speeds to minimize side loads on the landing gear. Fast taxiing should also be avoided to prevent excessive brake and tire heating and wear.

TURNING DIAGRAM

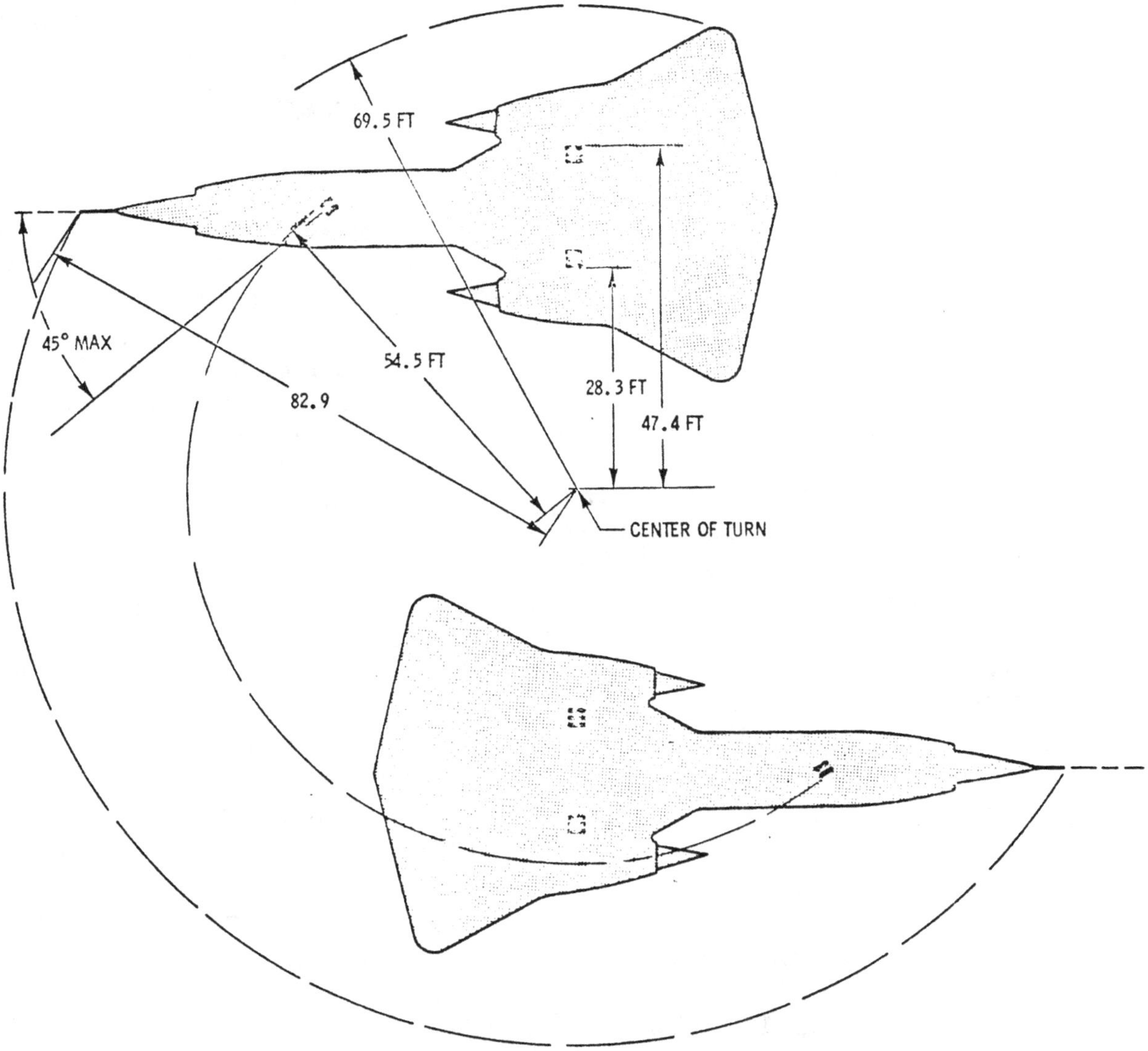

NOTE

- 101.9 FT MINIMUM RUNWAY WIDTH REQUIRED FOR 180-DEGREE TURN (MAIN GEAR WHEELS ON EDGE OF RUNWAY AT START OF TURN).
- SEE FIGURE 5-4 FOR MAXIMUM TURNING SPEEDS.

Figure 2-3

1. Braking and nosewheel steering - Checked.

   a. When clear of obstacles check for normal operation.

   b. Disengage nosewheel steering and check individual brake operation and dragging brakes. Re-engage NWS and check steering operation.

NOTE

Check STEER ON engage light operation when engaging and disengaging nosewheel steering.

2. Turn-and slip instrument - Checked.

   Check turn-and-slip indicator for turn needle deflection in the direction of turn while taxiing and for ball free in race.

BEFORE TAKEOFF

1. EGT trim - Checked

   a. Wheels - Chocked.

   b. Brakes - Apply.

   One engine at a time:

   c. Throttle - MILITARY.

   Move the throttle smoothly to the Military stop, observing ENP and EGT. EGT should increase toward the nominal trim band and the ENP indication should move toward zero. An EGT gage COLD flag will appear when the throttle reaches the Military position if EGT is below the nominal trim band.

NOTE

Automatic trimming does not occur until the throttle is positioned at or above the Military position.

   d. Throttle - Retard approximately one-half inch aft of the Military position and return it to Military rapidly. This removes hysteresis from the fuel control linkage.

   Note EGT gage COLD/HOT flag operation while the throttle is being cycled. If the throttle is retarded before EGT reaches the nominal trim band, disappearance of the COLD flag while the throttle is retarded confirms normal operation of the automatic EGT trim system permission circuit.

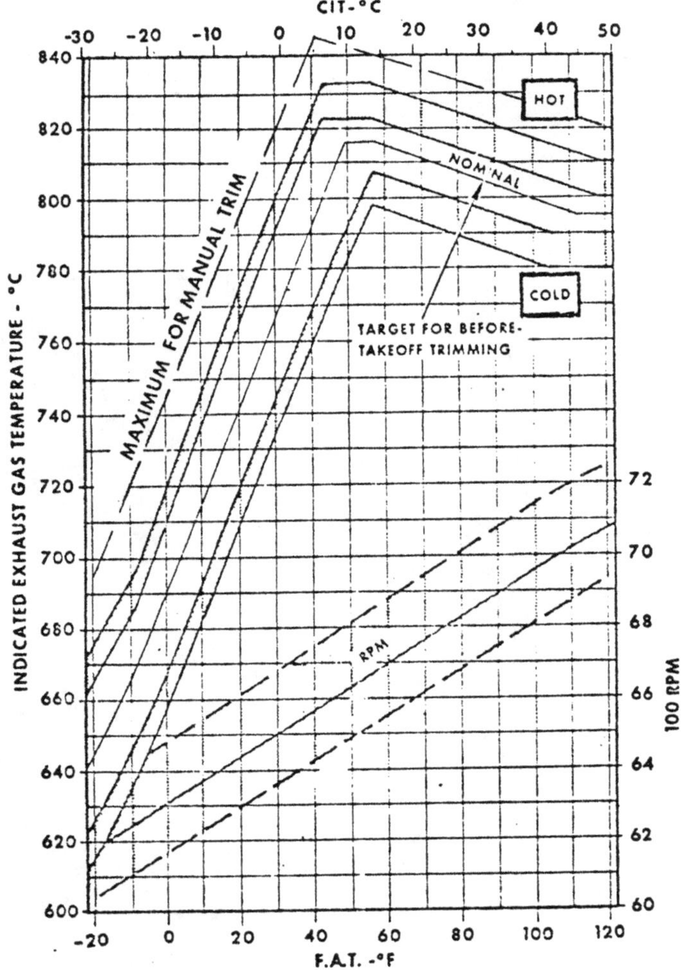

Figure 2-4

e. EGT trim - As required.

Trim to obtain appropriate value shown in Figure 2-4.

Note COLD/HOT flag operation.

If a HOT flag appears and EGT approaches an overtemperature condition, select MANUAL EGT trim and down trim, (estimate the time required to reach the proper Military operating range) then recheck trim with AUTO EGT selected.

In the event no HOT or COLD flag appears and EGT is normal at Military power, downtrim momentarily with MANUAL trim selected, then select AUTO EGT trim. At Military, the COLD flag should appear temporarily, and the engine should retrim automatically to the deadband EGT range.

f. RPM - Check engine speed vs FAT schedule shown in Figure 2-4.

g. Throttle - Retard smoothly to IDLE.

After retarding from Military power to idle, exercise care not to re-advance the power lever until the engine has stabilized at idle rpm. If this precaution is not observed, stall and "die-out" may occur. The stall may be inaudible but die-out will be evidenced by decreasing rpm and increasing EGT. If die-out occurs, move the power lever to cut-off to prevent overtemperature. The engine may be restarted as soon as a starter is available.

The stall and die-out occur only during ground static operation. Experience has shown that it is more apt to occur when relatively high ambient temperatures exist, as in summer.

2. Fuel system - Checked.

a. Check tank 1 fuel quantity 3000 lb minimum.

b. Check tanks 1, 2 and 4 ON and quantity decreasing.

3. Pitot heat switch - ON and checked

Ground crew confirm heat on.

4. Flight controls and trim setting - Checked.

Cycle and check hydraulic pressure, and recheck trim setting zero.

5. Fuel derich switch - Check ARM.

⚠ 6. No. 1 and No. 2 oxygen systems - Latched ON.

Check latched ON with pressure, and verbally confirm normal operation with other cockpit.

7. Warning lights - OFF (Except SAS).

8. Takeoff data - Reviewed.

a. Acceleration check.

b. Refusal speed.

c. Rotation speed.

d. Takeoff speed.

e. Single-engine speed.

NOTE

If a tire cooling period has been required, do not takeoff until the ground crew signals that tire condition is satisfactory.

TAKEOFF

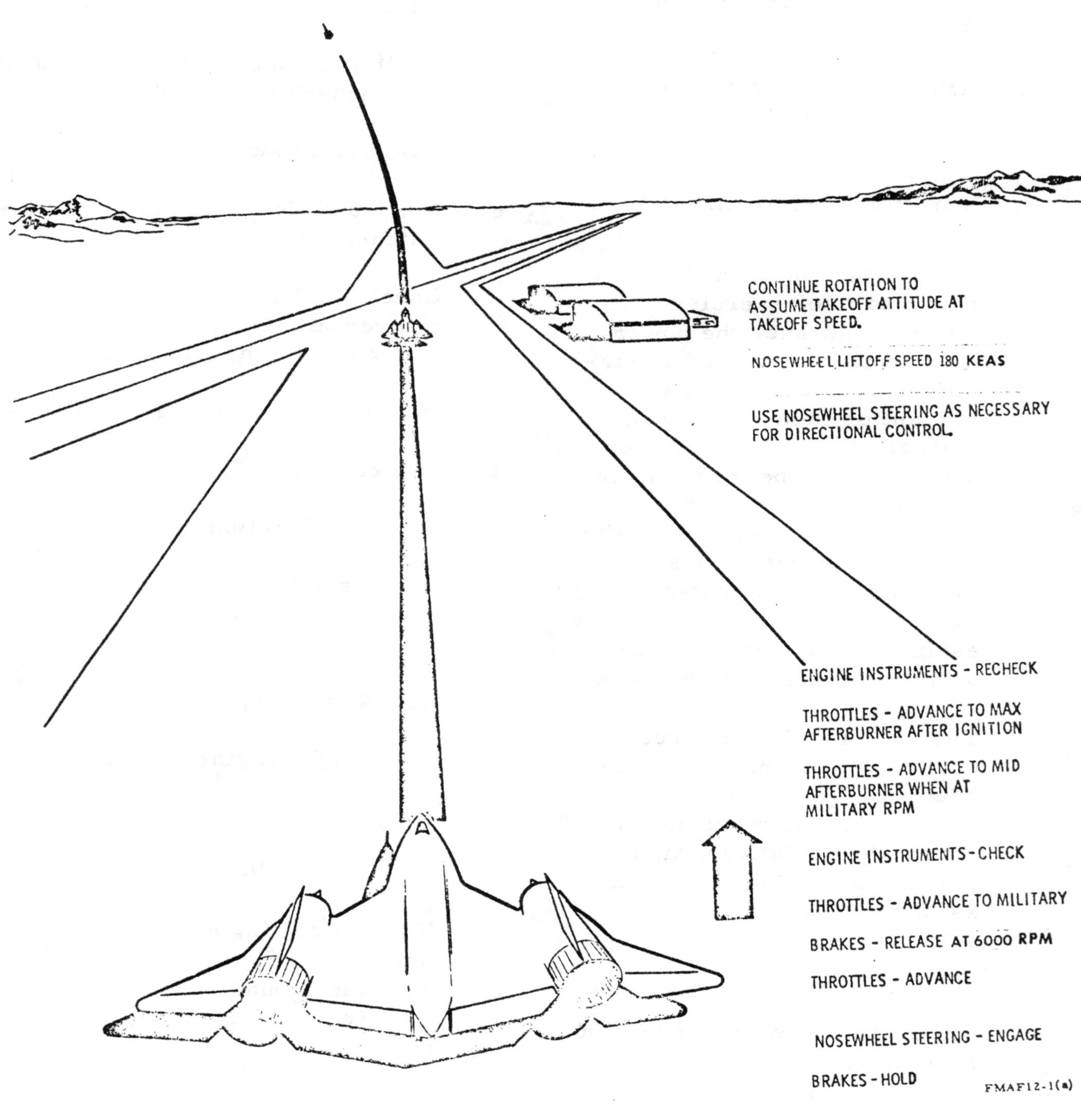
NOTE
ENGINE INSTRUMENT CHECKS SHOULD BE MADE DURING THE INITIAL PORTION OF TAKEOFF ROLL.
THE TIRES MAY SKID WITH THE BRAKES ON AT HIGH ENGINE RPM.
CONTINUE ROTATION TO ASSUME TAKEOFF ATTITUDE AT TAKEOFF SPEED.
NOSEWHEEL LIFTOFF SPEED 180 KEAS
USE NOSEWHEEL STEERING AS NECESSARY FOR DIRECTIONAL CONTROL.
ENGINE INSTRUMENTS - RECHECK
THROTTLES - ADVANCE TO MAX AFTERBURNER AFTER IGNITION
THROTTLES - ADVANCE TO MID AFTERBURNER WHEN AT MILITARY RPM
ENGINE INSTRUMENTS - CHECK
THROTTLES - ADVANCE TO MILITARY
BRAKES - RELEASE AT 6000 RPM
THROTTLES - ADVANCE
NOSEWHEEL STEERING - ENGAGE
BRAKES - HOLD
FMAF12-1(a)

Figure 2-5

TAKEOFF

①. IFF/SIF - NORMAL.

Set proper mode and code.

2. SAS - Engaged.

   a. Channel engage switches - ON.

   b. SAS recycle lights - OFF.

3. Crossfeed - ON.

4. Compasses - Checked.

Check and synchronize flight reference system. Check standby compass and FRS against runway heading.

5. STEER ON engage light - Check illuminated.

6. Throttles - Advanced.

   a. Brakes - Release at approximately 6000 rpm.

CAUTION

The tires may skid if the brakes are held at high thrust.

   b. Military power - Engine instruments - Check for values at or approaching those observed during trim.

      (1) Tachometer.

      (2) EGT.

      (3) Nozzle Position.

      (4) Oil Pressure.

   c. Throttles - Advance to mid afterburner range.

CAUTION

- To prevent overpseed, afterburner must not be ignited before engines reach Military rpm.
- Abort the takeoff if an afterburner fails to ignite within 3 seconds.
- Advancing the throttle will result in momentary nozzle excursion, and engine transient speed oscillation may approach 250 rpm.

   d. Maximum A/B - Check engine instruments.

Maximum afterburning thrust and fuel flow are obtained with the power lever between 108° and the quadrant forward stop (120° position).

CAUTION

The time for throttle advancement to Max A/B should be no less than 1 second.

Exact readout on these instruments is time-consuming. The readouts should be anticipated and needle position checked against a clock position. If there is any indication of deficient engine performance during throttle advancement, abort the takeoff. Monitor ground run distance and airspeed during the takeoff roll. If possible, any abort decision should be made before the aircraft has reached high speed. Refer to performance data, Appendix

I, for takeoff information. Directional control can be maintained with nosewheel steering up to nosewheel lift-off speed.

e. Acceleration - Check airspeed against computed acceleration check speed at selected acceleration check distance. Refer to performance data in the Appendix for takeoff information.

ROTATION TECHNIQUE

In general, the tires are more vulnerable to blowouts during takeoff than at landing because of the higher groundspeeds and gross weights involved. Wing lift quickly relieves the gear load as the nose is raised. Apply smooth, constant back pressure 15 to 25 knots below computed rotation speed. Lift the nosewheel off at rotation speed, using the rotation rate required to leave the ground at computed takeoff speed. Normal takeoff attitude is 8° to 10° nose high indication. The transition from start of rotation to takeoff requires approximately 5 seconds when using the normal takeoff technique. Refer to Takeoff Speed Schedule in Part II of the Appendix for rotation and takeoff speeds. Premature nosewheel liftoff should be avoided because the unnecessary drag extends the ground run. Delayed rotation also extends the ground run and may result in excessive tire speeds.

CROSSWIND TAKEOFF

During crosswind takeoffs the aircraft tends to weathervane into the wind. This will be noted when the nosewheel lifts off and nosewheel steering is no longer available. Rudder pressure must be held to counteract the crosswind effect. A definite correction must be made as the aircraft breaks ground. Apply lateral control as necessary for wings-level flight. Both the directional and lateral control applications are normal and no problems should be encountered when taking off during reasonable crosswind conditions.

AFTER TAKEOFF

When definitely airborne:

1. Landing gear lever - UP.

NOTE

- The gear will retract in approximately 12 seconds. Ventral fin extension time is approximately 10 seconds after gear is retracted. Observe landing gear limit speed until gear is retracted.
- With a small but positive rate of climb, accelerate to 400 KEAS for afterburning climb.

WARNING

Single-engine operation is critical immediately after takeoff. Increasing airspeed and decreasing angle of attack has greater benefits than gaining altitude at a maximum rate.

After gear retraction is complete:

2. Throttle - Set.

Establish climb power schedule. At approximately 275 KEAS, begin power reduction to minimum A/B. Retard throttles to Military power at 375 KEAS. Climb at 400 KEAS until Mach 0.9 intercepted. Climb at Mach 0.9 until reaching subsonic leveloff altitude.

NOTE

The Military power, 400 KEAS climb should be used on flights that require an intermediate level-off altitude; however, when turbulence is encountered or anticiapted, 300 KEAS to 350 KEAS may be desirable. This climb offers the pilot a more comfortable pitch attitude and conserves fuel. A/B thrust may be used on missions requiring immediate acceleration to supersonic cruise or if mission requirements dictate.

3. Engine instruments - Check.

At Mach 0.5:

4. Surface limiter - Engaged and surface limiter light off.

Rotate handle counterclockwise and push in to engage limiters.

5. Crossfeed - OFF.

If severe or extreme turbulence is encountered below 40,000 feet reduce the airspeed to 275-350 KEAS.

CLIMB

Climbing Acceleration Procedure

When acceleration climb procedure is used, accelerate from takeoff to 25 to 50 knots below initial climb schedule and rotate to climb attitude.

NOTE

Begin the rotation sufficiently in advance of reaching climb speed to avoid exceeding maximum recommended KEAS schedule. If rotation is delayed, it is possible to overshoot the airspeed by an appreciable amount.

Establish 450 KEAS at approximately 15,000 feet and climb at this speed to Mach 2.6, using normal climb procedure or continuous maximum afterburning thrust. Mach number will increase with altitude and Mach 1 should be reached at approximately 20,000 feet.

Level Transonic Acceleration Procedure

At approximately 30,000 feet, accelerate to 450 KEAS in a level or slightly descending flight path, and then climb at a constant 450 KEAS to Mach 2.6. For maximum acceleration, maintain limit EGT by use of the turbine inlet temperature switches. It is particularly important to have the EGT trimmed and stabilized prior to reaching approximately 200°C CIT. This will minimize rpm suppression as the compressor bleed valves open. There will be a slight but noticeable yaw if the compressor bleed bypass valves do not open at the same time on the left and right engines.

<u>Transonic Acceleration Techniques</u>

The normal procedure employs a subsonic climb to 36,000 feet which is followed by a dive at approximately 5500 fpm rate of descent and a round-out near 30,000 feet so as to intercept 450 KEAS at 1.25 Mach number while climbing. The pitch autopilot should be disengaged before pushover. The dive angle requires a carefully controlled pull-out in order to intercept 450 KEAS. During IFR or night operations, exercise extreme caution because of the dive angles and pull-out load factors involved. The level transonic acceleration technique may be more desirable during IFR or night operations.

<u>Supersonic Airspeed Schedules</u>

The optimum supersonic airspeed is 450 KEAS while between Mach 1.25 and Mach 2.6. Lower airspeed such as 400 KEAS or 375 KEAS may be used, as when turbulence is encountered or when directed by briefing to use airspeeds below 450 KEAS, but performance is considerably degraded.

<u>450 KEAS Supersonic Climb Procedures</u>

The following transition-to-climb procedure is recommended for supersonic missions after air refueling, intermediate altitude level-off, or after takeoff.

1. Throttles - Minimum A/B.

   For temperatures exceeding standard day conditions by more than approximately 10°C, it may be desirable to perform this subsonic portion of the climb in Maximum A/B.

2. Speed - Mach 0.9 (400 KEAS maximum).

   If the pitch autopilot is used, the Mach Hold function should be engaged during this constant Mach number portion of the climb.

At FL 300 and at Mach 0.9, correct airspeed indications should be 324 KEAS and 347 <u>KIAS</u> at 30,000 feet.

3. Throttles - Maximum A/B.

   Smoothly advance throttles to maximum thrust.

At FL 360:

4. Pitch autopilot - Disengaged.

   Disengage the pitch autopilot for the acceleration maneuver.

5. Pushover - Initiated.

   Pushover smoothly at approximately 0.8 g's to establish 5500 fpm rate of descent by Mach 1.05. It is most important to exceed Mach 1.05 early in the descent, and to avoid turning until the climb is established. After establishing the subsonic descent rate, maintain attitude until initiating round-out.

WARNING

Airspeed may increase rapidly after Mach 1.10 is reached. In the event of excessive KEAS increase, reduce power below Military if necessary while completing a transition to 450 KEAS. Do not use excessive load factors to avoid exceeding the schedule airspeed.

6. Climb KEAS - Intercepted.

The normal supersonic climb airspeed is 450 KEAS when above Mach 1.25. Start a gradual roundout at 420 KEAS. Avoid exceeding 450 KEAS by reducing power, if necessary, rather than by increasing load factor if airspeed is increasing too rapidly.

WARNING

In rough air, restrict climb speed as specified for turbulent air penetration instructions until smooth air is encountered.

In supersonic climb:

7. EGT trim - Check.

If necessary, trim to nominal operating temperature as shown by figure 2-4. Trim as necessary while accelerating to 60°C CIT to maintain EGT less than 845°C when above 60°C CIT. It is recommended that EGT be maintained between 775°C and 805°C after the climb is established.

At Mach 1.7:

8. Aft bypass controls - Position A or B.

Set position A at Mach 1.7. When the forward bypass exceeds 10% open, select position B.

NOTE

Unnecessary aft bypass should be avoided in order to reduce inlet air distortion at the engine face; however, if forward bypass must schedule in excess of 20%, drag increases noticeably.

9. RPM and CIP - Checked.

10. Exterior lights - Off.

11. Pitot heat switch - OFF.

At Mach 2.6:

12. KEAS - Checked.

For normal climb at 450 KEAS, decrease KEAS 10 knots per 0.1 Mach number increase above Mach 2.6.

At Mach 2.6:

13. Aft bypass controls - Position A.

At Mach 3.0:

14. Aft bypass controls - Set.

a. Set CLOSE position for scheduled cruise speed above Mach 3.05.

b. Maintain position A or use CLOSE setting for cruise near Mach 3.0.

The optimum setting for cruise near Mach 3.0 may be determined by setting the aft doors to CLOSE individually. If there is no change, or if an asymmetric thrust increase on the closed side is noted, cruise with aft bypass closed.

CRUISE

Observe limitations of Section V. Avoid sustained operation at speeds between Mach 2.5 and 2.7 when convenient. This area is normally more susceptible to inlet duct roughness than higher or lower speeds.

C.G. Control

Center-of-gravity control is important for optimum cruise performance. Fuel load distribution and the automatic tank sequencing provides a forward center of gravity for takeoff and climb. During cruise the automatic sequencing provides an aft center of gravity to minimize elevon deflection and resulting trim drag. Supplemental manual control of fuel usage is also permissible and may be used in the event of failure of the automatic fuel tank sequencing or if additional c. g. shifts are required.

Engine Operation

Exhaust gas temperature and engine speed limits vary with compressor inlet temperature. Refer to Engine Operating Limits, Section V, for limit schedules.

Effect of Engine Thrust Variation with EGT

For a given level of thrust, higher throttle settings and increased fuel flow are required as EGT is decreased. Full throttle ceilings in cruise and while turning are reduced; this occurs because combined burning efficiency of the engine and AB decreases with lowered EGT. The degradation in thrust for all throttle settings, at Mach 3.2 and 80,000 feet, is approximately 1.3 percent per 10°C of EGT decrease. Although only one flight condition is discussed, the trend is the same for other flight conditions.

Effect of RPM Suppression on Maximum Thrust

As EGT decreases, the engine nozzle opens to maintain scheduled rpm. At high Mach number and maximum power, low EGT may cause the nozzle to open fully and any further EGT decrease will result in rpm suppression below schedule. When this condition occurs, the engine speed will suppress approximately 50 rpm for each 10°C of EGT decrease. The airflow through the engine decreases due to the suppressed rpm, leading to a higher inlet duct bypass requirement and opening of the

forward bypass doors. At Mach 3.2 this results in a thrust degradation and drag increase of approximately 3.5 percent per 10°C of EGT decrease for each affected engine. If Mach number decreases as a result of the change in thrust and drag, the spikes schedule more forward and the forward bypass doors open further. Performance will deteriorate rapidly under these cumulative effects and it is recommended that cruise EGT be maintained between 775°C and 805°C to avoid the possibility of this situation occurring.

Autopilot Operation, Cruise or Climb

The aircraft may be manually trimmed in the yaw and roll axes while the roll autopilot is engaged in Heading Hold mode to minimize track error (course hang-off) or to correct wing-low conditions. This procedure may be used in any phase of flight. While engaged in attitude hold mode, (pitch or roll autopilot), use the respective trim wheel on the function selector panel to adjust the engaged attitude. (Manual pitch trim inputs at this time will only cause control transients.)

PRIOR TO DESCENT

Retrimming of EGT should not be required prior to start of descent unless manual uptrimming has been accomplished during climb or cruise. The amount of downtrim required will be approximately equal to the total prior uptrim. Pilot judgement must govern its use.

▲ 1. LN_2 quantity - Checked.

The total liquid nitrogen quantity available should be checked. If not sufficient for normal descent, refer to Fuel Tank Pressurization emergency procedure, Section III.

2. Inlet controls - AUTO & CLOSED.

Air inlet spike and forward bypass controls will be placed in AUTO and the aft bypass controls set at CLOSE unless manual inlet procedures must be used. If manual inlet procedures are required, follow schedule in Section III.

DESCENT

1. Throttles - Military power.

Slowly retard the throttles to the Military power position.

NOTE

Pause at minimum A/B approximately 5 seconds if retarding from a higher power setting.

2. EGT trim - Checked.

Trim if necessary.

3. Airspeed - 350 KEAS minimum.

Maintain cruise altitude while decelerating, or maintain cruise Mach number while descending, until the KEAS schedule is intercepted.

DESCENT PROFILE

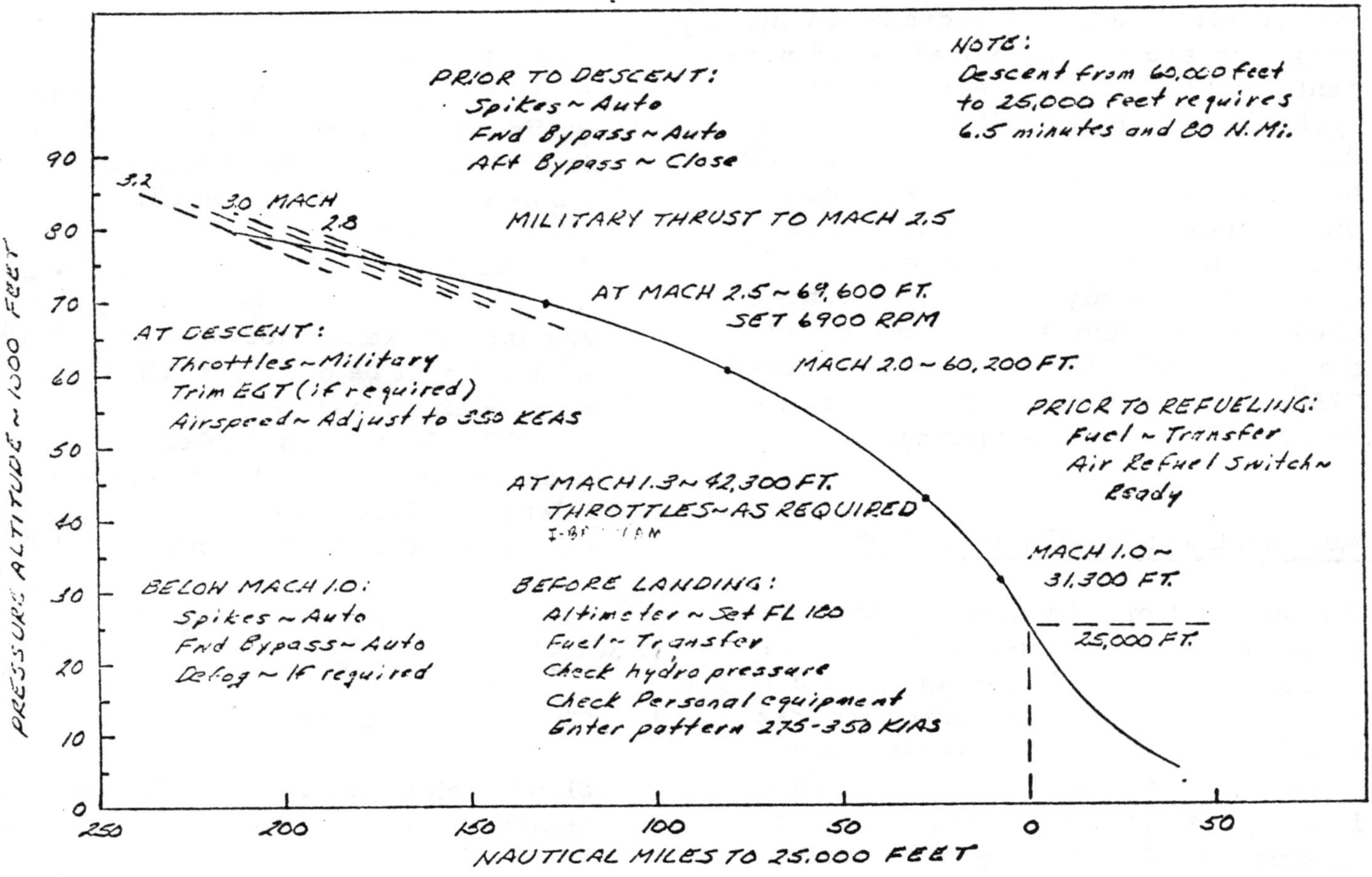

PRIOR TO DESCENT:
Spikes ~ Auto
Fwd Bypass ~ Auto
Aft Bypass ~ Close

NOTE:
Descent from 60,000 feet to 25,000 feet requires 6.5 minutes and 80 N. Mi.

3.2
3.0 MACH
2.8

MILITARY THRUST TO MACH 2.5

AT MACH 2.5 ~ 69,600 FT.
SET 6900 RPM

AT DESCENT:
Throttles ~ Military
Trim EGT (if required)
Airspeed ~ Adjust to 350 KEAS

MACH 2.0 ~ 60,200 FT.

PRIOR TO REFUELING:
Fuel ~ Transfer
Air Refuel Switch ~ Ready

AT MACH 1.3 ~ 42,300 FT.
THROTTLES ~ AS REQUIRED

MACH 1.0 ~ 31,300 FT.

BELOW MACH 1.0:
Spikes ~ Auto
Fwd Bypass ~ Auto
Defog ~ If required

BEFORE LANDING:
Altimeter ~ Set FL 180
Fuel ~ Transfer
Check hydro pressure
Check Personal equipment
Enter pattern 275-350 KIAS

25,000 FT.

PRESSURE ALTITUDE ~ 1000 FEET
0 10 20 30 40 50 60 70 80 90

250 200 150 100 50 0 50
NAUTICAL MILES TO 25,000 FEET

ALT	DISTANCE CRUISE MACH 3.2	3.1	3.0	2.9	2.8	2.5	2.4	2.0	PROFILE CHECK 350 KEAS DIST	KIAS	ETE
84M	273	266									
82M	264	257	251								
80M	255	248	242	237	231						:16
78M	247	240	232	227	222						
76M	238	230	222	218	212				219	441	:14.5
74M	230	221	213	208	203	191					
72M			204	199	194	182	178				
70M				189	185	173	169		175	437	:13
68M				180	175	164	160				
66M					166	155	151	148			
64M						145	142	138	138	432	:11.5
60M						128	124	120	120	428	:10.5
55M									101	422	:10
50M									86	416	:08.5
45M									74	409	:07.5
40M									63	401	:07
35M									54	391	:06
30M									46	376	:05

1. DISTANCES AND TIME INCLUDE 40 N. MI. LEVEL OFF PRIOR TO ARCP.
2. SUBTRACT 20 N. MI. AND :02 FOR DESCENT INTO IAF.

Figure 2-8

4. Fuel tank pressure - Monitor.

   Rate of descent must not result in negative fuel tank pressure.

At Mach 2.5:

5. Throttles - 6900 rpm set.

   Retard throttles simultaneously. Some throttle misalignment may be required. An rpm decrease of 400-500 rpm can be expected from 2.5 to 1.3 Mach. Maintain at least 6500 rpm while above Mach 2.0.

Below Mach 1.7:

6. Fuel forward transfer switch - FWD TRANS.

   Transfer fuel as necessary to obtain c.g. within subsonic limits. Monitor tank 1 quantity while transferring fuel.

7. Pitot heat switch - ON.

8. Exterior lights - On.

Below Mach 1.3:

9. Throttles - As required.

   Adjust descent profile as required after reaching Mach 1.3. Reduce rate of descent, if necessary, to avoid low fuel tank pressure below FL 400.

10. Fuel forward transfer switch - OFF.

11. Crossfeed switch - Set.

    Crossfeed OPEN should be selected if immediate penetration for landing is to be accomplished or if the FUEL QTY LOW light is on.

12. Inlet controls - Checked.

    Check spike and forward bypass controls AUTO and aft bypass controls at CLOSE unless manual inlet control procedures are required and emergency schedule is in use.

AIR REFUELING PROCEDURE

The pitot-static flight instruments should be crosschecked and the integrated display used for in-flight refueling procedures. Check that the altimeters are set at 29.92 in. Hg. The radar must be OFF or in STDBY.

FUEL MANAGEMENT PRIOR TO REFUELING

From 1500 to 3000 pounds of fuel should be transferred to tank 1 and forward transfer should be shut off prior to making a refueling contact. The fuel transfer improves the unaugmented pitch stability of the aircraft by moving its center of gravity forward. Forward transfer also replenishes tank 1 so that its boost pumps will supply both fuel manifolds automatically during refueling. With normal SAS operation, there is no marked change in aircraft handling characteristics after forward transfer. The forward transfer should be repeated if there is any appreciable delay in making the refueling contact, and between dry contacts during air-to-air refueling practice.

AIR REFUELING DIRECTOR LIGHTS

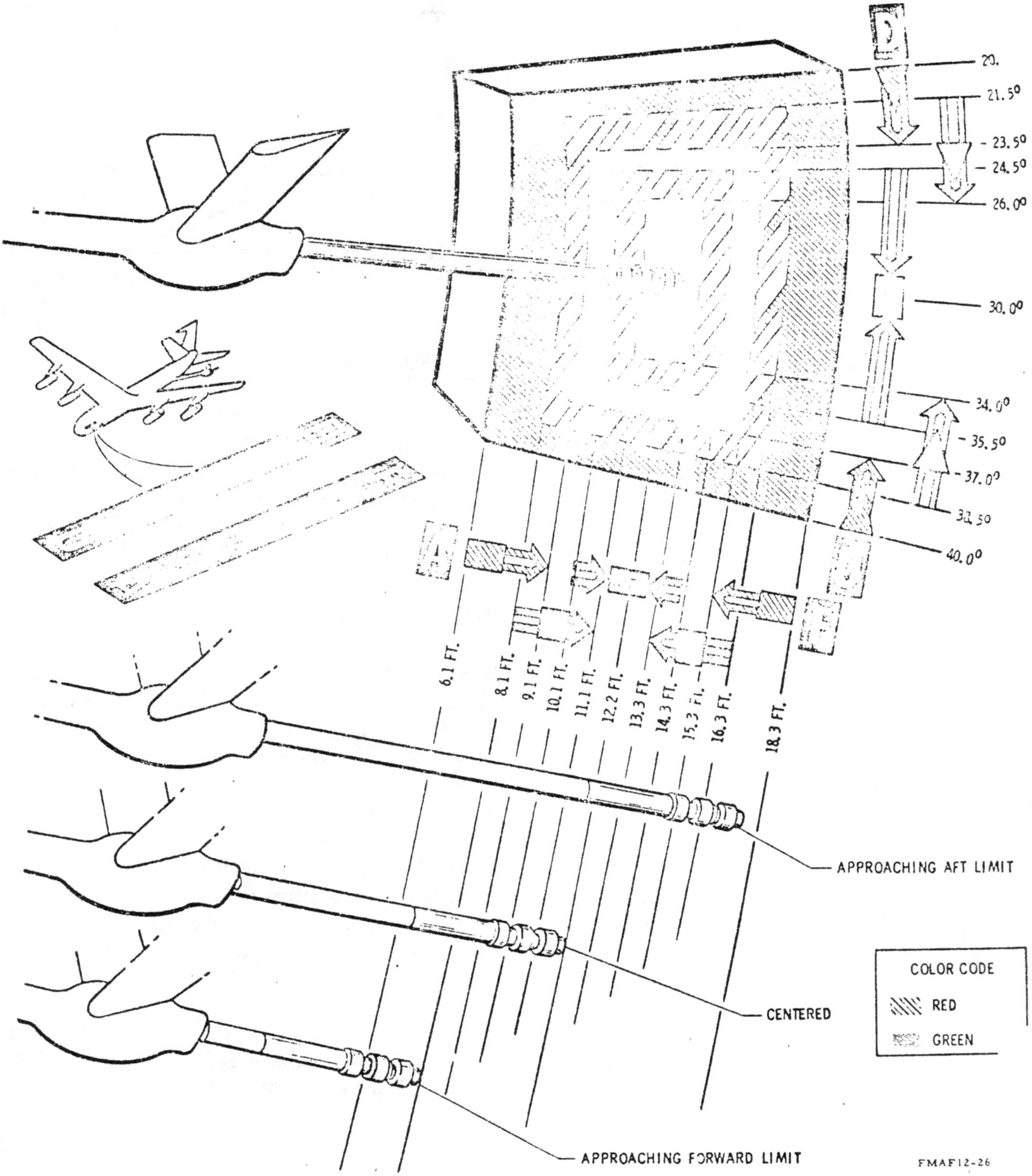

20.
21.5°
23.5°
24.5°
26.0°
30.0°
34.0°
35.5°
37.0°
38.5°
40.0°

6.1 FT.
8.1 FT.
9.1 FT.
10.1 FT.
11.1 FT.
12.2 FT.
13.3 FT.
14.3 FT.
15.3 FT.
16.3 FT.
18.3 FT.

APPROACHING AFT LIMIT

CENTERED

APPROACHING FORWARD LIMIT

COLOR CODE
RED
GREEN

FMAF12-26

Figure 2-7

AIR REFUELING BOOM ENVELOPE LIMITS

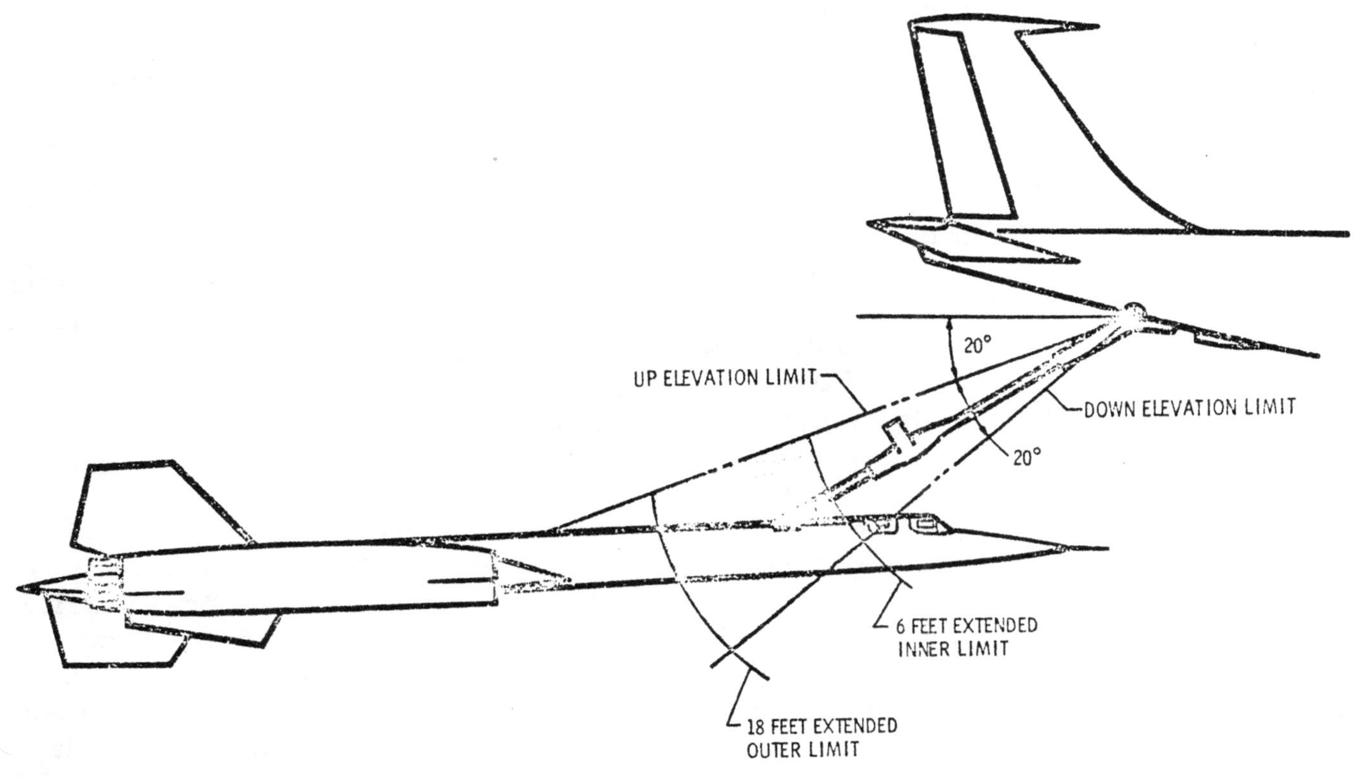

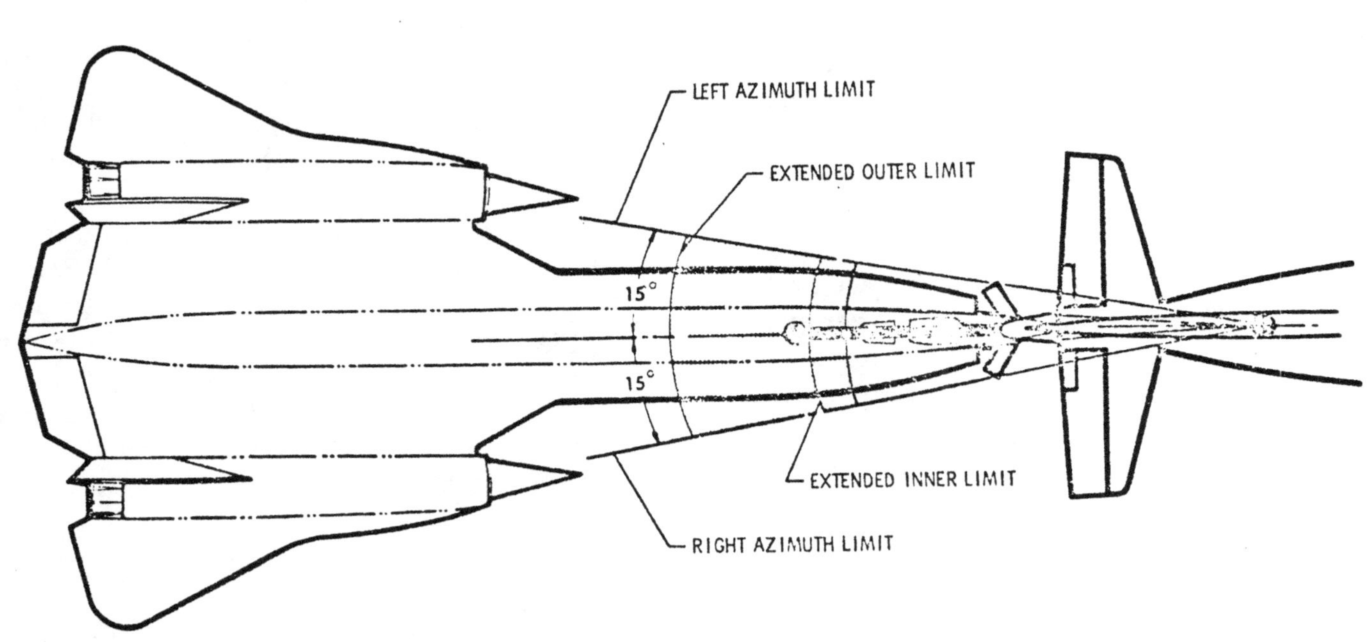

Figure 2-8

NOTE

Crossfeed should be used whenever the FUEL QTY LOW light is illuminated.

Approach For Refueling

Make the approach from behind, below, and to the right, observing position through the left windshield panel. The seat should be lowered prior to reaching the observation position. Upon completion of the Before Refueling procedures, maneuver to the precontact position about 100 feet to the rear and 50 feet below the refueling position. Stabilize the aircraft and trim at refueling speed for contact. Observe the tanker for director light signals. Maneuver as directed by the lights or by verbal instructions of the boom operator.

Refueling Hookup

A successful hookup is confirmed by steady illumination of the director light panel and extinguishing of the ready light. Slight maneuvering may be necessary at this point to illuminate the fore and aft and elevation neutral lights during fuel transfer. Hookup can be maintained between the aircraft and tanker during a turn or in a descent. No adverse flight characteristics result from tanker downwash. Radio contact between the aircraft and tanker is useful in establishing and maintaining position.

CAUTION

While refueling do not transmit on HF radio as the tanker HF transceiver may be damaged.

Manual EGT trimming may be used in lieu of the automatic trim system, if desired. This will avoid the possibility of "ratcheting". Manual trimming to temperatures above the Nominal Dead band schedule is permitted; however, EGT limits must be observed and, if the engine should stall, downtrimming must be accomplished immediately.

Overcontrol of the engines should be avoided while gaining and holding position. This is due to the non-linearity of throttle position vs engine thrust. A given throttle angle change near military power yields more thrust change than a similar change in the throttle mid range. The aircraft may become power limited if the afterburner-on technique is not used, and tobogganing descents of up to 1000 feet per minute can be requested as the military power throttle position is approached. Asymmetric thrust is easily controlled when the afterburner-on technique is used.

Turbulence encountered while in contact poses no particular problem with SAS operating normally, and shallow turns of up to 20° bank angle can be made without difficulty. However, if all pitch SAS including the back-up pitch damper are inoperative, it is recommended that refueling not be attempted except in an eme gency. The aircraft tends to be unstable without any pitch SAS, but control can be maintained under favorable conditions with fuel transferred to obtain a forward c.g. location.

All disconnects should be made with a rear-ward and slightly downward relative motion with wings level. This will insure separation of the boom from the receptacle with a straight line force. Side or rolling loads or excessive deviations from the desired elevation increase the possibility of boom and/or receptacle damage during disconnect.

Night refueling is essentially the same as for daytime operations except that added caution and effort is required to avoid overshoot, and the tendency toward throttle over-control while in contact is increased.

NORMAL AIR REFUELING PROCEDURE

When in the observation position, prior to air refueling:

1. Center of gravity - Checked.

Transfer fuel forward as necessary to position the c.g. corresponding to a fully fueled aircraft. (Otherwise set c.g. at 17%.) Confirm proper fuel transfer by FCO computation of c.g. position. The pitch trim should be approximately 0°.

CAUTION

Under normal conditions, if trimming is required, disconnect and retrim in the pre-contact position. It is recommended that trimming not be done on the boom unless it is impractical to disconnect. To avert the possibility of a sticking trim switch, the pilot should assure positive switch movement to the neutral position after each activation.

2. UHF radio - BOTH.
3. Anti-collision lights switch - FUS.
4. Air refuel switch - READY, ready light ON.
5. EGT trim switches - Manual, if desired.
6. Fuel forward transfer switch - OFF.
7. Fuel quantity switch - TOTAL.
8. Interphone - Set.

NOTE

- When stabilized, notify tanker when ready for hook-up.
- After tanker hook-up the air refuel READY light should extinguish.

The total fuel quantity indication and pitch trim should be monitored as conditions permit.

When refueling is complete and disconnect has been made as indicated by illumination of DISC light:

9. Air refuel switch - OFF.

WARNING

Carefully monitor airspeed and angle of attack after completion of refueling and adjust attitude as necessary to remain within limits. A minimum of 300 KEAS is required while making the transition to supersonic speeds after refueling.

10. Takeoff tanks - ON (Tanks 1, 2, 4).

11. Crossfeed - OFF.

12. EGT trim switches - Checked and set in AUTO position.

    If manual uptrim has been accomplished, downtrim EGT before selecting AUTO trim.

13. TACAN - T/R.

14. Interphone control panel - Set as desired.

15. Anti-collision lights - ANTI-COLLISION. (ON and EXTENDED).

POWER LIMITED REFUELING

Engines may be uptrimmed in accordance with operating limitations described in Section V. Uptrimming should be done prior to establishing tanker contact, but is permissible during contact as well. If CIT's are below 5°C, use caution if exceeding the nominal trim band limits. If Military power limit is reached prior to completion of fuel transfer, the receiver may request the tanker to initiate a "toboggan" maneuver or he may elect to use one afterburner, and maintain level flight. If one afterburner is to be used, disconnect from the tanker, light one A/B and set it in Minimum A/B. Left afterburner is recommended. Use the opposite throttle for variable thrust requirements, retrim the aircraft as necessary and return to the contact position.

BREAKAWAY PROCEDURE

Any crewmember, of either tanker or receiver, if aware of the existence of an emergency will transmit on the air refueling frequency the tanker aircraft tactical call sign followed by transmission of "breakaway, breakaway, breakaway". The boom operator may signal breakaway by turning the receiver director lights off and on rapidly. The receiver pilot will actuate the disconnect release. Retard throttles and drop aft and down from tanker until the entire tanker airplane is in sight.

NOTE

The pilot should use care not to overrun the tanker. If overrunning does occur, under no conditions should a turn either right or left be made until breakaway has been completed.

Engine Failure

After hookup, a tanker engine failure is more serious than a receiver aircraft engine failure. A receiver engine failure will result in extension of the boom and a probable disconnect; however, a tanker engine failure may cause the receiver aircraft to overrun the tanker. In the event of tanker engine failure, breakaway immediately.

Disconnect

A disconnect may be accomplished in any of the following ways:

1. Automatically.

   a. If boom envelope limits are exceeded.

b. When fuel pressure exceeds 70 psi.

c. When pull on boom becomes excessive.

2. Manually.

a. By the boom operator.

b. By depressing the A/R DISC trigger on the control stick grip.

CAUTION

Disconnect in an aft and downward direction, and avoid forward relative motion during disconnect. Disconnect maneuvers which tend to pry the boom out of the receptacle will result in receptacle or airframe damage.

AIR REFUELING ALTERNATE PROCEDURE

In case L hydraulic pressure is lost, R pressure may be utilized for refueling by moving the brake switch to the ALT BRAKE position.

CAUTION

Do not leave the brake switch in the ALT BRAKE position after refueling. R hydraulic pressure may be lost if L system fluid loss is due to a malfunction of the steering or refueling system.

Normal and manual boom latching air refueling procedures are applicable with ALT BRAKE selected. Follow normal procedures except for differences as described in the following paragraphs.

MANUAL BOOM LATCHING

A manual refueling procedure may also be used if a failure of the signal amplifier should occur. This procedure requires manual control of the refueling boom latches. The refueling boom can be latched in the receptacle by placing the air refuel switch in the MANUAL position and depressing the A/R DISC trigger on the control stick grip until the boom nozzle has bottomed in the receptacle. The receiver pilot can usually feel the boom nozzle bottom in the receptacle. When latching the boom manually:

a. Air refuel switch - MANUAL.

b. Air refuel disconnect switch - Depressed.

After nozzle is seated in receptacle:

c. Air refuel disconnect switch - Released.

When refueling is completed:

d. Air refuel disconnect switch - Press until boom is clear of receptacle.

Subsequent procedures are the same as for After Normal Refueling.

CAUTION

If the A/R DISC trigger switch is released when the nozzle is not in the bottom of the receptacle, it is possible for the nozzle to damage or break the extended nozzle latches, preventing any further refueling.

NOTE

On manual boom latching, there will not be a pressure disconnect, but the refuel manifold accepts tanker pressure with ample margin after tanks shut off automatically. The fuel vent manifold releases excess tank pressure which might occur due to malfunction of the tank shutoff valves.

CAUTION

When using the manual boom latching provisions the boom limit switches are deactivated. Consequently, the receiver pilot must initiate all disconnects before exceeding any of the boom extension limits as the boom operator will be unable to release the nozzle latches.

NOTE

- The latch toggle hydraulic system incorporates a pressure relief valve which will permit the boom to be pulled out when a pullout force of approximately 5400 pounds is applied. Therefore, if a malfunction occurs which prevents disconnecting the boom, place the air refuel switch in the MANUAL position, depress the A/R DISC trigger switch, and proceed with brute-force pullout.
- Boom interphone is inoperative when the tanker is in manual operation.

FUEL DUMPING

Fuel dumping provides a means of reducing gross weight rapidly. The nominal dump rate is 2500 pounds per minute, but the rate varies with the amount of fuel remaining and the number of boost pumps operating.

Normally, fuel is dumped in accordance with the automatic fuel usage sequence. An additional tank can be selected in each tank group to increase the dump rate.

To accomplish fuel dumping:

1. Fuel dump switch - DUMP.

   All tanks containing fuel except tank 1 will empty in normal fuel tank usage sequence. The boost pumps in tank 1 are inactivated while dumping. Fuel dumping terminates and tank 1 pumps can resume operation when the fuel level in tank 6 reaches 4000 pounds.

2. Fuel quantity - Alternately monitor total fuel and tank 6 fuel.

When 4000 pounds remain in tank 6:

3. Fuel dump switch - OFF.

   Fuel dumping can be terminated manually at any time by positioning the fuel dump switch to OFF.

In the event that the stop-dump switch in tank 6 has failed open, or the fuel level in tank 6 has reached the 4000 pound level prematurely and dumping is required (excessive fuel tanks 3 and 5):

1. Fuel dump switch - EMER.

   Selection of emergency fuel dumping overrides the stop-dump feature of the normal dump system (which operates to stop dumping automatically when tank 6 contains 4000 pounds or less) and will allow dumping to empty all tanks.

WARNING

Fuel dumping must be terminated manually when the required amount of fuel remains; otherwise, all except fuel in tank 1 may be inadvertently dumped.

2. Tank 3 and 5 fuel boost pump switches - Press on as required.

3. Fuel transfer switch - TRANSFER (if required).

   Transfer fuel forward as frequently as required.

4. Fuel quantity - Alternately monitor tanks 1 and 6.

When tank 1 quantity is above 4000 pounds

5. Fuel transfer switch - OFF.

When the required amount of fuel remains

6. Fuel dump switch - OFF.

BEFORE PENETRATION

The following procedures must be accomplished before landing. Accomplish All-Weather operating procedures in addition to the following when applicable.

1. Display mode selector switch - Set.

2. Attitude reference - Set.

3. Defog lever and cockpit temperature control rheostat - Set.

   The possibility of encountering windshield and cockpit fogging during approach and landing must be anticipated and appropriate settings selected. Maintain 75° to 85° F cockpit temperature.

Between FL 400 and FL 200:

4. Approach speed - Computed.

5. Altimeter - Set.

   Set station pressure on passing through FL 180, or just prior to penetration.

PENETRATION

Accomplish All-Weather operating procedures in addition to the following when applicable.

NOTE

Radar must be OFF or in standby during an ILS approach.

1. Crossfeed switch - Pressed ON.

CAUTION

Leave crossfeed on to assure fuel supply to both engines during landing and possible go-around operations.

2. Brake switch - ANTI-SKID.

Below Mach 0.5:

3. Surface limiter control handle - Pulled, light out.

Pull and rotate the surface limiter handle 90 degrees to disengage the surface limiters, lock the handle, and cause the SURFACE LIMITER warning light to extinguish.

BEFORE LANDING

See figure 2-9 for typical landing pattern altitudes and airspeeds. At heavy weights, increase airspeed if necessary to maintain a pitch angle of less than 8 degrees for turns to base leg and 9 degrees for turns to final approach.

1. Cockpit temperature - Set.

Check cockpit temperature and maintain approximately 80°F. Insure that there is no evidence of visible moisture coming from the air conditioning ducts.

2. Approach and landing speeds - Computed.

Final approach and landing speeds are based on landing weight. Angle of attack will be approximately 10 degrees for a normal final approach.

NOTE

Use the maximum performance landing speed schedule when operating conditions such as wet runway or short field length require minimum roll after touch down.

LANDING PATTERN - TYPICAL

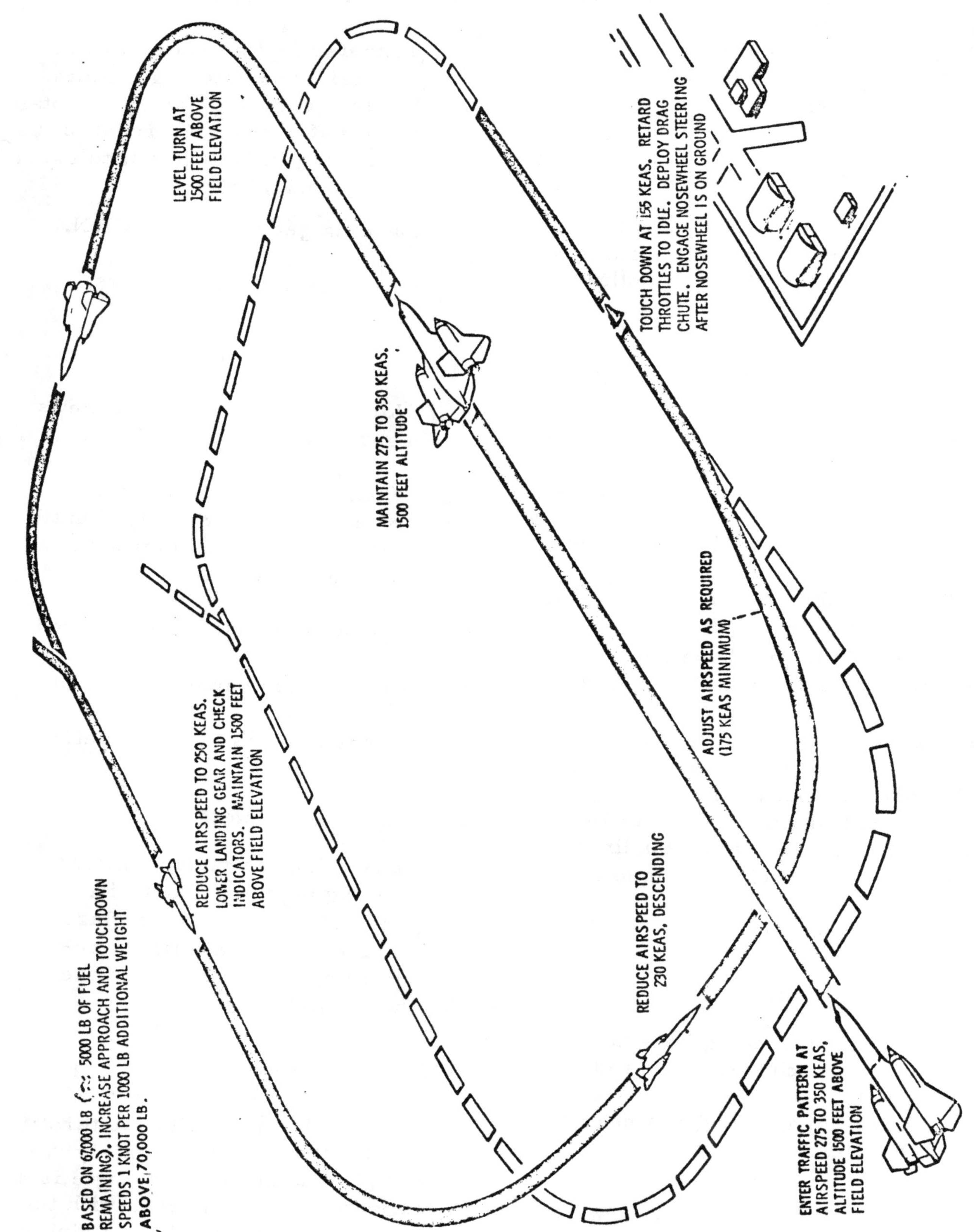
LEVEL TURN AT 1500 FEET ABOVE FIELD ELEVATION
TOUCH DOWN AT 155 KEAS. RETARD THROTTLES TO IDLE. DEPLOY DRAG CHUTE. ENGAGE NOSEWHEEL STEERING AFTER NOSEWHEEL IS ON GROUND
MAINTAIN 275 TO 350 KEAS. 1500 FEET ALTITUDE
ADJUST AIRSPEED AS REQUIRED (175 KEAS MINIMUM)
REDUCE AIRSPEED TO 250 KEAS. LOWER LANDING GEAR AND CHECK INDICATORS. MAINTAIN 1500 FEET ABOVE FIELD ELEVATION
REDUCE AIRSPEED TO 230 KEAS, DESCENDING
BASED ON 67,000 LB (5000 LB OF FUEL REMAINING). INCREASE APPROACH AND TOUCHDOWN SPEEDS 1 KNOT PER 1000 LB ADDITIONAL WEIGHT ABOVE 70,000 LB.
ENTER TRAFFIC PATTERN AT AIRSPEED 275 TO 350 KEAS, ALTITUDE 1500 FEET ABOVE FIELD ELEVATION

Figure 2-9

The design landing weight is 68,000 pounds with 10 fps sink rate. When landing at higher weights is required, the following speed and sink rate schedule applies:

NORMAL LANDINGS

Gross Weight	Final Appr. KEAS	Land Speed KEAS	Max. Sink rate allowable
70,000 lb or less	175	155	10 fps
75,000 lb	180	160	9.25 fps
80,000 lb	185	165	8.75 fps
90,000 lb	195	175	8 fps
100,000 lb	205	185	7.5 fps

Over 100,000 lb gross weight, observe Section V landing sink rate limits.

For heavyweight landings: At over 100,000 gross weight, use the normal final approach speed schedule and a landing speed which is 10 KEAS less than scheduled for normal landings. Use the minimum roll technique for stopping.

NOTE

Heavyweight landing speeds result in touchdown angles of attack 1/2 to 1 degree greater than for normal landings near 70,000 lb.

3. Center of gravity - Checked.

   Transfer fuel as necessary to maintain subsonic c.g. limits. Correct c.g. for landing is obtained by transferring 25 percent of fuel remaining in tank 6, not to exceed 3000 pounds, to tank 1.

4. Landing gear lever - DOWN.

   Check gear warning lights.

NOTE

- Normal gear extension time is approximately 12 to 16 seconds.
- When at heavyweight, gear extension may be delayed until after turn to final approach course, if desired.

5. Hydraulic pressure - Checked.

6. Annunciator panel - Checked.

7. Landing lights switch - Set.

NOTE

Lowering the vision splitter during night landings will materially reduce the glare caused by ground lights reflecting off the inside of the windshield.

NORMAL LANDING

Touchdown is made with the throttles in IDLE, and at approximately 9.5 degrees pitch angle. Pitch angle is approximately 10.5 degrees, with the nose on the horizon. A high sink rate

will develop if airspeed becomes excessively low, resulting in a hard landing. Refer to the Appendix for landing ground roll distances.

Use the maximum performance landing speed schedule when wet runway conditions exist. The lower touchdown speeds minimize braking distance in the event of drag chute failure.

NOTE

- Throttle movement should follow quadrant curvature so that the hidden ledge at the IDLE position can prevent inadvertent engine cutoff.
- Sudden fogging has occurred without warning as the throttles were retarded during the landing flare. Immediately turn both air system switches OFF if this occurs.
- Pitch angle at touchdown must not exceed 14 degrees to avoid scraping the tail.

AFTER TOUCHDOWN

1. Drag chute - DEPLOY.

   Deploy chute when main gear is on the runway and pitch angle is 10 degrees or less.

   CAUTION

   Deploying the drag chute at a pitch angle greater than 10 degrees may result in the chute canopy contacting the runway surface and receiving scuff damage.

   Chute deployment requires approximately 5 seconds.

   Start the nose down as the drag chute deploys. Aerodynamic braking is not needed with normal chute deployment. Excessive nose gear loads may result on contact if a high angle of attack is maintained to low airspeeds.

2. Nosewheel steering - Engage.

   Engage nosewheel steering when the nosewheel is on the runway. Steering is not effective until the rudder pedals and nosewheel are aligned and weight of the aircraft is on at least one gear.

   NOTE

   Check that STEER ON engage light is illuminated.

3. Brakes - Checked.

   Check for normal brake operation by light application prior to jettisoning drag chute. Brakes can be applied any time after the nosewheel is on the runway.

   The normal performance procedure should only be used on a dry runway. Apply brakes as required. Light braking is sufficient if the drag chute deploys normally. If the drag chute does not deploy, moderate braking is necessary at normal weights.

If the normal performance procedure is used on a wet runway, light to moderate braking pressure is required with the drag chute deployed. If the drag chute fails to deploy, shut down the right engine and use moderate up-elevon to increase drag and the load on the main gear. If braking is ineffective and a safe stop cannot be made, select ANTI-SKID OFF and skid the wheels.

4. Drag chute - JETTISON.

For normal landings, the drag chute should be jettisoned above 55 KEAS; however, do not jettison the drag chute if the crosswind component exceeds 12 knots, or if braking action is unsatisfactory.

CAUTION

If the drag chute is not jettisoned, the elevons should not be moved during taxiing as the shroud lines may jam between the inboard elevons and fuselage and cause structural damage.

NOTE

The drag chute cannot be jettisoned after using the emergency deployment system.

CROSSWIND LANDING

Refer to Section V for Landing Gear System limits which are applicable to crosswind landing conditions. Also refer to the Crosswind Component chart in the Appendix, figure A2-1.

Use a normal traffic pattern for a crosswind landing, making proper allowance for velocity and direction of the crosswind. Correct runway alignment on final approach can be maintained by crabbing or dropping one wing, or both. Remove the crab before touchdown, using the wing-low technique to prevent side drift.

Reduce sink rate to a minimum to accomplish a smooth touchdown. As crosswind components increase, sink rate must be minimized due to the increased side loads imposed on the landing gear.

CAUTION

It is essential to remove all crab before touchdown to minimize scuffing damage to the tires on contact.

Crosswind Condition With Dry or Grooved Runwa

Touchdown and try to remain on the upwind side of the runway. This provides more runway space on the downwind side, and puts the crosswind and runway "crown" effects in opposition. Deploy the drag chute early in the landing roll, as for a normal landing, but lower the nosewheel first and engage steering if the crosswind component is over 15 knots.

The chute's tendency to pull the aircraft off the runway in a crosswind will actually decrease as speed is reduced, and the effect will be easily controllable with nosewheel steering. Keep the stick forward to increase nosewheel steering effectiveness. Without nosewheel steering, increasing steering with the elevons or increasing rudder deflection will be required as the speed decreases.

Do not shut down either engine when on a dry runway or on a grooved runway which provides equivalent braking.

Crosswind Condition With Slippery Runway

For landing on a slippery runway with a crosswind, start the nose down immediately on landing and engage nosewheel steering before deploying the drag chute. After the nose is lowered, lateral stick deflection can be used instead of or together with aerodynamic steering with the rudders to increase directional control. Use roll inputs in the same direction as rudder and nosewheel steering. This also increases braking on that side when used in combination with neutral or aft stick.

With a slippery runway, shutdown of one engine is permissible in the event of drag chute failure if this is required to assist in stopping. Shutdown the upwind engine when under 100 KIAS, and select ALT BRAKE if continuing on the right engine alone. Shutdown is not recommended if barrier engagement is the alternate action available.

The nosewheel steering system provides adequate control in allowable crosswinds on slippery runways, even with damaged main gear tires. However, care must be taken not to overcontrol the aircraft and start a lateral skid. The nosewheel steering force can be very large and this force, combined with the reduced side reaction force capability of the main gear tires, may cause the main gear tires to "break away" and slide. The nosewheel steering force reaches a maximum at a 13 1/2 degree angle between the tires and the ground track. This corresponds to 6 degrees rudder deflection with the aircraft heading along the ground track.

MINIMUM ROLL AND HEAVYWEIGHT LANDING (Maximum Performance)

Use this procedure when minimum landing distance is desirable or for wet runway landings. With 100,000 pounds gross weight or less, reduce touchdown speeds 10 KEAS from normal values. Start drag chute deployment as soon as possible after the main gear is on runway, then start the nose down immediately, so that deployment is accomplished at less than 10° pitch angle. Apply maximum braking as soon as the nosewheel is on the runway. Engage nosewheel steering. The right engine may be shut down after landing to lessen the landing roll by reducing thrust.

WARNING

- Do not shut down both engines because this will result in the loss of nosewheel steering and normal brakes when engine rotor speed decays.
- Do not deploy the drag chute in flight.

Heavyweight landings with over 100,000 pounds gross weight should be avoided but can be accomplished if required. Use normal final approach speeds if possible, but do not exceed 11 degrees pitch angle. Intended touchdown speed should be 10 KEAS less than for the normal schedule. Observe rate of sink limits for touchdown from Section V. When touchdown speed is less than the limit for chute deployment, lower the nose and deploy the drag chute immediately. If touchdown speed is higher, lower the nose as the limit speed is approached and deploy the chute at or below limit airspeed. In order to minimize the possibility of tire failure at heavy weight, it is preferable to apply brakes early in the landing roll. Reduce the distance travelled at high speed as soon as possible rather than delay braking until the brake energy rating speeds are reached. Refer to Abort procedure in Section III for action required during heavyweight deceleration.

Dry Runway Procedure

Lower the nose while deploying the drag chute and apply moderate brake pressure if at light weight or maximum brake pressure if at heavy weight. Do not delay lowering the nose and applying brakes if the drag chute is slow to deploy.

Wet Runway Procedure

Lower the nosewheel to the runway after chute deployment, then use light to moderate braking pressure. Do not jettison the drag chute. If the chute is slow to deploy or does not deploy, use moderate up-elevon to increase the load on the main gear and for aerodynamic braking and shut down the right engine. Use ANTI-SKID OFF braking.

Icy Runway Procedure

Use same techniques as for wet runway landing procedure.

GO-AROUND

A go-around may be initiated at any time during the approach, or during the landing roll when sufficient runway remains for takeoff if no attempt to deploy the drag chute has been made. (See figure 2-10.) For go-around after touchdown, lower pitch attitude to approximately 5 degrees nose up, then adjust attitude to takeoff at 210 KEAS.

GO - AROUND

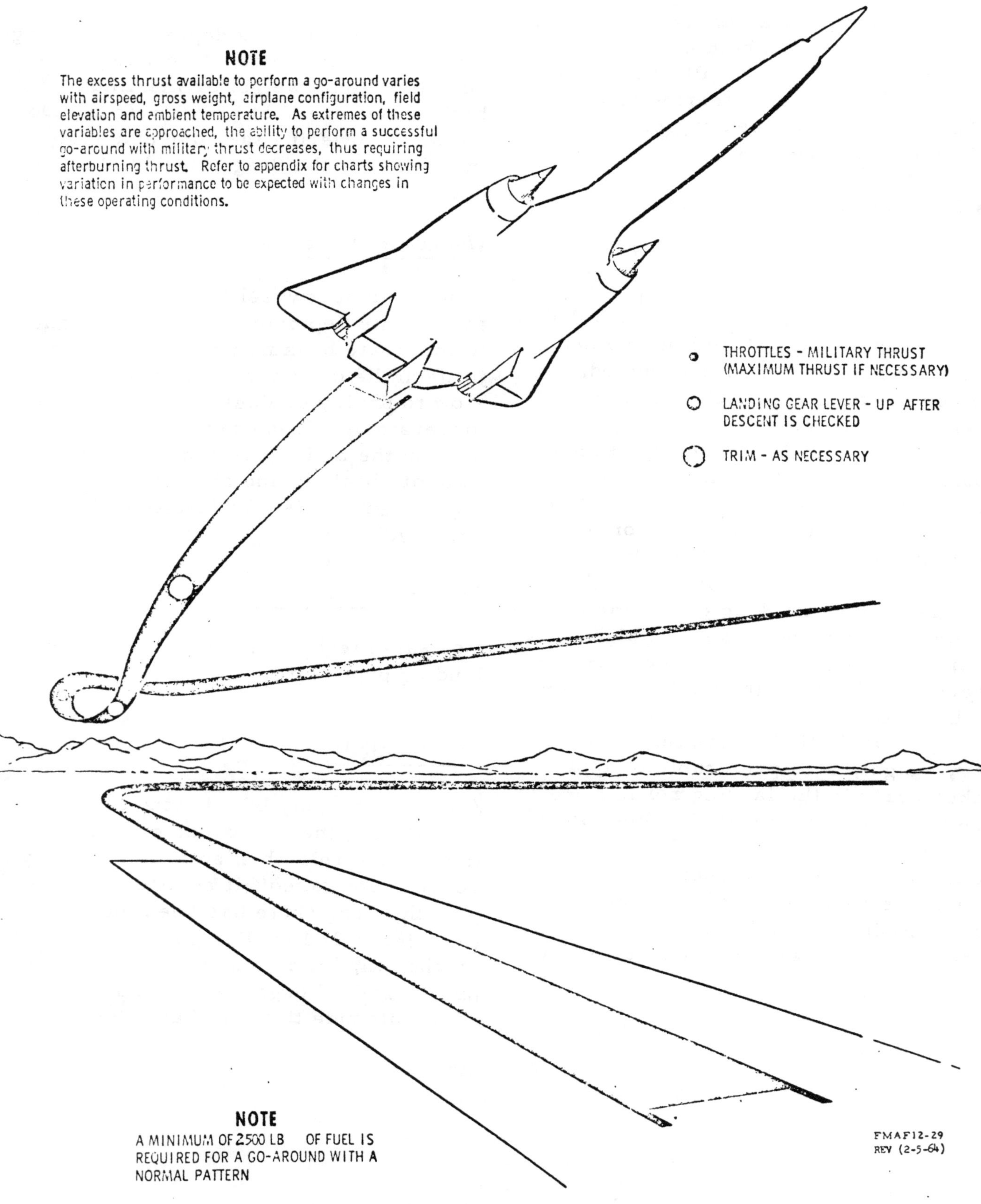

NOTE
The excess thrust available to perform a go-around varies with airspeed, gross weight, airplane configuration, field elevation and ambient temperature. As extremes of these variables are approached, the ability to perform a successful go-around with military thrust decreases, thus requiring afterburning thrust. Refer to appendix for charts showing variation in performance to be expected with changes in these operating conditions.
THROTTLES - MILITARY THRUST (MAXIMUM THRUST IF NECESSARY)
LANDING GEAR LEVER - UP AFTER DESCENT IS CHECKED
TRIM - AS NECESSARY
NOTE
A MINIMUM OF 2500 LB OF FUEL IS REQUIRED FOR A GO-AROUND WITH A NORMAL PATTERN
FMAF12-29
REV (2-5-64)

Figure 2-10

AFTER LANDING

1. SAS channel engage switches - OFF.
2. Landing lights - Set.
3. HF radio - OFF.
4. IFF - OFF.

When clear of the runway:

5. Pitot heat switch - OFF.
6. Right generator switch - OFF.

   Operate with generator off for at least three minutes in order to obtain adequate MRS record of electrical system performance.

7. Crossfeed - OFF
8. Right generator switch - ON

After R GEN OUT light extinguishes:

9. Left generator switch - OFF

   Continue through engine shutdown with left generator off.

ENGINE SHUTDOWN

1. Wheel chocks - Installed.
2. Exterior lights - OFF.
3. ▲ UHF and Nav radios - OFF.
4. Fuel derich systems - Checked and OFF.

   a. Set both engines 400 rpm above idle rpm.

   b. Derich test switch - Left.

   When EGT indication exceeds 860°C:

   c. Verify EGT gage and fuel derich lights on.

   d. Note engine speed decreases 50 to 400 rpm.

   e. Recycle fuel derich switch to ARM.

   f. Repeat b, c, and d for right derich check.

   Verify engine rpm returns to 400 rpm above idle rpm and EGT indication is normal.

   g. Fuel derich switch - OFF.

   h. Throttles - IDLE.

5. FCS liquid cooling switch - OFF.
6. MRS power switch - OFF.
7. Right generator switch - OFF.
8. Throttles - OFF.

   Ascertain that the compressor decelerates freely. The engine exhaust nozzle should achieve a position between 50% and 100% full open with the engine rotor at rest.

9. ▲ Canopy seal pressure lever - OFF.
10. ▲ Canopy - Open.
11. ▲ Seat and canopy safety pins - Installed.
12. Battery and appropriate electrical switches - OFF.

NORMAL OVER THE SIDE EGRESS - Stabilized Seat

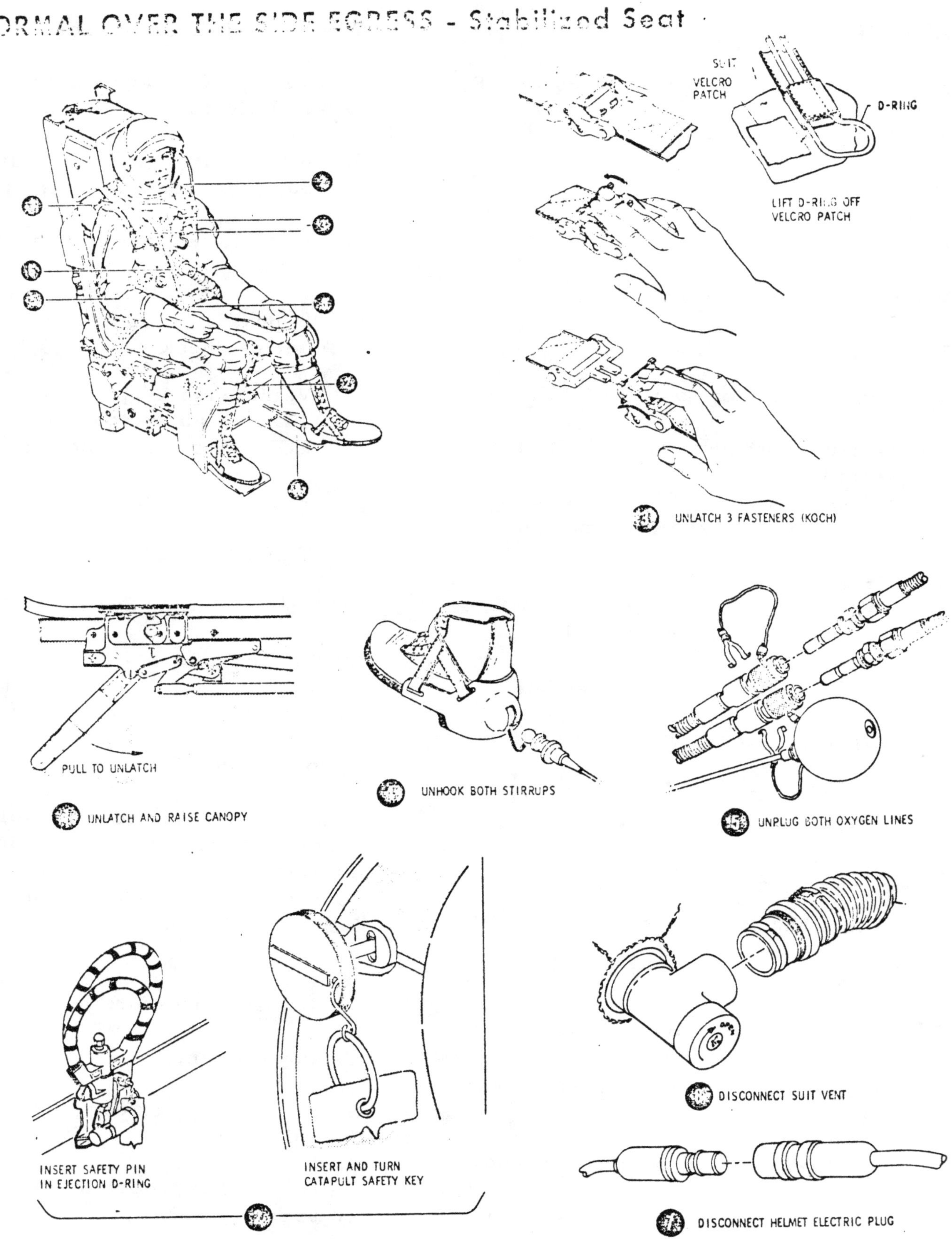
SUIT
VELCRO PATCH
D-RING
LIFT D-RING OFF VELCRO PATCH
UNLATCH 3 FASTENERS (KOCH)
PULL TO UNLATCH
UNLATCH AND RAISE CANOPY
UNHOOK BOTH STIRRUPS
UNPLUG BOTH OXYGEN LINES
INSERT SAFETY PIN IN EJECTION D-RING
INSERT AND TURN CATAPULT SAFETY KEY
DISCONNECT SUIT VENT
DISCONNECT HELMET ELECTRIC PLUG

Figure 2-11

Section III

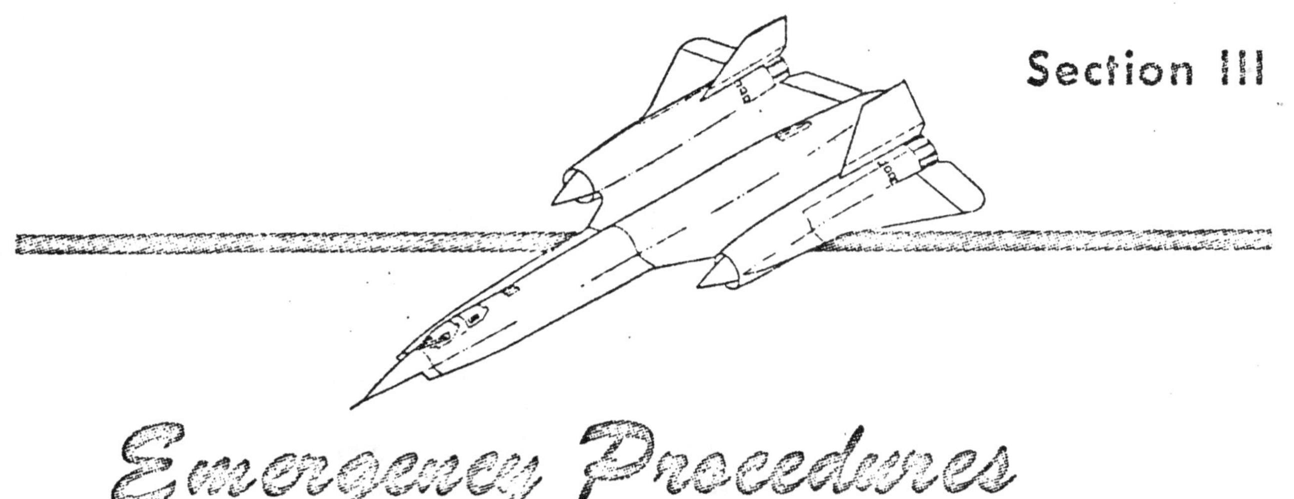

Emergency Procedures

TABLE OF CONTENTS

TABLE OF CONTENTS (Cont.)

INTRODUCTION

This section recommends procedures for use in the event of emergencies or other abnormal operating conditions. The procedures are designed to provide a margin of safety in the event of further failures. The safest course of action to recover the crew and aircraft will be recommended and should be followed unless circumstances such as weather, fuel, or other reasons dictate otherwise. The safest region for continued operation with most malfunctions is the subsonic speed range, provided altitude or aircraft range is not a factor.

Multiple Emergencies

These procedures are based on the assumption that each crewmember fully understands the normal operation of the aircraft and system and, generally, multiple emergency situations are not covered. Crewmembers must recognize that single malfunctions will often affect operation of other aircraft systems and will require corrective actions in addition to those contained in a specific emergency procedure.

Use of Checklists

Critical emergency checklist steps appear in capital bold print letters. They must be committed to memory and, in the event of an emergency, accomplished immediately without reference to the abbreviated checklist to prevent a delay that might aggravate the emergency. The remainder of the checklist steps should be accomplished using the challenge and response method as quickly as time and circumstance permit. The most important consideration is still to maintain control; therefore, checklist steps which are not bold print items should be accomplished only after aircraft control is definitely established. Where an emergency situation requires use of more than one of the listed procedures, a reference to the other procedure(s) is included on the abbreviated checklist.

Assumptions

In constructing the checklists, three basic assumptions have been made which have to be remembered in order to avoid reiteration in each individual procedure. These are: (1) Aircraft control is paramount. (2) Circuit breakers associated with a malfunctioning system must be checked. (3) The other crewmember must be alerted in the event of any emergency situation, and Hot Mike will be used, if advantageous, to facilitate crew coordination.

Symbol Coding

Symbols used to identify crew responsibility are the same as for the normal procedures checklists. These are:

Symbol	Action required by:
(none)	Pilot
▲	Both crewmembers
○	FCO

Definitions of Landing Situations

The terms "land when practicable" and "land as soon as possible" are not used interchangeably.

Land when practicable means land at home base or other suitable alternate. Air refueling as necessary to reach the selected base is allowed.

Land as soon as possible means land at the nearest suitable facility.

GROUND OPERATION

ENGINE FIRE

If a fire is evident during start, or on notification:

1. THROTTLES OFF.

2. FUEL OFF.

   Set both guarded emergency fuel switches to the fuel off position (up). During engine start, the ground crew should continue turning the engine if the fire is contained in the tailpipe. If the starter unit has disengaged, it can not be re-engaged until the engine has come to a complete stop.

3. Battery - OFF.

△ 4. Abandon the aircraft.

ABANDONING THE AIRCRAFT ON THE GROUND

In an emergency requiring ground abandonment, the primary concern is to leave the immediate area of the aircraft as soon as possible. The following procedure provides the fastest means of escape when fire or explosion are probable. The procedure may be initiated while the aircraft is in motion; however, the lap belt should not be released until the aircraft has stopped.

Aircraft On Fire

When the aircraft or the surrounding area is completely engulfed in flames, the crew may elect to use normal ejection procedures or abandon the aircraft, relying on the faceplate, helmet, and suit for protection.

Abandonment Procedure

△ 1. CANOPY OPEN OR JETTISON.

   Open or jettison the canopies first unless fire danger exists. Retain the canopies until all preparations for evacuation are completed if there is danger of fire engulfing the cockpit area.

   The recommended order for canopy jettison is pilot, then FCO, so that the pilot's canopy can not fall upon an open FCO cockpit and strike the FCO.

△ 2. SCRAMBLE HANDLE.

   Pull the scramble handle after the aircraft has come to a stop. This releases:

   (1) Lap belt.
   (2) Inertia reel shoulder harness.
   (3) Foot retention cables.

EMERGENCY OVER THE SIDE EGRESS-ON THE GROUND

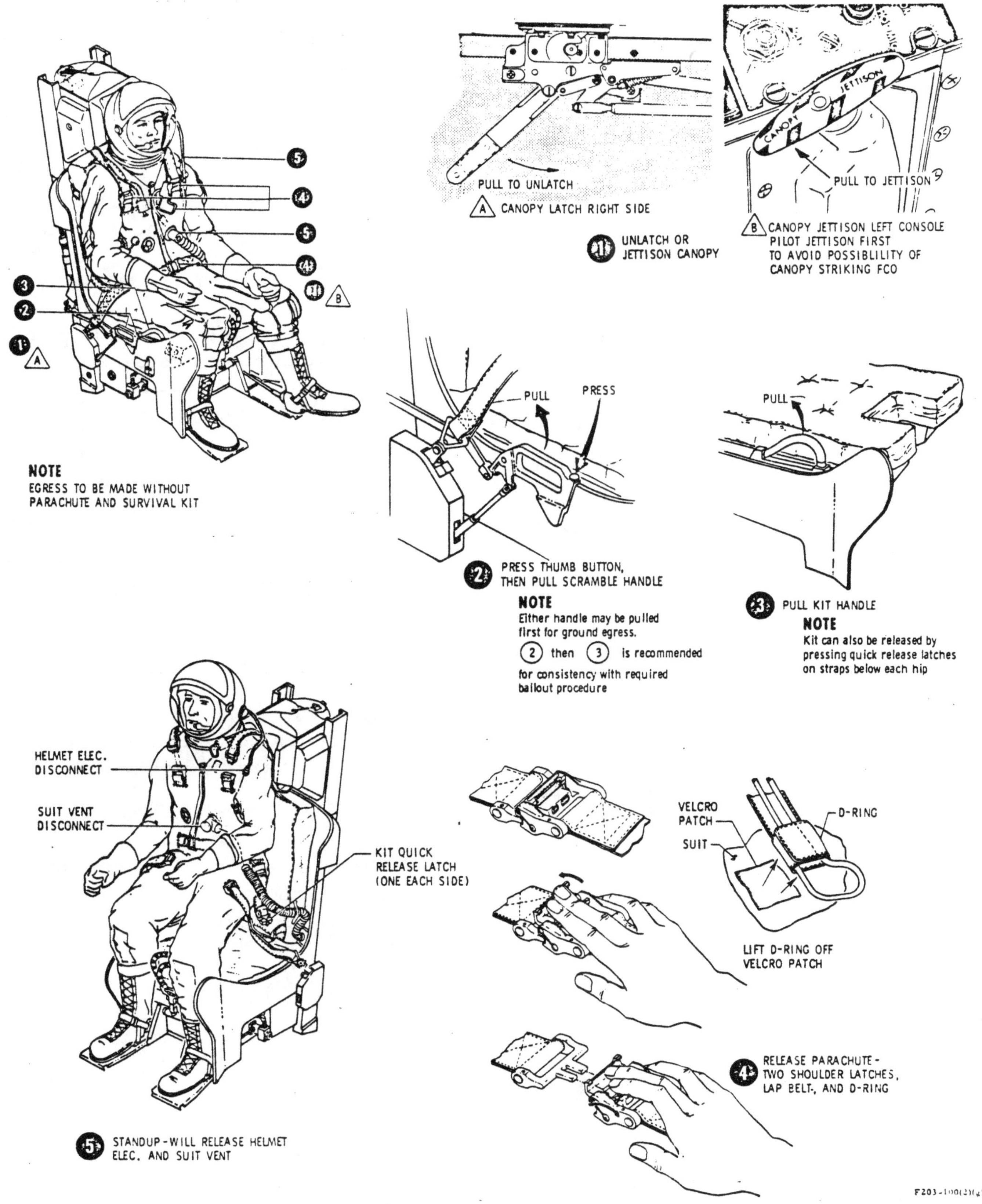

5
4
5
4
3
2
1 A
1 B

NOTE
EGRESS TO BE MADE WITHOUT PARACHUTE AND SURVIVAL KIT

PULL TO UNLATCH

A CANOPY LATCH RIGHT SIDE

PULL TO JETTISON

CANOPY JETTISON

B CANOPY JETTISON LEFT CONSOLE
PILOT JETTISON FIRST
TO AVOID POSSIBLILITY OF
CANOPY STRIKING FCO

1 UNLATCH OR JETTISON CANOPY

PULL

PRESS

2 PRESS THUMB BUTTON, THEN PULL SCRAMBLE HANDLE

NOTE
Either handle may be pulled first for ground egress.
(2) then (3) is recommended for consistency with required bailout procedure

PULL

3 PULL KIT HANDLE

NOTE
Kit can also be released by pressing quick release latches on straps below each hip

HELMET ELEC. DISCONNECT

SUIT VENT DISCONNECT

KIT QUICK RELEASE LATCH (ONE EACH SIDE)

VELCRO PATCH

SUIT

D-RING

LIFT D-RING OFF VELCRO PATCH

4 RELEASE PARACHUTE - TWO SHOULDER LATCHES, LAP BELT, AND D-RING

5 STANDUP - WILL RELEASE HELMET ELEC. AND SUIT VENT

F203-100(2)(a)

Figure 3-1

(4) Parachute arming lanyard and housing.
(5) Cable on ejection D-ring.

NOTE

After pulling the scramble handle, expect a loud report from the initiator firing.

WARNING

- The crewmember must remain seated until the survival kit handle is pulled.
- If the scramble handle does not function normally, the ejection seat safety pin should be installed to prevent inadvertent seat ejection. Then the harness, spurs, and lap belt must be released manually as required.

△3. KIT HANDLE.

Pull the survival kit handle next. This releases the kit from the torso harness, disconnects personal leads to the normal and emergency oxygen supplies and releases the parachute from the survival kit lid. It also detaches the kit lanyard from the torso harness if the kit has remained firmly in the seat.

The kit lanyard may remain attached to the torso harness if the crewmember does not remain seated until the kit handle is pulled.

The kit can also be released manually by pressing the quick relea latches below each hip. The right side latch also releases the kit lanyard.

△ 4. CHUTE RELEASE.

To release the parachute, open the quick disconnects at the shoulder and lap, and lift the chute D-ring from the suit.

Egress with the chute is possible if it can not be released.

WARNING

Mobility with the chute is limited.

Standing up will separate the helmet electrical connections and the suit ver hose.

BRAKE OR STEERING FAILURE

Brake or steering failure may or may not be indicated by illumination of the SKID OUT warning light.

Brake or Steering Failure Procedure

If normal brakes and/or nosewheel steering are not effective, or if L system hydraulic pressure is not available:

1. ALT BRAKE.

   Release brake pedal pressure, then move the brake switch to ALT BRAKE.

With ALT BRAKE selected, the power source for braking shifts to the R hydraulic system immediately. Steering shifts to the R hydraulic system after L system pressure decreases below 2200 psi. Nosewheel steering may remain inoperative after ALT BRAKE is selected if the nosewheel steering system has failed or if L hydraulic pressure remains above 2200 psi. In this event, steer by differential braking.

NOTE

In the event that both engines are shut down while the aircraft is moving, the brake switch should be set to the SKID OFF position and steady pressure applied until completely stopped; otherwise, anti-skid cycling or pumping the brakes could deplete the hydraulic system accumulator and result in loss of brakes.

If alternate brakes are ineffective:

2. SKID OFF.

Set the brake switch to the SKID OFF position. The power source for braking and steering is the L hydraulic system. The anti-skid system is disabled.

ANTI-SKID OUT

The ANTI-SKID OUT caution light illuminates while on the ground if the brake switch is positioned to SKID OFF OR ALT BRAKES or if the anti-skid system is disabled or fails.

With the ANTI-SKID OUT light on unaccountably:

1. Anti-skid - Recycle.

Attempt to recycle the anti-skid brake system by repositioning the brake switch if the situations permit and if there is no apparent reason for the system being disabled.

If the ANTI-SKID OUT light persists:

2. Skid - OFF.

Set the brake switch to the SKID OFF position.

Without anti-skid operating, extreme caution must be used while braking to prevent wheel skid.

Skidding is hard to detect due to aircraft size and weight. Tires may fail before a skid condition can be recognized and corrected. A main landing gear tire blow-out may be sensed by the pilot as a thump or muffled explosive sound.

TIRE FAILURE

At heavy weight, taxi only as necessary to clear the runway if one or more tires have failed. Taxiing to clear the runway is permitted with all tires failed.

EMERGENCY ENTRANCE

In the event that qualified ground personnel are not available, emergency entrance to the aircraft can be accomplished using the procedures illustrated by figure 3-2.

TAKEOFF EMERGENCIES

PROPULSION SYSTEM

The components considered as parts of the propulsion system include the main engines, afterburners, inlets, nozzles, tailpipes, fuel controls, and fuel-hydraulic, lubrication, and ignition systems. If abnormal operation of any of these components is indicated prior to reaching the acceleration check distance, the takeoff should be aborted. Refer to ABORT procedure, this section.

ENGINE FAILURE

If conditions permit and gear retraction has not been initiated:

1. ABORT.

Abort if abnormal operation of any of the propulsion system components is indicated before reaching the acceleration check distance.

CRASH RESCUE PROCEDURES

INSERT TOOL INTO ONE-HALF INCH SQUARE DRIVE OPENING AND ROTATE CLOCKWISE TO OPEN (BOTH COCKPITS)

TWO MEN ON EACH SIDE OF EACH CANOPY GRASP CANOPY AT THE MOST FORWARD POINT AND ROTATE CANOPY AFT ABOUT HINGE LINE SO COMPLETE ACCESS TO COCKPITS IS POSSIBLE FOR PILOT AND FCO REMOVAL. USE CAUTION.

OR

REMOVE JETTISON ACCESS COVER (A) BY PRESSING QUICK DISCONNECT. REMOVE PULL HANDLE. UNCOIL EXCESS CABLE, APPROX. 6 FEET

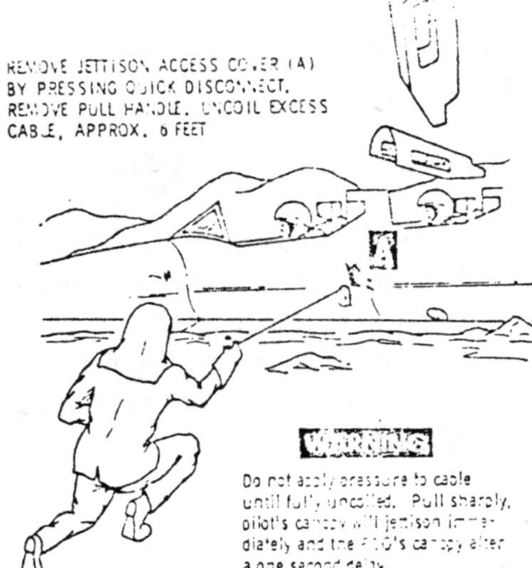

WARNING

Do not apply pressure to cable until fully uncoiled. Pull sharply, pilot's canopy will jettison immediately and the FCO's canopy after a one second delay.

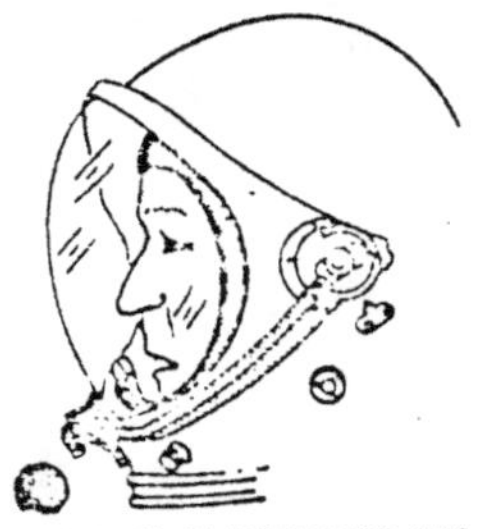

J SUIT-RAISE BAYLOR BAR

UNHOOK PILOT'S PERSONAL EQUIPMENT

WARNING

Do not puncture the nacelle in the TEB tank access area, in order to gain access to the interior of the nacelle. A violent fire and or explosion may result if the TEB tank is ruptured.

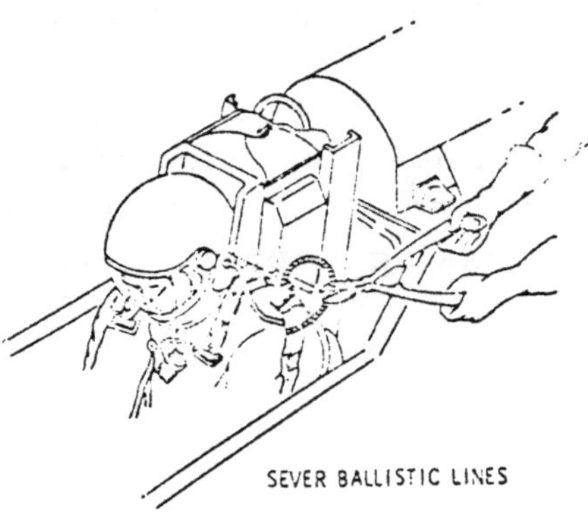

SEVER BALLISTIC LINES

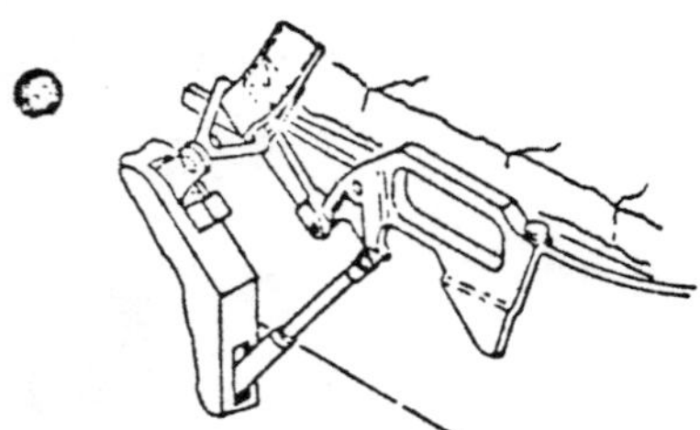

PRESS BUTTON AND PULL SCRAMBLE HANDLE

THREE MEN ARE REQUIRED TO REMOVE PILOT OR FCO, ONE ON EACH SIDE AND ONE ASTRIDE THE COCKPITS IN FRONT OF PILOT OR FCO

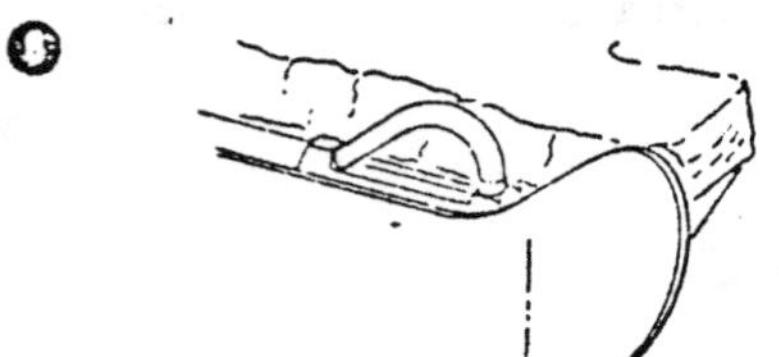

PULL KIT HANDLE

F203-103(d)

Figure 3-2

Abort if the acceleration check is unsatisfactory, or if a fire warning occurs before the refusal speed is attained.

Abort if the thrust of either engine decays to the point that minimum single engine flight speed can not be attained, provided that conditions permit and landing gear retraction has not been initiated.

In the event that both engines fail immediately after takeoff, decay of engine rpm will result in rapid loss of A and B hydraulic system pressure and subsequent loss of aircraft control. LAND STRAIGHT AHEAD IF THE GEAR IS DOWN AND SUFFICIENT RUNWAY IS AVAILABLE.

After takeoff, if unable to hold altitude and accelerate:

△ 1. EJECT.

   If gear retraction has been initiated, eject rather than attempt to land with the gear partially retracted or up.

If able to maintain altitude or accelerate:

1. THROTTLES MAX THRUST.

   If an engine fails immediately after takeoff and the decision is made to continue, maintain Maximum thrust on the operating engine. Lateral and directional control can be maintained when airspeed remains above the minimum single engine control speeds shown on figure 3-3; however, ability to maintain altitude and accelerate or climb depend on weight, drag, altitude, airspeed, and temperature. The appendix provides performance information.

2. GEAR UP.

   Initiate gear retraction if not already accomplished.

When the aircraft is under control:

3. Fuel dump switch - DUMP.

   Fuel dumping in addition to consumption by the operating engine lightens the aircraft at an appreciable rate. When at heavy weight for the existing air temperature, dumping fuel may reduce weight sufficiently to remain airborne. If turning at a sufficient speed, the inoperative engine will also discharge fuel from its afterburner.

   Monitor c.g. carefully if dumping with crossfeed open.

4. Rudder trim - As necessary.

   Bank and sideslip toward the operating engine as necessary to maintain directional control and minimize drag. 7 to 9 degrees of rudder trim, with bank and sideslip as needed to maintain course, yields minimum drag in the critical speed range from 220 to 250 KEAS.

Failed engine:

5. Complete shutdown or airstart, as appropriate.

WARNING

Positively identify the failed engine before retarding the throttle.

SINGLE ENGINE MINIMUM AERODYNAMIC CONTROL SPEED

YJ-1 ENGINES

ONE ENGINE - MAX THRUST
ONE ENGINE - WINDMILLING
20° RUDDER DEFLECTION

BASIS-ESTIMATED DATA -
FROM REVISED W/M DRAG ESTIMATE,
FLT TESTS, AND JJ ENGINE DATA

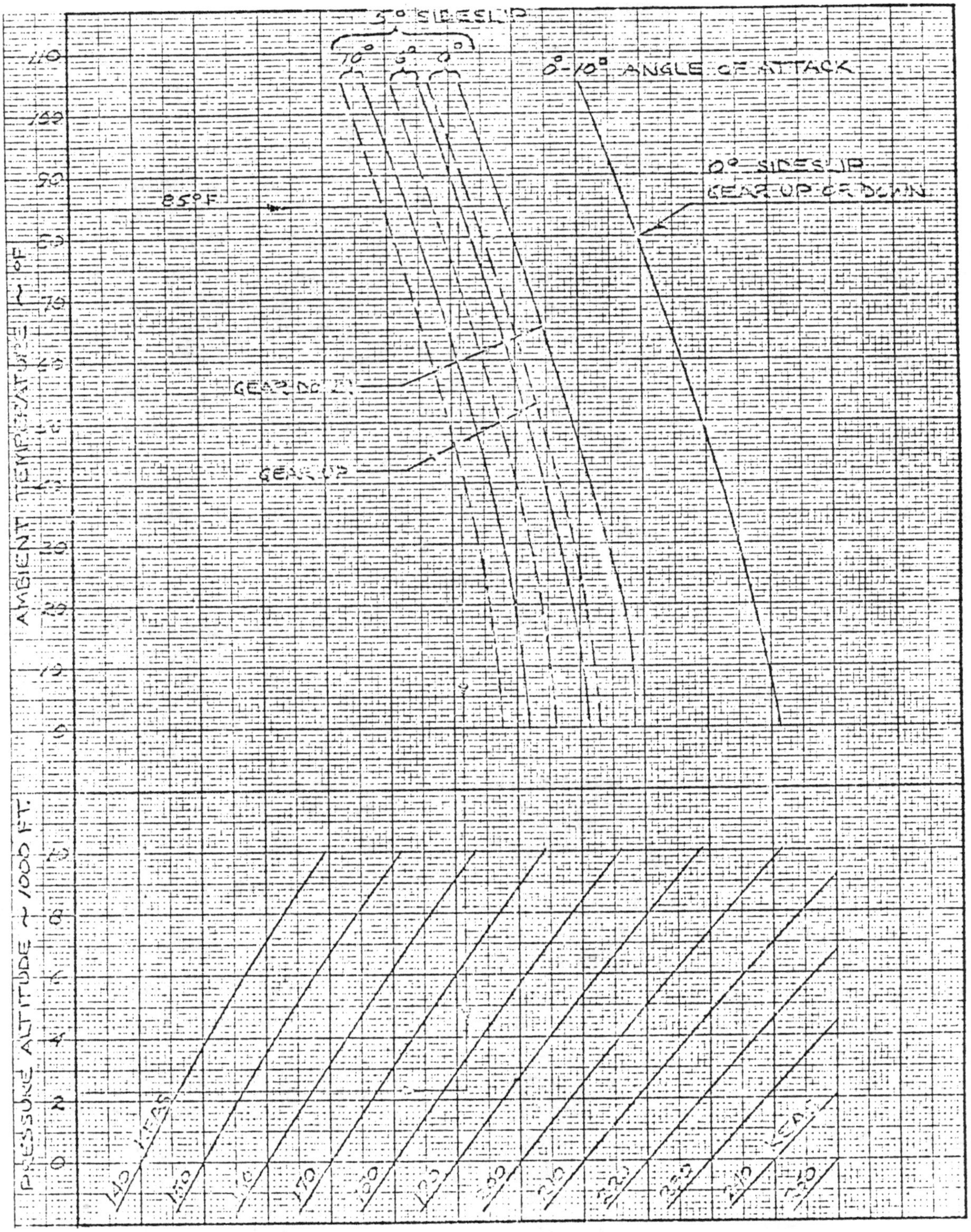

Figure 3-3

AFTERBURNER FAILURE DURING TAKEOFF

Abort if an afterburner fails prior to reaching the acceleration check speed. Refer to the Abort Procedure, this section.

In the event that an afterburner fails after reaching the acceleration check speed, confirm that both throttles are at the maximum afterburner position and continue the takeoff. Check EGT and for derichment. When safely airborne positively identify the affected engine and then retard that throttle below the afterburner operating range. Pause at the Military power position if the nozzle position indication is near closed, then check nozzle operation by retarding the throttle until the nozzle starts to open. A relight may be attempted if engine instrument indications and observation of the nacelle with the periscope disclose normal conditions; however, a malfunction should be assumed and a landing made as soon as possible.

AFTERBURNER NOZZLE FAILURE

A landing should be made as soon as possible in the event of a nozzle failure.

Nozzle failure may be indicated by nozzle position, excessive rpm fluctuations, or failure of the engine to control to scheduled speed. This may be accompanied by compressor stall and exhaust gas overtemperature. Engine shutdown may be necessary.

Also refer to the Afterburner Nozzle Failure discussion under In-Flight Emergency Procedures, this section.

Nozzle Failed Open

If an engine exhaust nozzle fails open and takeoff is continued, keep the throttle in the maximum afterburner position until a reduction in thrust is possible. Anticipate possible engine overspeed and be prepared to reduce throttle position below Military as soon as flight conditions permit.

Nozzle Failed Toward Closed

In the event of a nozzle failed toward closed condition, expect EGT rise, compressor stall, and the possibility of engine flameout.

Affected engine:

1. Throttle - Military or below, as required.

   Do not attempt to relight the afterburner, as the engine may flameout. (If engine flameout occurs, it can not be restarted due to the reduced rpm which will result.)

2. RPM and EGT - Maintain within limits.

   Compressor stall is likely, and EGT will probably rise.

FIRE

Assume that an abnormal condition exsists in the propulsion system and abort if either fire warning light illuminates before the refusal speed is reached; otherwise, continue the takeoff and proceed as for fire warning in flight and then land as soon as possible.

ABORT

The abort procedure assumes that a decision to abort will be made before rotation speed is reached. Aborts from above rotation speed are not prohibited, but the risks associated with aborting from such a high initial speed at takeoff weight must be balanced against those of continuing a takeoff when making the decision. In general, after rotation speed is reached, the most reasonable course of action is to continue rather than abort unless the emergency is such that the aircraft can not fly.

Engine Management

Both throttles should be retarded to IDLE and the brakes applied with the nose down as soon as the decision to abort is made. The planned rotation speed may be exceeded; however, the nosewheel should be kept on the runway to take advantage of nosewheel steering in combination with rudder control.

NOTE

In the event of chute failure, shutdown the right engine after both are idling or complete the shut down of a failed or flamed out engine. This reduction in thrust is necessary to improve braking and to reduce the possibility of tire failure due to extended ground roll distances.

WARNING

Wait until rpm and EGT show that both engines are idling or that one engine is failing before selecting the engine to shutdown. Loss of both engines may result in loss of hydraulic pressure for braking.

Aircraft Attitude, With Decision to Abort

Lower the nose and energize the brakes simultaneously with nosewheel contact. When rotation is well advanced, the aircraft may accelerate beyond takeoff speed and lift off before rotation can be checked. In this case, hold the aircraft off sufficiently to regain control and then touchdown without sideslip if possible. Fly the aircraft back to the runway, attempting to regain the center.

Chute Deployment

The drag chute requires 4 to 5 seconds for deployment after drag chute control actuation. If above 210 KEAS, it is permissible to actuate the deploy switch while decelerating in anticipation of reaching the limit airspeed for chute

deployment; however, premature deployment can result in destruction of the chute. Actuation of the chute system so as to reach limit airspeed simultaneously with loading of the chute is not recommended unless the risk is justified by a very marginal distance remaining situation.

Braking On Wet Runways

Unless hydroplaning is encountered, good nosewheel and rudder steering characteristics can be expected. Well controlled stops have been demonstrated on wet runways with and without the drag chute, with all main gear tires blown and wheels locked, and with one engine shut down.

Hydroplaning in various forms is a limiting factor with wet runway conditions and, although nosewheel and rudder steering remain effective, wheel braking force is nil until the tires can make contact with the runway. The aircraft tends to follow a trajectory and will drift with a crosswind.

Except for the extended stop distance involved, skids across or into dry runway areas are the chief hazard of wet runway stops. The wheels tend to lock-up and cause blown tires while sliding on a wet surface. Dry areas tend to destroy the tires due to increased friction or wheel spin-up. This allows the wheels to make runway contact and may ultimately destroy the wheels and then the brake assemblies. Even so, the aircraft can probably survive on the landing gear struts so long as it remains on the main runway, or on a hard surface overrun if there is a smooth transition from runway to overrun.

Drag Chute Failure

If the drag chute does not deploy, shut down the failed engine (or shut down the right engine if there has been no engine failure) in order to reduce thrust and increase braking effectiveness. Also use moderate up elevon so as to provide as much drag as possible without lifting the nosewheel. The increased gear load may cause tire failure at heavy weight; however, tire failure may be acceptable since the tires will not necessarily disintegrate. Braking deceleration available is nearly the same for braked tire rolling and blown tire locked conditions with a smooth wet surface. Locked wheel skids on a wet ungrooved runway of up to 7000 feet have left the wheels undamaged.

ABORT PROCEDURE

WARNING

- Do not unfasten the lap belt, shoulder harness, or pull scramble handle until the aircraft has come to a stop.
- The landing gear should be left in the extended position.

1. THROTTLES IDLE.

Retard both throttles to IDLE. Do not attempt to shut down either engine immediately unless failure to do so would vitally endanger the aircraft, such as for engine fire.

2. BRAKES.

Lower nose and -

For dry runway: Use moderate to heavy brake pressure until stop is assured. Do not use up elevon, risk of tire fatigue failure is increased.

For wet runway: Use light to moderate brake pressure. Up elevon for additional drag may be used if braking is marginal or if the drag chute fails.

NOTE

- Rated brake energy capacities and associated maximum braking speeds may be disregarded during aborted takeoffs. It is considered better to use the brakes at high speed, as tire failure may occur if the roll is extended by delayed braking.
- On wet runways without grooves, deceleration is nearly the same with blown tires locked as with braked tires rolling.

CAUTION

Hard braking may result in brake seizure after stopping, increasing time to clear the runway. If possible, keep the aircraft moving at slow speed until clear of the runway. Taxiing at low speed to clear a runway is permitted with all tires failed on a main gear. The massive tire bead tends to protect the wheels for a short distance at heavy weight.

3. CHUTE DEPLOYED.

Deploy the chute on all aborts. The maximum airspeed for drag chute deployment is 210 KEAS.

If tire failure occurs:

4. SKID OFF

Select SKID OFF prior to brake application. Brake with steady application of pressure.

If tire failure occurs with either wet or dry conditions, increased brake pressure will be required on that side to maintain braking force on the remaining tires. Maintain enough brake pressure to prevent spin-up of the wheel and, possibly, wheel and/or tire disintegration at high rotational speeds.

For L hydraulic system failure:

5. Brake switch - ALT BRAKE

Set the brake switch to ALT BRAKE when the L hydraulic system is below normal pressure, with left engine failure, or with drag chute failure on a wet runway.

CAUTION

Selection of ALT BRAKE changes the source of brake pressure from the L to the R hydraulic system, disables the anti-skid system and causes the ANTI-SKID OUT light to illuminate.

For fire or drag chute failure:

6. Throttle - OFF.
7. Fuel - OFF.
   Shut down one engine and operate its fuel shutoff in the event of fire or drag chute failure.
   a. Shut down the right engine if both engines are idling or if the right engine has failed or if the right side is involved in a fire.
   b. Shut down the left engine if it has failed or if the left side is involved in a fire.

WARNING

- Positively identify the failed engine before attempting engine shutdown.
- Shut down both engines immediately after stop if the aircraft is on fire and abandon the aircraft.

   c. Shut down both engines after clear of the runway if no fire is apparent.

WARNING

If there is no fire, delay engine shutdown until arrival of fire fighting equipment due to hazard of dumping fuel from nacelles on hot brakes.

If a choice of action permits, attempt to turn into the wind when stopping.

Prepare to engage the barrier if a suitable barrier is available and it appears that a reasonably safe stop can be made. If a safe stop is obviously impossible, ejection prior to reaching a runway overrun area or barrier is considered to be the most reasonable course of action.

If an ejection is to be accomplished, it should be initiated early enough to minimize risk of descent into a fire area.

WARNING

If the aircraft is involved in a major fire, ejection prior to barrier engagement is recommended. Burning fuel can be expected to engulf and spread ahead of the aircraft as it stops in the restraining cable.

BARRIER ENGAGEMENT

RUNWAY OVERRUN BARRIER

The following is applicable only to installations with the BAK-11 cable engaging systems installed with modified dual BAK-12 arresting engines.

Barrier Operation

The barrier is controlled by the control tower operator, and is armed by him prior to all takeoffs and landings. The pilot may call for disarming of the barrier if it is apparent that a safe stop can be made without it in the overrun area.

When armed, the barrier is operated by the aircraft nosewheel as it rolls over pressure-sensitive switchmats, located in the runway ahead of the main cable. The switches energize a timing computer which causes that length of the arresting cable under the vehicle to be thrown up to engage the main gear struts by a popup device. On engagement, the arresting cable is payed out with a relatively constant restraining force so as to stop the vehicle within 2000 feet.

Operating Restrictions

The maximum recommended groundspeed for barrier engagement is 180 knots at all gross weights. The minimum groundspeed is 30 knots with the model 8200, BAK-11 installation and 15 knots with the AMF BAK-11F installation. The barrier cable will not eject below these speeds.

The optimum barrier engagement is obtained when engagement occurs within the runway side stripe markings. A successful engagement can be expected, however, if the aircraft centerline is no closer than approximately 40 feet from the edge of the runway at the barrier. The probability of a successful engagement when closer than 40 feet from the edge of the runway may be marginal, especially at high speeds.

The nosewheel must be on the runway when crossing the switchmats and it is most desirable to contact the barrier squarely.

BARRIER ENGAGEMENT PROCEDURE

1. NOSE DOWN.

   Barrier switchmats must be crossed in a three-point attitude.

2. BRAKES RELEASED

To prevent exceeding structural limits of the struts, release brakes at the barrier. Continue steering as necessary. Do not jettison the drag chute.

3. Throttles - OFF.

4. Fuel - OFF.

WARNING

- If the aircraft is not on fire, delay engine shutdown until arrival of fire fighting equipment due to hazard of dumping fuel from nacelles in the vicinity of hot brakes.
- Do not unfasten the lap belt, shoulder harness, or pull the scramble handle until the aircraft has come to a stop.

LANDING GEAR AND TIRES

TIRE FAILURE

Long runs during taxiing or takeoff at heavy weight can result in blown tires due to crown or sidewall failures. A critical temperature for these tires is reached in the tire bead, approximately 455°F. Failure of a main gear tire during takeoff will overload the remaining tires on that side when takeoff weight exceeds 92,500 pounds; however, tire testing shows that the remaining tires, if they have been properly maintained, should be capable of sustaining a 50% overload for the remaining period required to take off at maximum weight or stop <u>if required cooling procedures are completed before starting the takeoff run</u>. (See figures 5-5 and 5-6.) As each main gear tire loss decreases the available brake energy capability by one-sixth, ability to stop from high speed is largely dependent on effectiveness of the drag chute.

Nosewheel Tire Failure

Failure of a nosewheel tire is not expected to generate a second tire failure, but it may not be possible to determine immediately whether a nose or main gear tire has failed. In either case, engine or structural damage may be sustained from tire fragments.

Tire Failure Procedure

Depending on the airspeed attained and whether or not engine damage is indicated, a takeoff may be preferable to aborting. The decision speed corresponds closely with speeds reached at the acceleration check point. Attempt to determine if engine damage has been sustained before reaching the refusal speed. The following procedure is recommended when a main or nose gear tire failure is suspected during the takeoff run.

In the event of tire failure before reaching acceleration check speed:

1. ABORT.

If tire failure occurs after reaching acceleration check speed,

1. CONTINUE TAKEOFF.

   Continue with normal takeoff procedure.

2. DON'T RETRACT GEAR.

   Leave the gear extended to minimize the possibility of damage in the wheel well.

3. SKID OFF.

   The brake switch SKID OFF position must be selected in order to stop the wheels rapidly after takeoff, as braking is disabled with ANTI-SKID or ALT BRAKE selected while the gear is down if there is no weight on the gear.

4. BRAKE WHEELS.

   The blown tire(s) must be stopped in order to minimize the possibility of damage to the aircraft.

5. Request confirmation of tire and aircraft conditions.

The gear should not be retracted until a visual check has been made by personnel in another aircraft or on the ground.

EMERGENCY GEAR RETRACTION

If the gear lever cannot be moved to the UP position after takeoff:

1. Gear override button - Press and hold.

   This overrides the solenoid which is normally actuated by the landing gear switch.

2. Landing gear lever - UP.

IN-FLIGHT EMERGENCIES

BAILOUT

Escape from the aircraft in flight should be made with the ejection seat. The following is a summary of ejection expectations:

a. At sea level, wind blast exerts only minor forces on the body up to 525 KIAS; appreciable forces from 525 to 600 KIAS; and excessive forces above 600 KIAS. The aircraft limit airspeed is below the speeds for excessive forces; however, when using an oxygen mask and regulator without a pressure suit, ejection should be delayed until below Mach 1.0 and 420 KEAS, slower when conditions permit.

b. Successful chute deployment should result after ejection from zero speed.

c. The free fall from high altitude down to 15,000 feet with drogue chute stabilization will result in stabilized descent in the quickest manner.

During any low altitude ejection, the chance for success can be greatly increased by zooming the aircraft to exchange excess airspeed for altitude. Ejection should be accomplished while the aircraft is in a level or climbing attitude. A climbing or level attitude will result in a more nearly vertical trajectory for the seat and crew members, thus providing more altitude

and time for seat separation and parachute deployment. The zero altitude capability of this aircraft should not be used as a basis for delaying ejection if ejection is necessary. Aircraft accident statistics emphatically show a progressive decrease in successful ejections as ejection altitude is decreased below 2000 feet. Whenever possible, eject above 2000 feet.

Before Ejection

Before ejection, when time and conditions permit:

1. Altitude - Reduce so that the pressure suit is not essential to survival.

2. Airspeed - Reduce to subsonic and as slow as conditions permit.

3. Head aircraft toward unpopulated area.

4. Transmit location and intentions to nearest radio facility.

⑤. IFF - EMER position.

▲6. Lower helmet visor.

▲7. Green apple - Pull if above 15,000 feet.

To Bailout

Accomplish as many of the following steps as is necessary to clear the aircraft. Refer to figure 3-4.

To eject, using the stabilized seat emergency escape system:

1. ALERT FCO.

Call "bailout, bailout, bailout" on interphone and set FCO BAILOUT switch to GO.

▲ 2. BODY POSITION.

Assume the proper body position. Sit erect with head against headrest. To pull ejection D-ring, cross arms (if possible) to assist in keeping arms close to the body.

▲ 3. EJECTION D-RING.

The FCO should eject first. The pilot should wait for the FCO CLEAR light to illuminate before ejecting if it is possible to do so without impairing his own ejection capability.

If the seat fails to eject, continue with the following after a normal delay:

▲ 4. CANOPY JETTISON.

Pull the canopy jettison handle. If the canopy still does not jettison, pull the canopy latching handle aft and then open the canopy and allow it to blow off in the air stream.

▲ 5. EJECTION T-HANDLE.

WARNING

Do not pull the T-handle with the canopy still in place.

Keep elbows close to sides and feet firmly against seat while pulling the ejection T-handle since the foot retractors and shoulder harness haul-back may not have actuated.

If an ejection seat is inoperative, the following procedure should be used to roll and drop out of the aircraft if sufficient control remains. If the FCO's seat has failed, the pilot should remain with the aircraft, assist the FCO to leave the aircraft, and then eject.

6. Airspeed - 250 to 300 KEAS.

△7. Green apple - Pulled.

△8. Scramble handle - Pulled.

The scramble handle is the outboard handle on the right side of the seat. It releases:

(a) Lap belt.

(b) Inertia reel shoulder harness.

(c) Foot retention cables.

(d) Parachute arming lanyard and housing.

(e) Cable on ejection D-ring.

WARNING

Do not pull the survival kit release (inboard) handle while in the seat, as this disconnects the emergency oxygen supply and then releases the kit and kit lanyard.

△9. Suit vent hose - Disconnected.

10. Trim full nose down, roll inverted.

△11. Lean forward and drop out.

After manual bailout, when clear of the aircraft and below 20,000 feet:

△12. Pull parachute <u>manual</u> deploy ring.

WARNING

- THE CREWMEMBER MUST DEPLOY THE MAIN CHUTE MANUALLY, USING THE CHUTE D-RING, AFTER SEPARATING FROM THE SEAT MANUALLY.
- The crewmember is not stabilized by the drogue chute and may be subjected to spinning or tumbling until the main chute is deployed.
- A free fall to a reasonably safe altitude should be attempted to avoid the possibility of serious chute damage due to high speed deployment.
- Visor heat will not be available.

After Ejection

After ejection, descent is normally made to approximately 15,000 feet while in the seat and with drogue chute stabilization. (Refer to figure 3-4.)

NOTE

Rotation or spinning of the seat may be experienced while descending with the drogue chute deployed. If it occurs, it may be possible to arrest such motions by using the arms and hands in the air stream.

In the event that the automatic man-seat separation sequence is inoperative, or if the crewmember elects to separate from the seat before automatic separation can occur at approximately 15,000 feet, the crewmember can initiate separation manually by pulling the scramble handle. This is the outboard handle on the right side of the sea[t]. It releases:

(a) Lap belt.

(b) Inertia reel shoulder harness

(c) Foot retention cables.

(d) Parachute arming lanyard an[d] housing.

(e) Cable on ejection D-ring.

EJECTION

1 POSITION FOR EJECTION

1. Sit erect with head against headrest
2. Feet firmly against seat
3. If possible cross arms, and pull ejection "D" ring

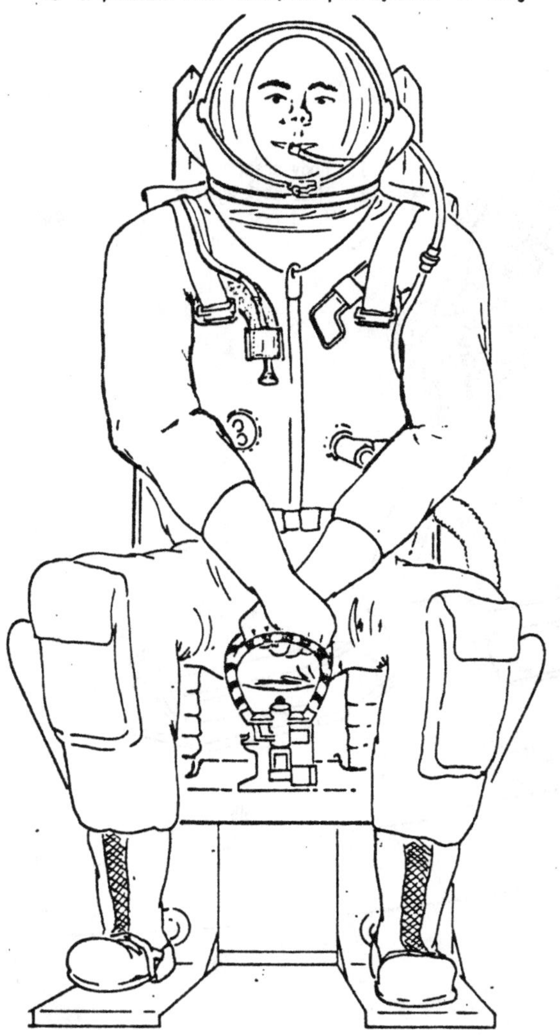

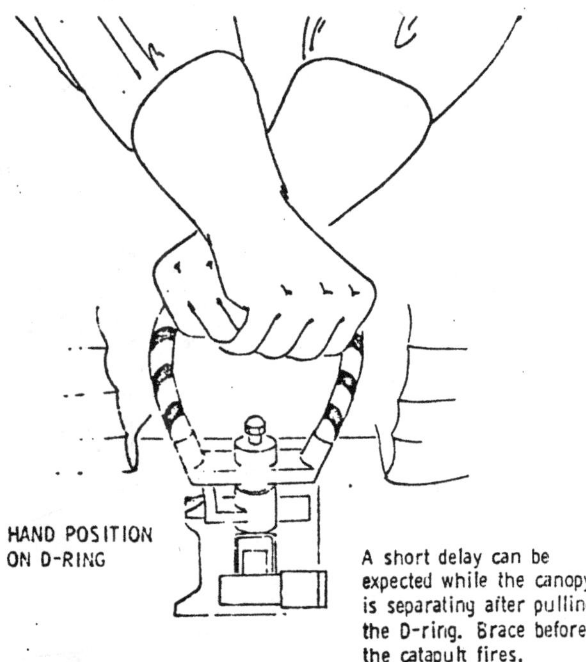

A short delay can be expected while the canopy is separating after pulling the D-ring. Brace before the catapult fires.

WARNING

If D-ring should fail to eject the seat the canopy must be jettisoned (**2**) before operating Secondary Ejection Handle (**4**) If canopy does not jettison (**2**), open manually (**3**)

2

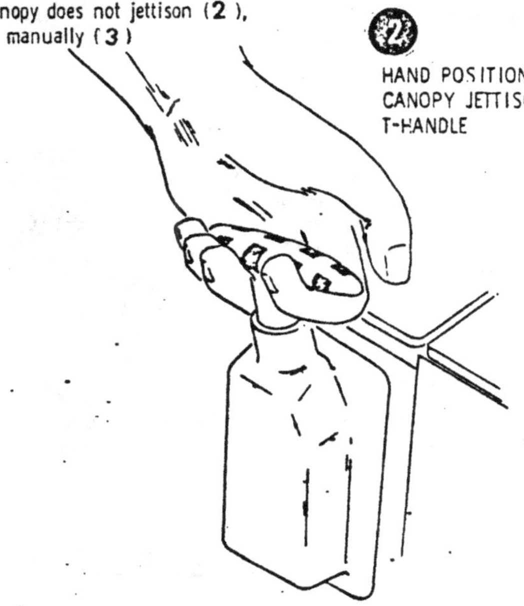

WARNING

Do not pull the Secondary Ejection T-handle with the canopy still in place

4

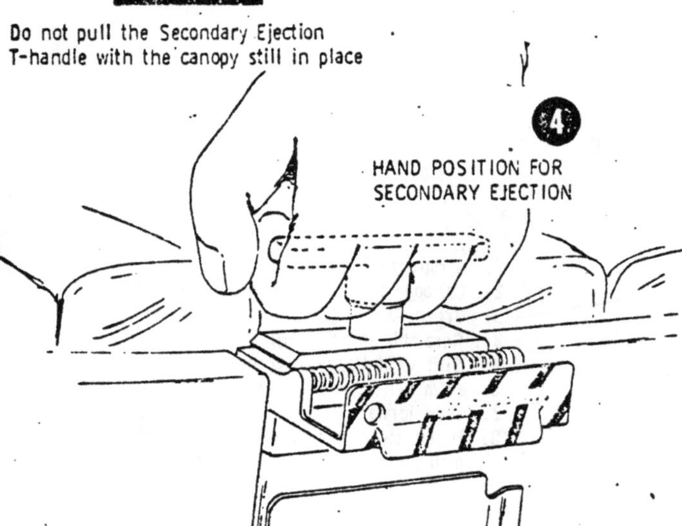

3

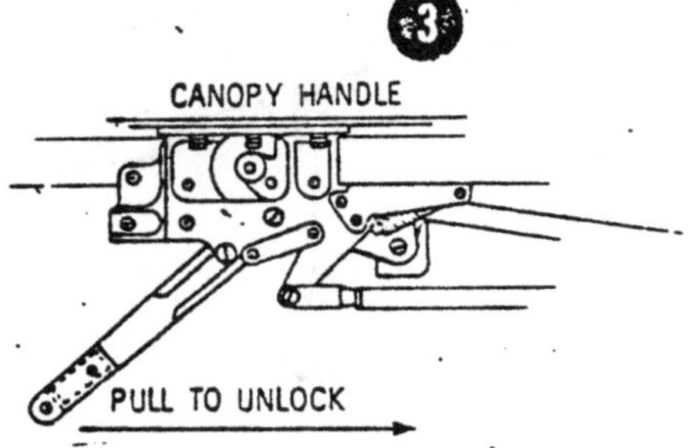

F203-96(2)(h)

Figure 3-4 (Sheet 1 of 7)

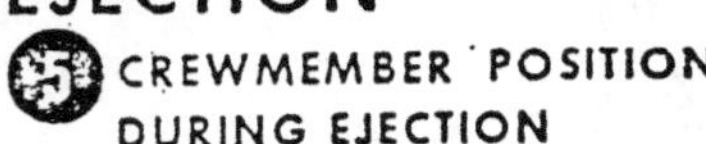

EJECTION

15 CREWMEMBER POSITION DURING EJECTION

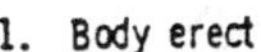

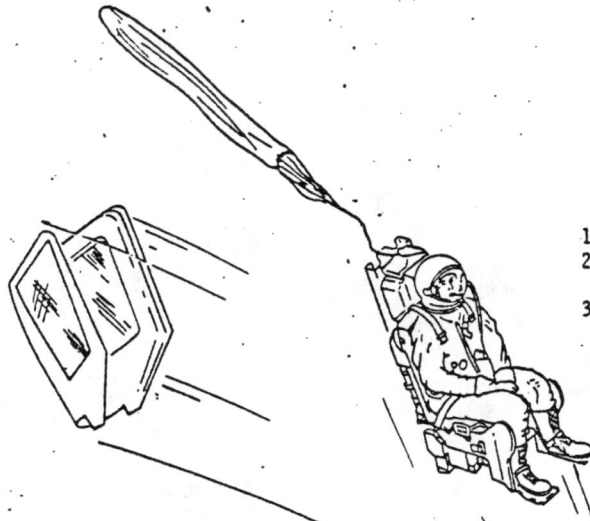

1. Body erect
2. Head back against head rest
3. Arms crossed

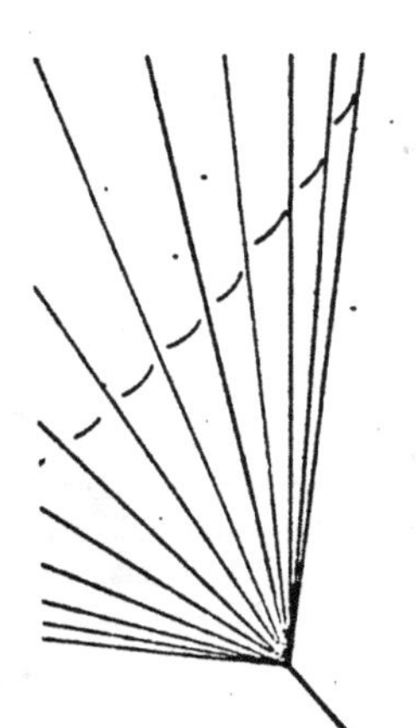

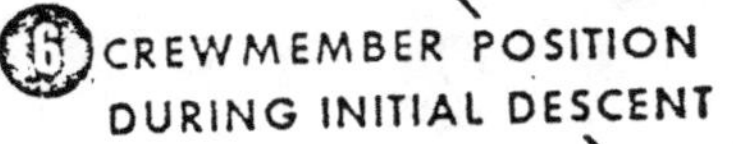

16 CREWMEMBER POSITION DURING INITIAL DESCENT

1. Maintain erect body position
2. Ride seat to automatic man-seat separation at 15,000 feet, or manually separate above 15,000 feet if desired

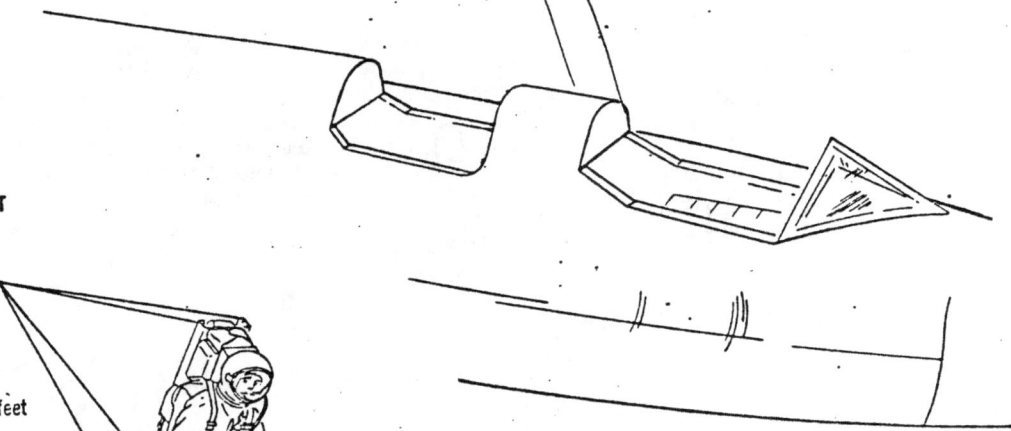

NOTE

Lower risers of drogue chute are severed 10 sec. after ejection

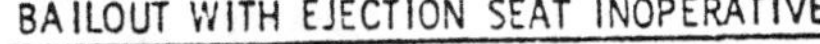

BAILOUT WITH EJECTION SEAT INOPERATIVE

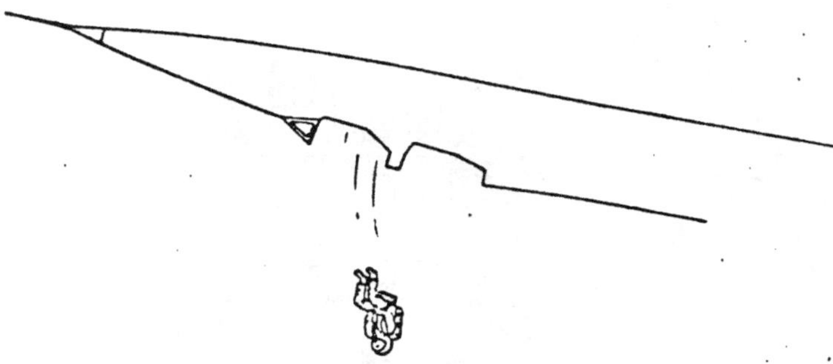

1. Airspeed - 250 to 300 KEAS
2. Canopy - Jettison or manually release
3. Green Apple - Pull
4. Scramble Handle - Pull
5. Suit Vent Hose - Disconnect
6. Trim full nose down, roll aircraft inverted
7. Lean forward and drop out
8. When clear of aircraft and below 20,000 feet pull parachute manual deploy ring

UNCLASSIFIED

Figure 3-4 (Sheet 2

EJECTION

17 SEPARATION OF PERSONNEL FROM SEAT(AUTOMATIC)

NOTE

Man-seat separation should occur at approximately 15,000 feet pressure altitude. Upper risers are cut 0.3 sec. after man separates from seat. Main chute deploys 0.2 sec. after separation. Rotation or spinning of the seat may be experienced while descending with the drogue chute deployed. If it occurs, it may be possible to arrest such motion by using the arms and hands in the airstream

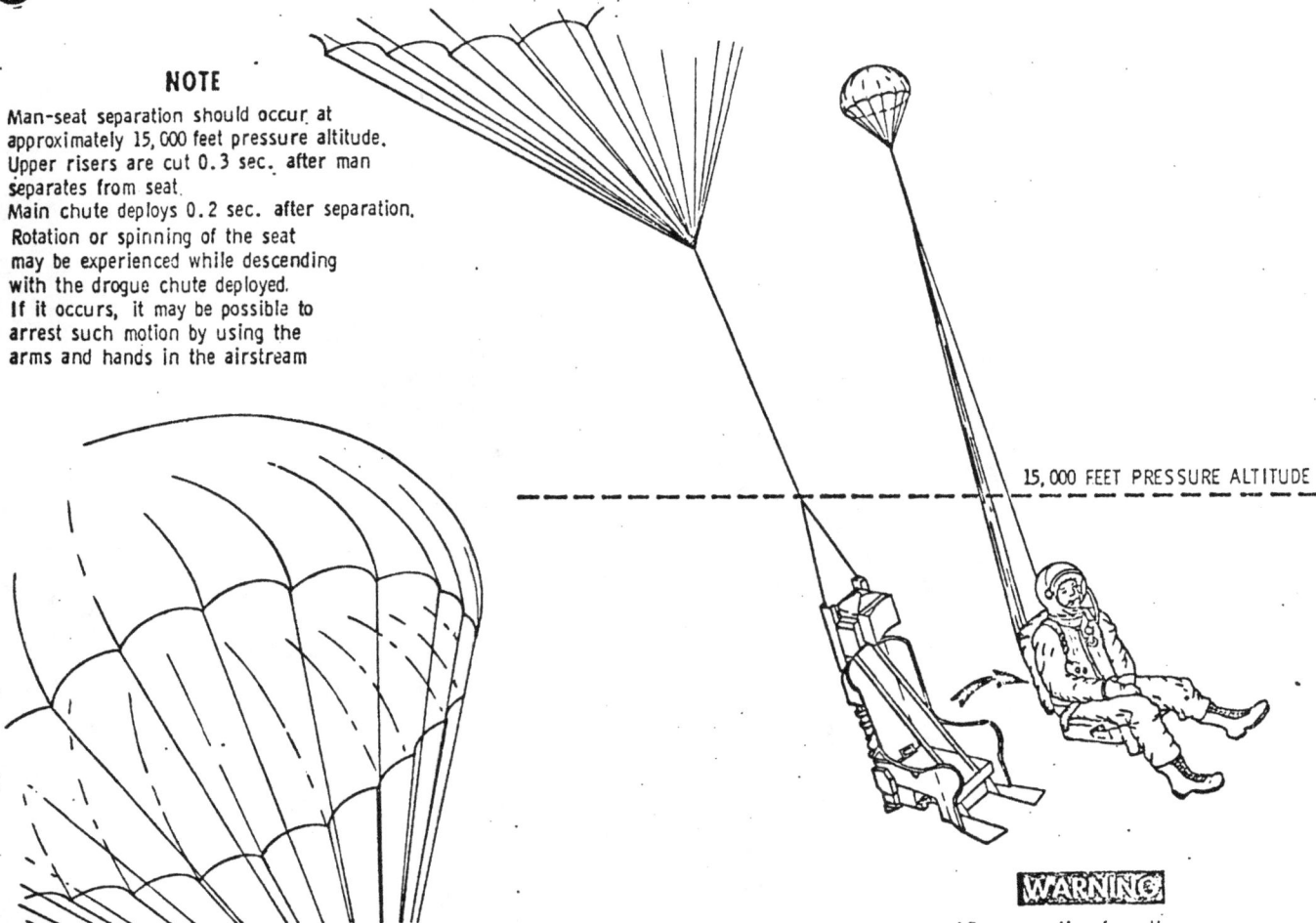

15,000 FEET PRESSURE ALTITUDE

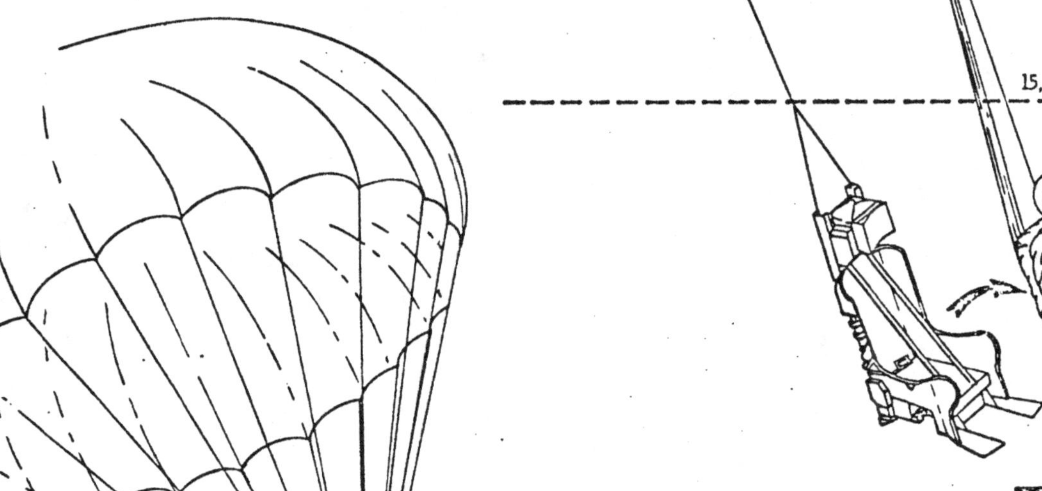

WARNING

After separation from the seat manually, the crew member must deploy the main chute, using the manual deploy ring

18 SEPARATION OF PERSONNEL FROM SEAT(MANUAL)

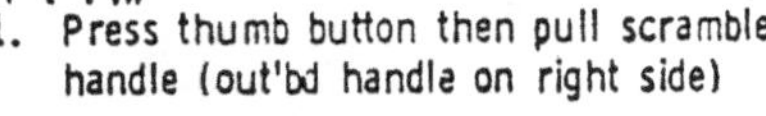

1. Press thumb button then pull scramble handle (out'bd handle on right side)
2. Crew member must forcibly separate himself from seat
3. The crewmember must deploy the main chute manually, using the manual deploy ring

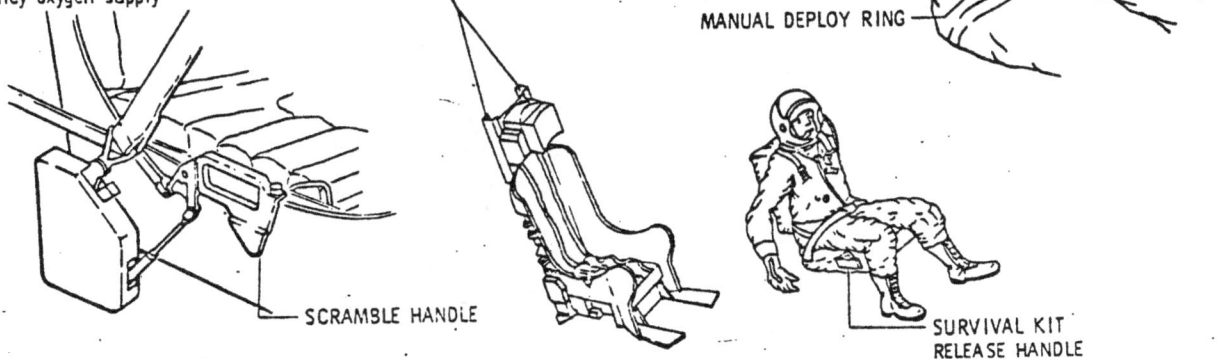

WARNING

Do not pull the survival kit release (inboard) handle while in the seat as this would release the kit and kit lanyard and disconnect the emergency oxygen supply

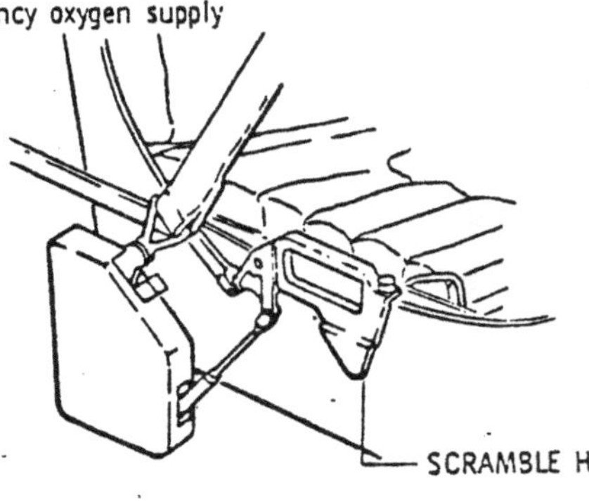

Figure 3-4 (Sheet 3 of 7)

EJECTION

9 PREPARATION FOR LANDING

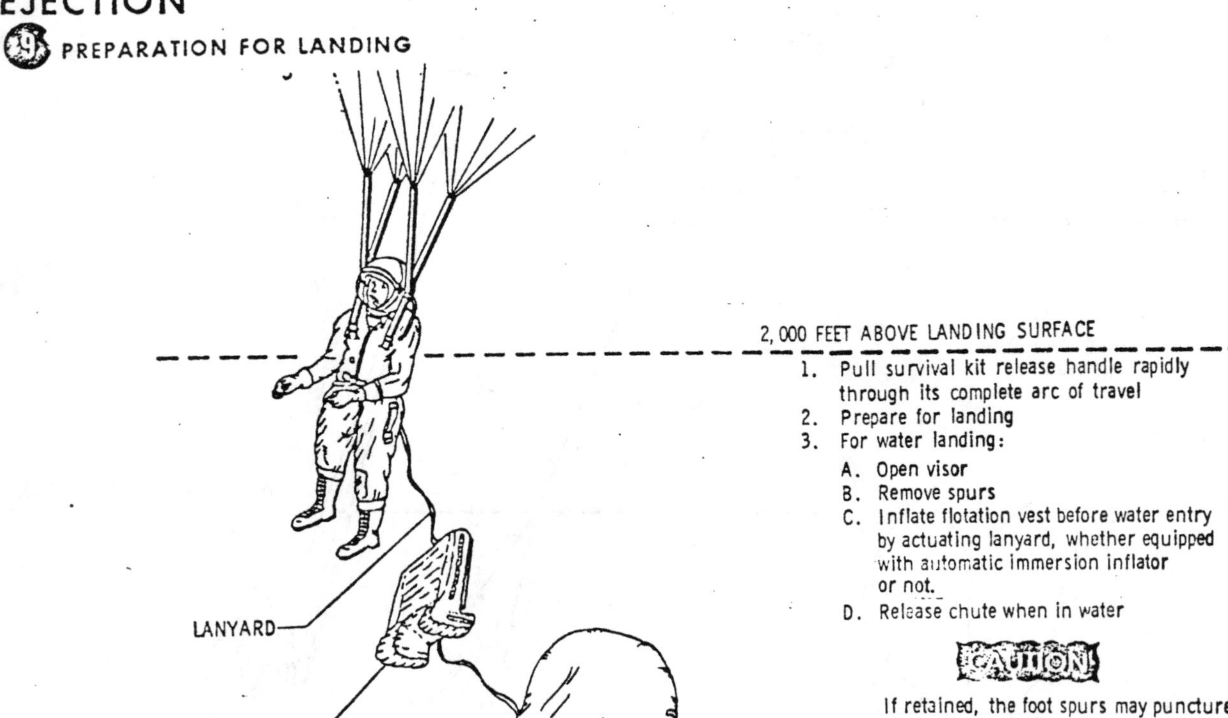

2,000 FEET ABOVE LANDING SURFACE
1. Pull survival kit release handle rapidly through its complete arc of travel
2. Prepare for landing
3. For water landing:
A. Open visor
B. Remove spurs
C. Inflate flotation vest before water entry by actuating lanyard, whether equipped with automatic immersion inflator or not.
D. Release chute when in water
CAUTION
If retained, the foot spurs may puncture the dinghy if care is not exercised
LANYARD
CUSHION
DINGHY

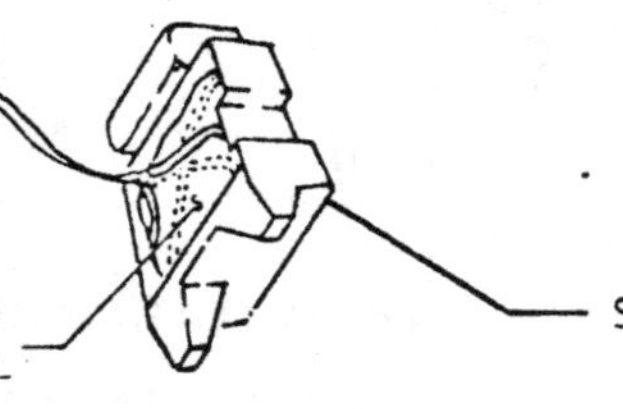

10 SURFACE CONTACT POSITIONS

Kit touchdown relief can be felt prior to crewmember landing

Retain helmet liner if possible, it can be used as a cap

If signal fire built, stand away from fire area to simplify helicopter rescue

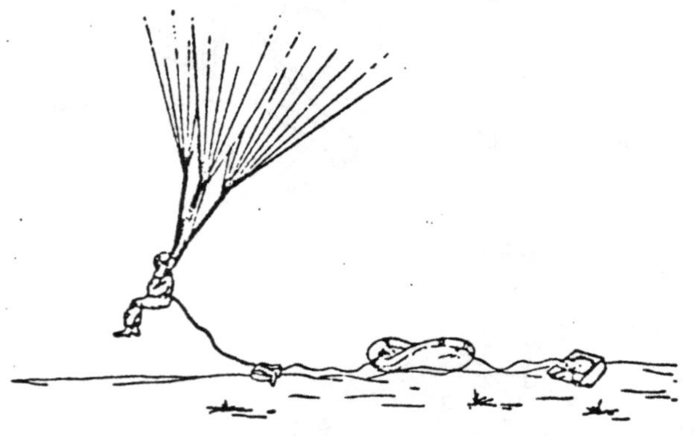

Figure 3-4 (Sheet 4 of 7)

EJECTION

11 RETRIEVAL OF SURVIVAL GEAR (WATER LANDING)

12 SUMMARY OF SEQUENCE OF EVENTS

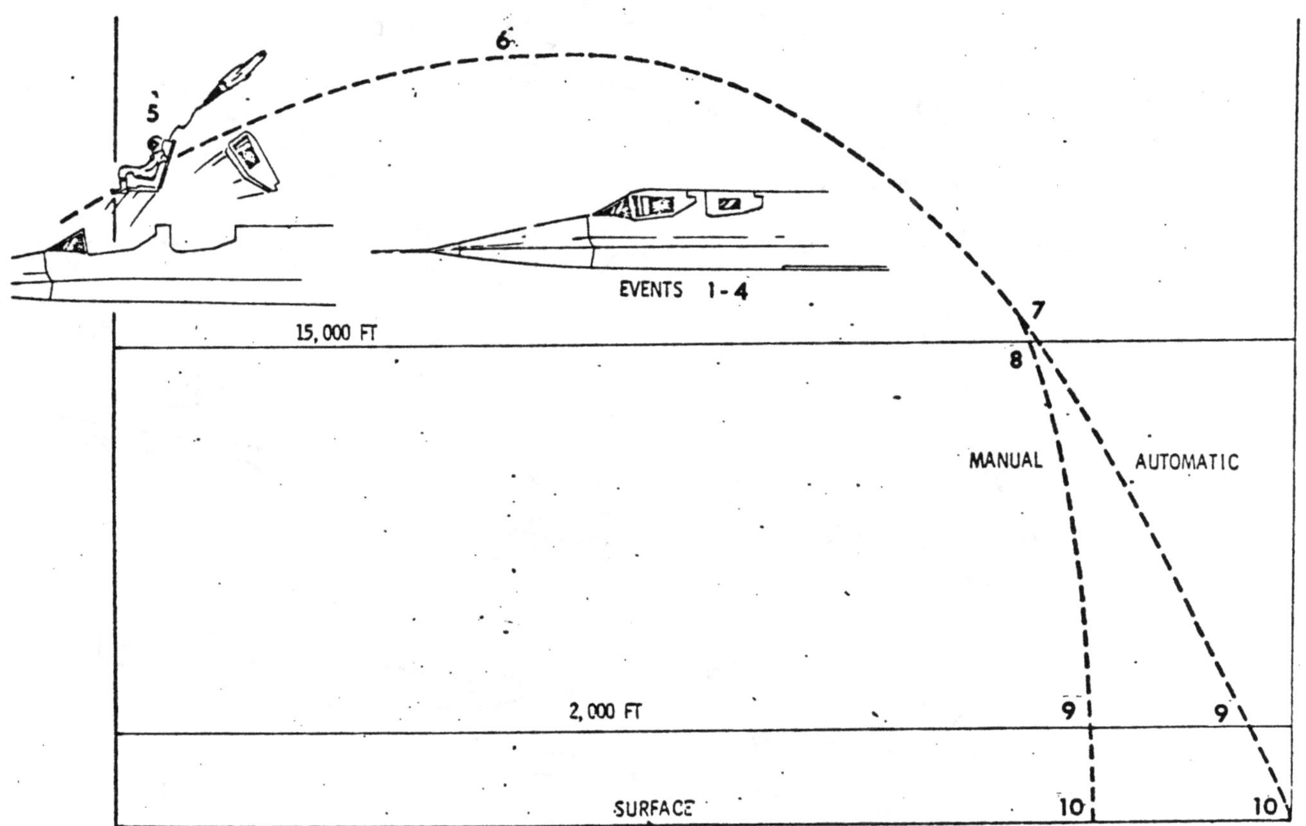

Figure 3-4 (Sheet 5 of 7)

EJECTION

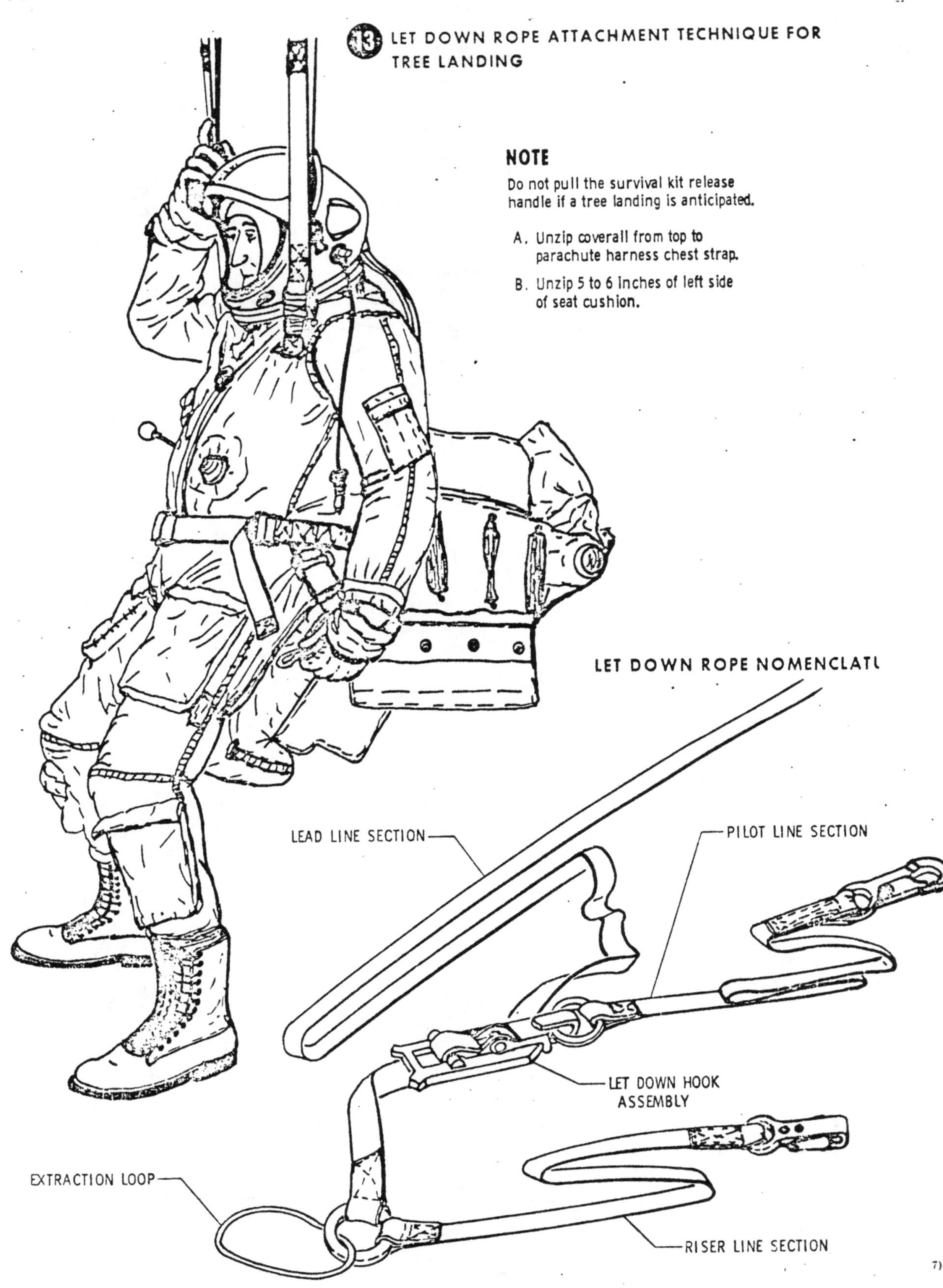
13 LET DOWN ROPE ATTACHMENT TECHNIQUE FOR TREE LANDING
NOTE
Do not pull the survival kit release handle if a tree landing is anticipated.
A. Unzip coverall from top to parachute harness chest strap.
B. Unzip 5 to 6 inches of left side of seat cushion.
LET DOWN ROPE NOMENCLATU
LEAD LINE SECTION
PILOT LINE SECTION
LET DOWN HOOK ASSEMBLY
EXTRACTION LOOP
RISER LINE SECTION

Figure 3-4 (Sheet 6 of 7)

EJECTION

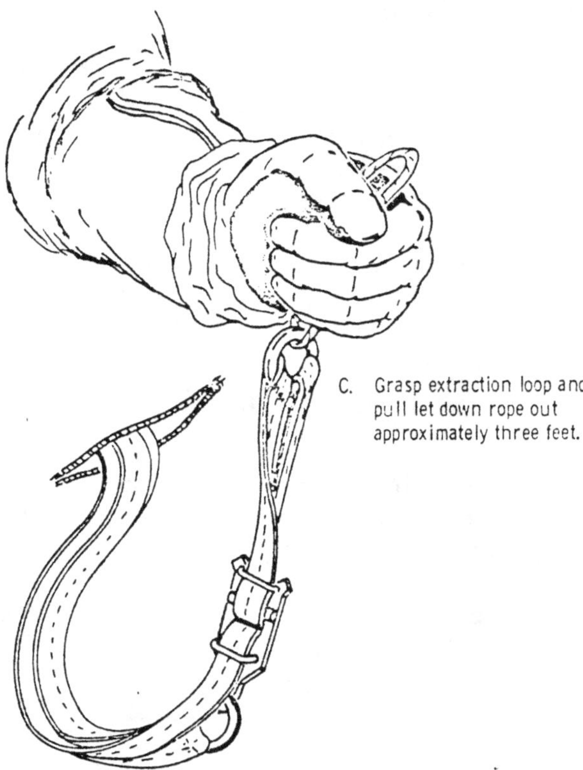

C. Grasp extraction loop and pull let down rope out approximately three feet.

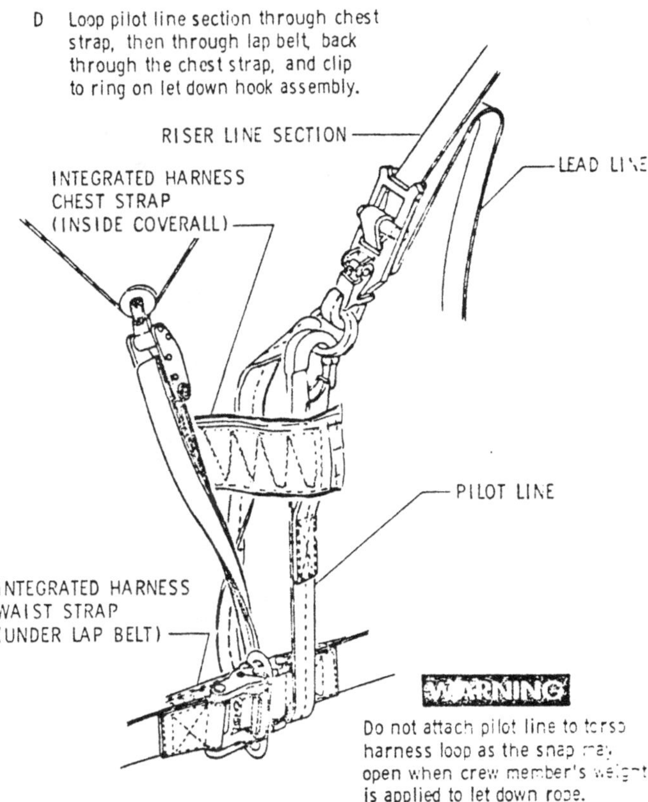

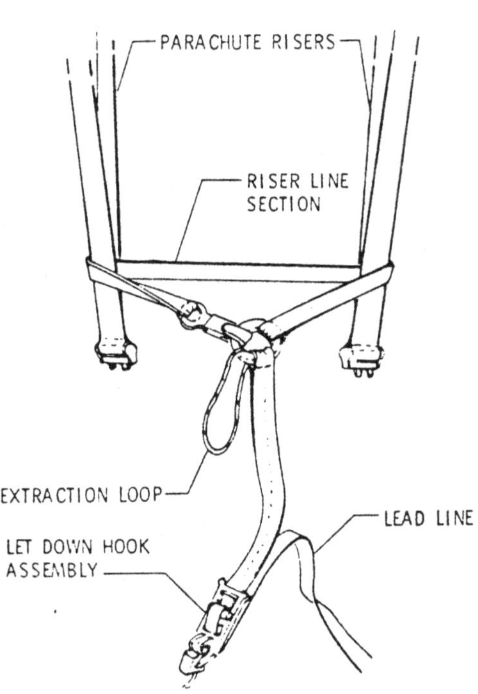

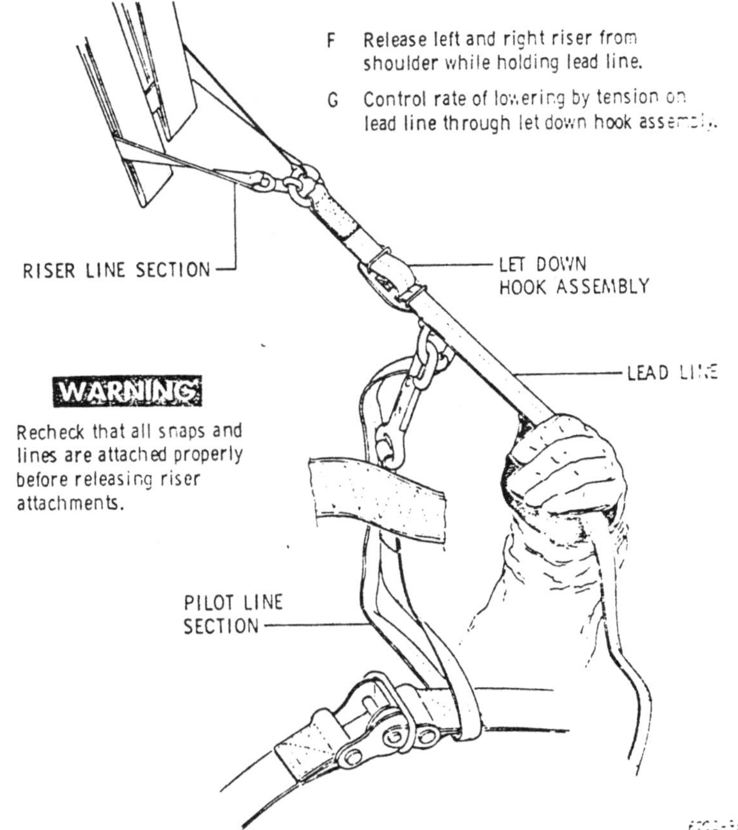

D Loop pilot line section through chest strap, then through lap belt, back through the chest strap, and clip to ring on let down hook assembly.
RISER LINE SECTION
LEAD LINE
INTEGRATED HARNESS CHEST STRAP (INSIDE COVERALL)
PILOT LINE
INTEGRATED HARNESS WAIST STRAP (UNDER LAP BELT)
WARNING
Do not attach pilot line to torso harness loop as the snap may open when crew member's weight is applied to let down rope.

PARACHUTE RISERS
RISER LINE SECTION

LET DOWN HOOK ASSEMBLY
LEAD LINE
E Loop riser line section through front straps of left and right risers and snap to ring as shown.

F Release left and right riser from shoulder while holding lead line.
G Control rate of lowering by tension on lead line through let down hook assembly.
RISER LINE SECTION
LET DOWN HOOK ASSEMBLY
LEAD LINE
WARNING
Recheck that all snaps and lines are attached properly before releasing riser attachments.
PILOT LINE SECTION

Figure 3-4 (Sheet 7 of 7)

WARNING

- Do not pull the survival kit release (inboard) handle while in the seat, as this disconnects the emergency oxygen supply and then releases the kit and kit lanyard.
- THE CREWMEMBER MUST DEPLOY THE MAIN CHUTE MANUALLY, USING THE CHUTE D-RING, AFTER SEPARATING FROM THE SEAT MANUALLY.
- The crewmember is not stabilized by the drogue chute and may be subjected to spinning or tumbling until the main chute is deployed.
- A free fall to a reasonably safe altitude should be attempted to avoid the possibility of serious chute damage due to high speed deployment.
- Visor heat will not be available.

Before Landing

Unless a "tree landing" is anticipated, pull the survival kit release handle ("Banana handle") after the main parachute has opened and when approximately 2000 feet above the landing point. The release handle should be pulled rapidly through its complete arc of travel in order to effect a clean release. Refer to figure 3-4.

NOTE

Do not pull the kit release handle in the event a "tree landing" is anticipated. This is to avoid possible entanglement of the kit, lanyard, and gear while landing.

Preparations for landing should be completed prior to touchdown. If a water landing is anticipated:

(a) Open visor and bend the microphone boom so as to hold the visor open. This should prevent the helmet from filling with water.

(b) Inflate the flotation gear by pulling the CO_2 inflator lanyard down firmly.

NOTE

The flotation gear cannot be inflated orally without actuating the CO_2 lanyard first.

(c) Remove spurs if possible.

CAUTION

If retained, the foot spurs may puncture the dinghy if care is not exercised.

(d) Release both Koch parachute riser releases when in the water.

(e) Release the chute bag by opening the lap belt Koch fastener if desired. Attempt to salvage the radio beacon.

FIRE

FIRE WARNING IN FLIGHT

Illumination of a FIRE warning light indicates a nacelle compartment temperature above approximately 565°C. An immediate check should be made for abnormal EGT, for trailing smoke or any other indication of fire by means of the rear view periscope. In case of doubt, assume that a fire does exist. Proceed in accordance with Engine Fire/ Engine Shutdown & Descent procedure as described under Propulsion System Emergencies, this section.

If at supersonic speed, retarding the throttle of the affected engine to Military power or less will require descent and deceleration to subsonic speeds.

SMOKE OR FUMES

The crew cannot detect fumes when wearing pressure suits. Each helmet oxygen system is independent of the cockpit and suit air supply. Smoke can be eliminated promptly by dumping cabin pressure unless smoke is entering the cockpit from the air conditioning system.

WARNING

When cabin pressure is dumped, cockpit depressurization will occur at an extremely rapid rate and the pressure suits may inflate.

Air-Conditioning System Smoke

If smoke is entering the cockpit from the air conditioning system:

1. L and R refrigeration switches - Cycle individually.

   Attempt to isolate the source of smoke, if time permits, by operating either L or R refrigeration switch to OFF for a few moments. If the smoke does not begin to clear, operate the switch back to ON and then set the other refrigeration switch to OFF.

With source isolated to one system:

2. Refer to L or R Air System Out Operation.

With smoke from both systems:

3. L and R refrigeration switches - OFF.

   If it is determined that smoke is entering from both systems, both systems should be shut down. This shuts off all vehicle air, a condition which should not be maintained while at supersonic speeds without initiating an emergency descent.

   WARNING

   - Shutting off both systems will depressurize the cockpits rapidly, just as in using the cabin pressure dump switch.
   - Continuing at supersonic speeds with both systems off will result in rapid overheating of the cockpit and equipment areas.

4. Initiate emergency descent.

Electrical Fire

The pilot and FCO must depend on visual detection of electrical fire when wearing pressure suits since they cannot smell cockpit air.

1. Isolate the malfunction.

   Turn off electrical systems in order to isolate the malfunction(s). If necessary, deactivate suspected systems by pulling circuit breakers. The battery and one generator may be turned off without adverse effect on essential systems; however, both generators should not be off simultaneously unless absolutely necessary as this would shut down all fuel boost pumps.

2. Leave failed system off.

If required:

3. Cockpit pressure dump switch - ON.

4. Land as soon as possible.

EMERGENCY DESCENT

This procedure may be used when extreme circumstances exist or are expected to develop, such as crew emergency, impending loss of all fuel or control system hydraulic power, etc., and minimum descent time is absolutely required.

Aircraft Control and Attitudes

A minimum use of flight controls is recommended for rapid descents during which aircraft control has become or may become critical (i.e., crew emergency, aft c.g. location with boost pumps inoperative). This may include non-turning flight until lower speeds are attained. If aircraft control is not critical (i.e., low oxygen quantity) a spiral descent is very effective in providing a rapid loss of altitude.

CAUTION

Turns causing appreciable load factors should be avoided when descending through the 50,000 foot level, as the pitch SAS gain switching will cause a transient "bump" which may increase the load factor to near limit value.

The nose will be between 10 and 30 degrees below the horizon while descending through the transonic speed region.

Power Setting and Inlet Configuration

The recommended power setting and inlet configuration provides high drag, the least probability of asymmetric unstart, and the best means of avoiding compressor stall and possible flameout. Inlet unstarts may be encountered when near Mach 2.0 if the engine rpm

is scheduling at the bottom of the idle band; in this event, set the aft bypass open to increase the unstart boundary margin.

CAUTION

Some damage to the engines can occur during an emergency descent if initial CIT is high and rate of deceleration exceeds 1.0 Mach number in three minutes. However, continued subsonic operation is permissible if the engines appear to operate normally after a descent where damage may have occurred.

Use Of Landing Gear For Drag

The landing gear may be extended at 400 KEAS when subsonic in order to maintain maximum rate of descent; however, the gear doors may be damaged if the gear is extended while above 300 KEAS or 0.7 Mach number. Gear extension at supersonic speeds is forbidden. Extending the landing gear while at speeds above Mach 2.3 may cause heat damage to the tires and result in a hazardous landing condition. With gear extended, a large nose-up pitching moment occurs in the speed range from 1.6 to 0.9 Mach number. Full nose-down elevon will be insufficient to maintain 1-g flight at high KEAS and/or aft c.g. in this area.

EMERGENCY DESCENT PROCEDURE

If extreme conditions require a rapid descent:

1. INLETS TO RESTART SIMULTANEOUSLY.

   Select the RESTART ON positions of both emergency spike-inlet restart switches simultaneously.

2. THROTTLES IDLE.

3. Aft bypass controls - CLOSE.

4. KEAS - 350 to 400.

   WARNING

   Do not exceed 400 KEAS or 1.5 g load factor.

   If necessary, reduce rate of descent to maintain positive fuel tank pressure.

   Increase rpm of one engine if high suit inflow temperatures are experienced.

5. C.G. - Forward of 24% MAC.

   Transfer fuel as necessary to set and maintain c.g. forward of 24% MAC.

When below Mach 1.7:

6. Pitot heat - On.

7. Exterior lights - On.

When subsonic:

8. Inlets - Normal.

   Select the OFF position of the emergency spike - inlet restart switches.

   NOTE

   Continued subsonic operation is permissible if engine operation appears normal.

For continued descent:

9. Landing gear lever - DOWN.

WARNING

Gear extension at supersonic speeds is forbidden.

CAUTION

In flight, gear door strength limits the airspeed with gear down to 300 KEAS or Mach 0.7, whichever is less, with a maximum permissible sideslip angle of 10°. Maximum permissible speeds are 300 KEAS or Mach 0.9, whichever is less, with gear down when sideslip angle does not exceed 5°.

FUEL DUMPING

Refer to Section II for Fuel Dumping Procedures.

FORCED LANDING OR DITCHING

Ditching, landing with both engines inoperative, or other forced landing should not be attempted. Ejection is the best course of action. All emergency survival equipment is carried by the crewmember; consequently, there is nothing to be gained by riding the airplane down.

If an ejection seat fails to fire, manual bailout is preferable to ditching or forced landing, since the aircraft will probably break up on touchdown.

PROPULSION SYSTEM

The following procedures are to be accomplished in the event of engine fire or abnormal operation or failure of a propulsion system component, i.e., inlet, engine, afterburner, nozzle, fuel control, or lubrication, fuel-hydraulic, or ignition system.

INLET UNSTART

Inlet duct unstarts can only occur after supersonic speeds are reached and an inlet has been "started", that is, supersonic flow conditions established inside part of the inlet. Unstart, or shock expulsion may be caused by inlet airflow becoming greater than engine requirements and duct bypass capability, spike position too far aft, or abrupt aircraft attitude changes. Improper spike or door positions can result from inlet control error, loss of hydraulic power, or electrical or mechanical failure. Unstarts are usually associated with climb or cruise operations above Mach 2 when at normal engine speeds; however, they may be encountered during reduced rpm descents at speeds above Mach 1.3.

Between Mach 1.3 and 2.2, when near Military rpm, recovery procedures using the restart switch ON position may result in compressor stall.

The characteristics associated with inlet unstarts are very similar to those of compressor stalls and, in fact, stalls and unstarts may be intermingled. The term "aerodynamic disturbance" or "A/D", as used in the inlet unstart procedures, refers to either condition -- regardless of whether it has or has not been identified as an unstart or compressor stall.

This page intentionally deleted.

Flight Characteristics During Unstart

Unstarts are generally recognizable by airframe roughness, loud "banging" noises, aircraft yawing and rolling, and decrease of compressor inlet pressure toward 4 psi. Fuel flow decreases quickly and the afterburner may blow out. EGT usually rises, with the rate of increase being faster when operating near limit Mach number and ceiling altitudes. A distinct increase in drag and loss of thrust occurs at the unstarted inlet because of increased air spillage around the inlet and reduced airflow through the engine.

The aircraft yaws toward the unstarted inlet during an unstart. This yaw causes a roll in the same direction. A pitch-up tendency may occur due to yaw and roll rates developed during the inlet unstart. Pitch control problems can also occur during associated maneuvering and will be accentuated by low KEAS and/or high angles of attack, maximum altitude operation, aft c.g., high Mach numbers, and any pitch rate which existed prior to inlet unstart. During the unstart, primary emphasis must be placed upon maintaining pitch control in order to prevent nose-up pitch rates and angles of attack in excess of eight degrees. Thrust asymmetry should be reduced as soon as possible.

Aileron effectiveness is reduced at high altitudes and high angles of attack. Roll control may become critical if the unstart occurs on the inboard inlet during a bank. At altitudes above 75,000 feet, aileron control may be ineffective in controlling roll during an unstart unless the angle of attack is immediately reduced. Aileron effectiveness increases rapidly as the angle of attack is reduced and only moderate aileron inputs will be required to control the roll. An excessive nose down attitude may result in an over speed in KEAS and Mach if the inlets are restarted during a recovery maneuver. Therefore, if manual restart is actuated the restart switches should remain on until speed and attitude are fully under control.

The roughness usually clears after the forward bypass doors open and the spikes are started forward manually or automatically. With automatic restart the spikes move toward full forward for approximately 3.75 seconds then return to the scheduled position at a reduced rate. If the roughness associated with unstart does not clear then the manual restart switches should be set in the RESTART position. As much as five to eight seconds may be required for the spikes to reach the full forward position. Roughness may persist until the spikes are fully forward during restarts at design Mach number when aft bypass open has been required.

Inlet pressure should be checked during recovery. Moderate CIP increases will occur as the inlet "clears" or restarts, and when the spike retracts to form the inlet throat farther aft. Return of the forward bypass doors to their normal operating schedule should result in a further CIP increase to normal operating values.

In automatic operation, unstarts which are caused by improper spike scheduling limit aircraft speed to Mach numbers below that for the unstarted condition. Manual scheduling procedure is necessary if the aircraft is to be accelerated further. If an unstart results from marginal bypass scheduling however, it may be possible to continue at speed by adjusting the forward or aft inlet bypass doors to positions which maintain stable flow conditions. In general, if engine speed is maintained, less bypass area is required as limit Mach number is approached.

Unstart Boundary Charts

Figure 3-5 shows the operating conditions where airframe roughness will occur due to unstable inlet airflow conditions. The unstart boundaries are a function of Mach number, engine speed, and spike and bypass door positions. The smallest roughness area below the idle rpm range with spike full forward occurs with the forward and aft bypass doors open. A more extensive area occurs with the bypass doors open but with the spike moving in accordance with the automatic schedule. In both cases, the onset of inlet airflow instability occurs earlier, i. e., at higher engine speeds, with the bypass doors closed. At windmilling rpm, heavy roughness will occur in the speed range above Mach 1.3 unless the spike is positioned fully aft.

Inlet Unstart Procedure

In the event of an unstart, accomplish only those of the following steps which are necessary to clear the inlet and return to normal operation.

1. ANGLE OF ATTACK within limit.

   Apply pure pitch correction (stick forward) first to eliminate the nose up pitch rate and to maintain angle of attack within limit. Rudders may be used to assist in roll correction if necessary. Delay roll correction with the stick until pitch angle is controlled.

   WARNING

   Start pitch correction first. High angles of attack can develop if pitch rate is not controlled and this can result in pitch-up. Maintain angle of attack below 8°. Decreasing angle of attack first assists roll control and makes recovery of attitude more positive.

If auto-restart not effective:

2. INLETS TO RESTART.

   The automatic inlet restart system will begin the inlet restart automatically. Both inlets should respond while above Mach 2.3. Operation of the system can be observed by CIP change and by spike and forward bypass position indications.

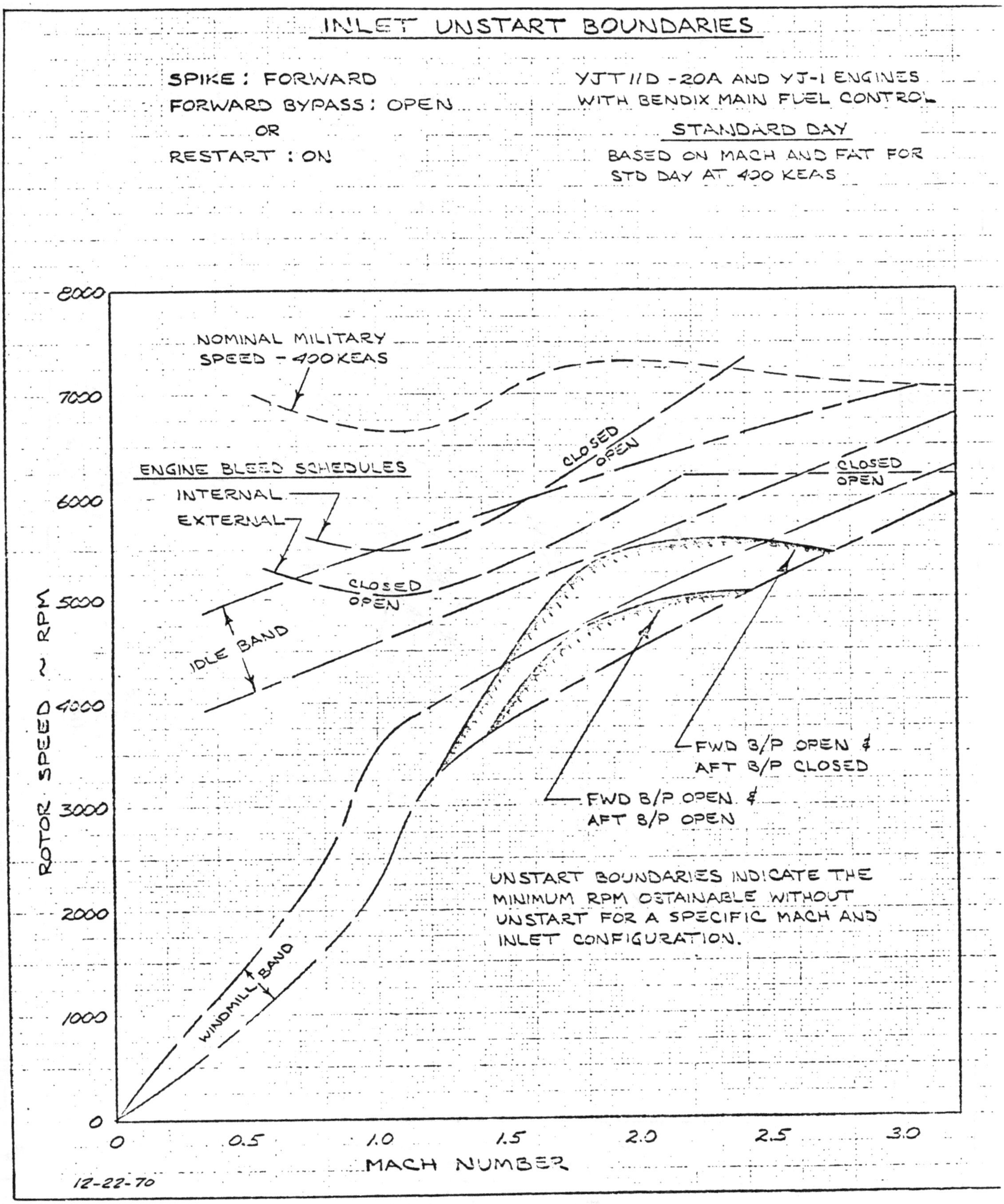

INLET UNSTART BOUNDARIES

SPIKE : FORWARD
FORWARD BYPASS : OPEN
OR
RESTART : ON

YJT11D-20A AND YJ-1 ENGINES
WITH BENDIX MAIN FUEL CONTROL

STANDARD DAY

BASED ON MACH AND FAT FOR
STD DAY AT 400 KEAS

NOMINAL MILITARY
SPEED - 400 KEAS

ENGINE BLEED SCHEDULES
INTERNAL
EXTERNAL

CLOSED
OPEN

CLOSED
OPEN

CLOSED
OPEN

IDLE BAND

WINDMILL BAND

FWD B/P OPEN &
AFT B/P CLOSED

FWD B/P OPEN &
AFT B/P OPEN

UNSTART BOUNDARIES INDICATE THE
MINIMUM RPM OBTAINABLE WITHOUT
UNSTART FOR A SPECIFIC MACH AND
INLET CONFIGURATION.

ROTOR SPEED ~ RPM

8000
7000
6000
5000
4000
3000
2000
1000
0

0 0.5 1.0 1.5 2.0 2.5 3.0

MACH NUMBER

12-22-70

Figure 3-5 (Sheet 1 of 2)

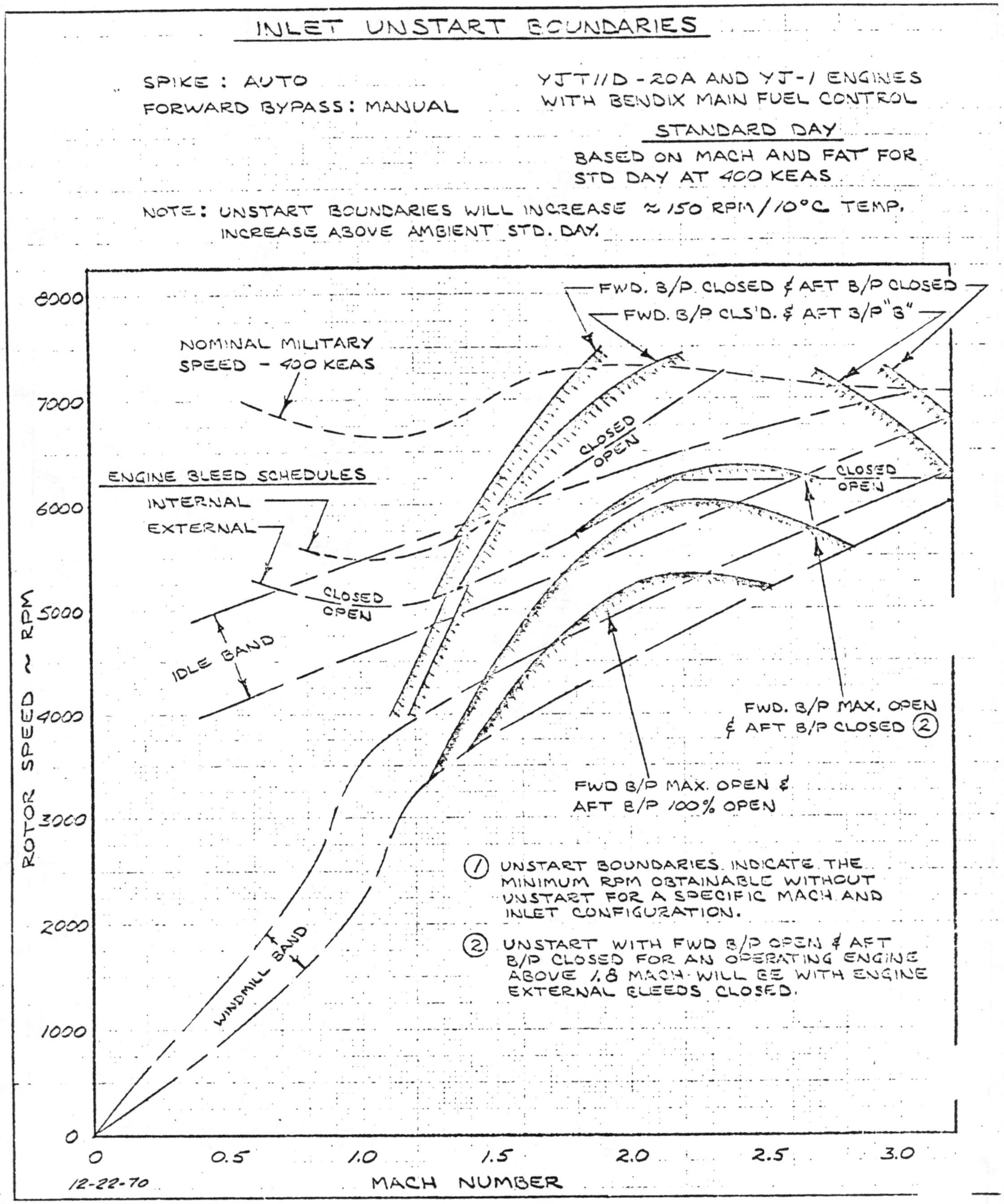

INLET UNSTART BOUNDARIES

SPIKE : AUTO
FORWARD BYPASS : MANUAL

YJT11D-20A AND YJ-1 ENGINES
WITH BENDIX MAIN FUEL CONTROL

STANDARD DAY

BASED ON MACH AND FAT FOR
STD DAY AT 400 KEAS

NOTE: UNSTART BOUNDARIES WILL INCREASE ≈ 150 RPM/10°C TEMP. INCREASE ABOVE AMBIENT STD. DAY.

FWD. B/P CLOSED & AFT B/P CLOSED
FWD. B/P CLS'D. & AFT B/P "B"
NOMINAL MILITARY
SPEED - 400 KEAS
CLOSED
OPEN
CLOSED
OPEN
ENGINE BLEED SCHEDULES
INTERNAL
EXTERNAL
CLOSED
OPEN
IDLE BAND
WINDMILL BAND
FWD. B/P MAX. OPEN
& AFT B/P CLOSED (2)
FWD B/P MAX. OPEN &
AFT B/P 100% OPEN

(1) UNSTART BOUNDARIES INDICATE THE MINIMUM RPM OBTAINABLE WITHOUT UNSTART FOR A SPECIFIC MACH AND INLET CONFIGURATION.

(2) UNSTART WITH FWD B/P OPEN & AFT B/P CLOSED FOR AN OPERATING ENGINE ABOVE 1.8 MACH WILL BE WITH ENGINE EXTERNAL BLEEDS CLOSED.

ROTOR SPEED ~ RPM
0
1000
2000
3000
4000
5000
6000
7000
8000

0
0.5
1.0
1.5
2.0
2.5
3.0
MACH NUMBER

12-22-70

Figure 3-5 (Sheet 2 of 2)

If the auto-restart system does not operate or is ineffective, set both inlet restart switches ON (down). In the restart configuration, the forward bypass is open and the spike is full forward.

WARNING

Initially set both inlets to the restart configuration to avoid possibility of confusing identity of the unstarted inlet and to reduce control problems.

3. AFT BYPASS.

If roughness does not clear with the inlets in the restart configuration, cycle the aft bypass switches OPEN, then to normal schedule. Aft bypass cycling will not normally be required below Mach 3.0, and definitely should not be used if roughness is associated with compressor stall.

CAUTION

- If roughness can not be cleared, reduce throttles to minimum afterburner or Military, depending on the severity of the unstart.
- Check for engine failure or inlet system malfunction.

If deceleration is required with the inlets in the restart configuration, close the aft bypass of each inlet and set the affected engine throttle to 6500 upon reaching speed below Mach 2.5.

4. CHECK EGT.

Be prepared to shut down the affected engine(s). Exhaust gas overtemperatures up to 900°C may be tolerated for a maximum of 15 seconds.

WARNING

The affected engine(s) must be shut down if EGT exceeds 900°C for more than three seconds; otherwise, severe turbine damage can result.

Since the main fuel control can not reduce fuel flow below the minimum fixed schedule, manual or automatic trimming and throttle reduction to IDLE have no effect during a severe overtemperature condition with the inlet(s) unstarted. (See EGT Overtemperature, this section. Also refer to Shut-down, Flameout, Glide Distance, and Airstart.)

At high Mach number and high altitude conditions, inlet unstart can cause severe engine overtemperature if the Derich system is not effective, or if the OFF position of the Fuel Derich switch is selected before inlet restart is obtained.

CAUTION

The Fuel Derich switch should remain in the ARM position until the inlet has restarted. Otherwise, severe overtemperature may occur.

5. 350 KEAS.

Adjust airspeed toward 350 KEAS.

WARNING

If in a nose down attitude, leave inlets in restart configuration until speed and attitude are fully under control.

Below Mach 3.0, continued heavy roughness with the inlets in the restart configuration is indicative of compressor stall, regardless of the cause of the initial disturbance. Airspeed should be maintained above 350 KEAS and the compressor stall recovery procedure employed immediately to minimize the possibility of flameout.

If a compressor stall condition exists in the speed range from Mach 1.8 to Mach 2.6, affected engine:

6. Aft bypass switch - CLOSE.

7. Forward bypass control - OPEN.

8. Restart switch - OFF.

If the compressor stall does not clear, affected engine:

9. Throttle - Reduce rpm.

Decrease rpm until stall clears. Reduce throttle setting to idle if necessary.

When unstart or compressor stall clears:

10. Inlets - Automatic operation (individually).

If the Derich light is on:

11. Derich switch - Recycle to ARM.

Cycle to REARM then to ARM. Fuel derich OFF should not be selected while an inlet is unstarted.

In the event of derichment, the fuel derich system must be rearmed (after the inlet has restarted) in order to interrupt the derichment system flow and return the engine fuel system to normal operation. This must be done prior to lighting the afterburner to avoid engine speed suppression.

The catalytic igniters may sustain the afterburners or cause relights as soon as the inlet clears. If derichment has occurred, operation of the fuel derich system can result in engine speed suppression of up to 750 rpm.

After the inlet clears:

12. Afterburners - Check.

After the inlet clears, check for afterburner operation if the throttles remain in the afterburner range. Relight if necessary. If the throttles were retarded to Military or below, set afterburner power as desired.

If an inlet unstart repeats after its restart switch is OFF, set the affected inlet restart switch to ON and use the Manual Inlet schedule or the Inlet Malfunction procedure.

INLET MALFUNCTION

When at supersonic speeds, inlet system malfunction may be indicated by successive unstarts, abnormally high or low inlet pressures, or by engine stalls. Malfunction of an inlet system may be due to failure of the spike and/or forward bypass automatic controls,

aft bypass manual control, actuators, or electrical or hydraulic control power. The engine and hydraulic instruments and inlet control circuit breakers should be checked before employing the inlet malfunction procedure in order to determine that an inlet malfunction is not associated with some other abnormal condition. The respective emergency spike-inlet restart switch or spike and door position controls must be used to individually control the left or right inlet.

Failure To Schedule Normally

A combination of unsymmetrical thrust and fuel flow and low compressor inlet pressure on one side during acceleration indicates that a spike and/or forward bypass has failed to move automatically on the proper schedule. This may be caused by failure of the automatic control(s) or of the spike forward lock to disengage when above 30,000 feet, or by circuit breaker opening. Normal spike and forward bypass positions and compressor inlet pressures are provided by figure 1-5 for supersonic conditions. Inlet pressures can be cross-checked or compared with chart values if schedule failure is suspected. Spike and forward bypass position indications can also be used for malfunction analysis.

Inlet Spike Unstable

Spike instability is reflected by fluctuations of the respective L or R hydraulic (SPIKE) pressure gage and by the spike position indication. If spike oscillations are of large amplitude, the gage fluctuations will be several hundred psi and a "hammering" may be felt by the crew. If the instability persists, attempt to restore normal operation by cycling the manual spike control to match flight Mach number, then return to automatic control. If the condition persists, use the Manual Inlet Operation schedule until a different Mach number is reached, then return to automatic control. If the condition still continues, use the manual schedule.

Aft Bypass Control Failure

Malfunction of an aft bypass control is indicated by failure of the corresponding position indication light to extinguish after the aft bypass control position setting is changed. It may be possible to correct a malfunction condition by cycling the control setting. Control failure can result in reduced performance, inlet roughness or unstart, or engine stall, depending on the existing or subsequent flight conditions. Refer to Stall and Unstart Boundary charts, this section, for conditions to expect with the aft bypass, spike, and forward bypass in various positions.

Inlet Malfunction Procedure

This procedure may be used in the event that the unstart recovery procedures are unsuccessful or if an inlet bypass or spike control malfunction has made automatic operation impossible.

NOTE

- Use spike and forward bypass position indications for malfunction analysis. Use appropriate manual spike and/or bypass schedules if needed.
- In some cases when unstarts have occurred due to automatic spike control malfunction, automatic operation can be continued at a lower Mach number than that at which the unstart was encountered. Continuing at lower Mach number may be preferable to using the manual spike operating schedule.

Proceed as follows for the affected inlet:

1. Forward bypass control - OPEN.
2. Restart switch - OFF.

If unstart does not repeat:

3. Forward bypass control - Manual schedule.

   Assume that spike operation is normal and if aft bypass operation is normal, that there is a malfunction in the automatic forward bypass control.

If unstart repeats::

4. Restart switch - ON.
5. Spike control - Manual schedule.
6. Restart switch - OFF.

   Unstart should not repeat if malfunction is in the automatic spike control.

7. Inlet controls - Manual schedule.

   The normal aft bypass and manual forward bypass schedules must be used when operating with manual spike control in order to obtain near-normal inlet performance. Recycle the derich switch, if necessary, and observe altitude, speed, and bank angle restrictions for manual inlet operation.

MANUAL INLET OPERATION

The inlet spike and forward bypass may be positioned manually in the event that an automatic inlet control malfunctions. Manual control is also desirable in the event that engine shutdown is necessary while at high speed, as the spike aft position results in minimum inlet roughness during descent to subsonic speeds. (Refer to ENGINE SHUTDOWN procedure, this section.)

CAUTION

Observe altitude, speed, and bank angle restrictions during manual inlet operation.

Manual Control of Forward Bypass

Manual operation of the forward bypass is permissible with AUTO spike selected. The normal aft bypass position schedule should be used. During cruise, using the manual schedule as a guide, the forward bypass may be set to obtain 1.0 psi less than the opposite inlet CIP when that inlet is operating normally. Set the forward bypass controls to obtain 2.0 psi less than the CIP gage "barber pole" hand when both inlets are being controlled manually.

Manual Control of Spike

Manual operation of a spike is permissible; however, the effect on forward bypass positioning must be recognized. When an inlet forward bypass control is set at AUTO and the spike control is set in the manual range (FWD to Mach 3.2), the manual spike control overrides automatic bypass operation and causes the forward bypass to open 100%. The forward bypass must be controlled manually to obtain variable positioning during manual spike operation. When both are set in the manual control range, spike settings between Mach 1.4 and Mach 3.2 bias the actual forward bypass position toward a more open condition than the bypass control position settings. The maximum bias is approximately 25% of the manually selected bypass opening when the spike setting is Mach 2.3, and there is at least 10% bias toward open when spike control settings are between Mach 1.5 and Mach 2.8.

NOTE

Set spike position first when manual spike and forward bypass setting changes are scheduled. Then reset the forward bypass after allowing several seconds for the spike to reach its new position.

Manual Inlet Schedule

The following schedule may be used if automatic forward bypass or spike operation causes unstarts or results in low CIP while using the Inlet Malfunction Procedure. (Checklist emergency procedures include an abbreviated form of this table.) During manual inlet descent, with spike and/or forward door malfunctioning, a combination of restart on and high rpm can result in engine compressor stall. Set 6500 rpm (idle power) at Mach 2.5 and let rpm droop as speed decreases.

MANUAL INLET SCHEDULE (Affected side only)

SUPERSONIC CLIMB AND CRUISE			
SPEED	FWD BYPASS ONLY	SPIKE	AND FWD BYPASS
Below 1.7 Mach ①	0%	FWD	0% ②
Above 1.7 Mach ④	40%	Lag .1 Mach	20%
Cruise	1 psi low	Match Mach ③	1 psi low

SUPERSONIC DESCENT ⑤	
SPEED	ALL MANUAL INLET CONDITIONS
Above 1.3 Mach	Restart - ON
At 1.3 Mach ①	Restart off, Forward Bypass Closed and resume normal procedures.

NOTES:

① For all subsonic operation insure spike forward and forward bypass closed.

② % reflects door control setting, not indicator reading.

③ Adjust spike forward .1 Mach number before turns.

④ Adjust forward door 20% more open prior to shifting AFT doors or afterburner light or cutoff. Use normal AFT door schedule.

⑤ Above 2.5 Mach, military power schedule.
Below 2.5 Mach, set 6500 rpm - let rpm decrease.

RESTRICTIONS:

1 3.0 Mach number

2 20° maximum bank angle above 70,000 feet

3 80,000 feet maximum altitude

Figure 3-6

COMPRESSOR STALL

Compressor stall is usually indicated by thumping pulsations. Other characteristic indications are a loss of thrust, fluctuating CIP, CIT, RPM, ENP, or EGT at fixed throttle position, or failure of rpm to increase during throttle advance. Momentary afterburner flameout with catalytic reignition can be expected. At low airspeeds, compressor stall frequently results in engine flameout. Some of these stall characters are also descriptive of inlet unstarts, hence accurate differentiation between stall and unstart is difficult. In addition, stalls and unstarts may be intermingled, making identification more difficult. A supersonic stall clearing procedure is incorporated within the Inlet Unstart emergency procedure.

Compressor Stall Regions - Supersonic

Stall regions are shown in figure 3-7 Maximum stall risk is at military rpm with internal bleeds closed, aft bypass, full open, and restart switch ON. Minimum stall risk is near idle rpm with engine internal bleeds open, and aft bypass closed. Stalls may be caused by transient airflow conditions resulting from compressor bleed shift, or by unstable or manual inlet operations. Other causes may be abrupt or erratic throttle movement, failure to momentarily delay throttle advancement during afterburner light, or improper scheduling of engine bleeds. Recovery from stall accompanying inlet unstart at high altitude is aided by reducing altitude. Below Mach 2.5, recovery is more consistently obtained through throttle reduction to lower rpm.

Effect of Open Aft Bypass on Stall - Supersonic

Open aft bypass generally has slight effect on engine stall at supersonic speeds while the forward bypass continues scheduling. Stall risk is significantly increased when excessive aft bypass opening results in closed, non-scheduling forward bypass. While below Mach 2.5 with an inlet restart switch ON, appreciable stall risk exists while near Military rpm and in Idle with engine start (external) bleeds open. During airstart, the aft bypass should be closed as rpm increases.

Effect of Manual Inlet Operation - Supersonic

While at supersonic speeds, stall risk is increased during manual inlet operation due to greater inlet distortion and reduced inlet efficiency.

Subsonic Engine Stalls

The causes of engine compressor stall in subsonic flight are not well defined, but the major parameters have been identified. In addition to the more readily recognized abnormal rpm, EGT, nozzle position, and fuel flow conditions (for which emergency procedures are described in this section), the engine stall parameters include angle of attack (α), compressor inlet pressure (CIP), turbulence and wind shear and rapidly changing ambient air temperatures which result in corresponding changes of inlet temperature (CIT), and -- to a lesser degree -- changes in α.

Engine stalls that occur during throttle advance are usually the result of an excessive uptrim condition. Engine stalls may also occur as a result of excessive

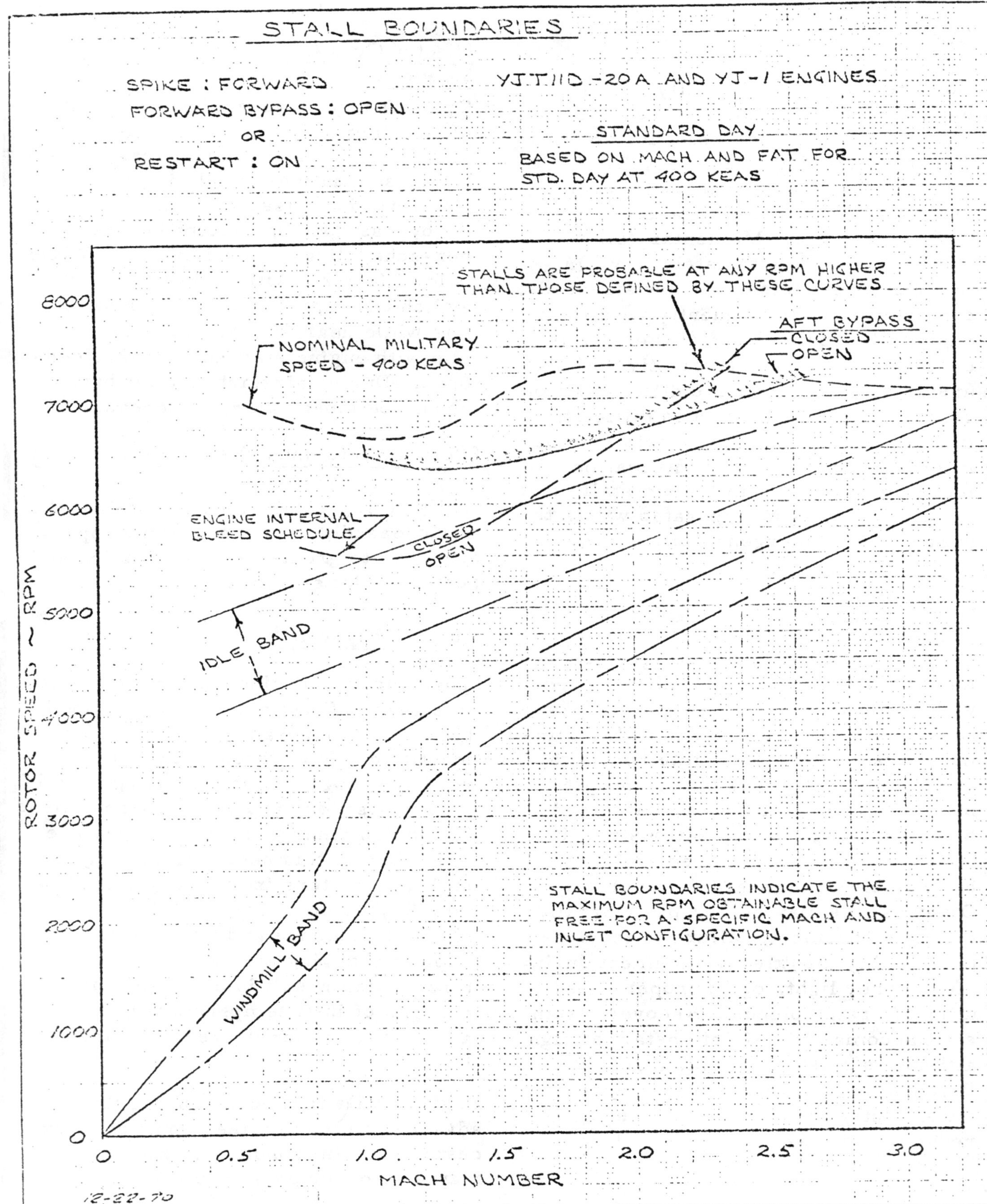
STALL BOUNDARIES
SPIKE : FORWARD
FORWARD BYPASS : OPEN
OR
RESTART : ON
YJT11D -20A AND YJ-1 ENGINES
STANDARD DAY
BASED ON MACH AND FAT FOR
STD. DAY AT 400 KEAS
STALLS ARE PROBABLE AT ANY RPM HIGHER
THAN THOSE DEFINED BY THESE CURVES
AFT BYPASS
CLOSED
OPEN
NOMINAL MILITARY
SPEED - 400 KEAS
ENGINE INTERNAL
BLEED SCHEDULE
CLOSED
OPEN
IDLE BAND
WINDMILL BAND
STALL BOUNDARIES INDICATE THE
MAXIMUM RPM OBTAINABLE STALL
FREE FOR A SPECIFIC MACH AND
INLET CONFIGURATION.
ROTOR SPEED ~ RPM
8000
7000
6000
5000
4000
3000
2000
1000
0
0
0.5
1.0
1.5
2.0
2.5
3.0
MACH NUMBER
12-22-70

Figure 3-7

overtrimming while at constant throttle settings, or because of malfunctioning nozzles, fuel controls, guide vanes, or engine bleeds. However, engine stalls that occur when engine operating conditions are otherwise normal and when control parameters are not being changed may indicate an approach to potentially dangerous flight situations. A low airspeed or high angle of attack condition, or both, may have been reached inadvertently. Then the stalls can be a result of low CIP (which is associated with low airspeed when at moderate to high subsonic operating altitudes) or result from high distortion in the inlet (which occurs at high angles of attack). Either can be dangerous when associated with operation beyond established flight limits.

When subsonic, angles of attack larger than 10° seem to provide the greatest susceptibility to engine compressor stalls; however, there have been many recorded instances of higher angles of attack where no stalls occurred. Stall free operation has been obtained at angles of attack as high as 15°.

Engine stalls are more probable when at Military rpm with CIT's below +10°C than when operating at higher CIT's. The condition corresponds to the operating area which is in the rapidly changing portion of the EGT trim band. The probability of stall is increased if at low CIP (high altitude, low airspeed), or if there is a sudden change (decrease) in air temperature, or if there is clear air turbulence, aircraft maneuvering, or open bypass door conditions. The effects can be additive when more than one of these factors are present.

High angles of attack do not affect CIP directly; rather, they cause non-uniform pressure distribution and disturbed airflow directions at the engine face. The uneven distribution is called inlet distortion. An engine can survive large amounts of distortion if low altitude or high KEAS provides a relatively high CIP and if other conditions are favorable. If there is very little distortion, operation may continue at CIP's as low as 2.5 psi. If at moderate to high subsonic operating altitudes and at low KEAS or CIP, and if other conditions are marginal, a small amount of distortion can result in compressor stall.

At angles of attack for the normal subsonic flight speed range (Mach 0.75 to 0.90), the linearity is such that increasing load factor (or lift coefficient) results in an almost equal factor of increase in angle of attack. Thus, if a gust produces a change in load factor from one-g to two-g's, angle of attack will also increase to an almost double value.

Sudden increases in angle of attack, such as from gusts, do not change CIP except for a very small and insignificant amount. However, such sudden increases do increase distortion of the airflow received at the face of the engine. Gusts can contribute to engine stall probability in this manner if of sufficiently large magnitude.

Subsonic Engine Stall Procedure

In the event of compressor stall while at subsonic speeds:

1. α WITHIN LIMIT

Reduce angle of attack and then maintain angle of attack and airspeed within limits.

WARNING

If a high angle of attack warning occurs, or if angle of attack and airspeed are not within limits, make angle of attack and speed corrections before adjusting the throttles. These actions alone may clear engine stall conditions, and are mandatory to avoid pitch-up if at high angle of attack and/or low airspeed.

After angle of attack and KEAS within limits:

2. RPM - Reduce.

Retard both throttles until stall clears.

Retarding the throttles should result in stall recovery if the condition persists while within angle of attack and KEAS limits.

3. KEAS - Adjust toward 350 KEAS.

Apply sufficient pitch correction to compensate for thrust loss. Airspeeds near 350 KEAS are generally favorable to normal engine operation.

4. EGT - Downtrim manually if necessary.

Downtrim EGT manually with both EGT trim switches for from one to three seconds if engine stalls are due to high EGT conditions.

5. Derich - Checked.

Rearm the fuel derich system if the fuel derich warning light is on.

When the engine stall clears:

6. Engine settings - As desired.

Readjustment of the throttles should restore correct operation. Retrim EGT manually to the normal operating range if necessary. If auto-EGT trimming is resumed, monitor EGT and HOT/COLD flag indications until assured that no malfunction persists.

If stall persists; affected engine:

7. Shutdown and accomplish airstart.

Compressor Stall in Descent

Compressor stalls may occur during descent at internal bleed shift. Often these stalls are self clearing through reopening bleeds. After a few of these cycles of shift followed by stall, the bleed shift is completed and stalls then do not recur. With the restart switch ON, compressor stalls are likely near military rpm while above Mach 1.2. Maximum risk is with the aft bypass full open and restart ON where stalls are probable before the internal bleeds shift. Closing the inlet aft bypass delays the probable stalls until the internal bleeds shift closed. Refer to figure 3-7.

As stated previously, stall and unstart characteristics are very similar and accurate identification of which condition exists is difficult. Use of the wrong corrective procedure can result in continued stall or unstart with eventual flameout. The following procedure is designed to clear severe or protracted stalls or similar inlet roughness conditions which can not be positively identified as to cause.

In the event of severe or protracted compressor stall in descent:

Affected engine:

1. Restart switch - ON.
2. Throttle - IDLE immediately.

   Retard the throttle to idle immediately after setting the inlet restart switch ON. When near the internal bleed shift points, setting the restart switch ON without throttle reduction can induce engine stall and possible flameout. Advance the other throttle to Military power.

If stall persists:

3. Airspeed - Increase.

   Higher KEAS will help clear the stall.

When subsonic:

4. Restart switch - OFF.
5. Throttles - As desired.

CAUTION

If an engine stall cannot be cleared, shut down that engine and accomplish airstart.

ENGINE FLAMEOUT

Engine flameout characteristics are a loss of thrust, a drop in EGT and rpm. Fuel flow may or may not decrease, depending on the operating condition prior to flameout. Engine flameout can result from interruption of fuel supply, component malfunction, or continued unstable inlet conditions with the compressor stalled.

Engine flameout with afterburners on or off should be treated identically except for initial throttle positioning. If flameout occurs with the afterburners on the throttles should be retarded to minimum afterburning position to reduce thrust asymmetry. If the afterburners are off at flameout, the operating engine should be set to the thrust required by flight conditions.

In the event of engine flameout, as confirmed by cross-checking EGT, fuel flow, rpm, and ENP, either accomplish the engine shut down procedure or employ the airstart procedure.

DOUBLE ENGINE FLAMEOUT

The possibility of both engines flaming out is greater at high speeds because it is possible for the second engine to flameout as a result of yaw angles induced by flameout of the first engine. (If this happens, it may be easier to restart the engine that flamed out due to the yaw maneuver first.)

With both engines out, the hydraulic pumps provide sufficient flow for satisfactory control surface rates at engine windmilling speeds above 3000 rpm. Control capability is progressively reduced as speed decreases, becoming marginal at approximately 1500 rpm. The ac generators furnish electrical power at engine windmilling speeds down to 2650 rpm, then the battery and emergency inverter provide power for SAS operation and other essential equp- equipment. See figure 3-7 for a comparison of engine rpm with airspeed.

When altitude permits:

1. ATTEMPT AIRSTART.

When altitude is critical or engines will not start:

△2. EJECT.

GLIDE DISTANCE - BOTH ENGINES INOPERATIVE

The glide distance chart, figure 3-8 shows zero wind glide distances with both engines windmilling. The 375 KEAS glide speed is the same as is recommended for airstarts. Somewhat slower airspeeds provide greater range but reduced airstart capability. There is sufficient engine rpm for adequate hydraulic pressure to approximately 10,000 feet.

WARNING

Landing with both engines inoperative should not be attempted.

WINDMILLING GLIDE DISTANCE

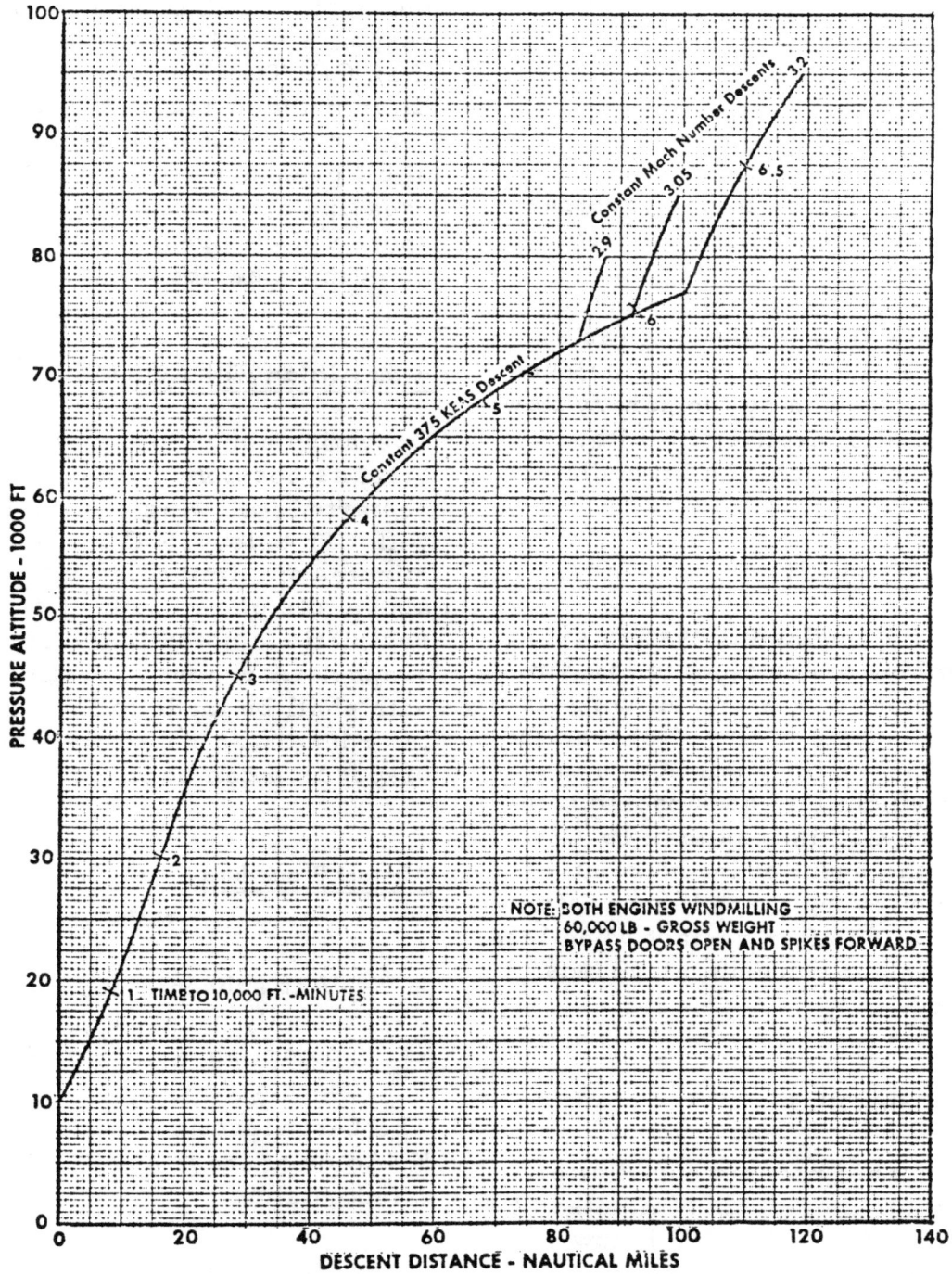

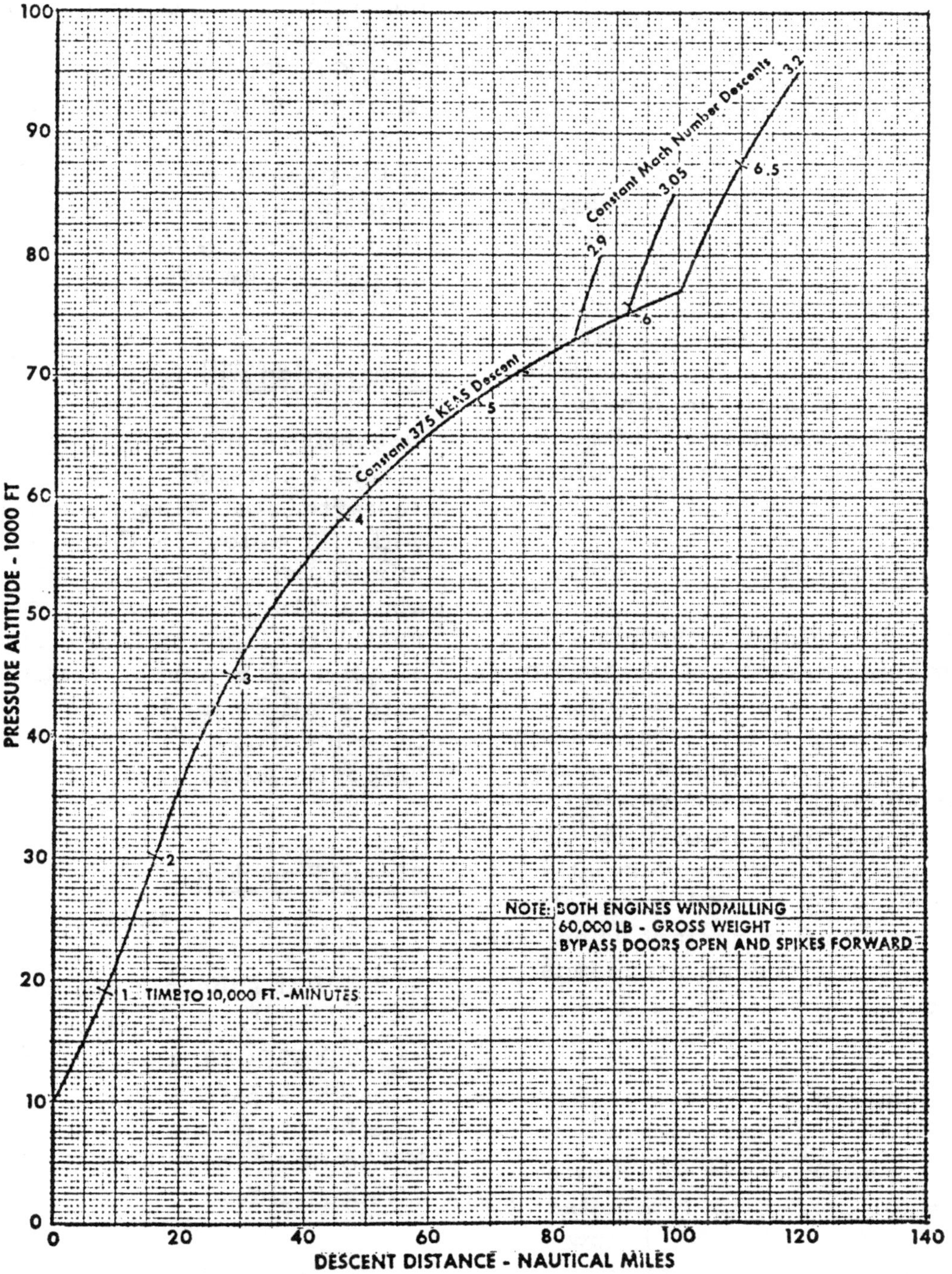

PRESSURE ALTITUDE - 1000 FT

DESCENT DISTANCE - NAUTICAL MILES

Constant Mach Number Descents

2.9

3.05

3.2

6.5

6

Constant 375 KEAS Descent

5

4

3

2

1 - TIME TO 10,000 FT. - MINUTES

NOTE: BOTH ENGINES WINDMILLING
60,000 LB - GROSS WEIGHT
BYPASS DOORS OPEN AND SPIKES FORWARD

Figure 3-8

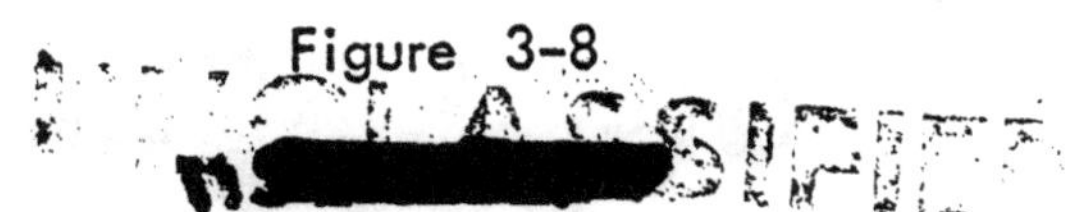

AIRSTART

Airstart procedures should be initiated immediately after flameout when possible; however, the reason for the flameout or shutdown must be considered prior to initiation of restart.

Use of Crossfeed

While it is expeditious to use crossfeed during an airstart attempt, crossfeed should not be left on after the start is obtained. Turn an additional tank on to the side where flow interruption is suspected before crossfeed is discontinued. If flameout is caused by temporary flow interruption, the throttle should be moved to the OFF position immediately.

Restriction with Low CIT

Portions of the engine oil system external plumbing are subject to low ambient air temperatures and oil congealing during subsonic windmill operation. Therefore, an engine shall not be intentionally windmilled at subsonic conditions where CIT is less than 15°C (60°F). In the event of inadvertent flameout, however, there is no restriction on operation after restart when less than 5 minutes at subsonic airspeeds have elapsed.

Airstart with Cold Oil

An airstart is not recommended after windmilling for more than 5 minutes at subsonic conditions with CIT below 15°C; however, if it is necessary to restart after exposure to such conditions, maintain CIT as high as possible prior to restarting. If an OIL TEMP warning light is illuminated, the throttle should remain at IDLE until the warning light is extinguished and a clear response of oil pressure is observed. If an OIL TEMP warning light is not illuminated at start, the throttle should remain at IDLE until a clear response of oil pressure is observed. When increasing above idle power, normal response of oil pressure to throttle movement must be observed.

After Airstart with Cold Oil

Operation above idle power shall be as brief as possible if an engine must be restarted following windmill operation with CIT below 15°C for more than 5 minutes and immediately operated above idle power without observing the restrictions noted in the above paragraphs. A report of pertinent flight data shall be made following the flight in order that a determination of engine disposition can be made.

Airstart Procedure

NOTE

> If at subsonic speeds, it is only necessary to accomplish ⊞ items for airstart.

The best airstart conditions are 375 KEAS or at least 7 psi CIP. On the affected side:

1. SMOOTH INLET.

Set the restart switch ON and open the aft bypass to attempt smoothing the inlet. This procedure may not smooth the inlet, and roughness will become severe near Mach 2.8. Although airstarts have been obtained while in roughness (inlet unstarted), there is a higher probability of restarting the engine when smooth operation has been restored.

2. DERICH.

Cycle the derich switch to REARM, then to ARM if the fuel derichment system has been actuated by high EGT.

⊛ 3. X-FEED OPEN.

Selecting crossfeed is the fastest method of assuring a positive supply of fuel to the engine before attempting an airstart.

⊛ 4. THROTTLE OFF, THEN 1/3 to 1/2 MILITARY.

Cycle the throttle to OFF, pause 2 to 3 seconds to assure cycling of the TEB chemical ignition system, then set the throttle at the position for 1/3 to 1/2 of the non-A/B range.

While supersonic, allow 15 seconds for rpm to increase (indicating that an airstart is being accomplished). Repeat procedure as necessary. Do not expend all TEB during airstart attempts while supersonic.

While subsonic, an airstart can usually be obtained in from 15 to 30 seconds at almost any allowable flight condition; however, 375 KEAS and at least 7 psi CIP are the recommended conditions. A period longer than 30 seconds may be required for starting. If repeated airstarts are not successful, leave the throttle in the 1/3 to 1/2 Military position several minutes until positive that no start can be obtained.

After engine starts:

5. AFT BYPASS SET.

Set the inlet aft bypass closed as rpm increases. Between approximately Mach 1.3 and Mach 2.3, compressor stall may be encountered when the internal bleeds shift as engine speed increases. If compressor stall in this speed range results in flameout, repeat the procedure, except no more than 6000 rpm should be maintained after start in this case. After starting and with the aft bypass closed, set the throttles and cockpit switches as required and reset the crossfeed switch and inlet controls.

If the engine will not start:

⊛ 6. Complete the Engine Shutdown procedure.

If unable to restart the engine after reaching subsonic speeds, shutdown the engine.

ENGINE SHUTDOWN

Engine shutdown must be accomplished in the event of complete engine failure, such as seizure or explosion, or in order to avoid or delay complete engine failure in the event of mechanical failure within the engine or engine accessories. Mechanical failure situations include uncontrollable rpm or EGT and unaccountably abnormal oil pressure, fuel flow, or vibration. Refer to emergency procedures related to the engine oil, EGT, fuel, and nozzle systems, and to information in this section relating to operation with one or both engines inoperative.

Complete failure probably will not permit normal windmilling operation but, if the engine continues to rotate, cooling fuel will circulate through the engine and aircraft cooling loops with the throttle OFF. An airstart should not be attempted, since doing so could result in fire or explosion.

Normal windmilling speeds can be expected after shutdown for mechanical failure. Fuel cooling will continue unless the fuel is shut off or drive shaft power to the fuel circulating pumps is lost. In some cases, an airstart may be attempted when operating conditions are favorable for control of oil pressure or EGT.

WARNING

Positively identify the failed engine before employing the engine shutdown procedure.

A descent from supersonic cruise altitudes to subsonic single engine cruise altitudes will be necessary if airstart attempts are unsuccessful or if engine failure has occurred. Initially, the spike should be forward and aft bypass open on the windmilling side. This delays the onset of inlet roughness. However, severe roughness will be encountered unless the spike is moved fully aft during the transition to low supersonic speeds. During the deceleration, the fully aft spike will result in mild buffet that will increase somewhat in intensity until a speed is reached where the buffet ceases abruptly. This speed is approximately Mach 1.3.

Note the effect of Mach number and engine rpm on inlet roughness as shown by the Inlet Unstart Boundaries charts, figure 3-5. Also refer to Single Engine Flight Characteristics, this section.

WARNING

With the spike forward, roughness intensity increases during deceleration between Mach 2.8 and Mach 1.3. Very severe roughness should be anticipated in this speed range if spike aft positioning is not completed. Maximum structural loads imposed are severe, but are well below design limits.

Descent distance can be extended by decelerating with maximum afterburning on the good engine. Overall economy can be improved by decelerating with minimum afterburning or Military power set on the operating side. Base the choice of A/B on or off on the power condition to be used for single engine cruise. When no airstart is to be attempted, descend at 350 KEAS until subsonic cruise altitude is reached.

ENGINE FIRE/ENGINE SHUTDOWN & DESCENT PROCEDURE

If a FIRE warning light illuminates, affected engine:

1. THROTTLE MIL/IDLE.

   Positively identify the affected engine. Then retard its throttle to Military when operating at a higher power setting. Retard the throttle toward IDLE if the warning light remains on or if operating with afterburner off when the warning occurs. If at supersonic speed, retarding the throttle of the affected engine to Military power or less will result in descent and deceleration to subsonic speeds.

   The thrust required for level flight may govern the power reduction possible on the affected engine if at low airspeed and heavy weight, as for fire warning immediately after takeoff. During landing approach, minimum control speeds consideration may govern the amount of power advancement which can be used on the unaffected engine.

   An immediate check should be made for abnormal EGT, for trailing smoke, or any other indication of fire. Use the rear view periscope. Request confirmation of fire from other personnel if outside assistance is available. In case of doubt, assume that a fire does exist.

   If the FIRE light extinguishes while at reduced power, and if there is no confirmation of fire, the flight may be continued with power reduced on the affected engine until a landing can be made at the nearest suitable facility. (Land as soon as possible.)

   If the FIRE light remains on with the throttle at IDLE, or if a fire is confirmed, shut down the affected engine.

Proceed as follows on the affected side if shutdown is necessary as a result of engine flameout, failure, or fire warning.

NOTE

- If at subsonic speeds, only [*] items should be accomplished.
- If the fire warning light extinguishes while shutting down the engine, do not attempt a restart. Fire or explosion could result.

2. RESTART ON.

   The spike forward and forward bypass open configuration delays onset of roughness or unstart when the engine is shut down at speeds near Mach 3.

[*]3. THROTTLE OFF.

[*]4. AFT BYPASS OPEN.

5. FUEL OFF FOR FIRE.

Positively identify the emergency fuel shutoff switch for the affected engine and set it to the fuel OFF position if shutdown is a result of Fire Warning.

WARNING

Shutting off fuel to a windmilling engine while at high Mach numbers may cause additional emergencies due to loss of cooling fuel for the engine and aircraft systems.

Fuel shutoff stops flow through one fuel cooling loop system. If speed is above approximately Mach 2.2, shutting off the fuel may cause engine oil to overheat and result in engine seizure. Shutting off the fuel may also cause additional emergencies due to loss of the associated aircraft cooling systems. Reduced Mach number decreases cooling requirements because of lower environmental temperatures.

6. Manual Spike control - Set Mach 3.2 (aft) position.

7. Airspeed - 350 KEAS (recommended).

Adjust speed toward 375 KEAS if airstart attempt is intended.

When roughness is encountered:

8. Restart switch - OFF.

Turn the restart switch OFF at onset of roughness, approximately Mach 2.8.

The full clockwise position of the manual spike control provides full aft spike and forward bypass open positioning (with the forward bypass control in the AUTO position) after the restart switch is set to OFF. Expect mild buffet as the spike moves aft and restricts inlet airflow.

CAUTION

Do not attempt airstart with the spike positioned aft.

Operative Engine Conditions

1. Inlet Controls - AUTO & CLOSE.

Position the spike and forward bypass controls to AUTO and set the aft bypass controls at CLOSE unless manual inlet control procedures are required and the emergency schedule is in use.

2. Throttle - Min A/B above Mach 2.0.

3. Throttle - Mil or Min A/B below Mach 2.0.

A/B on is required while above Mach 2.0 to keep descent and deceleration rates within limit. Minimum afterburner or Military power is recommended below Mach 2.0 until subsonic. Use of maximum afterburner may result in greatest descent distance extension; however, maximum power should not be selected while

above Mach 3.0 (to avoid the possibility of unstarting the good inlet due to sideslip) and it is relatively uneconomical while below Mach 2.0.

[*] 9. Bay air switch - OFF.

Close the bay air valve to make the maximum amount of cooling air available to the cockpits.

Turn the affected side refrigeration switch OFF if the action is justifiable (such as for smoke entering the cockpit, fuel shut off, etc.)

] 10. C.G. - Forward of 24% MAC

Transfer fuel as necessary to obtain c.g. within subsonic limits.

] 11. Fuel - Dump as necessary.

When below Mach 1.7:

12. Pitot heat switch - ON.

13. Exterior lights - On.

When below Mach 1.3:

14. Restart switch - ON.

Set the restart switch to ON and reposition the aft bypass when necessary in order to minimize roughness.

When subsonic:

[*] 15. C.G. - Monitor and control.

Refer to use of forward transfer and crossfeed as described under fuel system emergency operating procedures to control c.g. during single engine operation.

[*] 16. SAS - Appropriate channels OFF.

Review SAS and hydraulic services available.

Refer to procedures for SAS, flight control system, and hydraulic system emergencies for operating procedures.

[*] 17. Land as soon as possible.

SINGLE ENGINE FLIGHT CHARACTERISTICS

The aircraft design is such that no flight system is dependent on a specific engine; therefore, the loss of an engine will not result in subsequent loss of all hydraulic or electrical systems. If an engine fails at low speed just after takeoff, the large amount of asymmetric thrust may require bank toward the good engine and full rudder for directional control. Refer to figure 3-3 for minimum single engine control speeds. After regaining control, however, 7° to 9° rudder trim with bank and sideslip toward the good engine provide minimum drag during acceleration to climbout speed. Charts showing single engine climbout capabilities are included in the performance data appendix. Acceleration to climb speed and climb to landing pattern altitude must be accomplished with Maximum thrust on the operating engine when a climb capability exists for the operating condition. During single engine cruise, or after climbout, reduction to zero rudder trim and use of bank and sideslip to maintain course provides minimum drag. A bank of up to 10 degrees is recommended, using no more than enough rudder trim to maintain course.

Trim Changes

Pitch trim changes can be expected while dumping fuel due to shifting center of gravity as the tanks empty. Directional trim is quite sensitive to changes in airspeed and power settings during landing pattern operation. At high speed, engine failure or engine flameout could cause yaw angle to become critical at high rates if an effective damper were not operating. Temporary thrust reduction on the good engine helps to counteract the asymmetric thrust condition. Follow-up rudder action is necessary. If large yaw angles develop, inlet duct airflow disturbances may cause the other engine to stall or flame out.

Fuel System Management

Fuel system management during protracted engine out operation should be directed toward maintaining optimum center of gravity conditions, making all of the fuel available to the operating engine and, when possible, continuing the fuel cooling of necessary systems. Improper c. g. conditions will be indicated by abnormal pitch trim requirements.

Single Engine Air Refueling

Single engine air refueling procedures are provided by figure 3-9.

SINGLE ENGINE CRUISE

Single engine cruise performance for Military, Minimum A/B, and Maximum A/B thrust settings are tabulated in figure 3-10. These performance data conservatively reflect preliminary flight test results.

SINGLE ENGINE AIR REFUELING

ACTUAL	SIMULATED
Receiver weight and altitude variations may result in conditions where military power is inadequate while afterburning power is excessive. Flight tests have shown that single engine rendezvous and refueling can be accomplished satisfactorily with approximately 10,000 pounds of fuel aboard near 27,000 feet aircraft altitude. Approximately same control trim and forces as for single engine cruise may be used with bank angles up to 10 degrees. After completing rendezvous the following procedures are recommanded:	Reasonable practice of single engine refueling techniques can be accomplished at normal refueling altitudes with a fuel load of 25-30,000 pounds, one engine in IDLE, and use one engine in afterburner for thrust requirements with the following procedures:
1. Adjust EGT trim and throttles as necessary to stabilize behind tanker in mid-burner position.	1. Same as for "ACTUAL".
2. Turn both roll SAS channels OFF.	2. Same as for "ACTUAL".
3. Trim in roll and yaw axes as required to reduce effects of asymmetrical thrust and provide best stick pressures at stabilized speeds.	3. Same as for "ACTUAL".
4. Turn forward transfer ON if left engine is being used. Turn crossfeed ON if right engine is being used.	4. Same as for "ACTUAL".
5. Establish tanker contact using appropriate L or R hydraulic system.	5. Establish tanker contact using L hydraulic system.
6. Uptrim EGT to nominal as fuel is being transferred. Trimming beyond nominal band exposes the engine to stall condition.	6. Same as for "ACTUAL".
7. Initiate a toboggan maneuver prior to reaching maximum afterburner.	7. Same as for "ACTUAL" or: Initiate disconnect when power is limited at maximum afterburner. Do not intentionally exercise an outer limit disconnect.
	8. Re-establish normal 2-engine configuration and resume refuel operations if desired.

Figure 3-9

SINGLE ENGINE CRUISE PERFORMANCE Based on 61,000 Lb. Zero Fuel Weight Data Basis Preliminary Flight Test								
MILITARY THRUST								
Fuel Remain Lbs.	Mach	KEAS	Altitude Ft.	Fuel Flow Lb/Hr	N.Miles/ 1000 lb.	True Airspeed Knots	Time to 5000 lb. Hr:Min	Range to 5000 lb N. Miles
45,000	0.58	378	1,300	25,200	15.2	384	2:00	769
40,000	0.58	368	2,600	23,800	16.0	382	1:50	691
35,000	0.58	357	4,000	22,550	16.9	381	1:36	608
30,000	0.58	350	5,500	21,250	17.8	379	1:22	522
25,000	0.58	339	7,000	20,000	18.8	377	1:07	430
20,000	0.58	328	8,600	18,750	20.0	375	0:52	333
15,000	0.58	318	10,300	17,450	21.4	373	0:34	230
10,000	0.58	306	12,000	16,150	22.9	370	0:16	119
5,000	0.58	295	13,900	14,950	24.6	368	0	0
MINIMUM AFTERBURNER THRUST								
60,000	0.73	375	13,500	36,950	12.5	462	2:00	921
55,000	0.73	367	14,600	35,150	13.1	460	1:51	857
50,000	0.73	358	15,700	33,450	13.7	458	1:43	791
45,000	0.73	349	16,800	31,750	14.3	454	1:34	721
40,000	0.73	341	18,000	30,050	15.0	452	1:26	648
35,000	0.73	332	19,200	28,400	15.9	451	1:15	571
30,000	0.73	324	20,500	26,750	16.8	448	1:04	489
25,000	0.73	314	21,900	25,200	17.7	446	0:53	403
20,000	0.73	306	23,300	23,600	18.8	443	0:40	312
15,000	0.73	295	24,800	21,950	20.0	439	0:27	215
10,000	0.73	285	26,400	20,300	21.5	437	0:13	112
5,000	0.73	275	28,000	18,800	23.1	434	0	0
MAXIMUM AFTERBURNER THRUST								
60,000	0.85	361	22,500	51,700	10.0	518	1:28	737
55,000	0.85	353	23,500	49,050	10.5	515	1:22	685
50,000	0.85	346	24,500	46,900	10.9	513	1:16	632
45,000	0.85	338	25,600	44,800	11.4	511	1:09	576
40,000	0.85	330	26,700	42,250	12.0	509	1:02	517
35,000	0.85	323	27,900	39,950	12.7	507	0:54	456
30,000	0.85	313	29,000	37,700	13.3	503	0:46	390
25,000	0.85	304	30,300	35,400	14.1	500	0:38	322
20,000	0.85	295	31,600	33,150	15.0	498	0:29	249
15,000	0.85	286	33,000	30,800	16.0	494	0:19	171
10,000	0.85	276	34,400	28,600	17.1	490	0:09	89
5,000	0.85	267	36,000	26,450	18.4	487	0	0

Figure 3-10

The cruise altitudes given in figure 3-10 are also the aircraft constant throttle single engine ceiling capability. Since hot temperatures adversely affect aircraft ceiling, an altitude capability lower than shown on the charts must be expected on a hot day.

Minimum A/B thrust and Military thrust provide the best levels of single engine cruise performance. Military provides the best range performance, but penalizes the aircraft in altitude capability, especially at heavy gross weights. Minimum A/B provides good range performance with an ample altitude capability.

The maximum A/B single engine cruise has poor range performance and should be only used in cases where the required cruise altitude is higher than the minimum A/B cruise altitude capability.

AFTERBURNER FLAMEOUT

Afterburner flameout can be expected as a result of engine stall or abnormal inlet operation, or insufficient airspeed at altitude. Afterburner flameout may be detected by a loss of thrust and by comparison of nozzle position indicators. The flamed-out afterburner nozzle will be noticeably more closed. Fuel will continue to flow from the spray bars until the throttle is retarded to MILITARY. Correct the inlet, engine, or airspeed and altitude condition before attempting afterburner relight. At high Mach numbers, the minimum airspeed necessary for afterburner operation is lower with automatic scheduling than with spike forward.

In the event of afterburner flameout, attempt to relight as follows:

Affected engine:

1. Throttle - Retard to Military.
2. Nozzle operation - ENP checked.

   Retard the throttle below Military momentarily and observe ENP indication.

3. Throttle - A/B midrange.

   Note TEB shot counter number and fuel flow increase.

4. Nozzle position - Check.

   Check for more open nozzle position.

If relight not successful:

5. EGT - Increase trim.

   For CIT above 5°C, trim toward 805°C EGT.

   For CIT below 5°C, trim toward 830°C EGT.

CAUTION

Uptrim toward 830°C EGT carefully due to possibility of engine surge.

If relight by catalytic igniters is unsuccessful:

6. Igniter Purge switch - On for two seconds.

CAUTION

The TEB supply will be depleted rapidly if the igniter purge switch is held on for more than two seconds.

If relight not successful:

7. Throttle - Military.

AFTERBURNER CUTOFF FAILURE

If the afterburner does not cut off when the throttle is retarded to Military, an attempt can be made to vary the thrust by retarding the throttle below Military. The engine should be shut down if thrust cannot be modulated satisfactorily. The fuel may have to be shut off if the flowmeter indicates that the afterburner is dumping fuel.

AFTERBURNER NOZZLE FAILURE

Nozzle malfunctions may be indicated by the nozzle position indicator, excessive rpm fluctuations, or failure of the engine to control to schedule speed. This may be accompanied by compressor stall and exhaust gas overtemperature. Precautionary engine shut down may be necessary.

Nozzle Failed Open - High Altitude

A nozzle failed open condition will be more difficult to recognize at high altitude during afterburner operation near limit KEAS because open nozzle position is normal in these conditions. As altitude increases and KEAS decreases, the nozzle gradually closes to 60% to 80% open as limit altitude is approached. A failed open nozzle will result in abnormally high engine speeds under these conditions. An increase in afterburning throttle position or a reduction in cruise altitude while maintaining cruise Mach number (increasing KEAS) may permit cruise to be continued until the scheduled descent point is reached. Nozzle position and rpm of the normally operating engine can be used as a guide in selection of the lower cruise altitude range where an open nozzle position is normal. The Emergency Descent procedure should be used if rpm is not controllable. Be prepared to use less than Military throttle when the afterburner is shut down.

Nozzle Failed Open - Low and Intermediate Altitudes

At intermediate altitudes, the nozzle failed open condition may be recognized by reduction of thrust and an increase in rpm. At low altitude and Mach number it will be necessary to rapidly retard the throttle to a point midway between IDLE and MILITARY to keep rpm within limits. The same procedure will apply when altitude and Mach number are decreased and the nozzle

failure is detected. If the thrust requirement is critical, such as for takeoff, it may be practical to retain Maximum thrust, even with engine overspeed, until safe airspeed and altitude are attained.

Nozzle Failed Closed

A nozzle-closed failure can, in most cases, be detected by referring to the nozzle position indicator on the instrument panel and by analyzing engine symptoms. There are no obvious symptoms of a nozzle failed closed without afterburning because the nozzle is already closed, or nearly so, at Military thrust. EGT and rpm may fluctuate together. Either down trim the engine or retard the throttle slightly and check for rpm suppression or ENP change. A normally functioning nozzle will open slightly to maintain engine rpm. In the case of the nozzle failing closed, do not attempt to light the afterburner because the engine may flame out (after which it cannot be restarted due to reduced rpm). If the nozzle fails closed with afterburning, rpm supression will occur, probably unstarting the inlet. Compressor stall and afterburner flame out are extremely likely and EGT will probably rise.

Nozzle Failure Procedure

When a reduction in thrust or rpm is desirable:

1. Throttle - MILITARY or below, as required.
2. RPM and EGT - Maintain in limits.
3. Land as soon as possible.

ABNORMAL EGT INDICATIONS

EGT Overtemperature

An overtemperature condition may be encountered as a result of intentional or inadvertent uptrimming, failure of the main fuel control to regulate EGT as desired, malfunction of the automatic EGT trim system, nozzle failure, or airflow transients during engine stall or inlet unstart conditions. An overtemperature will be indicated by the EGT gage temperature display and, if over 860°C, by illumination of the EGT instrument red warning light. If the fuel derich system is armed, the Fuel Derich warning light will also illuminate when EGT exceeds 860°C and it will remain on until the Derich switch is reset, even though EGT returns to normal values. A relatively small rpm and/or fuel flow change may be observed.

Downtrimming EGT or throttle repositioning below Military are generally sufficient to clear an overtemperature condition unless the condition is caused by inlet airflow disturbances. In this latter case, appropriate compressor stall or inlet unstart procedures should be used.

The Derich system will tend to reduce EGT by decreasing fuel flow to the affected engine(s) if 860°C is exceeded while the system is armed. The fuel derich system should be rearmed <u>after</u> normal inlet or engine operation has been restored and EGT is within limit values.

Overtemperature situations tend to be more extreme when at the higher altitude and/or lower KEAS conditions. EGT may become uncontrollable in some cases when near maximum altitude and Mach number if the overtemperature is associated with compressor stall or unstart and if the Derich system is not sufficiently effective to reduce EGT.

WARNING

The affected engine(s) must be shut down if EGT exceeds 900°C for more than three seconds; otherwise, severe turbine damage can result.

Since inlet airflow is reduced during unstarts or compressor stalls, and since the main fuel control can not reduce fuel flow below the minimum fixed schedule, manual or automatic trimming and throttle reduction to IDLE have no effect during severe over-temperatures. The fuel flow reduction accomplished by the Derich system may not be sufficient. Airflow must be increased by decreasing altitude or increasing KEAS or by correcting the compressor stall or unstart in order to control EGT. Exhaust gas overtemperatures of up 900°C may be tolerated for a maximum of 15 seconds, then the affected engine(s) must be shut down. After the immediate action items of the shutdown procedure have been accomplished, the airstart procedure can be employed as soon as flight conditions are deemed favorable.

Auto EGT System - Malfunctioning Permission Circuit

The "permission circuit" allows automatic EGT trimming to occur when the corresponding engine throttle is positioned at or above the setting for Military power. Normally it is de-energized. It is energized by throttle positioning below the Military setting or by operation of the Fuel Derich system. When energized, the circuit receives dc power through the L or R EGT circuit breaker. Refer to figure 1-1B.

The Auto EGT Trim System is automatically disabled by a solenoid operated power interlock switch unless power is available to the permission circuit. The Auto EGT Trim System cannot function in the event of permission circuit malfunction. EGT trimming must then be accomplished manually.

EGT Gage Malfunction

In the event of an EGT gage malfunction, the indication may stick, fluctuate, operate erratically, or slew to zero or to its maximum indication of 1198°C. If the indication reaches or exceeds 860° C, the gage warning light will illuminate and derichment will occur if the Derich system is armed. The Derich system should be turned off if the EGT indication exceeds 860°C. If EGT tends to fluctuate erratically near this value, attempt to pull the EGT IND ac circuit breaker while the indication is below 860°C. Then recycle the Derich switch to ARM to protect the other engine.

If a malfunction of the EGT gage temperature display has occured, the Hot and Cold condition flags provided on the instrument will operate normally and not be displayed persistently while at or above Military power with the Auto EGT Trim system engaged. Temporary appearance from time to time indicates normal operation of the automatic trim system. EGT may be downtrimmed manually as a test to ascertain that a COLD flag will appear on return to automatic trimming. The system should be left on while the flight is concluded and an attempt made to match nozzle positions and to minimize sideslip (if using less than Military power) or to match fuel flow and to minimize sideslip (if the afterburners are on). A landing should be made as soon as practicable.

EGT Persistently Out of Auto-Trim Band With Flag Indication Visible

Auto EGT on Initially

With low EGT indicated and COLD flag persistently displayed --

Affected engine:

1. EGT Trim - Disengage the Auto EGT control.

WARNING

Do not attempt manual uptrim immediately. If an EGT harness has malfunctioned, uptrimming could result in severe EGT overtemperature.

2. Attempt to confirm low EGT indication.

   Check for thrust asymmetry (ball out of trim). Ball displacement toward the suspect engine is evidence of harness malfunction and false EGT indication and, possibly, a high EGT condition.

   CAUTION

   Downtrim EGT immediately for at least ten seconds if the suspect engine appears to have higher thrust than the normal engine. The trim motor may have reached its full uptrim condition while in Auto EGT.

   Ball displacement away from the suspected engine is evidence of a low EGT condition, particularly if the condition is noted at Military power or at Maximum thrust. Cross check fuel flow. Equalize thrust by throttle manipulation to center the ball, and then observe fuel flow again. A substantial fuel flow split (as much as 5000 to 6000 pounds per hour if afterburners are on) should be required to center the ball if there is an actual difference in EGT. A high fuel flow on the suspected engine indicates that a low EGT probably exists on that engine.

If low EGT condition appears to be confirmed:

3. EGT Trim - Trim with caution.

   Uptrim to equalize thrust (ball in center). Cross check nozzle positions and fuel flow indications.

   Operation may be continued in the manual EGT trim mode if engine response and parameter indications appear normal.

   NOTE

   Do not up-trim unless an asymmetrical low thrust condition on the suspected engine was readily apparent.

   Uptrimming will not be possible if the trim motor or motor switching has failed in a manner which prevents rotation in the uptrim direction.

   If uptrimming is not effective, attempt to downtrim for ten seconds. This should reduce EGT approximately 80°C, approximately half of the trim range available. Stop downtrimming if rpm decreases, as some rpm suppression can be expected if EGT is actually approaching the bottom of the trim band.

If low EGT is not confirmed, or if manual trim operates abnormally:

4. Land as soon as possible.

   Land as soon as possible if throttle manipulation checks and instrument cross-checks are inconclusive, or if attempts to trim EGT manually disclose abnormal or erratic trimmer operating characteristics.

   The EGT instrument and the Auto EGT Vernier Temperature Control both receive temperature signals from the same EGT harness. (See figure 1-1B) Therefore, un-

less the asymmetric thrust check is conclusive, it is not possible to confirm that a false signal is being supplied while in flight. A ground check with proper instrumentation is required. If a faulty EGT signal is being supplied, a low indication is more likely than one which is abnormally high. A landing should be made at the nearest suitable facility without delay, as the engine may have sustained severe damage due to high EGT.

With high EGT indicated and HOT flag persistently displayed --

Affected engine:

1. EGT Trim - Disengage the Auto EGT control and downtrim EGT.

   Attempt to downtrim manually to the normal EGT range.

   Operation may be continued in the manual EGT trim mode if engine response and parameter indications appear normal.

If manual downtrim is not effective:

2. Throttle - Retard below Military.

   Retard the throttle until EGT is within the normal trim range.

3. EGT Trim - Recheck downtrim capability.

   Auto-trim and manual uptrim signals to the EGT trim motor are disabled when below Military power or when deriched. Manual downtrim should be available if three-phase power is supplied.

   When the throttle is set below Military, effectiveness of downtrimming may not be reflected fully b the EGT gage. Readvance the throttle to Military momentarily to check effectiveness of the down trimming.

If downtrimming is effective:

4. EGT TRIM ac power circuit breaker - Pull.

   Continue with trim circuit disabled. Recheck EGT at Military power.

   Afterburner power is permitted if trim is satisfactory; however, subsequent trimming would require resetting power below Military and resetting the circuit breaker. TEB would be required for afterburner relight. Uptrimming could be accomplished by momentarily enga ing the EGT TRIM circuit breaker without reducing power if the discrepancy is a runaway uptrim signal to the trim motor.

   Continued operation with the engine above Military power is not recommended unless necessary in order to minimize risk during conclusion of a mission.

5. Land as soon as practicable.

If throttle control of EGT is ineffective:

6. Shutdown the affected engine.

EGT Persistently Out of Auto-Trim Band Without Flag Indication

Auto EGT On Initially

For abnormal high or low temperature indication without flag display --

If throttle above Military position, affected engine:

1. Manual EGT Trim - Selected.

To confirm EGT indication:

2. EGT downtrim response - Checked.

   Check for decreasing EGT indication (approximately 8°C per second).

3. Auto EGT trim - Set.

4. COLD flag response - Checked.

   The COLD flag should appear temporarily, while EGT is below the normal trim band. Flag appearance is confirmation of malfunctioning EGT gage temperature indication.

For possible electrical malfunction:

5. EGT TRIM ac circuit breaker - Check condition.

   The EGT TRIM circuit breaker provides three-phase ac power for trim motor operation through the Vernier Temperature Control (figure 1-1B) when in Auto EGT, or through the EGT Trim switch when trim is controlled manually. 15 volt dc power for the EGT instrument HOT and COLD flag operation is provided by a power supply within the Vernier Temperature Control.

6. EGT dc circuit breaker - Check condition.

   If in Auto EGT, permission circuit power must also be available through the EGT circuit breaker to operate the Auto EGT power interlock.

   The trim circuit, Vernier Temperature Control, or trim motor may have failed if EGT does not follow trim switch operation or if the COLD flag does not appear.

If malfunction of the EGT gage temperature indication is confirmed:

7. EGT IND ac circuit breaker - Pull

   Disable the EGT gage indication. Attempt to complete the action while the indication is below 860°C.

   Continue with Auto EGT on. Monitor HOT and COLD flag appearance. Periodic temporary appearance is evidence of EGT control.

If the Derich System has operated and EGT is stuck above 860°C:

8. FUEL DERICH dc circuit breaker - Pull

NOTE

This does not disable the Fuel Derich System for the opposite engine.

If EGT decreased without COLD flag appearance:

9. EGT Trim - Disengage the Auto EGT control.

   Monitor EGT and control temperature manually.

10. Land as soon as practicable.

EGT Gage COLD Flag Visible While Throttle Below Military Position

Auto EGT On Initially

Appearance of an EGT gage COLD flag while the corresponding engine throttle position is below Military indicates that its Auto EGT "permission circuit" is inoperative.

With the permission circuit inoperative, affected engine:

1. Auto EGT should be usable at or above Military power.

2. Select Manual EGT before operating below Military power.

AUTO EGT trim should not be selected while below Military power. EGT should be checked within the normal band while advancing power from below the Military position.

ACCESSORY DRIVE SYSTEM

Accessory drive system (ADS) failure is indicated by progressive loss of a generator and the corresponding A and L or B and R hydraulic systems. The rpm, oil pressure, and ENP indications should also be checked.

In the event of ADS failure, complete the appropriate hydraulic and/or electrical system emergency procedures. Shutdown the engine if there is excessive vibration or fire; otherwise, operate the affected engine as required.

Loss of Liquid Nitrogen Supply

Loss of LN_2 supply to the ADS does not affect ADS operation during flight, as contamination of the ADS by oxygen would not occur until the ADS were completely cooled. However, loss of LN_2 must be reported following flight so that proper maintenance can be accomplished.

OIL QUANTITY LOW

Cross check oil pressure indication when low engine oil quantity remaining is indicated by illumination of either OIL QTY LOW warning light.

The following procedure should be accomplished if either OIL QTY LOW warning light is illuminated:

1. Begin normal descent.
2. Oil pressure - Monitor.
3. Land as soon as practicable.

NOTE

Monitor engine operation closely for oil pressure fluctuations, high temperature, vibration, or other indications of imminent engine failure. Be prepared to shut the engine down.

OIL PRESSURE ABNORMAL

Refer to Oil Pressure Limits, Section V. Although not desirable, operation may be continued with oil pressure above the normal pressure range. Operation may also be continued with oil

pressure in the 35 to 40 psi range; however, engine operation should be monitored for indications of failure, which include engine roughness or rapidly increasing vibration. Oil pressures below 35 psi are unsafe and require either that the engine be shut down or that a landing be made as soon as possible, using the minimum thrust required to sustain flight until a landing can be accomplished.

OIL TEMPERATURE ABNORMAL

Abnormal high or low oil temperature is indicated by illumination of the OIL TEMP warning light. It is unlikely that low oil temperature will occur in flight with the engine operating, and a high oil temperature should be assumed. Abnormally high oil temperature could be caused by deficient lubrication flow or insufficient fuel/oil cooling. Abnormally low oil temp may be indicated before start or after extended windmilling operation at subsonic speeds.

In the event of an L or R OIL TEMP warning light indication in flight:

1. Oil quantity and pressure - Monitor indications.

   It may be possible to relieve a high temperature indication by reducing Mach number and altitude, or by maintaining fuel flow above 12,000 pph.

NOTE

> If L or R OIL TEMP warning light illuminates after extended windmilling operation, refer to AIRSTART procedure, this section.

FUEL CONTROL FAILURE

If a fuel control malfunction is suspected, minimize throttle movements and monitor rpm and EGT closely. Position the Emergency Spike - Restart switch to RESTART ON for the affected engine inlet if EGT becomes uncontrollable. This should reduce CIP and fuel flow. Shut the engine down if an uncontrollable EGT overtemperature condition continues.

FUEL-HYDRAULIC SYSTEM FAILURE

Fuel hydraulic system failure is usually the result of hydraulic pump failure or leakage from a broken line, connector, or actuator. Complete system failure will render the engine nozzle and compressor bleed systems inoperative.

The initial indication of hydraulic system failure would be an excessively open nozzle, accompanied by rpm overspeed. The nozzle will eventually remain full open, regardless of power setting. The main fuel control fuel flow schedule limits engine overspeed at all power settings to approximately 300 rpm above the normal Military schedule. Engine overspeed may not be nearly this great while at maximum afterburning and high Mach, since normal nozzle scheduling is nearly full open at these conditions. Some rpm control can be achieved while afterburning through EGT down-trim at maximum throttle position.

The engine internal bleeds tend to open or remain open, and the external bleeds drift toward closed. A closed external (start) bleed condition will probably result in engine stall and flameout when below 5000 rpm. Airstart attempts after flameout for this reason probably would be unsuccessful.

Continued engine operation is permissible with a failed hydraulic pump if engine speed is maintained above 5000 rpm and if maximum limits for rpm and EGT are observed; however, a landing should be made as soon as practicable.

If open nozzle and rpm overspeed indications are accompanied by abnormally high fuel flow indication, a fuel-hydraulic system line rupture or leakage is confirmed. If below Mach 1.5, engine shutdown should be accomplished and the fuel system of the affected engine should be isolated by closing the emergency fuel shut-off valve. If at higher supersonic speeds, delay shutdown and fuel shut-off until below Mach 1.5, if practicable, to avoid the possibility of engine damage.

To confirm suspected fuel-hydraulic system failure:

1. Engine nozzle - Check for full open indication.
2. Engine RPM - Check for overspeed from normal schedule.
3. Fuel flow - Check.

   If substantially above normal, a fuel hydraulic line has been ruptured. A fuel flow of approximately 8000 to 9000 pph above normal confirms a broken line. Check for fuel trail with periscope.

If hydraulic system failure confirmed:

4. Descend.

   Maintain 5000 RPM minimum while at subsonic speeds and minimize throttle movement. Observe max limits for RPM and EGT.

If fuel-hydraulic line ruptured or leaking:

5. Throttle - Off, when below Mach 1.5.
6. Emergency fuel shutoff switch - Fuel off.
7. Crossfeed switch - Closed.
8. Complete engine shutdown procedure.

SYSTEM EMERGENCIES

FUEL SYSTEM

Automatic operation of the fuel system provides center of gravity and trim drag control in addition to furnishing fuel to the engines. Fuel is also used to cool cockpit air, engine and accessory drive system oil, hydraulic fluid, and the SAS gyros. Abnormal operating conditions or emergencies affecting fuel system operation (such as generator failure) may affect c.g. control and operation of the fuel cooled systems in addition to engine operation.

Fuel System Manual Operation

Manual control of the fuel system is accomplished through use of the fuel panel crossfeed, tank pump and pump release switches, the forward transfer switch, and the essential dc bus circuit breakers for the ullage systems. Manual control, manual pump selection, crossfeed, and/or fuel transfer is necessary in the event of low fuel pressure, abnormal c.g. condition or c.g. trend, incorrect boost pump sequencing, or single engine operation. It is recommended as a precautionary measure after failure of one generator and as part of the SAS emergency procedure. Manual control of the fuel system is not recommended during normal flight operation except for takeoff and landing and for sustained subsonic flight. Note that manual operation of the fuel system supplements but does not terminate automatic fuel sequencing.

FUEL LOW QUANTITY WARNING

If the fuel quantity low warning light comes on with more than 6000 pounds of total fuel indicated on the quantity gage, compute total fuel from the individual tank quantities. Monitor tank 6 quantity and land as soon as practicable. Note that total and individual tank quantities are affected by pitch attitude and acceleration, tending to be low during decelerations or negative pitch angles, as during a descent, and high during accelerations or positive pitch angles, as during a climb.

Fuel Quantity Low Condition

If total fuel quantity remaining is near 3000 pounds, monitor fuel quantity and land as soon as possible. If immediate landing is not possible and a tanker aircraft is available, accomplish air refueling with any JP-type fuel. Subsequent operation should be restricted to a maximum speed of Mach 1.5 if refueled with other than standard JP-7 or 523E fuel.

Fuel Pressure Low Warning

If one or both FUEL PRESS LOW warning lights illuminate:

1. XFEED - ON.
2. PRESS TANK 6 ON.

   Fuel pressure should be restored with crossfeed and tank 6 on. Analyze the difficulty and attempt to resume sequencing.

3. Individual tank quantities - Check.

To restore automatic sequencing:

4. Pump release - Press.
5. Crossfeed - Press off.

If pressure can not be restored:

6. Land as soon as possible.

FUEL TANK PRESSURIZATION FAILURE

Fuel tank pressurization failure is indicated by the tank pressure gage and by the liquid nitrogen quantity indicator. The liquid nitrogen quantity low warning lights on the annunciator panel indicate impending failure.

NOTE

Do not continue flight above Mach 2.6 without nitrogen inerting of the fuel system.

No corrective action is possible after both liquid nitrogen systems are depleted except to limit maximum speed to Mach 2.6 and to reduce rates of descent to minimize the difference between fuel tank and ambient pressures. In descent, the fuel tank suction relief valve in the nosewheel well opens when slightly negative tank pressures occur. Rates of descent should be limited so that tank pressure does not become less than -1/2 psi in order to maintain normal maneuvering capability and structural safety factor.

Air refueling and normal climb procedures may be accomplished without nitrogen inerting of the fuel system. In climbs, the fuel tank vent relieves internal pressure when the tank to ambient differential pressure reaches 3.0 to 3.5 psi. Mach 2.6 must not be exceeded during climb and subsequent cruise.

WARNING

Limit tank pressures are -0.5 and +5.0 psi. The limits are based on structural capabilities of the fuselage tanks with design limit load factors.

To descend, in the event of fuel tank pressurization failure:

1. Descend so that minimum tank pressure limit is not exceeded.

   Adjust power and airspeed as required.

If flight included cruise over Mach 2.6:

2. Loiter subsonic at FL 330 to FL 300 for 10 minutes if possible.

   Descend from FL 400 to FL 300 as slowly as possible.

   Below FL 250 descend at approximately 1500 fpm.

If loiter not possible:

3. Continue descent so that minimum tank pressure limit is not exceeded.

NOTE

Cooling will be accelerated and pressure may be relieved faster after reaching subsonic speeds if the nose gear is extended.

FUEL SEQUENCING INCORRECT

Incorrect automatic fuel sequencing is indicated primarily by the fuel boost pump lights. (A light may illuminate out of normal sequence, or fail to illuminate on schedule.) In this event, control the boost pumps manually until

correct automatic sequencing resumes or a landing is made. It is possible that faulty fuel sequencing may manifest itself by secondary indications, such as a fuel low level light coming on prematurely, or an abnormal adjustment required in pitch trim (due to c.g. change by faulty fuel distribution). Note that forward c.g. requires increased power to maintain speed and altitude. If normal sequencing does not resume, and manual sequencing is either inconvenient or impossible, turn crossfeed on or transfer fuel to ensure that any available fuel will get to the engines.

CAUTION

Do not permit a manually selected fuel boost pump to continue running in an empty fuel tank. The boost pump will be damaged.

NOTE

Crossfeed may be required to provide fuel to both engines during fuel sequence malfunctions.

Fuel System Management with Engine Shutdown

Although automatic fuel sequencing continues during single engine operation, manual control of the fuel system is necessary.

With the left engine shutdown:

1. Manually select tanks 3 and 4 - ON.
2. Crossfeed - ON.
3. Forward transfer fuel as necessary to maintain the required c.g. position for subsonic operation.

With the right engine shutdown:

1. Manually select tanks 2 and 5 - ON
2. Crossfeed - ON.
3. Forward transfer fuel as necessary to maintain the required c.g. position for subsonic operation.

NOTE

Fuel transfer capability is lost when operating on battery power, and the crossfeed valve position can not be changed.

Fuel cooling will continue automatically and there will be an indication of fuel flow to the inoperative engine, if it is windmilling, unless its emergency fuel shutoff switch is actuated. This heat sink system fuel is either supplied automatically to the opposite engine's mixing valve, if the crossfeed valve is open, or returned to the air refueling manifold.

Crossfeed During Forward Transfer

Crossfeed should not be used during forward transfer when fuel remains in tank 2. If it were, most of the fuel transferred would come from tank 2 because of the aircraft nose up attitude and the lower fuel pressure head that these forward tank pumps would have to overcome. Only a small forward c.g. shift would result.

PARTIAL LOSS OF BOOST PUMPS

Partial loss of boost pumps may result from individual pump failure, sequencing failure, or loss of ac power at the generator bus. Partial boost pump failure may not be indicated by the fuel pressure low warning lights unless the condition is associated with some other system emergency (such as generator failure with bus tie split).

Incorrect fuel sequencing and center of gravity shift may be the first indication. Proceed as directed for Fuel Sequencing Incorrect. Crossfeed may be required; however, when crossfeed is on, more fuel will tend to feed from the forward tanks which have boost pumps operating than from the aft tanks. This could cause a hazardous aft c.g. shift.

COMPLETE LOSS OF BOOST PUMPS

Loss of all boost pumps can only result from multiple failures such as loss of both generators. The condition is indicated by illumination of both fuel pressure low warning lights, probably in conjunction with the generator out warning lights. If this occurs during takeoff, ground tests show that fuel tank pressurization will supply sufficient fuel to the engine-driven pumps to maintain engine and afterburner operation if all tanks are nearly full. Abort the takeoff if speed and runway length permit; otherwise, continue takeoff and then land immediately.

WARNING

Fuel can not be dumped with complete boost pump failure. Use caution and observe operating limits of Section V if a heavy weight landing is required.

Maintain a higher power setting on the right engine than on the left side in order to minimize aft c.g. shift as fuel is used.

Fuel Management Prior to Complete Pump Failure

When there is a possibility of complete pump failure, that is, after loss of one generator, make successive forward transfers to obtain and maintain as forward a c.g. as is allowed. Crossfeed should remain off (except when necessary for forward transfer) until tanks 4 and 5 are empty, then it may be selected on to equalize usage from the forward tanks.

Supersonic Cruise Capability

Without generator or battery power, the inlet spikes are expected to shift forward. Optimum cruise will be at subsonic speeds.

Supersonic Descent

Use normal descent speed and Military power in order to minimize the rate of descent and maintain a positive deck angle. If necessary, if the spikes are forward and forward bypass doors open (restart condition), use the manual inlet descent procedure. At some point, probably below 60,000 feet, the engines will be receiving insufficient fuel flow to maintain full power. The engines will throttle themselves, and may surge. If this occurs, reduce the throttle setting to eliminate "chugging". The rate of descent will increase as power is reduced.

Subsonic Cruise

Loiter speed and altitude schedules are recommended to minimize engine fuel flow requirements. The optimum loiter altitude is approximately 7000 feet below the altitude for long range cruise, 0.75 to 0.80 Mach number is suggested.

Landing Approach

A straight-in landing approach without abrupt nose over should be started approximately 20 minutes prior to intended landing, using approximately 250 KIAS in order to maintain a level or nose-up deck angle. The probability of engine failure during approach is increased if rpm is advanced during approach to the threshold.

WARNING

Be prepared to bailout immediately in the event of complete flameout. A dead-stick landing must not be attempted.

ELECTRICAL POWER SYSTEM

Each generator is capable of supplying the power requirements of the electrical system except for the FCS requirements.

SINGLE GENERATOR FAILURE

A single generator failure will be indicated by illumination of the L GEN OUT or R GEN OUT warning light. If the BUS TIE OPEN warning light also lights, a "split load" condition exists in which the operating generator output will not be connected to the primary ac bus of the failed generator. If either generator fails, the FCS radar will immediately shut off, since both generators must be operating to supply sufficient current to operate the Freon compressor motor and FCS radar. If the R generator fails, the Freon compressor motor will shut off as well. If the faulty generator is brought back on the line while the BUS TIE OPEN light is on, each generator will power only its own bus.

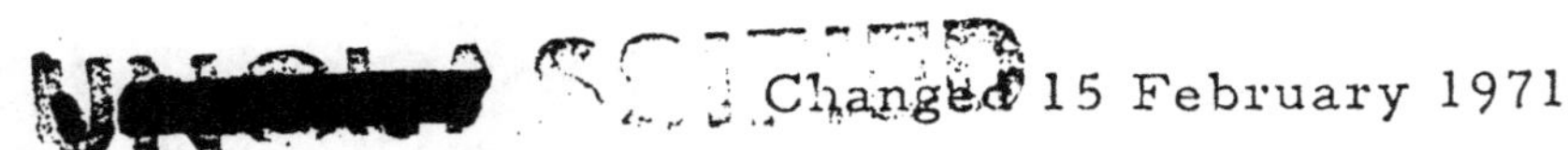

Single Generator Failure Procedure

If right generator has failed:

1. FCS liquid cooling switch - OFF

If either generator has failed:

2. FCS power switch - OFF.

3. Affected generator switch - OFF, then ON.

    This will recycle the automatic reset circuit up to three times. If the generator fault was a transient, the generator will reconnect to the bus and the GEN OUT warning light will extinguish and normal operation may be resumed.

CAUTION

When resuming normal operation after shutdown of the FCS liquid cooling system, allow at least ten seconds between liquid system turn-on and power application to the FCS. This will prevent excessive loads from being applied to the constant speed drive.

If the GEN OUT light remains on after two manual reset attempts:

4. Affected generator switch - OFF.

5. Crossfeed switch - Set.

    If the BUS TIE light is on, press the crossfeed switch on when tank 6 alone is supplying both engines or if a FUEL PRESS LOW light illuminates.

    Crossfeed may be required with one or both engines depending on tank 6 fuel as, with the generator buses split, the even or odd numbered tank 6 boost pumps corresponding to the failed generator are inoperable. A low fuel pressure warning may result when tank 6 is sequenced on automatically.

6. Speed - Subsonic operating speeds are recommended.

    Operational restrictions are 350 KEAS and Mach 2.8.

NOTE

Monitor hydraulic system warning lights and pressure instruments for indications of CSD failure.

7. C.G. - Maintain c.g. within subsonic limits.

    Transfer fuel as necessary and maintain c.g. within the subsonic limits. Confirm proper transfer by FCO computation of c.g. When subsonic, transfer fuel as necessary to maintain c.g. near the forward subsonic limit.

As soon as practicable after starting deceleration to subsonic speeds:

8. Mach trim c/b and pitch trim - Set and maintain pitch trim in normal subsonic cruise range.

    When the desired pitch trim setting is accomplished, pull the Mach trim circuit breaker.

NOTE

- All flight control trim systems will be inoperative in the event of a second generator failure.
- The pitch autopilot should not be used during descent to subsonic speeds.

When subsonic:

9. Affected generator - Attempt to reset.

   It may be possible to reset the generator while in the more favorable subsonic environment.

10. Mach trim c/b and pitch trim - As required.

    If desired, reset the Mach Trim circuit breaker and use normal pitch trim procedure.

11. Land as soon as possible.

    The only source of electrical power will be the batteries in the event that both generators fail.

DOUBLE GENERATOR FAILURE

If both generators fail, the only source of power will be the batteries. When the battery switch is ON, the batteries will supply the dc essential bus and the emergency inverter automatically, or manually by actuation of the instrument power switch if the automatic feature does not operate. With reduced usage, the batteries will last approximately 40 minutes. In order to reduce battery drain, systems not essential for flight or not usable on battery power should be deenergized. Some dc systems not always essential for flight can only be turned off by their circuit breakers. The following are some representative circuit breakers that may be pulled to conserve battery power, flight conditions permitting:

MACH TRIM
RATE OF TURN
MA-1 COMP
OIL QTY & TEMP R and L
INTERPHONE
NITRO QTY NO. 1 and NO. 2

Fuel System Capability Remaining

With both generators inoperative, the fuel boost pumps will be inoperable. Fuel availability is reduced under this condition and the engines will be fed by the gravity head plus LN_2 pressure existing. As the fuel quantity indicator is inoperative, it will not be possible to monitor tank quantities remaining. Fuel cannot be dumped. Approach and landing speed schedules will have to be based on estimated weight and attitude indications.

Flight Control System Capability Remaining

All flight control trim systems will be inoperative. SAS will remain usable until battery voltage drops.

Double Generator Failure Procedure

If both generators have been disconnected automatically, as indicated by illumination of both GEN OUT warning lights:

1. SAS - Reset, if necessary.

2. Heading and Vertical Reference AUX-MAIN switch - AUX (up) position.

3. Battery switch - Check ON.

   The EMER BAT ON warning light illuminates when the switchover to battery power is accomplished.

④ FCS liquid cooling switch - OFF.

⑤ FCS power switch - OFF.

6. Generator switches - Individually cycle to OFF and ON.

   This will recycle the automatic reset circuit to attempt to put the generators back on the line three times. If the generator faults were due to a transient and both generators reconnect to their buses, the generator out lights will extinguish and normal operation may be resumed.

If one generator resets:

7. Use Single Generator Failure procedure.

If both generators fail to reset:

8. Instrument power switch - EMER (up), if necessary.

   The INST ON EMER PWR annunciator light will illuminate when the emergency ac bus is energized by the inverter.

NOTE

In the event that one or both generators are reset, return the instrument power switch to the NORM position to avoid the possibility of autopilot control fluctuations occurring due to electrical phase differences in the ac voltage sources.

9. Control c.g.

   The c.g. will tend to shift aft as fuel is used. If possible, use a higher power setting on the right engine to maintain c.g. position.

▲10. Conserve batteries.

   One channel of pitch and yaw SAS and both channels of roll SAS may be disengaged while supersonic. Turn off the UHF, IFF, TACAN, and other non-essential systems where possible.

   When subsonic, select spikes forward and forward bypass doors open, aft bypass doors closed, and then pull all circuit breakers for the inlet system.

   Battery loads may be reduced further when subsonic by disengaging all SAS and pulling circuit breakers for non-essential systems and by turning the batteries off if flying under VFR conditions.

NOTE

SAS gyro c/b's should not be pulled. The gyros should remain operating in the event SAS reengagement is desired. Single channel pitch SAS reengagement may be required if a subsonic c. g. has not been attained.

11. Land as soon as possible.

WARNING

There is no assurance of continued engine operation if one or more fuel tanks are empty. If one of the tanks serving an engine is empty, engine flameout can occur at any time if flight attitude, engine flow requirements, and/or fuel level in the associated tank(s) is such that the remaining head of fuel in the supply tank(s) cannot keep that manifold clear of nitrogen gas or vapor.

Attempt to maintain fuselage attitude level to nose-up by airspeed control during descent.

Immediately prior to landing:

12. SAS - Attempt to engage a pitch channel.

BUS TIE OPEN LIGHT ON

Do not take action if the BUS TIE OPEN warning light illuminates while the generators continue to operate. Each generator will then power only its own bus. In the event either generator subsequently fails, the essential ac bus will be powered automatically by the operating generator.

TRANSFORMER-RECTIFIER FAILURE

One transformer-rectifier will supply the normal electrical demands.

A double failure of the transformer-rectifiers, as indicated by illumination of both XFMR RECT OUT warning lights, removes generator power from the essential dc bus. Proceed as for double generator failure, steps 1, 2, 3, 8 through 12.

Loss of emergency ac bus power is indicated by loss of SAS and one or more of the following as indicated by appearance of power OFF warning flags: Triple Display Indicator, Attitude Director Indicator, Altitude-Vertical Velocity Indicator and the Standby Attitude Indicator. Battery power is provided after inverter warm-up by moving the instrument power switch to the EMER (up) position. With reduced usage, the battery will last approximately 40 minutes.

When at subsonic speed and there are not abnormal control problems, turn all SAS channels off before reacquiring ac power. Then, after ac power is on,

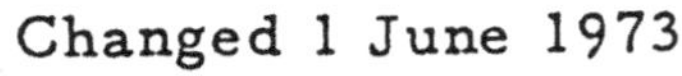

wait 30 seconds before re-engaging the SAS channels individually. Expect transients in the respective axes on engagement. If at high speed and conditions are such that SAS is required immediately, reacquire ac power with SAS engaged. Expect a large attitude transient as ac power is turned on.

HYDRAULIC POWER SYSTEM

With both engines out, the hydraulic pumps provide sufficient flow for satisfactory flight control system operation at windmilling speeds above 3000 rpm. Reduced control system capability is available down to a windmilling speed of approximately 1500 rpm. With one engine windmilling, all primary and most utility services are supplied by the operating engine hydraulic systems. The windmilling engine utility system pressure and flow may be sufficient to supply service until the engine is almost stopped.

ABNORMAL HYDRAULIC SYSTEM PRESSURE

Steady hydraulic system pressures between 2200 and 3000 psi, and above 3500 psi, are considered abnormal. An abnormal pressure condition should be corrected before continuing when observed prior to flight. Although not desirable, operation may be continued without restriction if an abnormal pressure condition is observed during flight; however, system operation should be monitored for indications of low quantity or degraded performance for the remainder of the flight, and the condition must be reported to ground personnel after landing. Transient pressure fluctuations into the abnormal ranges are not considered abnormal when they can be associated with system usage.

L & R (UTILITY) HYDRAULIC SYSTEMS

Illumination of the utility hydraulic system L-R HYD QTY LOW warning light indicates a low fluid quantity condition for one or both system(s) (not low quantity and/or low pressure as for the A and B hydraulic system HYD LOW warning lights). A low pressure condition is indicated by the corresponding hand of the L and R hydraulic system (SPIKE) pressure gage. If the L system fails or pressure falls below 2200 (nominal) psi, crossover to the R system for gear retraction is automatic and control of ventral fin retraction and extension and nosewheel steering can be transferred to the R hydraulic system. The Gear Emergency Extension procedure must be used to lower the landing gear if the L system fails.

Items which are affected by failure of the L hydraulic system are normal brakes, ventral fin, nosewheel steering, aerial refueling, and the left inlet spike and bypass doors. Items which are affected by loss of the R hydraulic system are right inlet spike and bypass doors, alternate brakes, and missile bay doors. Control of missile bay doors crosses over to the L system in case of R system loss or pressure drop to 2200 (nominal) psi.

L and/or R Hydraulic System Failed

Consider that a utility hydraulic system has failed when its pressure indication remains below 2200 psi or the L-R HYD QTY LOW warning light remains illuminated. Subsonic operating speeds are recommended; however, if at high speeds when failure occurs, deceleration may be discontinued while supersonic, if necessary, providing cruise speed is held below Mach 2.3. Avoid sustained operation at speeds for engine bleed shifting.

At initial indication of system malfunction, while above Mach 2.3:

1. Begin descent.

If either system pressure indication drops below 2200 psi, (affected inlet):

2. Restart switch - ON.
3. Throttle - Military.

NOTE

Use normal descent procedure on good inlet. If desirable, when descent is to be continued to subsonic speeds, use the afterburner throttle range on the unaffected side while above Mach 2.0.

If spike not forward:

4. Forward bypass control - OPEN.
5. Spike control - FWD.
6. Emergency spike switch - ON.

CAUTION

To ensure that the spike will lock in the forward position select forward bypass OPEN and spike FWD before operating the emergency spike switch; otherwise, it is possible for the spike to become unlocked after gas pressure is depleted if hydraulic pressure is fluctuating.

7. Land when practicable.

Refueling is not recommended if the L hydraulic system has failed.

WARNING

With both L and R systems failed, wheel brakes and steering will not be available. Ejection may be necessary if a suitable landing area cannot be reached.

If L system failed:

8. Brake switch - ALT BRAKES.

Select alternate steering and brakes for air refueling and landing.

9. Gear- Emergency extension for landing.

The Gear Emergency Extension procedure must be used prior to landing with the L hydraulic system failed.

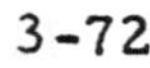

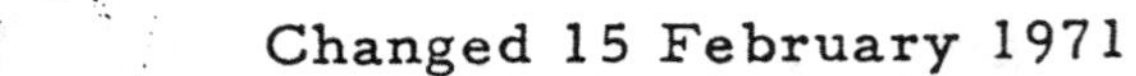

CAUTION

After landing with the L Hydraulic system failed or with the left engine inoperative, stop on the runway and have the gear downlocks installed before taxiing. This positively assures that the gear will remain locked in the down position when L Hydraulic system pressure is not available.

A AND B HYDRAULIC SYSTEMS

The loss of either A or B hydraulic system fluid quantity or pressure will be indicated by the HYD SYS LOW warning light on the pilot's annunciator panel. The A and B hydraulic pressure gage will indicate complete failure of a system. If the A or B hydraulic system fails, the control forces will not change. Either system will operate the control surfaces, but at a slower rate and with some reduction in available control at higher speed and Mach number combinations. Airspeed reduction with a single hydraulic system is a precautionary procedure which allows for the reduction in available hinge moment capability. Monitor all system operation closely and attempt to determine if a complete failure is imminent. Be prepared for ejection prior to complete failure.

Effect of System Loss on SAS

Disengagement of the failed A or B hydraulic system SAS channels is necessary to regain SAS damping capability in all channels, as a hydraulic system failure is not sensed directly by the SAS equipment. The signal gain of the operating yaw channel is doubled automatically by disengagement of the channel for the failed hydraulic system. It is necessary to recycle the operating roll SAS to regain damping in this axis.

A or B HYD SYS LOW Light On

In the event that an A or B HYD SYS LOW warning light illuminates:

1. A & B hydraulic pressure gage - Checked.

If at supersonic speed:

2. Begin normal descent.
3. Land as soon as possible.

A or B Hydraulic System Failed

In the event that an A or B hydraulic system has failed, as indicated by illumination of the HYD SYS LOW warning light and confirmed by indication of the hydraulic pressure gage:

1. Airspeed - 350 KEAS or less.

CAUTION

Do not exceed 350 KEAS with only one hydraulic system operative. If a system fails above this speed, reduce airspeed immediately. Set Idle power if at transonic speeds with high KEAS.

2. SAS channels - Respective SAS switches OFF.

   Set respective pitch, yaw, and roll SAS channel engage switches OFF.

3. Roll SAS - Cycle operating channel OFF, then ON.

   The roll SAS may be disengaged, if desired, to avoid pitch coupling.

4. Reserve hydraulic oil system - To operative A or B system.

   Monitor the operative system closely. Be prepared to eject if failure becomes imminent.

If at supersonic speed:

5. Make normal descent to subsonic speed.

6. Land as soon as possible.

A and B Hydraulic Systems Failed

If both the A and B hydraulic systems have failed:

1. EJECT.

WARNING

All control will be lost if both the A and B hydraulic systems fail.

PRIMARY FLIGHT CONTROL SYSTEM

In the event that control difficulties are encountered:

1. Hydraulic pressure - Check A & B hydraulic systems.

If both systems are normal and the autopilot is engaged:

2. Autopilot - Disengage and check control.

3. SAS - Check warning lights.

   If SAS failure has occurred, proceed as directed under SAS Emergency Operation, this section.

4. If unable to determine cause of malfunction - Proceed at subsonic speeds (recommended) and make a precautionary landing when practicable.

TRIM FAILURES

Pitch, yaw or roll trim malfunctions may be of the inoperative type or the runaway type. Runaway trim failures in pitch may occur at slow rate (0.113°/sec change in elevon deflection) if due to automatic trim motor operation or at fast rate if due to manual trim motor operation (1.13°/sec change in elevon deflection). A low rate runaway type of malfunction will be apparent by the need for constant manual pitch trimming. A high rate pitch trim runaway will result in a moderately rapid change in pitch attitude. If the cause is a sticky manual pitch trim switch, a rapid oscillation may develop if the pilot applies corrective pitch trim inputs. The possibility of this happening can be minimized by manually centering the trim switch following each trim application. The runaway yaw trim rate is approximately 0.90° per second trim change. The roll trim rate is approximately 0.95°/sec. Runaway yaw trim will be accompanied by rudder pedal deflections as the surfaces move. Runaway pitch or roll trim will not be accompanied by stick movement due to surface movement.

In the event trim runaway failure is suspected, proceed as follows:

1. Trim Power Switch - OFF.

If circumstances permit:

2. Reduce speed to below 350 KEAS and 2.5 Mach number.

With runaway nose up pitch trim:

3. Transfer fuel forward to reduce forward stick force requirement.

WARNING

Do not transfer fuel if nose down pitch trim has occurred.

When time and conditions permit, if the pitch and yaw trim switch malfunction is suspected:

4. Trim Selector Switch - AUX TRIM

5. Trim Power Switch - ON (momentarily)

   If trim difficulty has been corrected, proceed in normal flight, using the Auxiliary Trim Control Switch for pitch and yaw trim.

If trim difficulty persists:

6. Trim Power Switch - OFF

7. Autopilot - ON. Check for control improvement.

8. Affected trim circuit breakers - Pull.

NOTE

The manual pitch and yaw trim motors are powered from the same circuit breaker. Therefore, both are rendered inoperative when the pitch and yaw trim circuit breaker is pulled.

Trim Malfunctions:

a. If runaway slow rate pitch trim - Pull AUTO PITCH trim circuit breakers.

b. If runaway high rate pitch trim - Pull PITCH & YAW trim circuit breaker.

c. If runaway roll or yaw trim - Pull the ROLL or PITCH & YAW circuit breakers.

9. Trim power switch - ON.

With manual pitch trim inoperative and auto trim available, engagement of the pitch autopilot will gradually correct an out of pitch trim condition. This will relieve the pilot of a need for maintaining stick deflection to maintain attitude. The pitch autopilot can also be used when the auto trim motor is inoperative, but automatic pitch trim synchronization will not be available. In this situation the pitch autopilot should be used only in straight and level flight and turns should be hand flown.

CAUTION

Disengagement of the pitch autopilot when not in trim may be accompanied by a considerable transient.

If the trim malfunction is a runaway in the roll axis, right or left stick deflection will be required for the rest of the flight but stick force will not be more than normally required for the same amount of deflection. If the malfunction was a runaway in the yaw axis, rudder pedal force will be required to maintain neutral rudder pedal position.

STABILITY AUGMENTATION SYSTEM

SAS emergency operating procedures and the applicable flight limitations should be used whenever there has been a channel disengagement or a reduction in SAS effectiveness. Disengagements may result from failures of any of the following systems or components: SAS gyro or electronics circuitry, flight control servos, or electrical power supply. Disengagement or loss of effectivenss may occur as a result of complete or partial loss of A or B system hydraulic power. Disengagement of any channel is indicated by illumination of the master caution light, the SAS CHANNEL OUT light on the annunciator panel, and one or more of the recycle indicator lights on the SAS control panel.

SAS Emergency Operation

When a malfunction is indicated in any SAS axis by poor aircraft response or by SAS recycle light illumination:

1. A and B hydraulic systems - Check for normal pressure indication.

   If hydraulic pressure failure is indicated, follow A or B Hydraulic System Emergency procedures.

2. ▲ Electrical system - Check warning lights off and circuit breakers normal.

3. Recycle appropriate channels.

   For pitch or yaw SAS, press recycle light.
   For roll SAS, cycle either A or B Roll SAS Channel Engage switch OFF, then ON.
   If the malfunction was a transient condition, the channel will reengage.

4. Pulse the aircraft in the appropriate axis.

   Pulsing the aircraft will check for a dead or sticking gyro. Although this check is desirable, it is not a conclusive check. If the light remains extinguished, assume no failure. If the light re-illuminates, or will not recycle, assume a failure is indicated by the lights.

NOTE

Consider that no failure exists when all pitch and yaw recycle lights have been extinguished, regardless of previous combinations of illumination, if normal operation of the recycle lights is verified by depressing the SAS Lights Test button.

The SAS Failure Warning Lights charts, figures 3-11 and 3-12 illustrate the probable causes of failure, indications, remaining capabilities, procedures, and limits which apply after channel disengagement.

SAS FAILURE WARNING LIGHTS CHART

(LOGIC CIRCUITRY FAILURE INVALIDATES THIS CHART)

▒ = MOST LIKELY FAILURE ◇ = ANOTHER POSSIBLE FAILURE

SAS LIGHTS ON - ● OFF - ○		Servo / electronics failures	Two-channel failures	A / B / M	Logic power supply failures	Any failure combination	Low hydraulic pressure	NO LIGHTS BUT POOR SAS OPERATION
SERVO	A, B		TWO ELECTRONICS OR TWO SERVOS AND ONE ELECTRONICS		A or B LOGIC POWER SUPPLY	ANY FAILURE COMBINATION	LOW HYDRAULIC PRESSURE	A, B: LOGIC POWER SUPPLY FAILURE
ELECTRONICS	A, B, M	ANY ONE CHANNEL; AND ANY ONE ELECTRONICS CHANNEL			◇			A, B, M: AC or DC POWER FAILURE
ACTION		Recycle -- press light off. Make control transient in affected axis. / Do not use pitch or yaw logic over-ride or BUPD with bad servos	Try A or B logic over-ride. Use BUPD* if re-quired		Recycle - Press lights off	Logic over-ride on - A or B	Failed system SAS off - 350 KEAS max	
NOTES		If light stays on, or comes back on during transient, use good channel. Monitor hydraulic pressure. Servo may be usable with electronics failed.						

MAX airspeed 350 KEAS -- With LOGIC OVERRIDE or with 2nd pitch or yaw SAS channel failed.

MAX mach number 2.5 -- With 2nd yaw SAS channel failed.

Autopilot -- Do not use with BUPD.

With pitch SAS 2nd failure -- Transfer fuel forward.

*BUPD is limited to .85 mach and 330 KEAS.

Figure 3-11

COMBINATION PITCH, YAW, & ROLL AXIS FAILURES - M LIGHTS ON					
Recycle lights illuminated	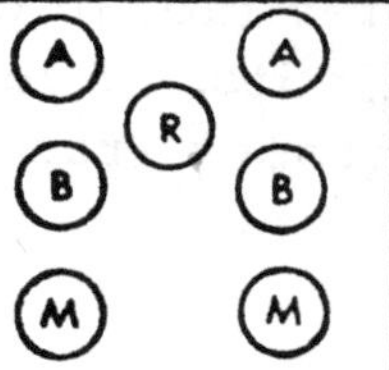 A A R B B M M	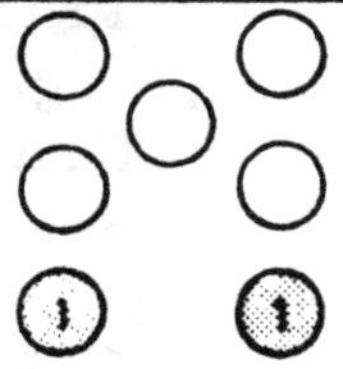	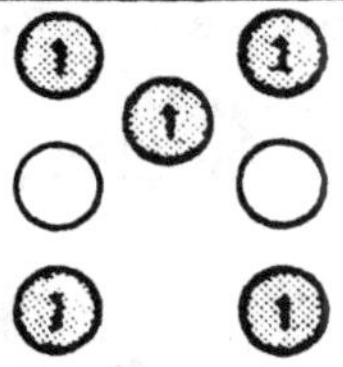	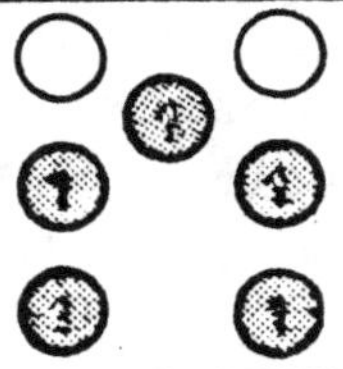	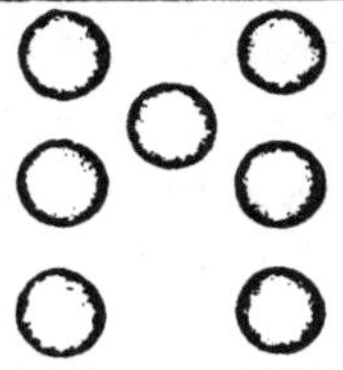
ACTION: Check - generators circuit breakers pilot: left console fco: right console Then try to recycle roll channels and press pitch & yaw lights off.	THEN: If lights stay on - failed channels off discontinue if practicable if pitch axis failed supersonic, transfer fuel to tank 1 to obtain subsonic c.g. limits. See electrical system emer. procedures.				for 2nd cond. To decel: Use Manual inlet descent procedure.
LIMITS: (See generator failed limits if applicable.)	Max speed: Mach 3.0 350 KEAS in cruise 400 KEAS in climb Land when practicable. Do not use logic override.				Mach 2.0 max 350 KEAS max forward c.g. Subsonic speed recommended.

(1) = first condition failure. (2) = second condition failure.

COMBINATION PITCH, YAW, & ROLL AXIS FAILURES - M LIGHTS OFF				
Recycle Lights illuminated	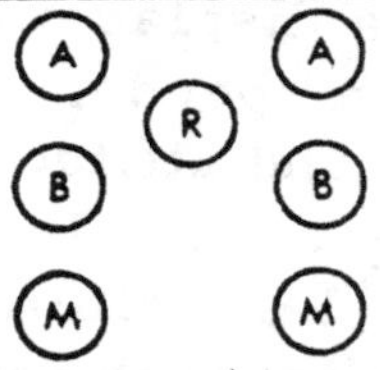 A A R B B M M	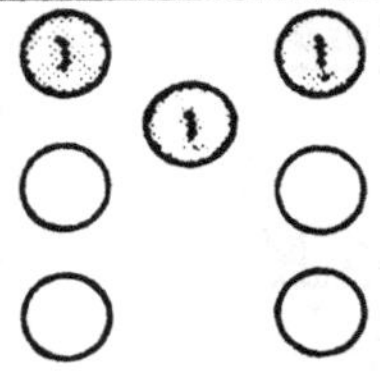	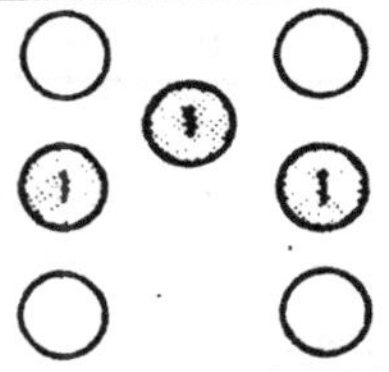	operation poor, but no lights on.
ACTION: check hydro pressure		A system	B system	both systems
If pressure normal, recycle a roll channel and press pitch & yaw lights off.	If pressure low, any combination of simultaneous or progressive A, B, and/or roll light illumination may occur. Failed channels off. Recycle roll channel on. See primary flight control emergency procedure.			
LIMITS:	Max speed: 350 KEAS			

Figure 3-12

PITCH OR YAW AXIS "FIRST" FAILURE

Aircraft flight characteristics do not change as a result of a single failure (or sequence of failures) in the pitch and yaw axes which leaves one A or B channel operating in each of these axes. However, a roll channel failure will cause some undesirable cross coupling in the pitch and yaw axes.

A "first" condition failure exists after attempts to extinguish one or more SAS recycle lights are ineffective and an A or B or M, or an A and M, or B and M light combination remains illuminated in an axis. Either an A or B channel must remain (light off) in each of the pitch and yaw axes. Different types of first condition failures may cause either simultaneous or progressive illumination of the A and M or B and M combination, as illustrated by the SAS Warning Light Charts, Figures 3-11 and 3-12.

Pitch or Yaw Axis "First" Failure Procedure

When a "first" failure exists in a pitch or yaw axis, A or B or M or A and M or B and M illuminated:

1. Maximum speed - 350 KEAS in cruise or 400 KEAS in climb, and Mach 3.0.

2. Failed channel switch - OFF.

3. Land when practicable.

While supersonic, if pitch axis channel failed:

4. Fuel - FWD TRANS to increase tank 1 quantity.

   Transfer fuel to obtain subsonic c.g. limits. Accomplish and confirm proper transfer by FCO computation of c.g.

PITCH "SECOND" FAILURE

After a second condition failure in the pitch axis, some longitudinal overcontrol probably will occur when at high Mach numbers. Observance of second failure limits is required, and descent to subsonic operating speed is recommended when practicable. If unable to use BUPD, air refueling and landing may present some difficulties in maintaining precise attitude control. With SAS off and a neutral stability condition, there is no tendency for the aircraft to return to a trimmed attitude when a displacement occurs at landing pattern speeds. However, divergent speed and attitude tendencies occur slowly enough to be completely controllable. Minimum airspeed limits with pitch SAS inoperative (figure 3-13) should be observed.

Pitch Axis Second Failure Procedure

1. Airspeed - 350 KEAS Maximum.

Failure analysis and corrective action depends on the sequence of first and second failure warning light indicatio Depending on failure analysis and fligh conditions, do the following:

MINIMUM AIRSPEED LIMITS WITH PITCH SAS INOPERATIVE

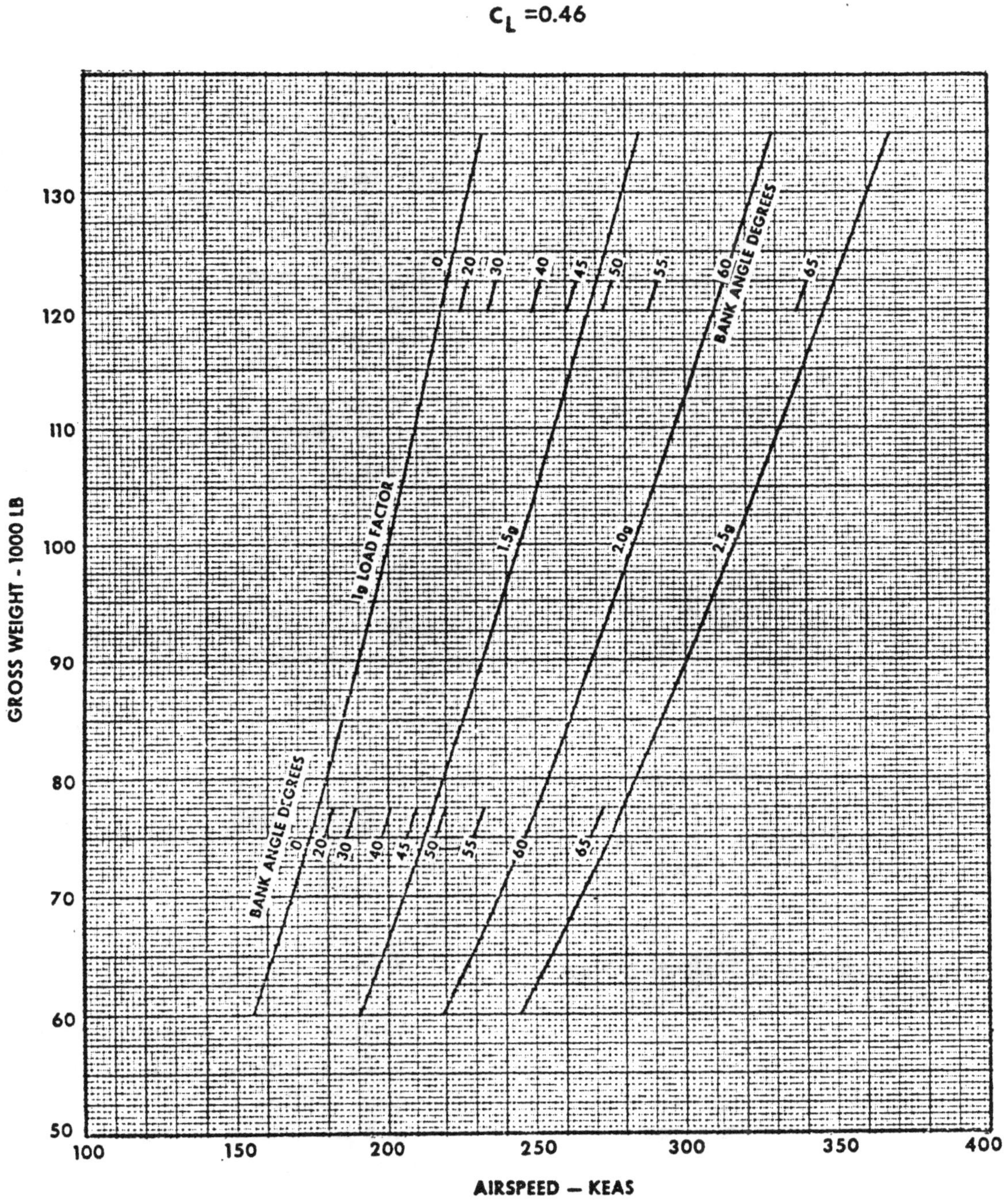

$C_L = 0.46$

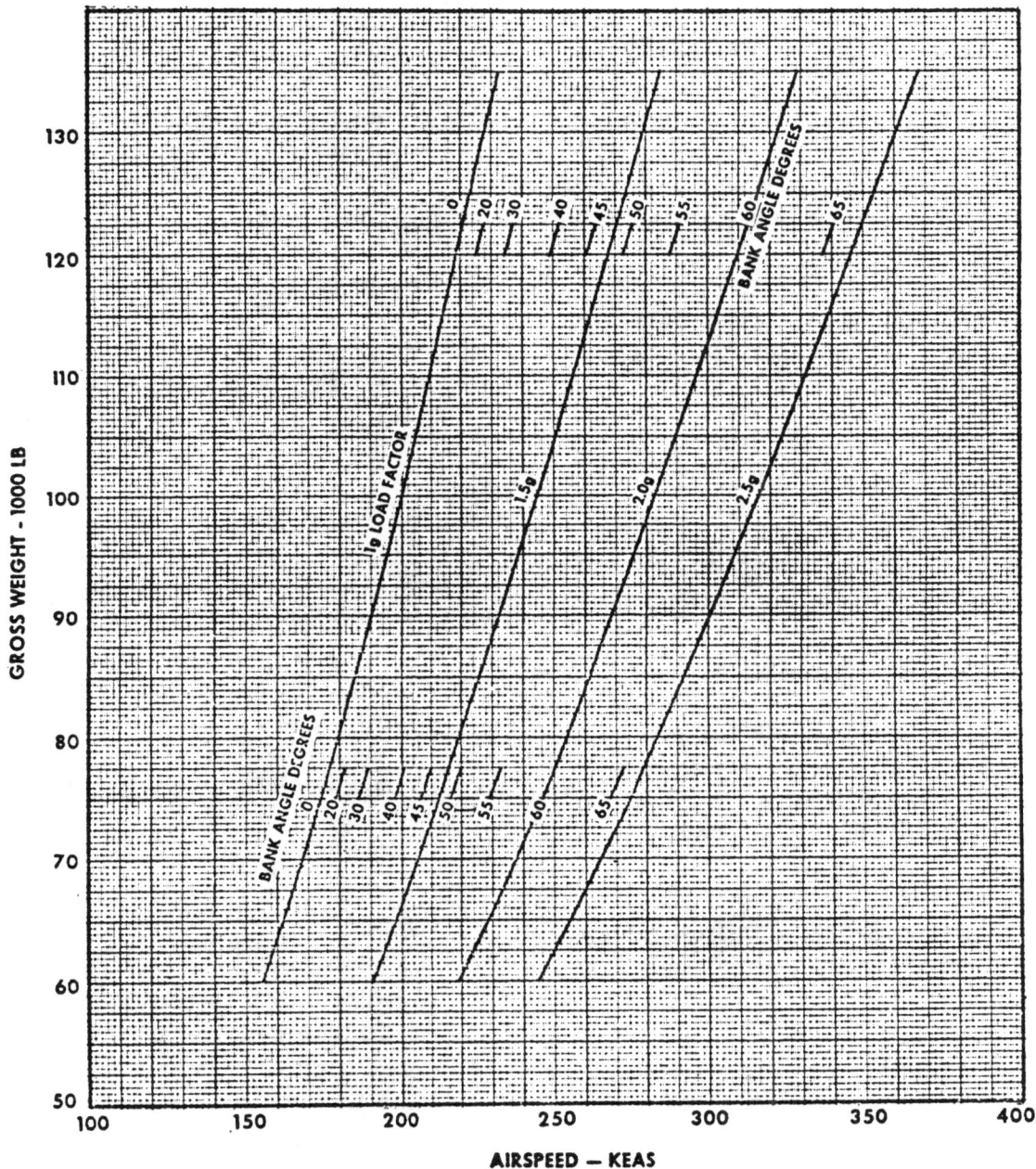

GROSS WEIGHT - 1000 LB

AIRSPEED — KEAS

1g LOAD FACTOR

1.5g

2.0g

2.5g

BANK ANGLE DEGREES

0, 20, 30, 40, 45, 50, 55, 60, 65

Figure 3-13

2. Remain at supersonic speed (if supersonic) and use logic override procedure, if appropriate.

If conditions permit:

3. Descend.

4. Fuel transfer switch - TRANSFER. Transfer and maintain 3000 pounds in tank 1 for remainder of flight.

5. Fuel dump switch - If close to base, DUMP until tank 6 boost pump light comes on.

6. Airspeed - Maintain Mach 1.3 minimum until fuel transfer is complete.

NOTE

Low supersonic conditions take advantage of available pitch stability margin. Make essential maneuvers at low load factor and bank angle.

7. Airspeed - Slow to best subsonic cruise speed at optimum altitude for remainder of flight. Maintain approximately 25:75 fuel ratio between tanks 1 and 6.

8. Refer to Backup Pitch Damper Operation, this section, and consider using.

9. If BUPD cannot be used, use caution to avoid abrupt maneuvers during landing approach.

Yaw Axis Second Failure Procedure

1. Airspeed - Mach 2.5 maximum.

Failure analysis and corrective action depends on the sequence of first and second failure warning light indications. Depending on failure analysis and flight conditions, do the following:

2. Remain supersonic and use logic override procedure, if appropriate.

If conditions permit:

3. Descend.

4. Airspeed - Slow to best subsonic speed and altitude for duration of flight.

ROLL AXIS FAILURES

Unless the failure can be associated with a specific hydraulic or electrical power supply, regain the use of one channel by the following arbitrary step sequence:

1. A and B channel switches - OFF.

2. A Channel switch - ON.

NOTE

- Be prepared to move the switch to OFF immediately if a hardover signal results, indicating that the failed channel was inadvertently selected.

- Operation with only one roll channel engaged results in overriding of logic circuitry. There is no automatic protection against inadvertent selection of a failed channel, or against subsequent failure of a properly operating channel which has been engaged.

If a hard-over signal is obtained on engagement or during subsequent operation, or if no improvement is noted in flight characteristics:

3. A Channel switch - OFF.

4. B Channel switch - ON.

NOTE

Be prepared to disengage the channel immediately if a hardover signal results.

If no improvement, or if a hard-over results:

5. Roll Channel switches - Both OFF.

Some undesirable cross-coupling may occur during single roll SAS channel operation. This appears as small amplitude oscillations in the pitch and yaw axes, as the elevons on only one side of the aircraft respond to roll damping signals during single channel operation and compensation for the asymmetric roll signals is provided by pitch and yaw axis control operation.

Schedule activity may be continued for the remainder of the flight with a single roll SAS channel operating. The roll auto-pilot may be engaged and the automatic navigation feature of the ANS used as desired.

NOTE

- Operation with both roll channels disengaged is permitted if cross-coupling prevents precise aircraft control with one roll channel engaged.

- In the event of single engine failure at low speed or while reducing airspeed, loss of hydraulic power from the windmilling engine may cause failure of that roll SAS channel and simultaneous automatic disengagement of the other roll channel.

Effect of Single Engine Landing on Roll SAS Operation

To avoid changes in pitch control characteristics at a critical time during single engine landings, either make the approach with both roll SAS channels disengaged or with the roll channel which is powered by the inoperative engine disengaged. All roll SAS off is recommended for single engine landings.

Roll SAS Failure At High Speed

A second roll SAS channel failure while at high speed will probably be indicated by abnormal pitch transients and small roll transients without illumination of either pitch or roll SAS indicator lights. The symptoms may be difficult to attribute to roll channel failure. When pitch transients occur with one roll channel engaged, disengage both roll SAS channels and check for control improvement. If no improvement is noted, the single roll channel may be reengaged if desired.

Operation With Roll SAS Disengaged

Failure or intentional disengagement of both roll SAS channels is expected to increase pilot fatigue, reduce mission effectiveness, and will disable the roll autopilot; however, no hazard to safety should result and there are no flight restrictions on continued operation.

LOGIC OVERRIDE

Logic override procedures are usable after a "second" condition failure when the sequence of light illumination indicates that a channel with operative servos is available. Refer to After Second Failures, SAS Warning lights chart. When use of logic override is effective, flight characteristics are the same as with SAS fully operational. However, as a precaution against subsequent hardover failure signals, the autopilot must not be engaged in that channel and second condition failure speed limits apply.

WARNING

- If logic override is recommended, use it only in the channels specified and only after decelerating to second condition failure limit speeds in order to prevent excessive structural loads which could result from a hardover failure at higher speeds.
- Neither logic override nor BUPD operation should be attempted with either channel known to have a failed servo, or hydraulic failure.

WARNING

Logic override eliminates the protection provided by the logic Circuit against a hard-over failure. Without this protection, a hardover failure at high airspeed and/or aft center of gravity can develop excessive structural loads. Logic override should only be used as indicated in the chart at speeds below 350 KEAS, and preferably after fuel forward transfer.

If the SAS warning lights indicate that a good channel is available and SAS operation in the failed axis is desired:

1. Airspeed 350 KEAS or less.

CAUTION

- The logic override function is to be used only if all lights of a pitch or yaw axis have illuminated. This switch locks-in a given channel and prevents automatic disengagement in the event of a failure.
- Do not use logic override if both A & B lamps have illuminated individually. No practical test exists for determining whether the failure was caused by the logic or was an actual servo loop failure.

If one channel in the inoperative axis is needed for flight, logic override may be used after both A and M electronics OR both B and M electronics fail sequentially, even if an A or a B lamp has previously illuminated. Double electronics failure, in sequence, is indicated by an M lamp followed by A and B lamps simultaneously.

CAUTION

If an A or B lamp has illuminated previously and that engage switch has been turned OFF, do not place that switch ON.

Use caution; proceed as follows:

2. Affected pitch or yaw SAS A and B channel engage switches - Both OFF.

3. Pitch or yaw logic override switch- Arbitrarily select A.

Depending on sequence and type of failure indication:

4. Either A or B or both engage switches - ON.

   All lamps will extinguish.

CAUTION

Be prepared to counteract a previously undetected hard-over condition when channel is engaged.

5. Evaluate damping by making a small stick or rudder transient.

If control does not improve:

6. Channel engage switch or switches - OFF.

7. Pitch or yaw logic override switch - B.

Depending on sequence and type of failure indication:

8. Either A or B or both engage switches - ON.

   All lamps will extinguish.

   CAUTION

   Be prepared to counteract a previously undetected hard-over condition when channel is engaged.

9. Evaluate damping as above.

If control does not improve, both A and B electronics have failed. Proceed as follows:

For Yaw axis:

10. Channel engage and logic override switch or switches - OFF.

For Pitch axis:

11. Refer to BUPD procedures.

BACKUP PITCH DAMPER (BUPD)

BUPD plus logic override procedures are available after a "second" condition failure in the pitch axis. The BUPD is optimized for operation at air refueling speeds. It may or may not improve flight characteristics at other flight conditions.

To engage the BUPD below 0.85 Mach number and 330 KEAS:

1. SAS channel engage switch - Affected channel OFF.

2. BUPD switch - ON.

3. Pitch logic override switch - Select appropriate channel.

4. SAS channel engage switch - <u>Beep ON.</u>

If control does not improve:

5. Appropriate switches - OFF.

AIR DATA COMPUTER

If malfunction or failure of the air data computer (ADC) is suspected, proceed as follows:

1. Integrated instruments - Check against standby airspeed and altimeter.

If integrated instruments are inaccurate:

2. Revert to standby instruments for aircraft control.
3. MACH TRIM circuit breakers - Pull.
4. Autopilot - Off.

PITOT-STATIC SYSTEM

Under some conditions both of the pitot-static operated systems may become inaccurate or inoperative from a common malfunction. Failure of the pitot heater may simultaneously affect both normal systems in icing conditions. The pitot probe could be plugged by a foreign body of sufficient size. If both systems fail, proceed as follows:

1. Maintain aircraft control by use of attitude and power indicating instruments.
2. Check pitot heat switch and circuit breaker.
3. Request escort aircraft for letdown and landing.

ENVIRONMENTAL CONTROL SYSTEM

L OR R AIR SYSTEM OUT

In the event one air conditioning system fails or is intentionally shut off, as due to engine failure, and the L or R AIR SYS OUT warning light illuminates:

1. Bay air switch - SHUTOFF.
2. All fire control system or chine bay equipment - OFF.

CAUTION

The equipment will be overheated during subsonic or supersonic flight if electrical power to the equipment is not turned off. If supersonic, the OVERHEAT warning lights may illuminate with the equipment off.

If at supersonic speeds:

3. Cockpit temperature control rheostat - COLD.
4. Descend when practicable.

   It may be possible to maintain supersonic cruise for an indefinite time at reduced cooling if speed is reduced.

5. Monitor FCS RACK and BAY OVERHEAT warning lights for indication of overheat.

NOTE

Use normal procedures before penetration and before landing for management of environmental control system.

COCKPIT OVERTEMPERATURE

1. Cockpit temperature indicator - Check.

If the temperature is too high:

2. Cockpit temperature control rheostat - Rotate toward COLD.

3. Defog - Check CLOSED.

If auto-temperature control is not effective and cockpit temperature remains too high:

4. Cockpit temperature override switch - Hold in COLD position.

NOTE

In this position the motor-driven valves take 7 to 13 seconds to travel from full hot to full cold.

If no decrease in cockpit temperature occurs in 30 seconds:

5. Bay air switch - SHUTOFF.

6. Turn off fire control system or chine bay equipment.

7. Reduce airspeed and descend as soon as possible.

SUIT OVERTEMPERATURE

1. Suit heat rheostat - Turn to COLD.

2. SUIT HEAT C/B - Pull (Left side pilot's cockpit).

3. Suit air flow - Restrict.

4. AIR SYS OUT lights - Check.

If overtemperature condition persists:

5. Descend as soon as possible.

FCS RACK OR BAY OVERHEAT

In the event that the FCS RACK or BAY OVERHEAT warning light illuminates:

1. Bay air temperature - Check.

   Set appropriate source with temperature indicator selector switch and check temperature indication for confirmation of overheat condition before taking further action. An indication in excess of 150°F is confirmation of the overheat warning.

2. Bay air switch - ON.

△3. Turn off nonessential bay equipment.

If overtemperature condition continues:

4. Descend when practicable.

NOTE

Use normal procedures before penetration and before landing for management of environmental control system.

COCKPIT TOO COLD

Cockpit air and suit vent air temperatures may become unbearably low, even when cruising at high Mach number, in the event of some types of air conditioning system malfunctions or if the FULL COLD position of the manifold temperature control switch is selected inadvertently. Temperatures may be substantially below -30°F. A landing may be necessary if the condition can not be corrected. In general, the aft cockpit will be colder. If the automatic and manual temperature controls are ineffective with AUTO manifold temperature selected, attempt to minimize suit vent air flow and increase suit air temperature. The pilot can use defog air to heat the forward cockpit. If these actions are not sufficient, it may be possible to rectify the condition by shutting off an air conditioning unit.

COCKPIT DEPRESSURIZATION

Cockpit depressurization above approximately 35,000 feet will be indicated by pressure suit inflation. If suit inflates, proceed as follows:

1. Cockpit altitude - Check.

△2. Canopy seal levers - Check ON.

3. Cockpit pressure dump switch - Check OFF.

WARNING

During this time the crewmembers will be depending on the pressure suit only for altitude protection.

If cockpit still does not repressurize:

4. Descend as soon as possible.

COCKPIT DEFOGGING

The possibility of cockpit fogging exists any time that ambient humidity is high. Then, when the ambient air is cooled below its dew point temperature in the air cycle refrigerator, moisture condenses from the air in the form of fog. On very humid days, fog will form when air is cooled just slightly below the ambient Dry Bulb Temperature. The amount of fogging will increase as the amount of cooling is increased. This effect will be most noticeable near sea level when at or near Military rpm. Under these conditions, the air is cooled to its minimum temperature and the rate of airflow into the cockpit is greatest. The resulting high entrained moisture content results in objectionable cockpit fog.

Cockpit Temperature Control

The procedure for eliminating cockpit fog is to raise the temperature of the cockpit air and re-evaporate the moisture. This can be accomplished by either manual or automatic control of cockpit temperature until there is sufficient increase to evaporate the fog; however, manual WARM should not be held so long as to become uncomfortably hot.

AUTO TEMP Control Positioning

If cockpit fog is anticipated due to high ambient humidity, but has not yet appeared, attempt to select an automatic

temperature control setting that will maintain the cockpit temperature at or slightly above ambient temperature while the engines are at a high rpm. This can be done prior to or during the pre-takeoff engine trim run. A little experimenting should establish a position on the AUTO TEMP selector that prevents excessive fogging on humid days. After takeoff, the position of the control should be changed to provide comfort during the remainder of the flight. The same control position should be re-established prior to landing in order to prevent fogging during the approach.

Defogging Procedure In Flight

If persistent fogging is encountered:

1. Both air system switches - OFF.

   Set both air refrigeration system switches to the OFF position. This shuts off the cockpit and bay air supplies.

   NOTE

   Defog hot air is not available with both refrigerator switches off.

LIFE SUPPORT SYSTEMS

WARNING

Steps of following procedures preceded by two asterisks (**) apply to the crewmember affected by loss of oxygen.

LOSS OF ALL OXYGEN SYSTEMS (Aircraft and Emergency Systems)

Complete loss of oxygen could occur with inadvertent separation of the oxygen quick disconnect block on the stabilized seat, or with multiple failures within the oxygen systems. Confirm situation by checking oxygen controls, lights, and quantity/pressure indications. Tug gently on oxygen hoses at seat to check security of connections. If disconnect has occurred, all oxygen (including Emergency system) is cut off.

If complete loss of oxygen is confirmed:

**1. ADVISE OTHER CREWMEMBER.

**2. CLOSE SUIT CONTROLLER VALVE.

**3. OPEN SUIT VENT-AIR VALVE.

CAUTION

Tighten helmet tiedown strap.

4. BAY AIR - OFF.

NOTE

Aircraft ventilation air is capable of increasing suit pressure approximately 2.35 psi above cabin pressure. This will reduce physiological altitude inside suit from 26,000 foot pressure level to approximately 17-18,000 foot level.

5. MAKE 400 KEAS EMERGENCY DESCENT.

Maintain as close to 400 KEAS as possible to maintain effective ventilation air pressures and flow.

**6. Loosen face seal.

Adjust face seal to a more open position with barrier control knob to aid flow of vent air into the oral/nasal cavity area of the helmet. Vent airflow into the oral/nasal cavity will exhaust through the anti-suffocation valve to provide some "flushing" of exhaled carbon dioxide.

▲7. Maintain communications.

Advise ground of situation and maintain interphone communications with each other to provide mutual assistance as necessary. Conditioned response by the affected crewmember to calls from the other crewmember may help in accomplishing necessary procedures.

⑧. Be Prepared to Eject.

If the pilot has experienced an oxygen loss, the FCO must closely monitor attitude and speed; and evaluate pilots capability of making a successful recovery. If the pilot does not respond and aircraft control is obviously lost, the FCO should eject before the aircraft enters a pitch-up condition or if it exceeds the design limit speed schedule.

9. Land as soon as practicable.

If either crewmember experiences symptoms of decompression sickness, land as soon as possible, in accordance with AFM 60-16.

LOSS OF AIRCRAFT OXYGEN SYSTEMS

With loss of both aircraft oxygen systems due to failures other than separation at the seat, the EMERGENCY oxygen system should be available. The EMERGENCY oxygen system provides sufficient duration to permit a normal descent.

WARNING

Attempt to positively ascertain that the ship oxygen supplies are not available or that they have been lost before pulling the green apple to activate the crewmember's emergency oxygen supply. If the ship systems have not actually been lost and the green apple has been pulled, there is the very remote possibility that the total emergency system might not be available if all ship systems are subsequently lost.

If loss of all aircraft oxygen systems is confirmed:

**1. Emergency Oxygen - Pull GREEN APPLE.

2. Descend to 10,000 foot cabin altitude, terrain and weather permitting.

3. Land as soon as practicable.

If either crewmember experiences any difficulty in maintaining oxygen flow:

4. MAKE EMERGENCY DESCENT.

LEAK IN PRESSURE SUIT

1. Descend below 35,000 foot aircraft altitude.

   Below 35,000 feet, the suit will not be pressurized if cabin pressure is lost or bailout is necessary.

LOSS OF FACEPLATE HEAT

Without faceplate heat, moisture could condense in helmet causing oxygen controller to freeze during bailout.

▲1. Check circuit breaker and communication cord.

2. If faceplate fogging begins to block vision - Descend below 10,000 foot cabin altitude, terrain and range permitting.

▲3. Raise faceplate.

LOSS OF SUIT VENT-AIR

▲1. Check vent hose connection.

2. If crewmember perspires excessively - Descend below 35,000 foot MSL.

LEAK IN HELMET (oral/nasal area)

1. Descend below 10,000 foot cabin altitude, terrain and range permitting.

2. If leakage is gross - MAKE EMERGENCY DESCENT.

3. Oxygen systems - Monitor quantity and pressures.

EMERGENCY WARNING SYSTEM

Illumination of any light(s) on the annunciator panels indicates the existence of an abnormal or emergency operating condition, or that a situation deserving special attention is occurring. The conditions and a summary of recommended actions represented by illumination of the lights is provided on Figures 3-14 and 3-15.

ANNUNICATOR PANEL LIGHTS
ANALYSIS CHART

LIGHT	CONDITION	RECOMMENDED ACTION
AIR SYS OUT L or R	Air conditioning system has failed.	BAY AIR switch OFF, temperature control COLD, reduce speed and descend, if practicable.
ANTI-SKID OUT	Anti-skid braking inoperative.	Select SKID OFF, or ALT BRAKE if L hydraulic pressure low.
CAN UNLKD	One or both canopies not locked or seals not inflated.	Check canopy latch handles forward and locked, canopy seal pressure levers ON. Land if canopy unsafe.
EMER BATT ON	Essential dc bus on battery power, both generators or transformer rectifiers inoperative.	If generator power can not be restored to dc bus, check INST ON EMERG PWR, conserve battery, and land.
FCO CLEAR	FCO has ejected.	
* FCS LIQUID COOL OUT	Liquid cooling compressor is inoperative.	Check FCS liquid cooling switch ON. Actuate REFRIG OVERHEAT RESET switch.
* FCS RACK OVERHEAT	Air surrounding FCS rack is too hot.	Check BAY AIR switch ON. If ON, turn off FCS equipment and descend as soon as practicable.
FUEL PRESS (LOW) L or R	Pressure has dropped below 7 psi.	Press Crossfeed and additional tank pumps on.
FUEL QTY LOW	Total in tanks 1 and 6 has decreased to between 3000 and 6000 lbs.	X-feed and tank 6 on. Check total fuel. If low, land or refuel. If sequence faulty, restore normal sequence.
BUS TIE OPEN	ac buses are split.	No action unless GEN OUT also on.
GEN OUT L or R	Generator is disconnected.	Check BUS TIE OPEN light, recycle generator. Stay below 350 KEAS and Mach 2.8 with one generator off.
HYD (LOW) A or B	A or B quantity below 1-1/4 gal or pressure below 2200 psi.	Descend. If pressure low - stay below 350 KEAS, check SAS, land.
HYD (LOW) L or R	L or R quantity below 1-1/4 gal.	Begin descent. Monitor pressure. Stay below Mach 2.3.
INER PLTFM OFF	INS is inoperative	Use MA-1 compass system for navigation.

*FCO cockpit.

Figure 3-14

ANNUNICATOR PANEL LIGHTS
ANALYSIS CHART

LIGHT	CONDITION	RECOMMENDED ACTION
INST ON EMERG PWR	Emergency ac bus on inverter power from essential dc bus, both generators and/or TR's out.	If generator power can not be restored to dc bus, conserve battery and land.
MANUAL INLET	Restart switch on, or one or more spike and door controls not in automatic position.	Set restart OFF, or use AUTO inlet controls unless manual operation of inlet desired.
* MISSILE BAY OVERHEAT	Air in one or more missile bays is too hot.	Check BAY AIR switch ON. If light stays ON, descend as soon as practicable.
* MTC BAY OVERHEAT	Air in the MTC bay is too hot.	Turn off all FCS equipment and non-essential radio and navigation equipment.
* MTC BAY PRESS LOW	MTC bay has de-pressurized.	Descend to lower altitude as soon as practicable.
N QTY LOW	Nitrogen quantity below 1 liter.	Check tank pressure. Stay below Mach 2.6 if both systems low.
OIL QTY LOW L or R	Quantity below 2-1/4 gal.	Begin descent, monitor oil pressure.
OIL TEMP L or R	Temp over 282°C or below 15°C.	Check pressure, reduce Mach number. Keep fuel flow over 12,000 pph. If temp not controlled, shutdown & land.
** OXY PRESS LOW SYS 1 or SYS 2	System pressure below 50 psi.	Check individual pressures and descend as soon as possible.
SAS CHNL OUT	SAS pitch, yaw, or roll channel disengaged.	Attempt to recycle disengaged channel. See SAS warning light chart if channel remains disengaged.
SURF LIMITER	Surface limiter position not correct for existing speed.	Engage or release SURF LIMIT RELEASE handle.
XFMR RECT OUT L or R	Transformer-rectifier not supplying power.	For double failure, check EMER BAT ON and INSTR ON EMERG PWR illuminated.
VENTRAL FIN	Ventral fin not in correct position relative to landing gear.	Land if fin remains retracted after gear up. Use emergency retraction procedure if fin remains extended with gear down.

*FCO cockpit.
**Both cockpits.

Figure 3-15

LANDING EMERGENCIES

SINGLE ENGINE PENETRATION AND LANDING

This procedure may be used in lieu of the normal penetration, before landing, and landing procedures when one engine is inoperative and the engine shut down procedure has been completed.

1. All FCS and related equipment - OFF.

2. Gross Weight & c.g. - Dump & transfer as required.

   When time and conditions permit, dump fuel to obtain normal landing weight. Monitor c.g. Transfer fuel as necessary to maintain subsonic c.g. limits.

3. Display mode selector switch - SET.

4. Attitude reference - Set.

5. Altimeter - Set.

   Set station pressure on passing through FL 180, or just prior to penetration.

6. Defog lever and cockpit temperature control rheostat - Set.

   The possibility of encountering windshield and cockpit fogging during approach and landing must be anticipated and appropriate settings selected. Maintain 75° to 85°F cockpit temperature.

NOTE

The pitot-static-operated flight instruments will be used for penetration.

Accomplish All-Weather operating procedures in addition to the following when applicable.

7. Crossfeed switch - XFEED ON

CAUTION

Leave crossfeed on to assure fuel supply to both engines during landing and possible go-around operations.

8. SAS channels - Set.

   a. Inoperative engine SAS channels- All OFF.

   b. Operative engine roll SAS - OFF.

   c. Operative engine pitch and yaw SAS - ON.

When below Mach 0.5:

9. Brakes & anti-skid - Set

   a. For left engine failed - ALT BRAKES.

   b. For right engine failed - ANTI-SKID.

10. Surface limiter handle - Pulled & light off.

    Pull and rotate the surface limiter handle 90 degrees to disengage the surface limiters, lock the handle. Check SURF LIMITER warning light extinguished.

△ 11. Approach and landing speeds - Computed.

The minimum approach is 200 KIAS.

A single-engine landing is basically the same as normal landing, except that the pattern may be entered at any point. Expand the pattern to avoid steep turns.

Establish a steeper than normal final approach. A rate of descent of 1500 fpm is recommended.

Attempt to dump fuel and avoid a heavy weight landing if an ILS approach is required.

When landing at heavy weight, the single engine performance available with maximum power may not be sufficient to allow a $-2\ 1/2^{\circ}$ ILS glide path to be maintained with the gear down.

12. Landing light switch - Set.

NOTE

Lowering the vision splitter during night landings will materially reduce the glare caused by ground lights reflecting off the inside of the windshield.

At least 90 seconds must be allowed for emergency gear extension and ventral fin retraction if the left hydraulic system is inoperative.

13. Landing gear lever - Down and checked.

    Check gear warning lights.

    a. For left engine failed, pull GEAR RELEASE handle.

       At least 90 seconds must be allowed for emergency gear extension if the left hydraulic system is inoperative.

    b. For right engine failed (if the left hydraulic system is operating) the landing gear may be lowered and the ventral fin retracted after lining up on final approach.

       Normal gear extension time is 12 to 16 seconds.

14. Hydraulic pressure - Checked.

15. Annunicator panel - Checked.

16. Rudder trim - Neutral.

    During approach in adverse weather conditions, delay setting neutral trim until runway is in sight.

    The outstanding feature of a single engine landing is the change in rudder trim as thrust is varied. A marked rudder trim change will occur as the throttle is retarded in the flare. This is reduced by setting the rudder trim to neutral on the trim indicator after established on final approach.

When landing is assured, retard throttle smoothly and make a normal landing.

CAUTION

After landing with the L Hydraulic system failed or with the left engine inoperative, stop on the runway and have the gear downlocks installed before taxiing. This positively assures that the gear will remain locked in the down position when L Hydraulic system pressure is not available.

SIMULATED SINGLE-ENGINE LANDING

Directional trim changes will be more pronounced during an actual single engine situation with one engine windmilling.

1. Retard one throttle to IDLE.

2. Follow Single Engine Landing procedure.

SINGLE ENGINE GO-AROUND

Make decision to go around as soon as possible and definitely prior to flare.

1. Throttle - As required.

2. Continue approach until go-around is assured.

3. Landing gear lever - UP, as appropriate.

   Delay gear retraction until there is no possibility of contacting the runway.

4. Accelerate to 275 KIAS.

LANDING GEAR SYSTEM EMERGENCIES

GEAR UNSAFE INDICATION

An unsafe indication could be caused by low L hydraulic system pressure or malfunction within the landing gear extension or indicating system. Upon detecting an unsafe gear indication, proceed as follows:

1. Land gear control and indicator circuit breakers - Check IN.
2. L hydraulic pressure - Check.
3. Recycle landing gear lever to down position, repeat as desired.

If landing gear still indicates unsafe:

4. Pull emergency gear release handle.
5. Landing gear position - Request visual confirmation.
6. If all landing gear appear fully extended, make a normal landing on side of runway away from suspected unsafe gear. Observe the following precautions:
   a. Shoulder harness - Manually lock.
   b. Hold weight off unsafe gear as long as possible then allow gear to contact runway as smoothly as possible. If nose gear is held off, lower nose at approximately 130 KIAS.
   c. Allow aircraft to roll to a stop straight ahead. Have downlocks installed prior to further taxiing or engine shutdown.
7. If any gear remains fully retracted, use Emergency Extension procedure.
8. If all gear are not fully extended, refer to Partial Gear Landing procedure, this section.

NOTE

- Increasing airspeed may assist in locking a partially extended nose gear.
- Yawing aircraft may assist in locking a partially extended main landing gear.

GEAR EMERGENCY EXTENSION

The emergency landing gear extension system unlocks the landing gear uplocks and allows the landing gear to free fall to the down and locked position. If R hydraulic system pressure is available, the landing gear handle must be placed in the DOWN position or the landing gear control circuit breaker must be pulled to permit emergency extension. The time required for emergency gear extension is 60 to 90 seconds. The emergency landing gear handle must be pulled approximately 9 inches for full actuation. If it is not fully actuated, one or more gear may fail to extend. Normal pull force may be up to 65 pounds.

If the L hydraulic system has decrease below 2200 psi or normal gear extensio is unsuccessful, proceed as follows:

1. Landing gear handle - DOWN.

If landing gear remains retracted:

2. Emergency landing gear release handle - PULL.

3. Verify gear down and locked.

If L hydraulic system pressure low:

4. Brake switch - ALT BRAKE.

NOTE

Alternate nosewheel steering is available when L system pressure decreases below 2200 psi.

If landing gear remains retracted or landing gear handle sticks in the UP position:

5. Landing gear control circuit breaker - PULL.

6. Repeat steps 2 and 3 of this procedure.

CAUTION

The landing gear must not be retracted if the manual release handle is being held in the free fall (full out) position as damage to the system can result. The GEAR RELEASE handle should be permitted to return to the stowed position before attempting to retract the gear with the landing gear lever.

WARNING

After landing, nosewheel steering will not be available while L system hydraulic pressure remains above 2200 psi if the CONT circuit breaker has been pulled. In this event, steer by differential braking. If necessary, select ALT BRAKE and shut down the left engine to cause L system hydraulic pressure to decrease and allow the R system to power the steering.

3. Verify gear down and locked.

If L hydraulic system pressure low:

4. Brake switch - ALT BRAKE.

NOTE

Alternate nosewheel steering is available when L system pressure decreases below 2200 psi.

If landing gear remains retracted or landing gear handle sticks in the UP position:

5. Landing gear control circuit breaker - PULL.

6. Repeat steps 2 and 3 of this procedure.

CAUTION

The landing gear must not be retracted if the manual release handle is being held in the free fall (full out) position as damage to the system can result. The GEAR RELEASE handle should be permitted to return to the stowed position before attempting to retract the gear with the landing gear lever.

WARNING

After landing, nosewheel steering will not be available while L system hydraulic pressure remains above 2200 psi if the CONT circuit breaker has been pulled. In this event, steer by differential braking. If necessary, select ALT BRAKE and shut down the left engine to cause L system hydraulic pressure to decrease and allow the R system to power the steering.

PARTIAL-GEAR LANDING

Do not make a partial-gear landing unless the nose gear only is down and locked and the main gear and ventral fin are up and locked. Because this configuration could only result from unlikely circumstances, partial-gear landings are in effect not approved.

MAIN GEAR FLAT TIRE LANDING

For a landing with one or more flat tires, turn the anti-skid system off in order to obtain effective braking with the good tires and to prevent spin-up of wheels with flat tires. (With anti-skid on, the brake system prevents spin-down of the inboard and outboard wheels of each truck. If either or both of these tires has failed, normal runway contact is not made and they tend to lock-up when braking is attempted. The anti-skid system defeats the braking attempt by releasing brake pressure as soon as the wheel starts to spin down.)

Plan the landing for minimum gross weight, and touchdown on the good truck first on the side of the runway away from the flat tire(s). Little danger exists when landing at light weight if only one tire on a truck has failed, as the remaining tires have sufficient strength to support the aircraft. If two tires on a truck have failed, the third may fail during the roll-out because of the overload condition imposed on it.

When all tires on a truck are known to have failed, apply enough brake pressure to lock all three wheels on that side. Maintain the pressure to prevent spin-up and the tire/wheel fragmentation that might result with rolling wheels. Use asymmetric braking (modulating the opposite side), nosewheel steering, and rudder and elevon roll inputs for directional control and stopping effort.

Engine shutdown is generally not advantageous, and is not recommended if a crosswind component exists which could be a critical factor. Left engine shutdown would require selection of the alternate brake system and result in the anti-skid system being on. Do not shut down the engine on the downwind side if an appreciable crosswind exists.

With all tires failed on both trucks, lock the wheels and skid to a stop rather than risk wheel fragmentation. Use the nosewheel and aerodynamic steering for directional control.

1. ANTI-SKID OFF.
2. Touch down on good tires.
3. Drag chute switch - DEPLOY, as soon as possible.
4. Nosewheel - Lower.
5. Nosewheel steering - Engage.

   Check that STEER ON engage light is illuminated.
6. Hold weight off bad side as long as possible using full aileron.
7. Brakes - Use differential braking to maintain directional control.

WARNING

If there is no fire, maintain IDLE rpm until fire-fighting equipment arrives. Engine shutdown allows fuel to vent in the vicinity of the wheel brake area, thus creating a fire hazard.

NOSE GEAR FLAT TIRE LANDING

If it is necessary to land with a flat nosewheel tire or tires, avoid c.g. forward of 22 percent, if possible, and proceed as follows after making a normal touchdown.

1. Drag chute switch - DEPLOY.

2. Nose gear - Hold off.

   Hold the nosewheel off as long as practicable (approximately 130 KIAS) and then lower gently to runway.

3. Use nosewheel steering and differential braking to maintain directional control.

After stop, before shutdown:

4. Fuel - TRANSFER.

LANDING WITHOUT NOSEWHEEL STEERING

It should be possible to stop safely without nosewheel steering when landing on a dry or grooved runway with crosswind components within recommended values. However, crosswinds combined with a slippery runway or damaged tire condition can present a hazardous situation. The effects of the weathercock force and aerodynamic steering with the elevons and rudders is shown by figure 3-16 for a heavy crosswind combined with a slippery runway and ineffective nosewheel steering condition. The following procedures are recommended for this type of situation.

NOTE

> Take care not to overcontrol the aircraft on a slippery runway and start a lateral skid, either with main gear tires intact or damaged. The reduced side reaction force capability of the main gear tires may result in a "Break out" and slide.

(a) Distinguish between aircraft heading and ground track direction. A lateral skid can develop if the main gear loses traction because of a slippery runway and/or damaged main gear tires. The ground track will probably diverge downwind. First, attempt to regain traction by "steering into the skid". If this is successful, bring the heading back parallel to the runway centerline, then steer to the upwind side of the runway. Steer by applying roll control first. After full elevon deflection is reached, use rudder as necessary. Lateral stick deflection toward the desired direction of heading change produces the steering characteristics of the rudders to effect a turning moment, but without the side force in the opposite direction which the rudders must always produce. Corrective upwind rudder deflection results in a downwind rudder force (in addition to the desired turning moment) which compounds the downwind skid problem.

(b) Jettison the drag chute if roll control and use of the rudders fails to correct the ground track and it appears that the aircraft is in danger of sliding off the runway (assuming that unfavorable terrain exists adjacent to the runway). Retain the chute until a safe stop is assured if a track on the runway can be maintained.

EFFECTS OF WEATHERCOCK FORCE AND AERODYNAMIC STEERING ON CROSSWIND LANDING WITHOUT NOSEWHEEL STEERING

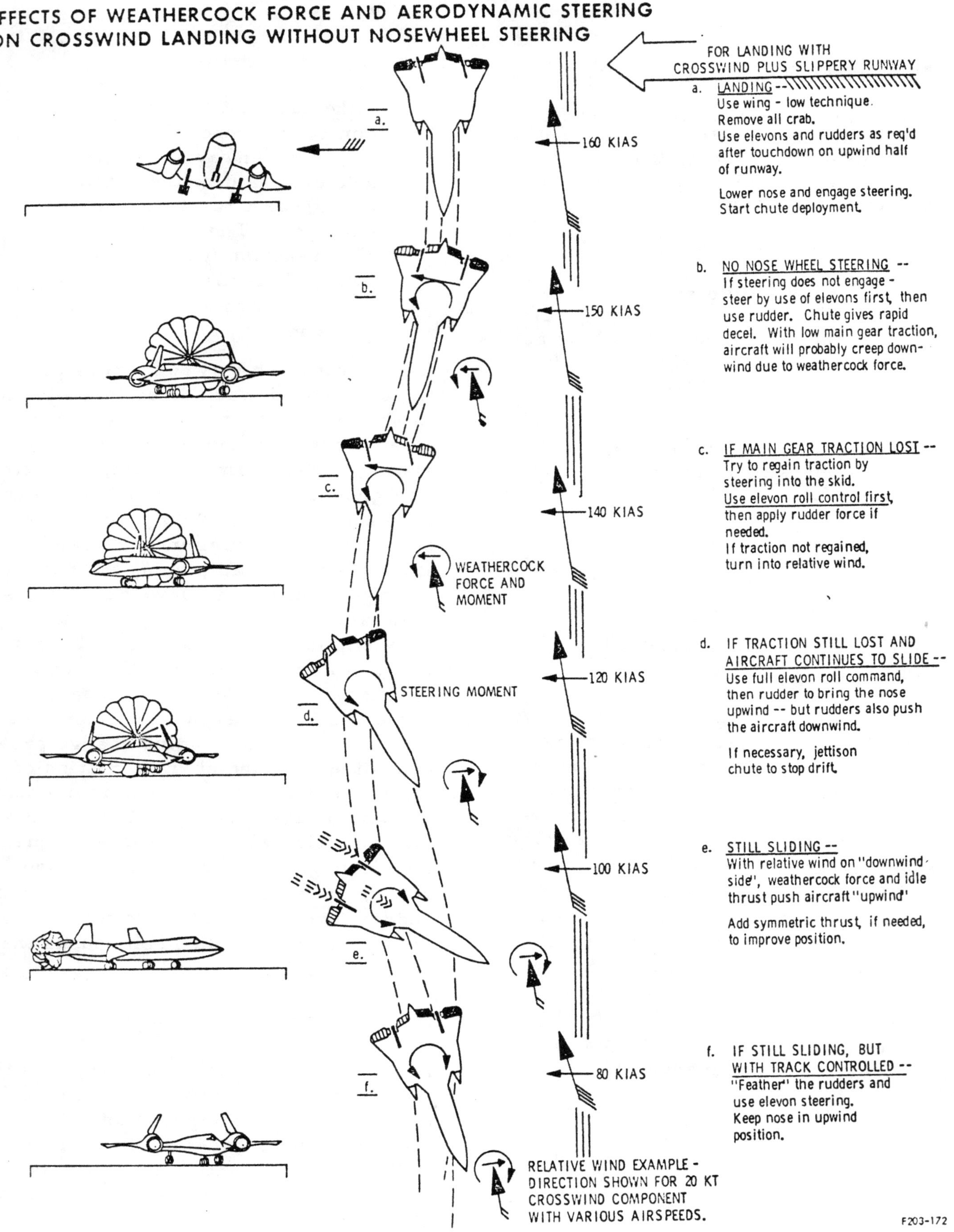

FOR LANDING WITH
CROSSWIND PLUS SLIPPERY RUNWAY
a. LANDING --
Use wing - low technique.
Remove all crab.
Use elevons and rudders as req'd after touchdown on upwind half of runway.
Lower nose and engage steering.
Start chute deployment.
b. NO NOSE WHEEL STEERING --
If steering does not engage - steer by use of elevons first, then use rudder. Chute gives rapid decel. With low main gear traction, aircraft will probably creep downwind due to weathercock force.
c. IF MAIN GEAR TRACTION LOST --
Try to regain traction by steering into the skid.
Use elevon roll control first, then apply rudder force if needed.
If traction not regained, turn into relative wind.
d. IF TRACTION STILL LOST AND AIRCRAFT CONTINUES TO SLIDE --
Use full elevon roll command, then rudder to bring the nose upwind -- but rudders also push the aircraft downwind.
If necessary, jettison chute to stop drift.
e. STILL SLIDING --
With relative wind on "downwind side", weathercock force and idle thrust push aircraft "upwind"
Add symmetric thrust, if needed, to improve position.
f. IF STILL SLIDING, BUT WITH TRACK CONTROLLED --
"Feather" the rudders and use elevon steering.
Keep nose in upwind position.
a.
b.
c.
d.
e.
f.
160 KIAS
150 KIAS
140 KIAS
120 KIAS
100 KIAS
80 KIAS
WEATHERCOCK
FORCE AND
MOMENT
STEERING MOMENT
RELATIVE WIND EXAMPLE -
DIRECTION SHOWN FOR 20 KT
CROSSWIND COMPONENT
WITH VARIOUS AIRSPEEDS.
F203-172

Figure 3-16

(c) If traction cannot be regained by turning into the skid, and if the ground track continues to diverge, try bringing the aircraft heading around so that it is well upwind of the ground track direction. This moves the "relative wind" to the other side of the nose, and puts the weathercock force in the upwind direction. This tends to correct the ground track. The forward thrust component of the idling engines also tends to correct the track. If the aircraft continues to slide, the weathercock force will now tend to rotate the heading back into the relative wind. Use roll control first, then the rudders, to keep the heading upwind and maintain a track on the runway. Revert to normal steering techniques whenever main gear traction can be regained.

NOTE

> Asymmetric thrust is not recommended for directional control because of the difficulty expected in obtaining a controlled and timely response. Lateral stick and rudder steering should be adequate to control aircraft heading. However, if these means fail and an otherwise irretrievable drift has developed, short bursts of equal thrust on both engines may be used to correct the ground track and stay on the runway for barrier engagement. The added thrust is in the proper direction to help regain runway position.

FLAT STRUT LANDING

The initial indication of a flat strut condition on landing is a wing low condition after touchdown and a directional control characteristic similar to that for crosswind condition. Directional control may be difficult until the nose wheel is lowered. The wings not level attitude is similar to that for all tires blown on one side.

Lower the nose immediately, engage nosewheel steering and select SKID OFF. Deploy the drag chute in a normal manner and keep the chute until a stop is assured. After stop, clea the runway and wait for assistance. If inflation of the strut is not possible, th other strut should be deflated to 1" clearance (to reduce side loads during aircraft movement). Slow taxiing is permitted; however, towing is recommended.

The wings not level attitude during landing roll causes uneven loading of the inboard and outboard tires on each truck. This may result in wheel spindown for the more lightly loaded tires during braking. Protracted cycling of the anti-skid system can occur and result in reduced braking effectiveness if the anti-skid system is not turned off.

GEAR DOWN AIR REFUELING, LOW ALTITUDE

During air refueling rendezvous, airspeed must be monitored closely to prevent exceeding landing gear down limits. Throttle response is very rapid; therefore, throttle movements just prior to and during contact should be less than those used at normal refueling altitudes.

VENTRAL FIN EMERGENCY RETRACTION

Any time the ventral fin does not retract after the landing gear is down and locked, the fin may be retracted by an electrical circuit, paralled to the normal fin retraction circuit, provided the malfunction is electrical. The parallel circuit is energized by pulling the manual landing gear release handle to its travel limit and holding it until the VENTRAL FIN warning light goes out. This procedure applies whether the landing gear has been extended normally or with the manual landing gear release handle.

1. Main landing gear locked down - Recheck.
2. Manual landing gear release handle - Pull to limit stop after main gear downlock lights are illuminated. Hold handle extended until ventral fin is retracted, as indicated by warning light going out.

HEAVY WEIGHT LANDING

Use normal procedure as described in Section II for minimum roll and heavyweight landings. Observe rate of sink limits of Section V.

COCKPIT FOG WHILE LANDING

1. Both air system switches - OFF.

NOTE

Windshield defog hot air is not available with both refrigeration switches off.

Section V

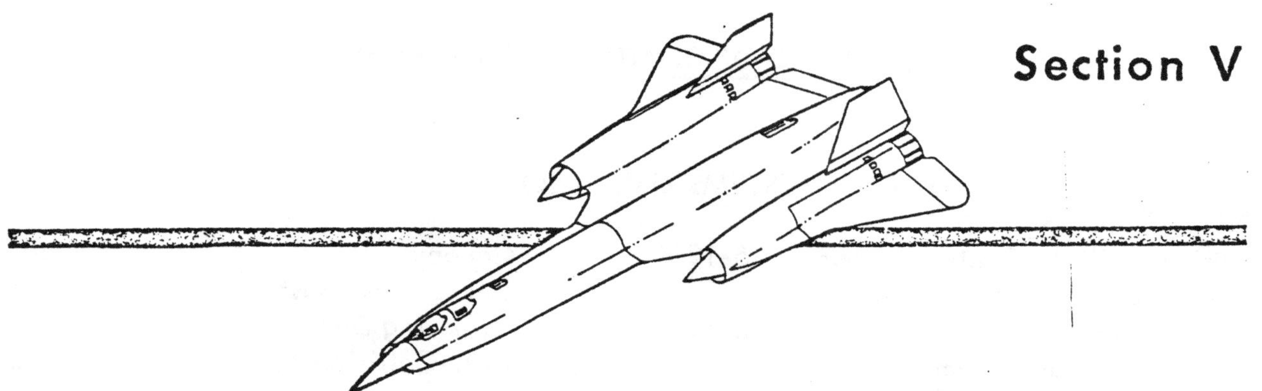

Operating Limitations

TABLE OF CONTENTS

TABLE OF CONTENTS (Continued)

INTRODUCTION

This section provides general aircraft and engine operating limits and restrictions that must be observed in normal operations.

The terms "limit (or limitation)" and "restriction" are not used interchangeably. A limit refers to a boundary which was a primary consideration during design of the aircraft, or which has been established as a maximum operational capability of the aircraft from the standpoint of flight safety. Mach 3.2 is the design speed limit in the sense that the design was optimized for operation at this speed. It must not be exceeded intentionally. The term restriction refers to a boundary which has been established as a result of current testing or service experience, etc., to enhance operating safety of the aircraft. An imposed restriction has all the force of a limitation, but does not necessarily represent a capability boundary. For example: the 310 KEAS minimum supersonic airspeed restriction does not necessarily represent the minimum airspeed which can be exploited; but it must be observed until satisfactory tests reveal that a lower safe speed boundary can be established.

Any deviation from the following limits or restrictions which is scheduled for test purposes should be coordinated with the vehicle manufacturer prior to flight. Limits and restrictions which are associated with system malfunctions are provided in Section III with the appropriate emergency procedures.

INSTRUMENT MARKINGS:

The instrument markings tabulated below are self-evident and are not necessarily repeated elsewhere in this section.

Standby airspeed indicator:

Red line (minimum) at 150 KIAS

Tachometer:

Red line at 7450 rpm.

Oil Pressure:

Red line (minimum) at 35 psi.

Green arc from 40 to 55 psi.

Compressor inlet static pressure:

Red arcs 0 to 4 psia and 27 to 30 psia.

Compressor inlet temperature:

Red line at 427°C.

Fuel tank pressure:

Red lines at -0.5 psi and 5.0 psi.

Hydraulic pressure gages - A & B, L & R systems:

Red lines at 2200 psi and 3800 psi.

Green arcs 3000 to 3500 psi.

FUEL

The approved fuel is JP-7. This fuel corresponds to PWA 535 with PWA 536 lubricity additive which has been premixed in the ratio of one gallon of additive per 5200 gallons of fuel.

Emergency Fuels

Fuels such as JP-4, JP-5 and JP-6 may be used only for emergency requirements when refueling must be accomplished or risk loss of the aircraft. Operation with emergency fuels is restricted to speeds below Mach 1.5. Rate of climb is not restricted. If fuels other than approved fuels are ever used, the applicable inspection publication should be consulted for post flight instructions.

OIL

The approved oil is PWA 524B. If necessary because of low ambient temperatures, it may be diluted with Trichloroethylene, Federal Specification O-T-634, Type 1, in accordance with the Maintenance Manual procedures.

ENGINE OPERATING LIMITS

General operating limits for YJT11D-20A and YJ-1 engines are summarized in figure 5-1.

In-Flight Shutdown

Except for authorized testing, an engine shall not be intentionally windmilled at subsonic conditions when CIT is less than 15°C. (59°F.)

TIME LIMITS

YJT11D-20A and YJ-1 engines may be operated continuously at all ratings when within the normal exhaust gas temperature limits.

Emergency Operation

Continuous or accumulated operating time in the emergency EGT operating zone for more than 15 minutes may require engine removal. No more than one hour may be accumulated with EGT in excess of the normal limit schedules. EGT must be reduced immediately if an emergency limit temperature is exceeded.

EXHAUST GAS TEMPERATURE LIMITS

The nominal operating bands, limits for continuous operation, and emergency operating zones for YJT11D-20A and YJ-1 engines are prescribed as a function of compressor inlet temperature as shown in figure 5-1.

Ground Operation

If an engine surges during pre-takeoff trimming, down-trim to eliminate the surge but do not trim lower than 60°C below the desired trim point for the ambient temperature. After takeoff, engines down-trimmed for surge protection should be up-trimmed to 775°C EGT when CIT reaches 0°C.

NOTE

> When EGT is within the nominal operating band shown by figure 5-1, surging is a ground run problem only.

The EGT limit for starting is 565°C.

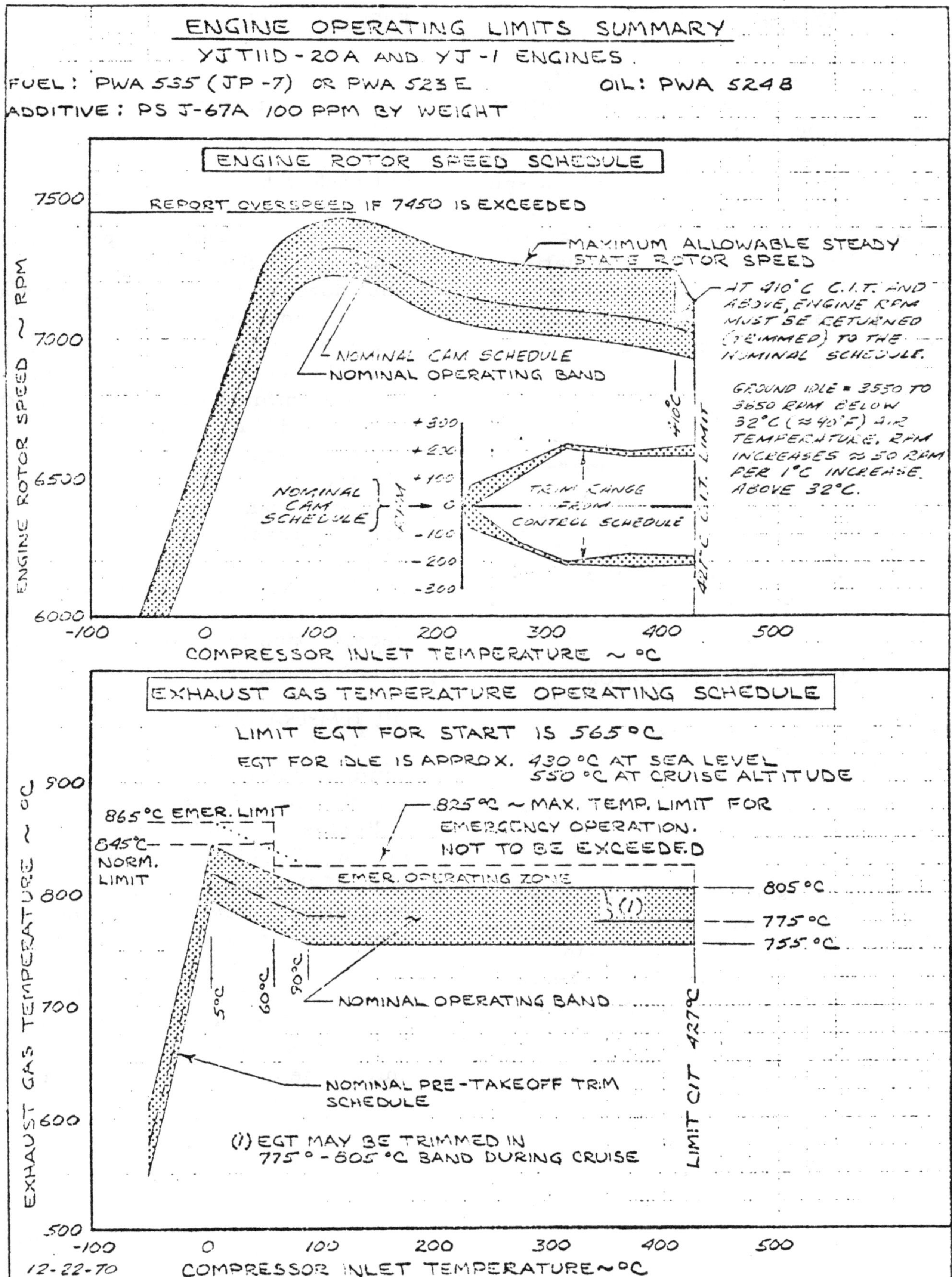
ENGINE OPERATING LIMITS SUMMARY
YJT11D-20A AND YJ-1 ENGINES
FUEL: PWA 535 (JP-7) OR PWA 523E
OIL: PWA 524B
ADDITIVE: PS J-67A 100 PPM BY WEIGHT
ENGINE ROTOR SPEED SCHEDULE
REPORT OVERSPEED IF 7450 IS EXCEEDED
MAXIMUM ALLOWABLE STEADY STATE ROTOR SPEED
AT 410°C C.I.T. AND ABOVE, ENGINE RPM MUST BE RETURNED (TRIMMED) TO THE NOMINAL SCHEDULE.
NOMINAL CAM SCHEDULE
NOMINAL OPERATING BAND
GROUND IDLE = 3550 TO 3650 RPM BELOW 32°C (≈ 90°F) AIR TEMPERATURE. RPM INCREASES ≈ 50 RPM PER 1°C INCREASE ABOVE 32°C.
410°C
427°C C.I.T. LIMIT
NOMINAL CAM SCHEDULE
RPM
TRIM RANGE FROM CONTROL SCHEDULE
ENGINE ROTOR SPEED ~ RPM
COMPRESSOR INLET TEMPERATURE ~ °C
EXHAUST GAS TEMPERATURE OPERATING SCHEDULE
LIMIT EGT FOR START IS 565°C
EGT FOR IDLE IS APPROX. 430°C AT SEA LEVEL
550°C AT CRUISE ALTITUDE
865°C EMER. LIMIT
845°C NORM. LIMIT
825°C ~ MAX. TEMP. LIMIT FOR EMERGENCY OPERATION. NOT TO BE EXCEEDED
EMER. OPERATING ZONE
805°C
775°C
755°C
5°C
60°C
90°C
NOMINAL OPERATING BAND
LIMIT CIT 427°C
NOMINAL PRE-TAKEOFF TRIM SCHEDULE
(1) EGT MAY BE TRIMMED IN 775° - 805°C BAND DURING CRUISE
EXHAUST GAS TEMPERATURE ~ °C
12-22-70
COMPRESSOR INLET TEMPERATURE ~ °C

Figure 5-1

Emergency Operation

An accurate accounting of operating time in the emergency EGT operating zone must be maintained, as any operation in or above the emergency operating zone (figure 5-1) requires special maintenance action. The permissible emergency EGT when at low CIT's is above the derich system actuation point; therefore, the derich system must be disarmed if the emergency limit EGT is to be attained.

NOTE

> At compressor inlet temperatures below 5°C, the possibility of engine stall exists at EGT's between the maximum permissible value and the nominal operating band.

COMPRESSOR INLET TEMPERATURE

The maximum compressor inlet temperature is 427°C.

ENGINE SPEED

Engine speed should not exceed the higher value shown for the nominal operating band, figure 5-1. Engine speeds above 7450 rpm should be reported as an engine discrepancy and include maximum rpm attained.

FUEL PRESSURE

The fuel pressure low warning lights must be extinguished for takeoff.

OIL PRESSURE

Oil pressures below 35 psi are unsafe and require that the engine either be shut down or that a landing be made as soon as possible, using the minimum thrust required to sustain flight until a landing can be accomplished. 35 psi is the minimum pressure permitted while at idle rpm. In flight oil pressures between 35 and 40 psi are undesirable but acceptable and the normal operating band is 40 to 55 psi. Gradually increasing oil pressures up to 60 psi are acceptable at high Mach numbers provided the indication returns to normal values after aircraft deceleration to subsonic speeds.

OIL TEMPERATURE

Illumination of the oil temperature warning light for an operating engine indicates an unsafe condition, and a landing should be made as soon as possible if the temperature can not be controlled. Refer to Section III for airstarting of a windmilling engine with its oil temperature warning light illuminated. An engine can not be started safely on the ground when its oil temperature warning light is illuminated unless the oil has been diluted.

NOTE

Engine operation is unsafe with oil temperature above 290°C, and ground starting is unsafe with undiluted oil temperature below 15°C. Illumination of an engine oil temperature warning light occurs when oil temperature of the respective engine is either above 282°C (± 11°C) or, below 15°C (± 3°C).

MAXIMUM WEIGHT LIMITS

The maximum gross weight is restricted to 130,000 pounds. The maximum gross weight capability for single engine flight is presented in Appendix I, Part II.

FUEL LOADING LIMITATIONS

Fuel loading tables are supplied in Chart E of the Weight and Balance Handbook, T.O. 1-1B-40.

CENTER OF GRAVITY LIMITS

The aircraft shall be fueled to operate within the center-of-gravity range from 17 to 27 percent Reference Chord. However, the center-of-gravity range for subsonic flight and landing should be maintained forward of 24 percent Reference Chord by forward transfer of fuel. Abnormal fuel usage sequences are not recommended except for forward fuel transfer to maintain the aircraft c.g. forward of 24% Reference Chord. This action may be necessary prior to landing, for subsonic flights, or inflight refueling.

LOAD FACTOR LIMITS

Load factor limits are shown by sheets 1 and 2 of figure 5-2. The diagrams presented for symmetrical flight are also applicable to entry into turns with normal bank angles. Allowable load factors applicable to rolling flight (abrupt maneuvers as a result of rapid control displacement) are especially identified and have a minimum limit of 1 g. Specific limits are shown for flight at Mach 3.2, 2.6, and at Mach 2.0 or less. Limit speeds and load factors applicable to operation at intermediate Mach numbers are obtained by linear interpolation between the limits shown on sheets 1 and 2 of Figure 5-2.

The following rule of thumb may be used as operational limit load factor values for symmetrical flight:

Mn 2.0 or less:	
65,000 to 124,000 lb	-.2* to 2.5 g
124,000 to 130,000 lb	-.2* to 2.0 g
80,000 to 90,000 lb	
below 50,000 ft	-.2* to 3.5 g
above 50,000 ft	-.2* to 2.5 g

*-1.0 g at subsonic speeds

Mn 2.0 to 2.6	
all weights	-.1 to 2.0 g

Mn 2.6 to 3.2	
all weights	-.1 to 1.5 g

MACH NUMBER, AIRSPEED AND ALTITUDE LIMITS AND RESTRICTIONS

Refer to figure 5-3 for summarized Mach, airspeed, and altitude limits and restrictions.

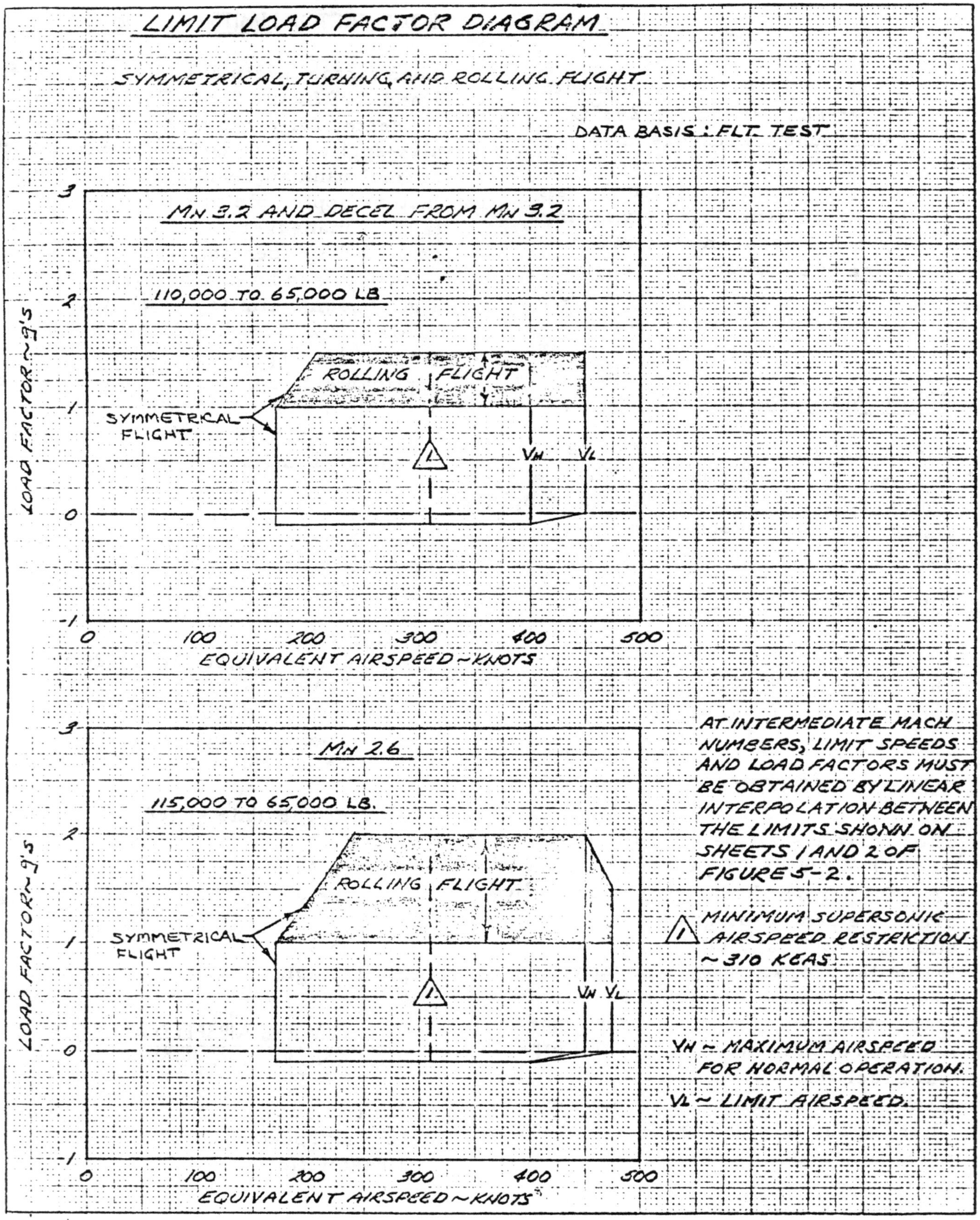

LIMIT LOAD FACTOR DIAGRAM
SYMMETRICAL, TURNING, AND ROLLING FLIGHT
DATA BASIS: FLT TEST
MN 3.2 AND DECEL FROM MN 3.2
110,000 TO 65,000 LB
ROLLING FLIGHT
SYMMETRICAL FLIGHT
VH
VL
LOAD FACTOR ~ g's
EQUIVALENT AIRSPEED ~ KNOTS
0
100
200
300
400
500
-1
0
1
2
3
MN 2.6
115,000 TO 65,000 LB.
ROLLING FLIGHT
SYMMETRICAL FLIGHT
VH VL
AT INTERMEDIATE MACH NUMBERS, LIMIT SPEEDS AND LOAD FACTORS MUST BE OBTAINED BY LINEAR INTERPOLATION BETWEEN THE LIMITS SHOWN ON SHEETS 1 AND 2 OF FIGURE 5-2.
1 MINIMUM SUPERSONIC AIRSPEED RESTRICTION ~ 310 KEAS
VH ~ MAXIMUM AIRSPEED FOR NORMAL OPERATION.
VL ~ LIMIT AIRSPEED.

Figure 5-2
(Sheet 1 of 2)

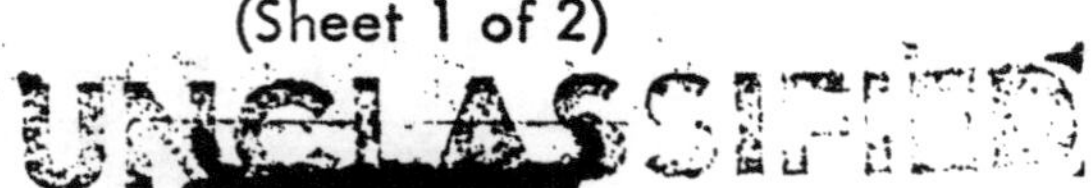

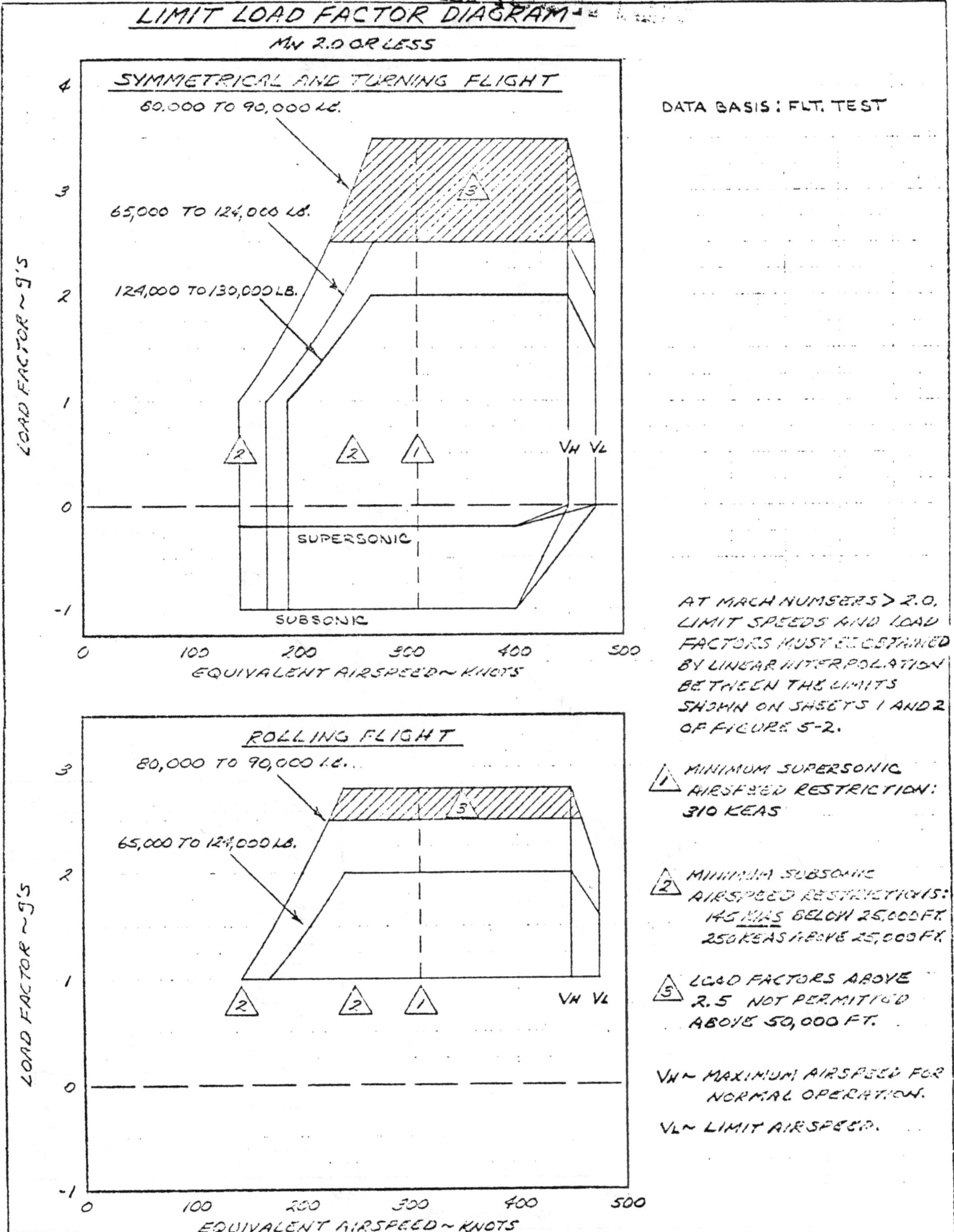
LIMIT LOAD FACTOR DIAGRAM
Mn 2.0 OR LESS
SYMMETRICAL AND TURNING FLIGHT
60,000 TO 90,000 LB.
65,000 TO 124,000 LB.
124,000 TO 130,000 LB.
SUPERSONIC
SUBSONIC
VH VL
LOAD FACTOR ~ g'S
EQUIVALENT AIRSPEED ~ KNOTS
DATA BASIS: FLT. TEST
AT MACH NUMBERS > 2.0, LIMIT SPEEDS AND LOAD FACTORS MUST BE OBTAINED BY LINEAR INTERPOLATION BETWEEN THE LIMITS SHOWN ON SHEETS 1 AND 2 OF FIGURE 5-2.
ROLLING FLIGHT
80,000 TO 90,000 LB.
65,000 TO 124,000 LB.
1 MINIMUM SUPERSONIC AIRSPEED RESTRICTION: 310 KEAS
2 MINIMUM SUBSONIC AIRSPEED RESTRICTIONS: 145 KIAS BELOW 25,000 FT. 250 KEAS ABOVE 25,000 FT.
3 LOAD FACTORS ABOVE 2.5 NOT PERMITTED ABOVE 50,000 FT.
VH ~ MAXIMUM AIRSPEED FOR NORMAL OPERATION.
VL ~ LIMIT AIRSPEED.

Figure 5-2

(Sheet 1 of 2)

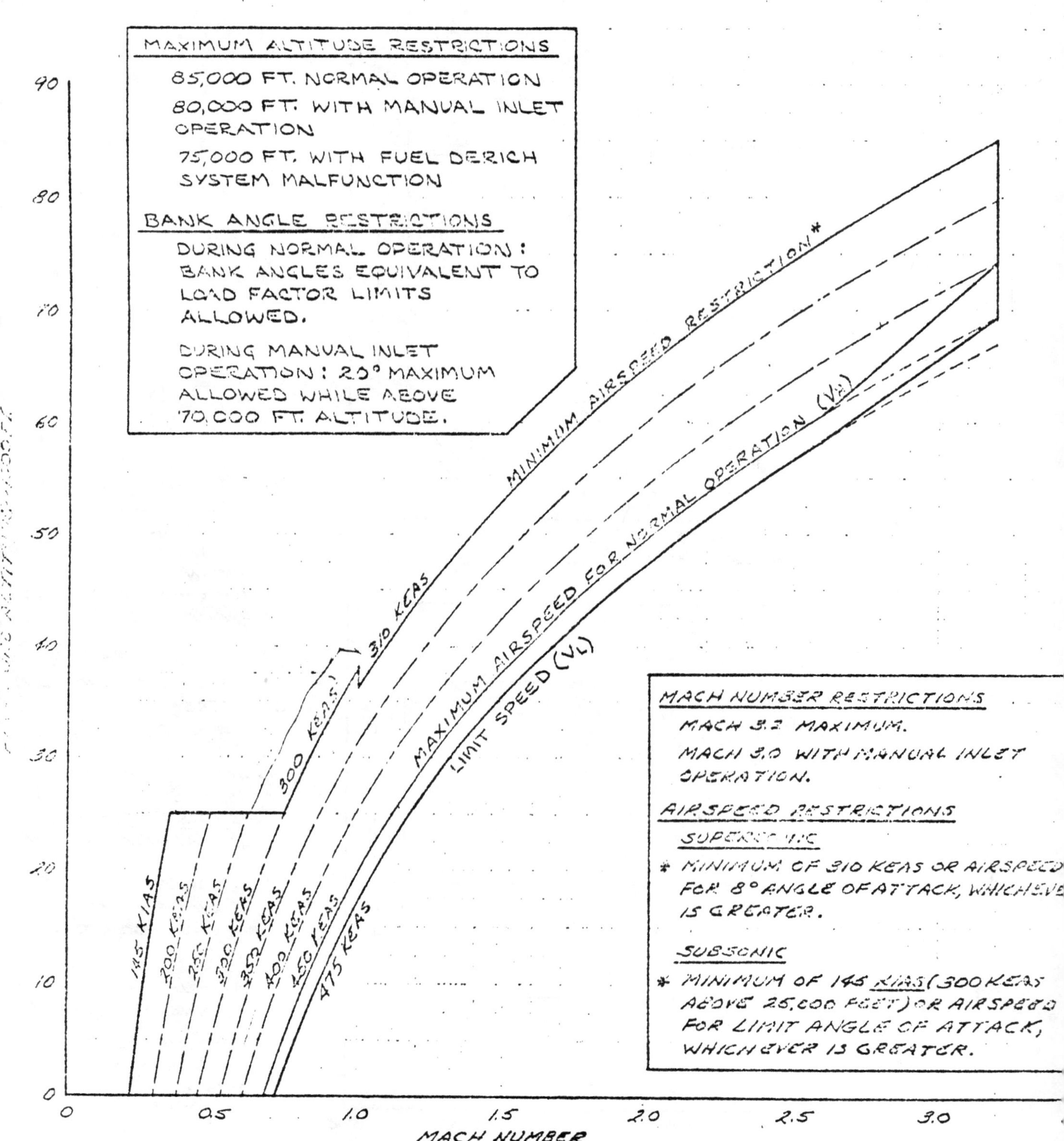
LIMIT SPEED AND ALTITUDE ENVELOPE
DATA BASIS: FLT. TEST
MAXIMUM ALTITUDE RESTRICTIONS
85,000 FT. NORMAL OPERATION
80,000 FT. WITH MANUAL INLET OPERATION
75,000 FT. WITH FUEL DERICH SYSTEM MALFUNCTION
BANK ANGLE RESTRICTIONS
DURING NORMAL OPERATION: BANK ANGLES EQUIVALENT TO LOAD FACTOR LIMITS ALLOWED.
DURING MANUAL INLET OPERATION: 20° MAXIMUM ALLOWED WHILE ABOVE 70,000 FT. ALTITUDE.
MINIMUM AIRSPEED RESTRICTION*
MAXIMUM AIRSPEED FOR NORMAL OPERATION (VH)
LIMIT SPEED (VL)
310 KEAS
300 KEAS
145 KIAS
200 KEAS
250 KEAS
300 KEAS
350 KEAS
400 KEAS
450 KEAS
475 KEAS
MACH NUMBER RESTRICTIONS
MACH 3.2 MAXIMUM.
MACH 3.0 WITH MANUAL INLET OPERATION.
AIRSPEED RESTRICTIONS
SUPERSONIC
* MINIMUM OF 310 KEAS OR AIRSPEED FOR 8° ANGLE OF ATTACK, WHICHEVER IS GREATER.
SUBSONIC
* MINIMUM OF 145 KIAS (300 KEAS ABOVE 25,000 FEET) OR AIRSPEED FOR LIMIT ANGLE OF ATTACK, WHICHEVER IS GREATER.
0
10
20
30
40
50
60
70
80
90
0
0.5
1.0
1.5
2.0
2.5
3.0
MACH NUMBER

Figure 5-3

WARNING

Flight instruments must be monitored closely during autopilot operation, particularly when using attitude hold, to prevent airspeed from exceeding the normal operating envelope.

MAXIMUM MACH NUMBER

Mach 3.2 is the limit speed. Mach 3.17 is the maximum scheduled cruise speed recommended.

NOTE

Mach 3.2 is the maximum speed limit providing the maximum engine inlet temperature (CIT) of 427°C is not exceeded. It is possible to reach the CIT limit before attaining 3.2 Mach number during extremely hot day operation.

WARNING

Mach overshoots may occur, especially when using Mach Hold, during turns and/or when the aircraft is heavy such as after leveloff. The excursions when using Mach Hold may be severe (± 0.05 Mach number) while flying through rapid small temperature changes at altitude. Use of the basic pitch autopilot instead of Mach Hold is strongly recommended when excessive speed and altitude excursions are experienced.

Without Inert Tank Atmosphere

Mach 2.6 is the maximum speed restriction without an inert atmosphere in the fuel tanks

AIRSPEED

Refer to figure 5-3, Limit Airspeed and Altitude Envelope. The maximum airspeed limit is 475 KEAS. Except when specifically authorized, the minimum airspeed restrictions are 310 KEAS while supersonic and, while subsonic, 250 KEAS above 25,000 feet and 145 KIAS below 25,000 feet pressure altitude except where higher airspeed is required to stay within α limits.

ANGLE OF ATTACK

Except for authorized testing, maximum angles of attack are restricted to 8° while supersonic, 14° while subsonic with landing gear extended, 12° while subsonic with landing gear retracted, when operating within the c.g. limits specified in this section.

WARNING

Because of the reduced stability that can occur at high angle of attack, maneuvers should not be attempted which intentionally approach limit load factor or angles of attack at minimum airspeed or high pitch rate. Recovery from inadvertent maneuvers should be initiated before limit conditions are reached.

Note

Engine stalls can be experienced at high angles of attack.

ALTITUDE

The maximum altitude restriction is 85,000 feet except when higher altitudes are specifically authorized. Do not exceed 80,000 feet with manual inlet operation. Do not exceed 75,000 feet unless both fuel derichment systems are operative and on.

High Altitude Turns

Except for specific phugoid testing, cruise altitudes within 1000 feet of the maximum afterburner ceiling should be avoided at turn entry. Turns must be anticipated and a descent of approximately 2000 feet should be accomplished prior to turn entry when cruising near the supersonic ceiling for the existing Mach number, gross weight, and ambient temperature.

Rate of Descent

Rate of descent must be limited so as to maintain positive fuel tank pressure when sustained speeds have exceeded Mach 2.6. While above Mach 1.8, the maximum rate of descent should be such that rate of deceleration does not exceed 1.0 Mach number in 3 minutes. There is no limitation on rate of deceleration while below Mach 1.8.

PROHIBITED MANEUVERS

Spins, full stalls, and inverted flight are prohibited in this aircraft. Except for authorized testing, intentional inlet unstarts are prohibited.

Simulated Single Engine Flight

Simulated single engine approaches at less than 200 KEAS or 300 feet terrain clearance or with more than 20,000 pounds of fuel remaining are prohibited.

AIRCRAFT SYSTEMS LIMITATIONS

PRIMARY FLIGHT CONTROL SYSTEMS

The control surface limiter shall be engaged whenever flight speed exceeds either 330 KEAS or 0.7 Mach number.

AUTOMATIC FLIGHT CONTROL SYSTEMS

Stability Augmentation System

The SAS shall be on for all takeoffs. Landings with normally functioning SAS channels intentionally disengaged are not permitted except that the roll SAS may be disengaged prior to simulated and actual single engine landings. Operation with all pitch and yaw SAS intentionally disengaged is not permitted; however, either all pitch or all yaw SAS channels may be disengaged for training demonstrations provided Mach 1.0 is not exceeded and gross maneuvers are not attempted.

Autopilot

The c.g. shall be within recommended limits during use of the autopilot. The pitch autopilot shall not be used at speeds in excess of 300 KEAS while below 25,000 feet. The altitude hold mode shall not be used above 60,000 feet. As a precaution against subsequent hardover signals, the autopilot must not be used with the BUPD engaged.

Mach Hold

Use of the Mach Hold feature of the autopilot is not restricted.

LANDING GEAR SYSTEM

The main landing gear is designed for landing sink speeds at touchdown which decrease from 10 feet per second at 68,000 pounds to 6 feet per second at 125,000 pounds gross weight. Side loads during takeoff, landing, and taxiing must be kept to a minimum, as landing gear side load strength is critical during ground maneuvering. Rapid turns while taxiing can overstress the airframe at relatively low speeds. Figure 5-4 illustrates the relationship of forward speed and turning angle. Note that the turn speeds indicated are limits at which design load factor is developed; operatinally, turns should be made at speeds less than these to permit a structural safety factor.

In flight, gear door strength limits the airspeed with gear in transit or down to 300 KEAS or Mach 0.7, whichever is less, with a maximum permissible sideslip angle of 10°. Maximum permissible speeds are 300 KEAS or Mach 0.9, whichever is less, with gear down when sideslip does not exceed 5°. Operation at supersonic speeds with gear extended is forbidden.

Tires

The maximum ground speed rating of Goodrich 27.5 x 7.5 x 16 "silver" tires is 239 knots (275 mph). The conversion from 239 knots to KEAS with various combinations of temperature and altitude is shown by figure 5-5 for various wind conditions. Limit indicated airspeed on the ground decreases by the amount of tailwind component along the runway, and increases by the headwind component. Takeoff after an abort is not permitted if taxiing restrictions have been exceeded during the takeoff run.

Taxiing Restrictions

a. The taxi speed limit is 60 mph (52 knots). 40 mph (35 knots) is the maximum taxi speed recommended.

b. A cooling period between the end of taxi and start of takeoff may be required. See figures 5-6 and 5-7 as guides to recommended cooling time vs tax distance. Figure 5-6 is based on 31% deflected tires and figure 5-7 is based on 400 psi tire pressure at all gross weights. Under normal taxi and takeoff conditions at maximum gross weight, the tire limit capability is not exceeded if the 400 psi tire pressure is maintained. See figure 5-7.

c. A check of tires, wheels and brakes is required when clear of the runway after an aborted takeoff or a heavy weight landing.

LIMIT TAXI TURN SPEEDS

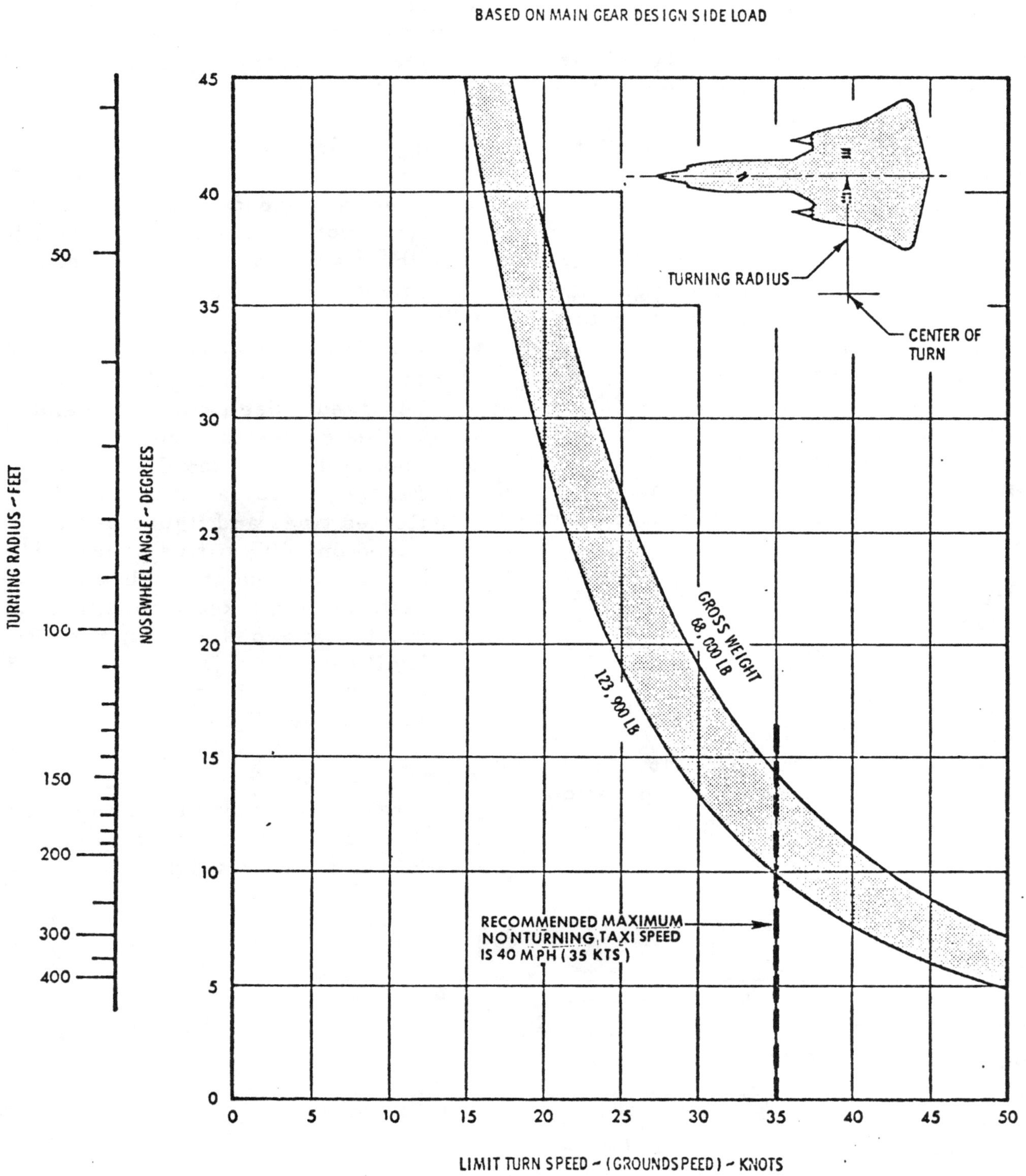

BASED ON MAIN GEAR DESIGN SIDE LOAD

TURNING RADIUS ~ FEET

50
100
150
200
300
400

NOSEWHEEL ANGLE ~ DEGREES

0
5
10
15
20
25
30
35
40
45

TURNING RADIUS

CENTER OF TURN

GROSS WEIGHT 68,000 LB

123,900 LB

RECOMMENDED MAXIMUM NONTURNING TAXI SPEED IS 40 MPH (35 KTS)

0 5 10 15 20 25 30 35 40 45 50

LIMIT TURN SPEED ~ (GROUNDSPEED) ~ KNOTS

Figure 5-4

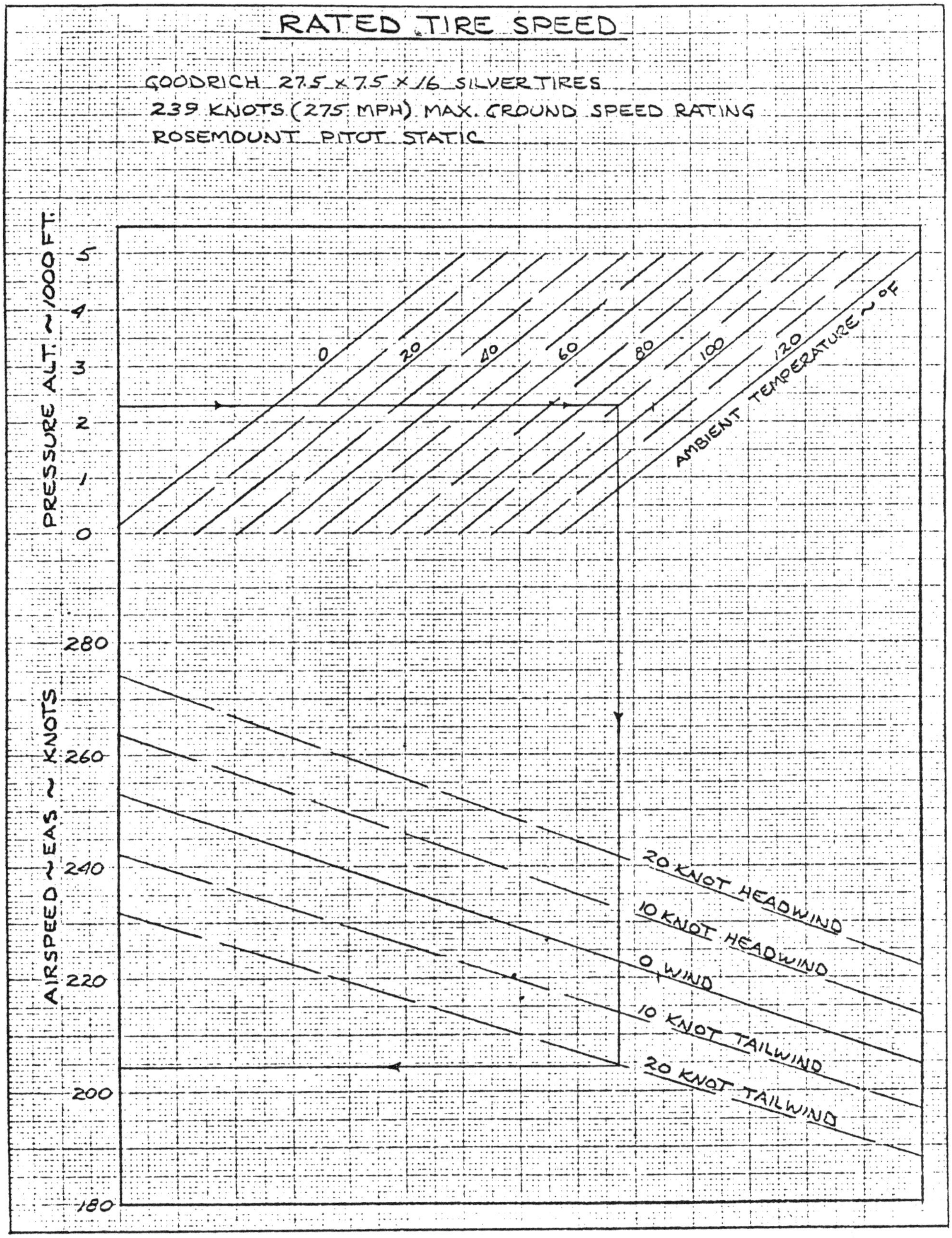

RATED TIRE SPEED

GOODRICH 27.5 x 7.5 x 16 SILVER TIRES
239 KNOTS (275 MPH) MAX. GROUND SPEED RATING
ROSEMOUNT PITOT STATIC

PRESSURE ALT. ~ 1000 FT.

0 1 2 3 4 5

AMBIENT TEMPERATURE ~ °F

0 20 40 60 80 100 120

AIRSPEED ~ EAS ~ KNOTS

180 200 220 240 260 280

20 KNOT HEADWIND
10 KNOT HEADWIND
0 WIND
10 KNOT TAILWIND
20 KNOT TAILWIND

Figure 5-5

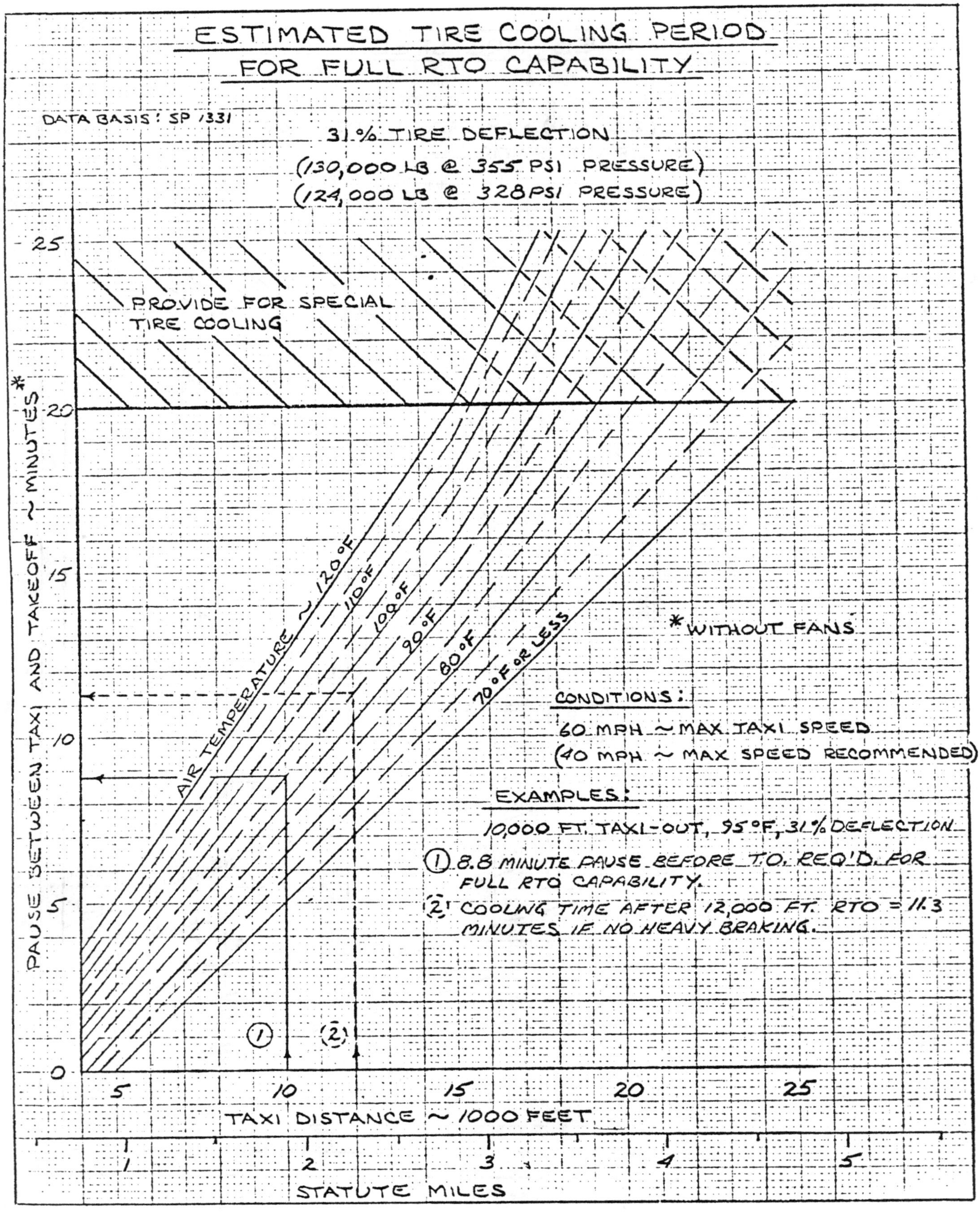

ESTIMATED TIRE COOLING PERIOD FOR FULL RTO CAPABILITY

DATA BASIS: SP 1331

31% TIRE DEFLECTION

(130,000 LB @ 355 PSI PRESSURE)

(124,000 LB @ 328 PSI PRESSURE)

PROVIDE FOR SPECIAL TIRE COOLING

PAUSE BETWEEN TAXI AND TAKEOFF ~ MINUTES*

0
5
10
15
20
25

AIR TEMPERATURE ~ 120°F

110°F

100°F

90°F

80°F

70°F OR LESS

*WITHOUT FANS

CONDITIONS:

60 MPH ~ MAX TAXI SPEED

(40 MPH ~ MAX SPEED RECOMMENDED)

EXAMPLES:

10,000 FT. TAXI-OUT, 95°F, 31% DEFLECTION

(1) 8.8 MINUTE PAUSE BEFORE T.O. REQ'D. FOR FULL RTO CAPABILITY.

(2) COOLING TIME AFTER 12,000 FT. RTO = 11.3 MINUTES IF NO HEAVY BRAKING.

(1)
(2)

5
10
15
20
25

TAXI DISTANCE ~ 1000 FEET

1
2
3
4
5

STATUTE MILES

Figure 5-6

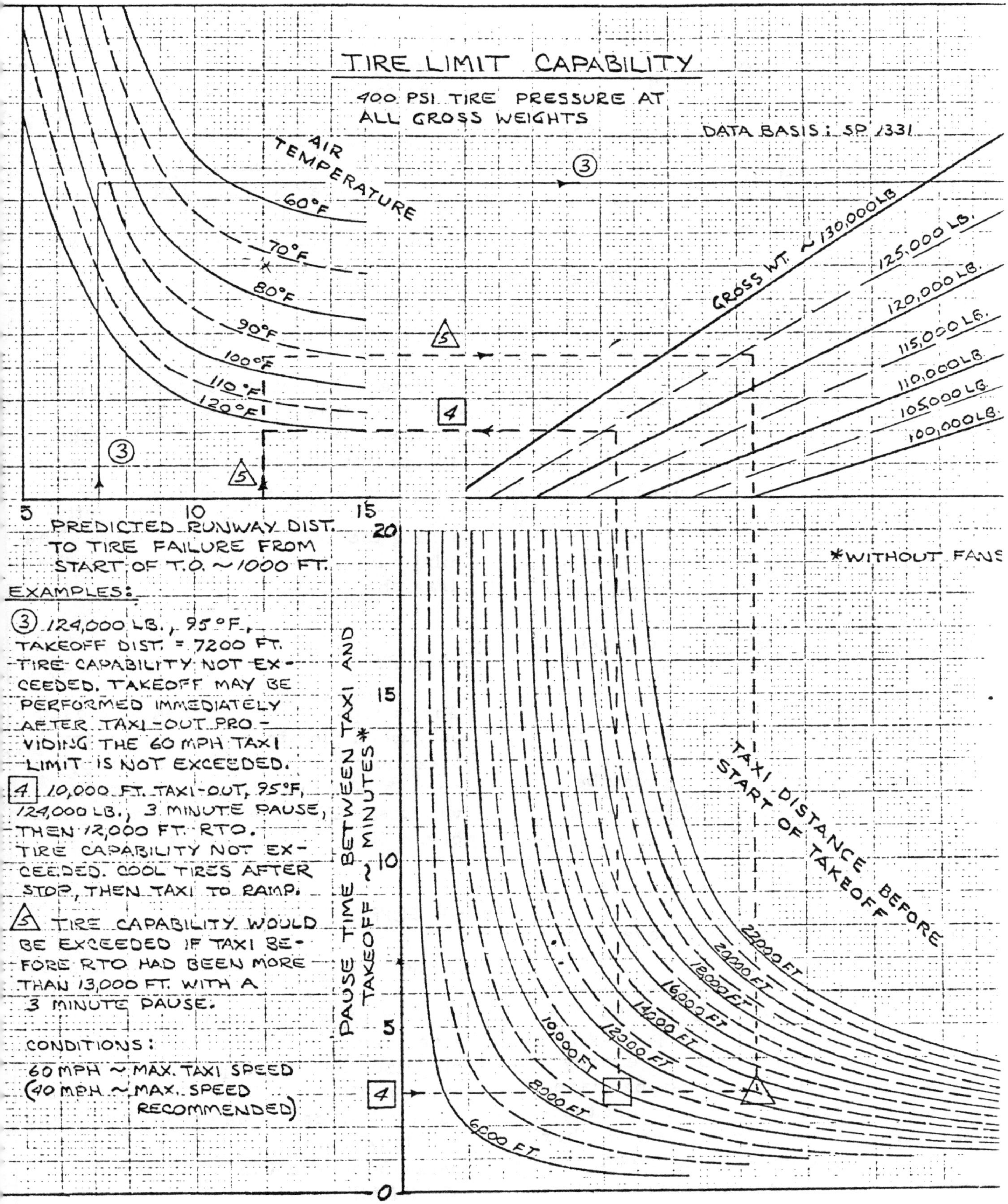

UNCLASSIFIED

TIRE LIMIT CAPABILITY

400 PSI TIRE PRESSURE AT ALL GROSS WEIGHTS

DATA BASIS: SP 1331

AIR TEMPERATURE

60°F
70°F
80°F
90°F
100°F
110°F
120°F

GROSS WT. ~ 130,000 LB
125,000 LB.
120,000 LB.
115,000 LB.
110,000 LB.
105,000 LB.
100,000 LB.

5 10 15

PREDICTED RUNWAY DIST. TO TIRE FAILURE FROM START OF T.O. ~ 1000 FT.

*WITHOUT FANS

PAUSE TIME BETWEEN TAXI AND TAKEOFF ~ MINUTES*

0 5 10 15 20

TAXI DISTANCE BEFORE START OF TAKEOFF

6000 FT.
8000 FT.
10000 FT.
12000 FT.
14000 FT.
16000 FT.
18000 FT.
20000 FT.
22000 FT.

EXAMPLES:

(3) 124,000 LB., 95°F, TAKEOFF DIST. = 7200 FT. TIRE CAPABILITY NOT EXCEEDED. TAKEOFF MAY BE PERFORMED IMMEDIATELY AFTER TAXI-OUT PROVIDING THE 60 MPH TAXI LIMIT IS NOT EXCEEDED.

[4] 10,000 FT. TAXI-OUT, 95°F, 124,000 LB., 3 MINUTE PAUSE, THEN 12,000 FT. RTO. TIRE CAPABILITY NOT EXCEEDED. COOL TIRES AFTER STOP, THEN TAXI TO RAMP.

⚠5 TIRE CAPABILITY WOULD BE EXCEEDED IF TAXI BEFORE RTO HAD BEEN MORE THAN 13,000 FT. WITH A 3 MINUTE PAUSE.

CONDITIONS:

60 MPH ~ MAX. TAXI SPEED
(40 MPH ~ MAX. SPEED RECOMMENDED)

Figure 5-7

> **WARNING**
>
> Extreme caution should be exercised when making tire and wheel check after a heavy weight landing, an aborted takeoff, or after any heavy braking. Overheated tires may explode and cause injury or loss of life. The check should be delayed until reasonable cooling has been accomplished if there is evidence that an overheated condition exists.

A ground vehicle may be used to monitor taxi distance and speed. If a tire and/or brake cooling period is necessary, it should be continued until ground inspection reveals that each individual tire and wheel is sufficiently cooled for continued operation (temperatures relatively tolerable to the touch).

> **NOTE**
>
> Tire and brake cooling may be accelerated by use of fans.

Brakes

Brakes in a new condition have a capacity for one hard stop from speed and weight conditions for the maximum rating. The one-stop energy rating of the brakes is 198,700,000 foot-pounds. Speeds corresponding to this energy rating from which stops can be made vary with gross weight, ambient temperature, altitude, wind, and whether or not the drag chute is deployed. Corresponding indicated airspeed, altitude, temperature, and weight conditions are shown by figure 5-8 for the above brake rating for stops on dry runways with zero wind component. Headwind components may be added to values shown, and tailwind components must be subtracted. Refer to Part II of the Appendix for detailed information related to maximum refusal speeds and heavy weight landings with various operating conditions.

DRAG CHUTE

The maximum airspeed for drag chute deployment is 210 KEAS. The drag chute shall not be deployed while in flight except in the event of a drag chute unsafe emergency indication. The maximum crosswind component for jettisoning the drag chute is 12 knots. The minimum airspeed for jettisoning the drag chute is 55 KEAS.

CANOPY

The canopy shall be opened or closed only when the aircraft is stationary. Maximum taxi speed with a canopy open is 40 knots. Gusts or strong winds should be considered as a portion of the 40-knot speed limit.

LIFE SUPPORT SYSTEM

Do not operate the aircraft if the pressure suit neck ring prevents normal support of the helmet by the headrest; also, verify proper access to the D-ring and adequate support in the seat prior to flight. Flights without pressure suits, using the oxygen mask and regulator, are restricted to altitudes below 50,000 feet and speeds below 420 KEAS.

MAXIMUM INITIAL BRAKING SPEED FOR STOP USING RATED BRAKE CAPACITY

ONE STOP CAPABILITY

(4×7) + (2×5) ROTOR BRAKES
198,700,000 FT-LB CAPACITY
ROSEMOUNT PITOT STATIC

WITHOUT DRAG CHUTE
NOSE DOWN
DRY AND HARD RUNWAY
ZERO WIND, ZERO SLOPE

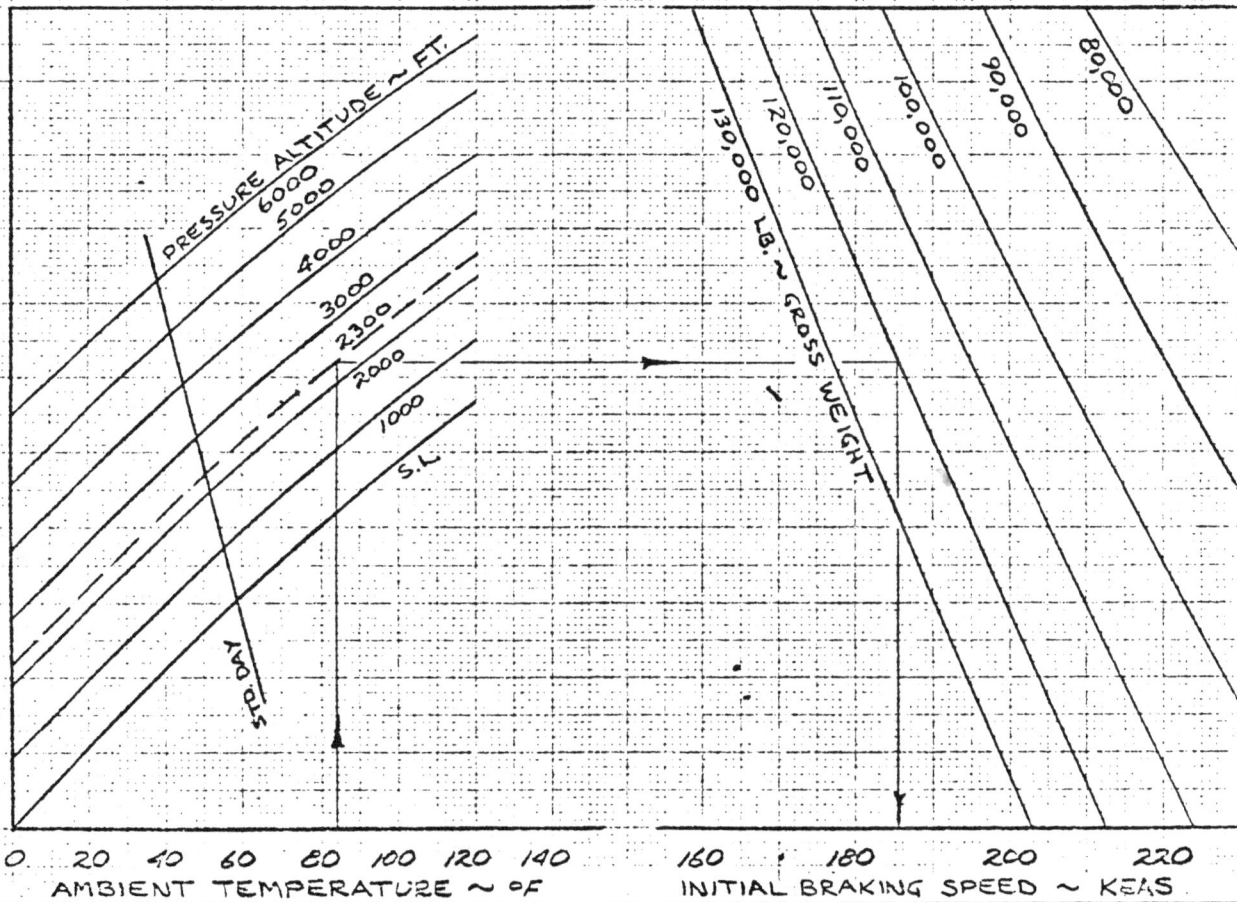

PRESSURE ALTITUDE ~ FT.
6000
5000
4000
3000
2300
2000
1000
S.L.
STD. DAY

130,000 LB. ~ GROSS WEIGHT
120,000
110,000
100,000
90,000
80,000

0 20 40 60 80 100 120 140
AMBIENT TEMPERATURE ~ °F

160 180 200 220
INITIAL BRAKING SPEED ~ KEAS

Figure 5-8

REFUELING DOOR

If flight has involved operation at high temperatures and the air-refueling door has not been closed, ground crew personnel must be advised after landing of temperature encountered and duration of exposure.

ANTI-COLLISION LIGHTS

If flight has involved operation at high temperatures and the anti-collision lights have not been retracted, ground personnel must be advised after landing of temperature encountered and duration of exposure.

Appendix I

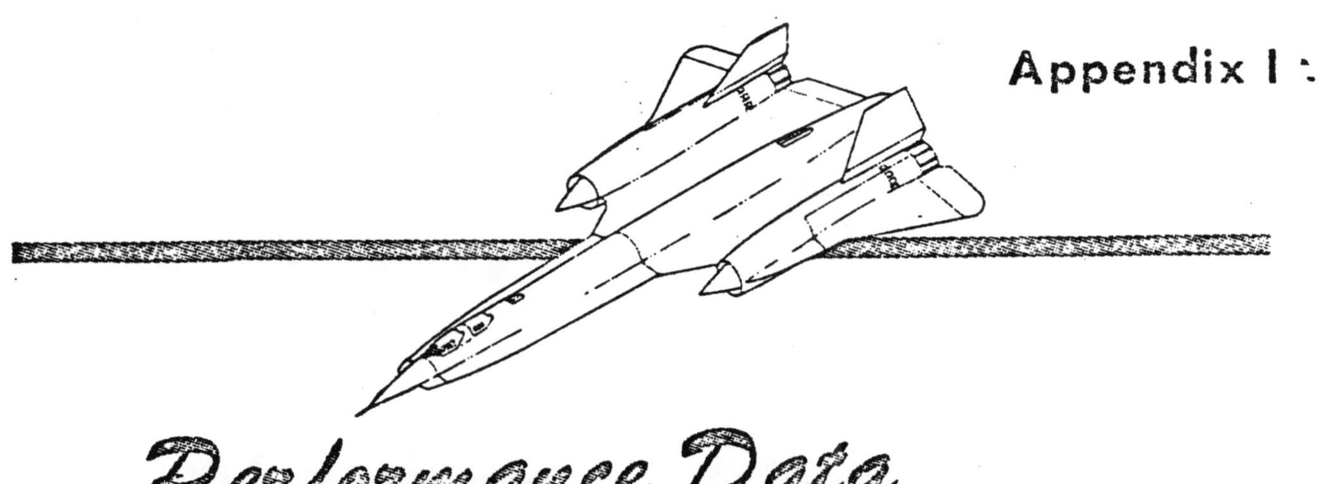

Performance Data

$M_0 = 2$ CRUISE RANGE ≈ $M_0 = .9$ RANGE

TABLE OF CONTENTS

PART I

INTRODUCTION

TABLE OF CONTENTS

LIST OF ILLUSTRATIONS

PART I

INTRODUCTION

DATA BASIS

The performance charts are based on flight test data from YJT11D-20A (J-58) engine equipped aircraft, supplemented by SR-71 flight test and wind tunnel performance data. Additional material will be incorporated as the information becomes available.

FUEL AND FUEL DENSITY

These data are applicable to aircraft fueled with PWA 535 with PS J-67A additive. Deviations from the nominal fuel density of 6.57 pounds per gallon have negligible effect on performance as long as actual aircraft gross weight and fuel load are known. However, with all tanks filled to capacity, the maximum fuel load changes about 1000 pounds for each 0.1 pound per gallon change in fuel density. This effect on operational capabilities must be considered.

AIRSPEED SYSTEMS AND POSITION ERROR CORRECTIONS

Air Data Computer System

The integrated instrument display supplies the primary airspeed and altitude information for all operations. Values shown are almost completely compensated for position error and compressibility effects. There is no appreciable lag in the indications.

The air data computer system converts pitot-static pressures to airspeed, pressure altitude, and Mach number signals for the pilot's integrated instrument display and for the fire control officer's triple display indicator (TDI). The pitot-static boom also supplies air pressures directly to the ship standby system indicated airspeed and altimeter instruments.

Standby System

Figures A1-1 and A1-2 show position error corrections for the standby altimeter and airspeed indicator. Corrections for operation at subsonic speeds were obtained by calibrations in flight and during ground runs. Corrections provided for supersonic speeds were obtained from wind tunnel results. Standard corrections for compressibility effects on airspeed must be applied (subtracted) after the position error corrections are made. The compressibility correction is supplied on figure A1-3. The differences between indicated airspeed (KIAS) and equivalent airspeed (KEAS) are a function of speed, altitude, and ship system position error. A comparison of KIAS from the ship system instrument with KEAS from either the integrated display or the TDI is shown by figure A1-4. For example: at 400 KEAS, the ship system will indicate 422 KIAS at 20,000 feet indicated pressure altitude, and 476 KIAS at 50,000 feet indicated altitude. Other combinations can be de-

termined from figure A1-4 for use in the event of air data computer system failure.

TRUE MACH NUMBER VS EQUIVALENT AIRSPEED

Figure A1-5 shows the relationship between true Mach number, pressure altitude, and equivalent airspeed, based on a τ of 1.4, the standard atmospheric parameter.

MACH-AIRSPEED-TEMPERATURE CHART

From the TDI Mach and CIT gage, ambient air temperature and true airspeed can be obtained, as shown, on the Mach-Airspeed-Temperature Chart, figure A1-6. For example, at a TDI Mach of 3.05 and CIT of 300°C, the ambient air temperature is 72°C and the true airspeed is 1680 knots (28 nmi/minute). The effect of adiabatic compression and temperature rise on atmospheric characteristics has been included by using a variable τ parameter.

STANDARD ATMOSPHERE TABLE

The standard atmosphere table, figure A1-7, provides reference temperature pressure, air density, and sonic speed information which may be of assistance in overall flight planning.

STANDARD UNITS CONVERSION

The standard untis conversion chart, figure A1-8, provides a means for direct conversion of temperature, distance, and speed between English and metric units.

W/δ AS A FUNCTION OF GROSS WEIGHT AND ALTITUDE

Figure A1-9 presents the relationship between the cruise parameter W/δ and the airplane's gross weight and altitude. The altitude required to attain a desired W/δ cruise condition for any normal supersonic cruise gross weight can be obtained from this chart.

TURNING PERFORMANCE

Figure A1-10 presents generalized turning performance at constant Mach numbers for various ambient temperatures and bank angles. Turn radius, distance, and time are plotted for a range of Mach numbers, ambient temperatures, bank angles, and degrees of turn.

Example:

For a turn of 180° at Mach 3.00 a forecast ambient temperature of -56.5°C at a bank angle of 30° find the turn radius, distance, and time. Enter figure A1-10 at Mach 3.00 and -56.5°C ambient temperature and note that true airspeed is 1720 knots. Proceed horizontally to 30° bank angle and read turn radius as 74.5 nautical miles. Proceed downward to 180° of turn and read turn distance as 235 nautical miles flown. Proceed horizontally to 1720 KTAS and read the turn time as 8.1 minutes.

POSITION ERROR CORRECTIONS VS MACH NO - STANDBY SYSTEM

Basis: Limited flt. tests
26 Oct. 1965

PITOT - STATIC SYSTEM ALTITUDE AND AIRSPEED INSTRUMENTS
ROSEMOUNT PITOT STATIC TYPE NO. 855

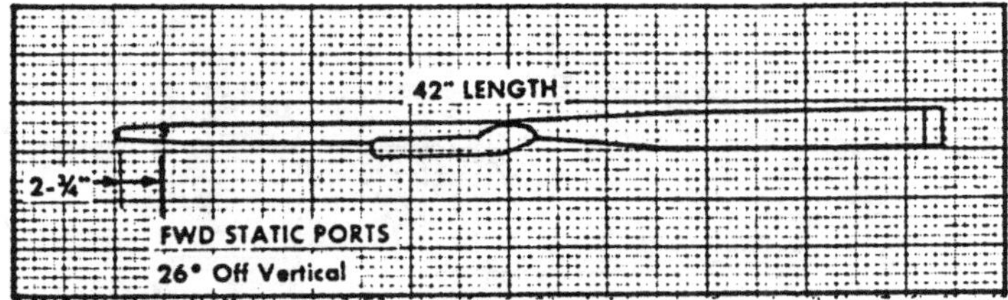

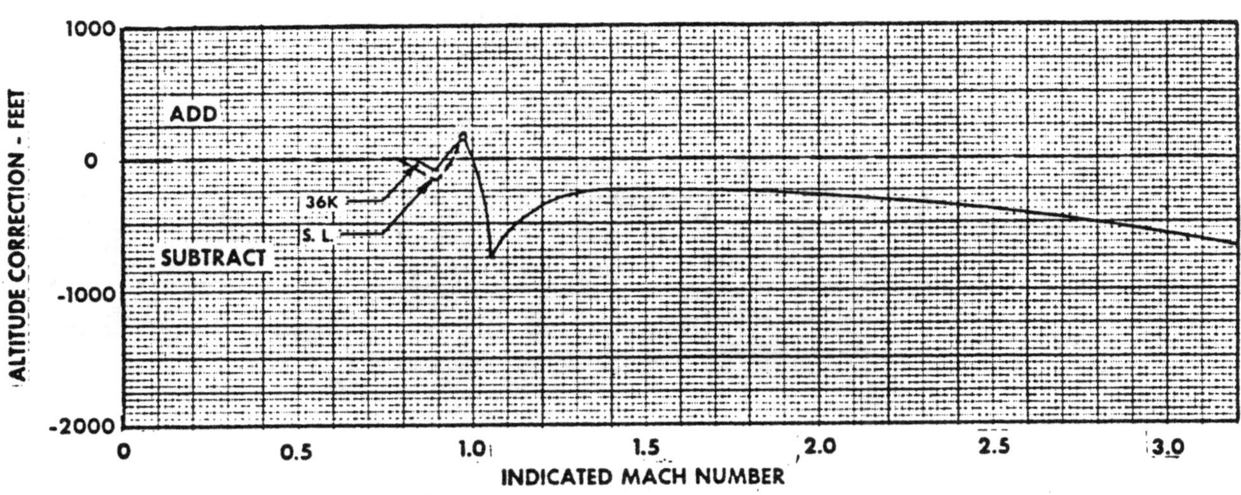

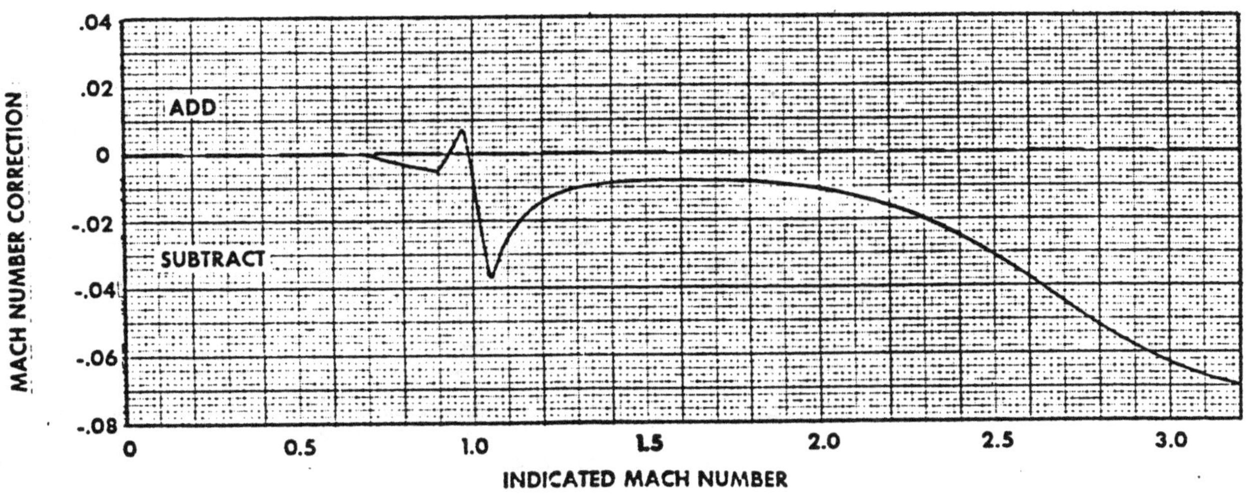

Figure A1-1

POSITION ERROR CORRECTIONS VS IAS - STANDBY SYSTEM

Basis: Limited flt. tests
26 Oct. 1965

PITOT - STATIC SYSTEM ALTITUDE AND AIRSPEED INSTRUMENTS
ROSEMOUNT PITOT STATIC TYPE NO. 855

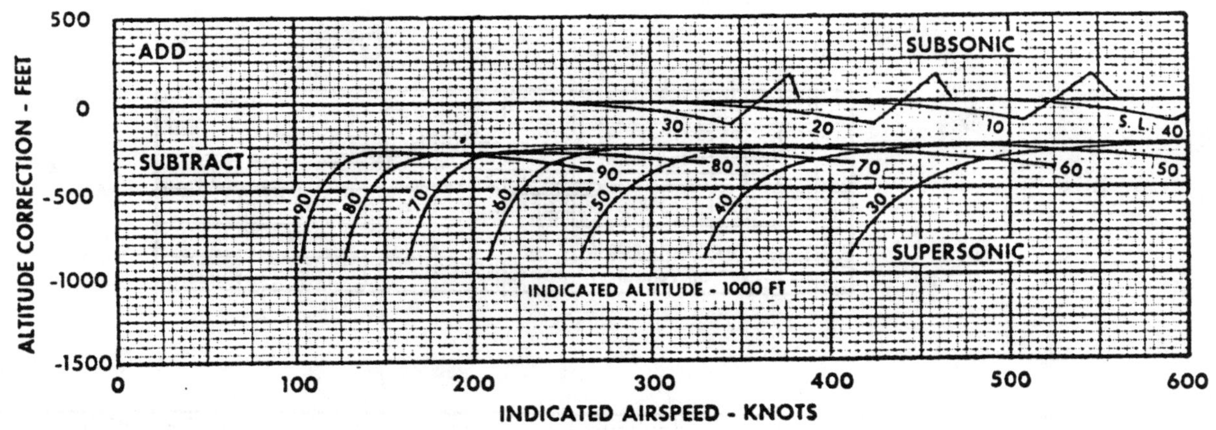

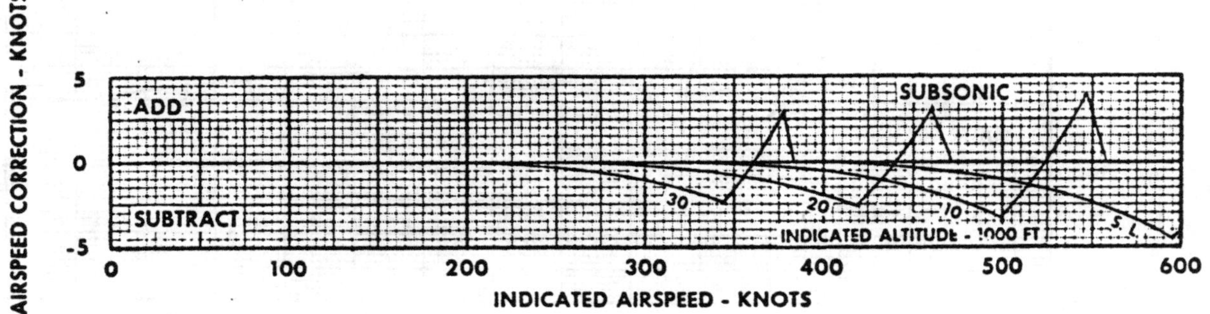

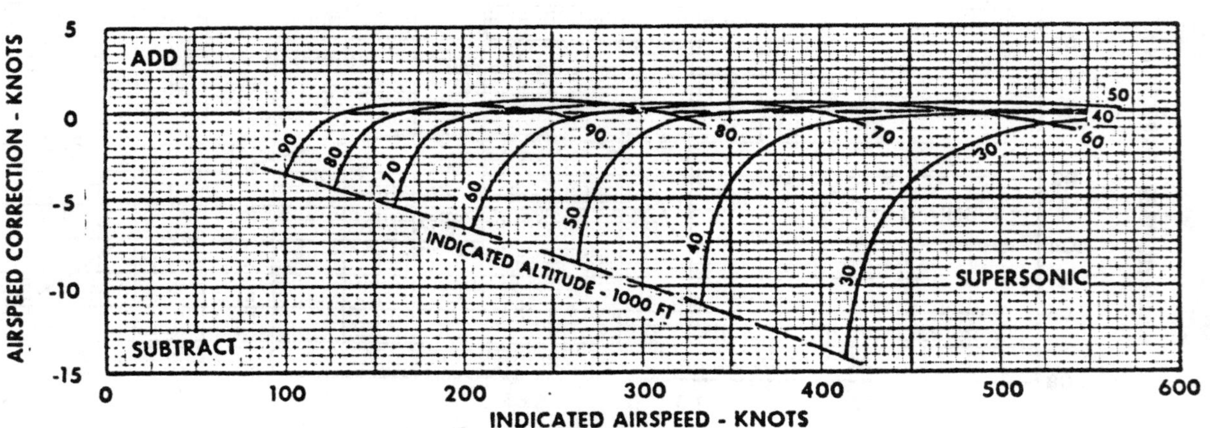

Figure A1-2

AIRSPEED COMPRESSIBILITY CORRECTION CHART

APPLICABLE TO SHIP STANDBY SYSTEM AFTER POSITION ERROR CORRECTION

SUBTRACT ΔV_C FROM KCAS TO OBTAIN KEAS

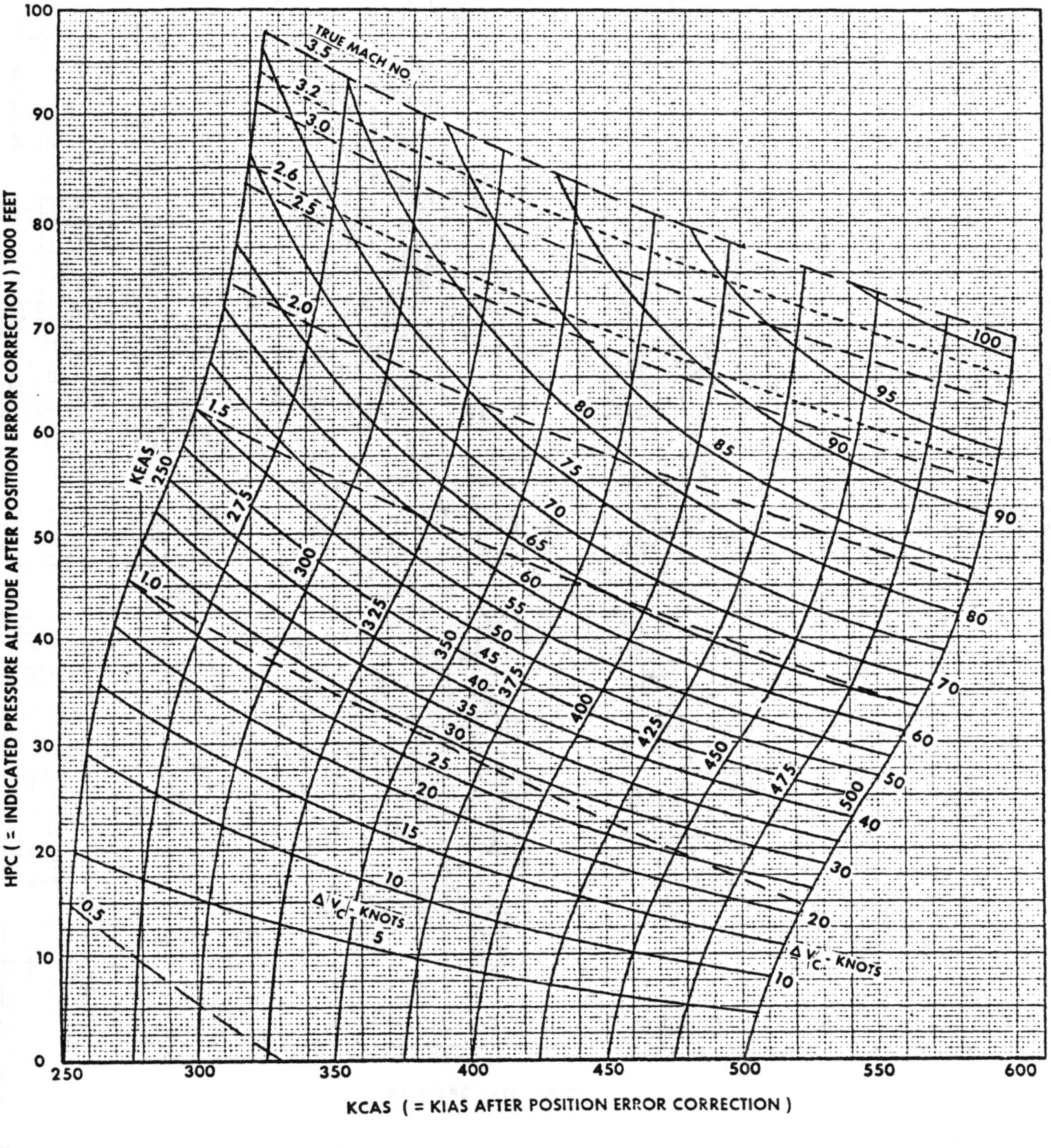

Figure A1-3

APPROXIMATE DIFFERENCES BETWEEN IAS AND EAS INDICATIONS

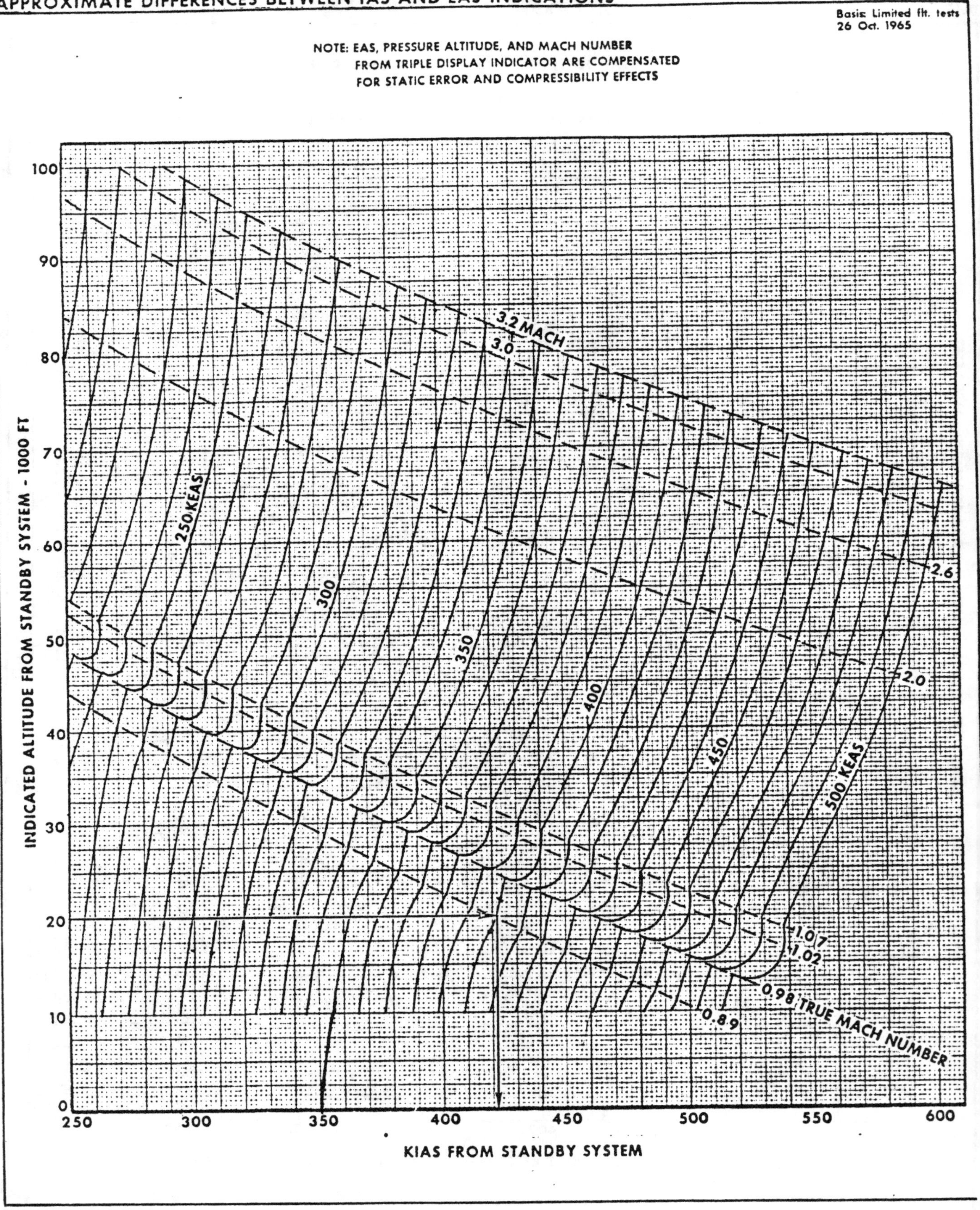

Basis: Limited flt. tests
26 Oct. 1965

NOTE: EAS, PRESSURE ALTITUDE, AND MACH NUMBER FROM TRIPLE DISPLAY INDICATOR ARE COMPENSATED FOR STATIC ERROR AND COMPRESSIBILITY EFFECTS

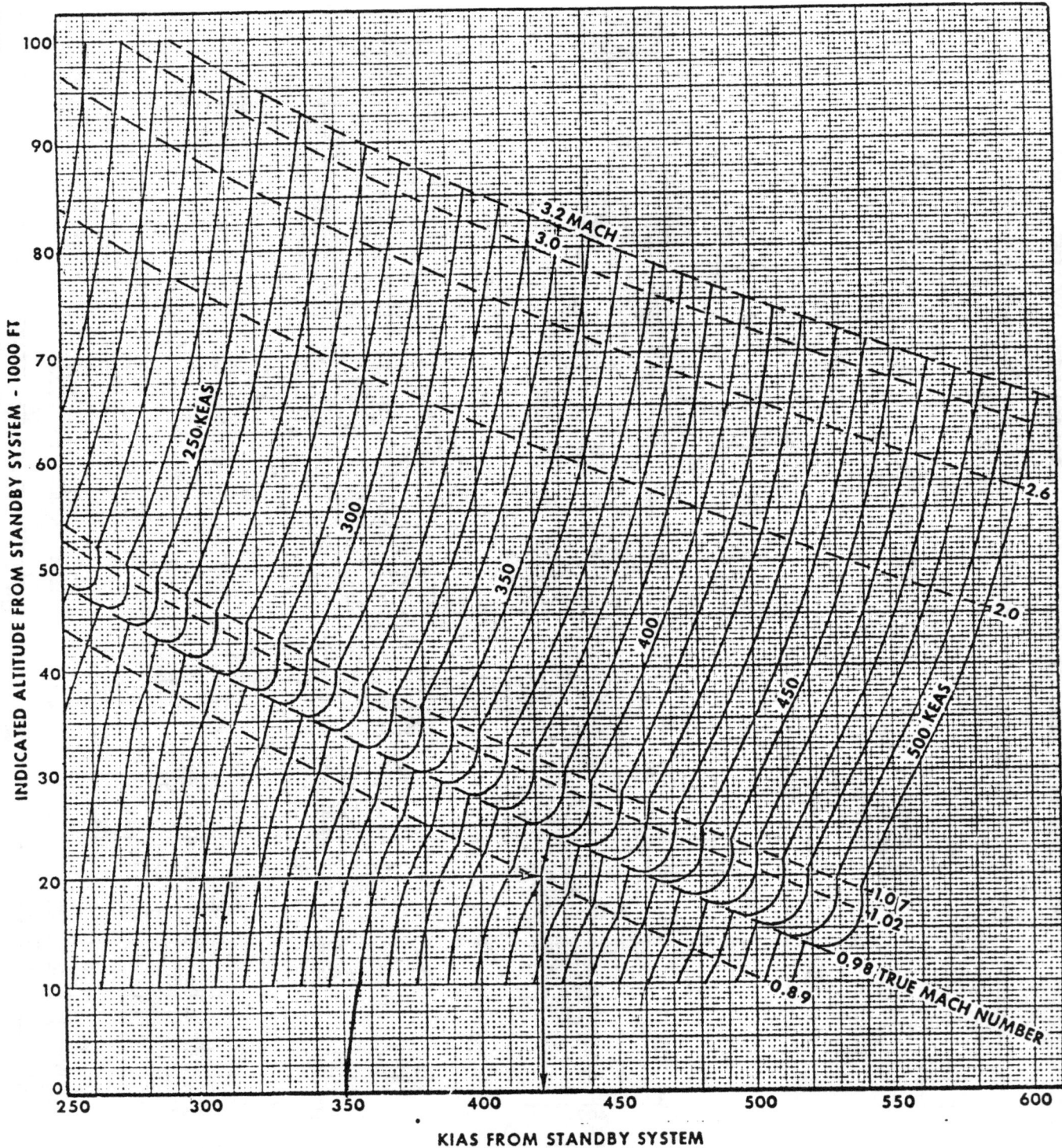

INDICATED ALTITUDE FROM STANDBY SYSTEM - 1000 FT

0 10 20 30 40 50 60 70 80 90 100

250KEAS 300 350 400 450 500KEAS

3.2 MACH 3.0 2.6 2.0 1.07 1.02 0.98 TRUE MACH NUMBER 0.89

250 300 350 400 450 500 550 600

KIAS FROM STANDBY SYSTEM

Figure A1-4

TRUE MACH NUMBER VS EQUIVALENT AIRSPEED

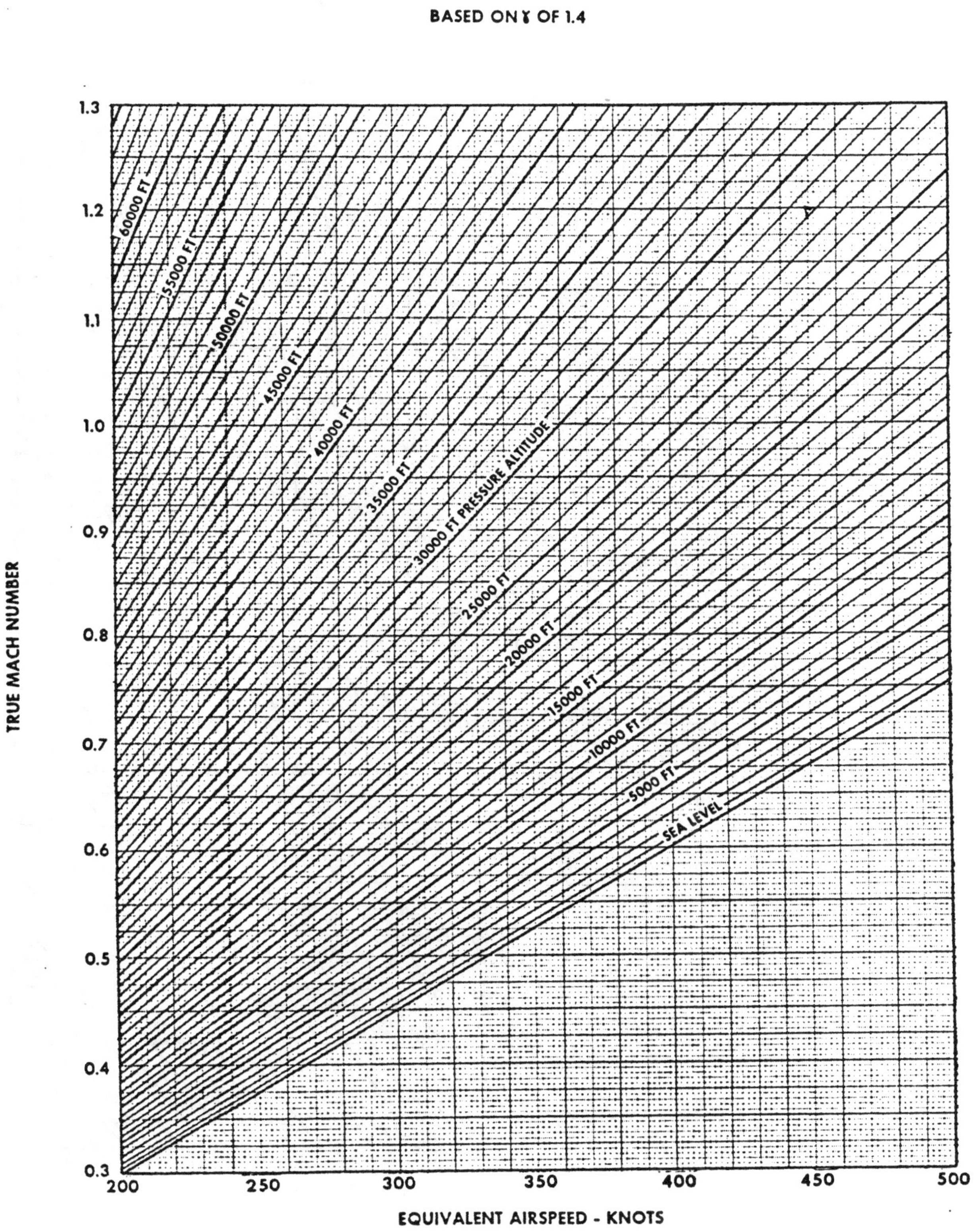

Figure A1-5 (Sheet 1 of 3)

TRUE MACH NUMBER VS EQUIVALENT AIRSPEED

TRUE MACH NUMBER

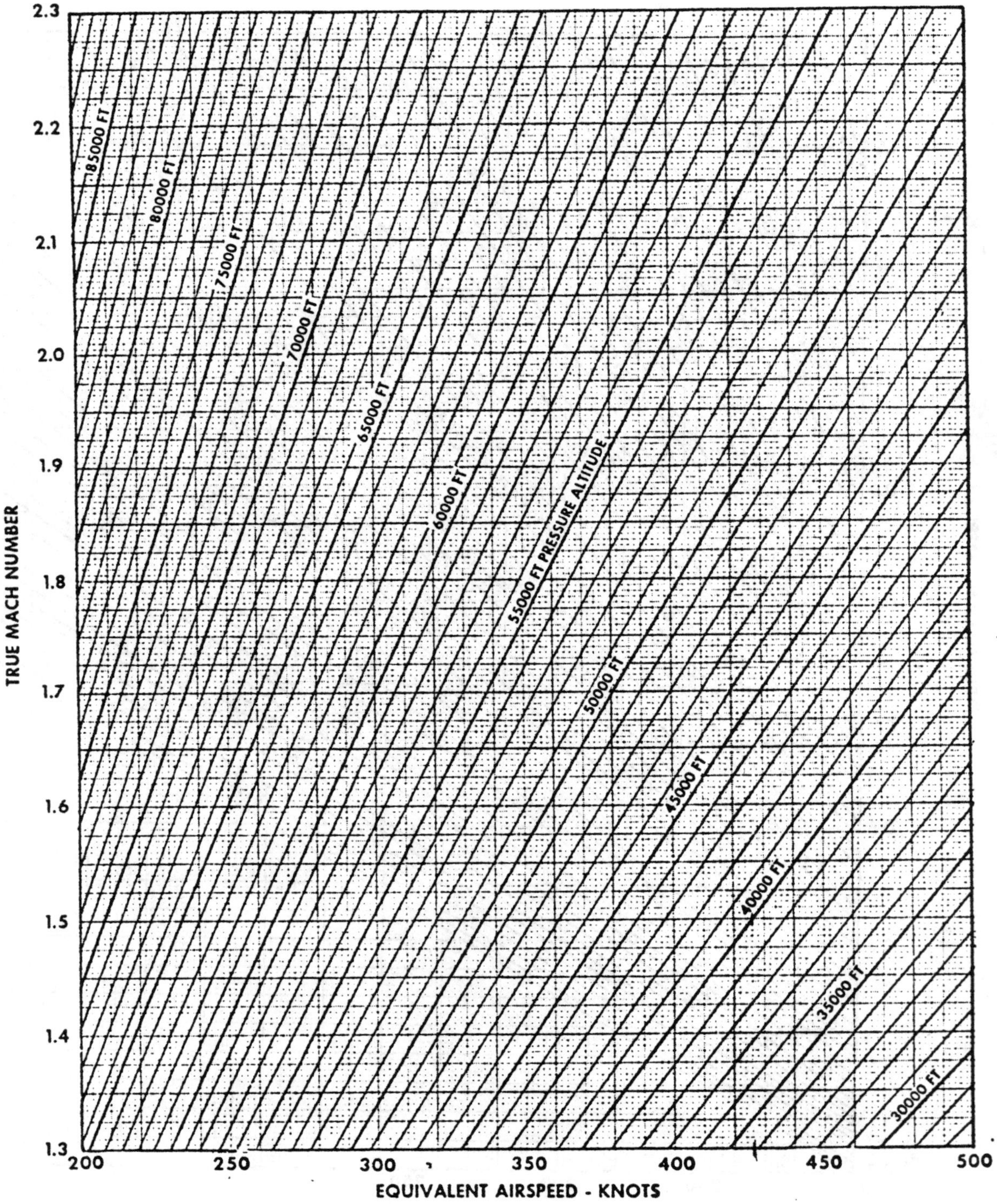

2.3
2.2
2.1
2.0
1.9
1.8
1.7
1.6
1.5
1.4
1.3

85000 FT
80000 FT
75000 FT
70000 FT
65000 FT
60000 FT
55000 FT PRESSURE ALTITUDE
50000 FT
45000 FT
40000 FT
35000 FT
30000 FT

200 250 300 350 400 450 500

EQUIVALENT AIRSPEED - KNOTS

Figure AT-5 (Sheet 2 of 3)

UE MACH NUMBER VS EQUIVALENT AIRSPEED

BASED ON γ OF 1.4

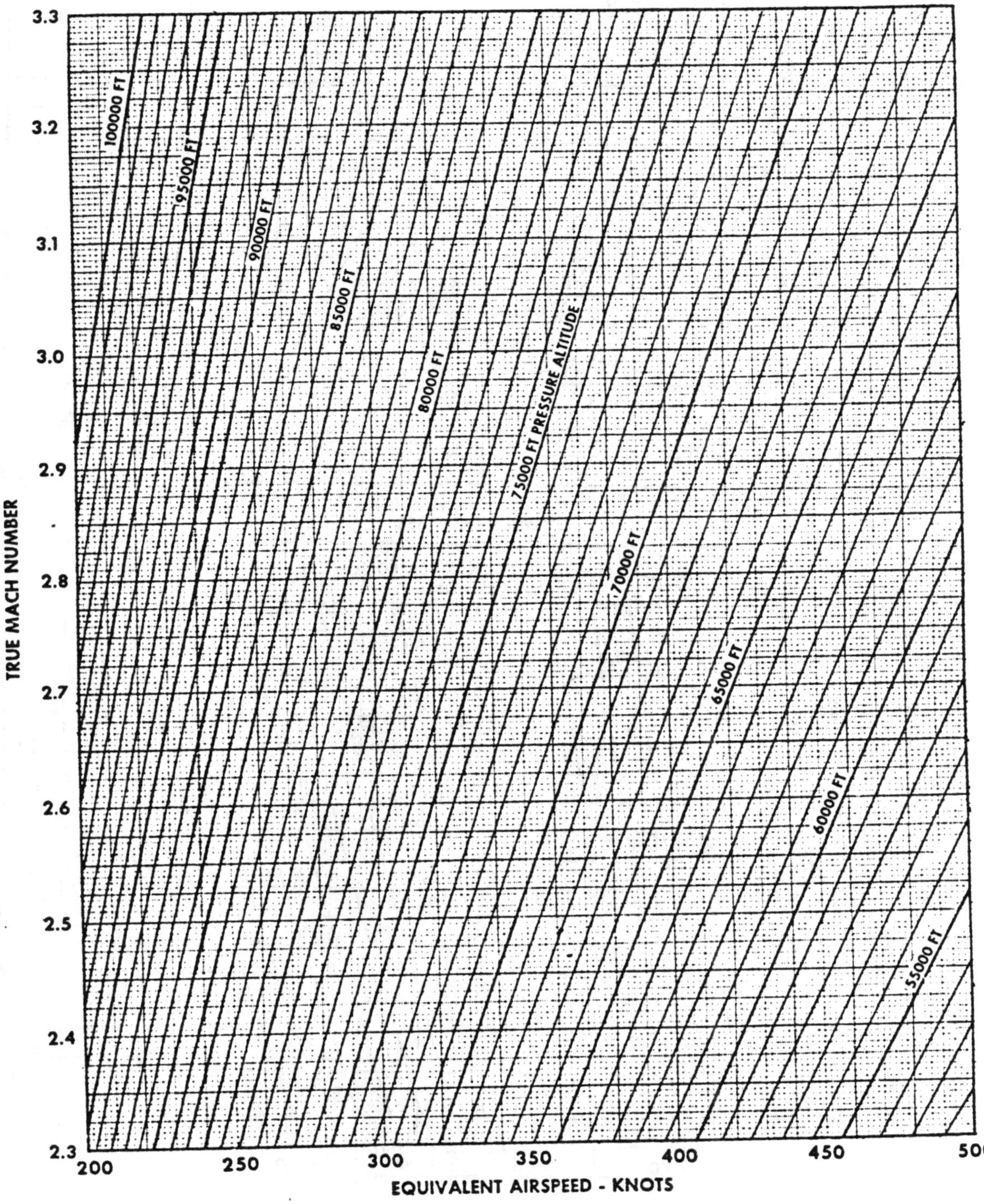

Figure A1-5 (Sheet 3 of 3)

MACH-AIRSPEED-TEMPERATURE CONVERSION

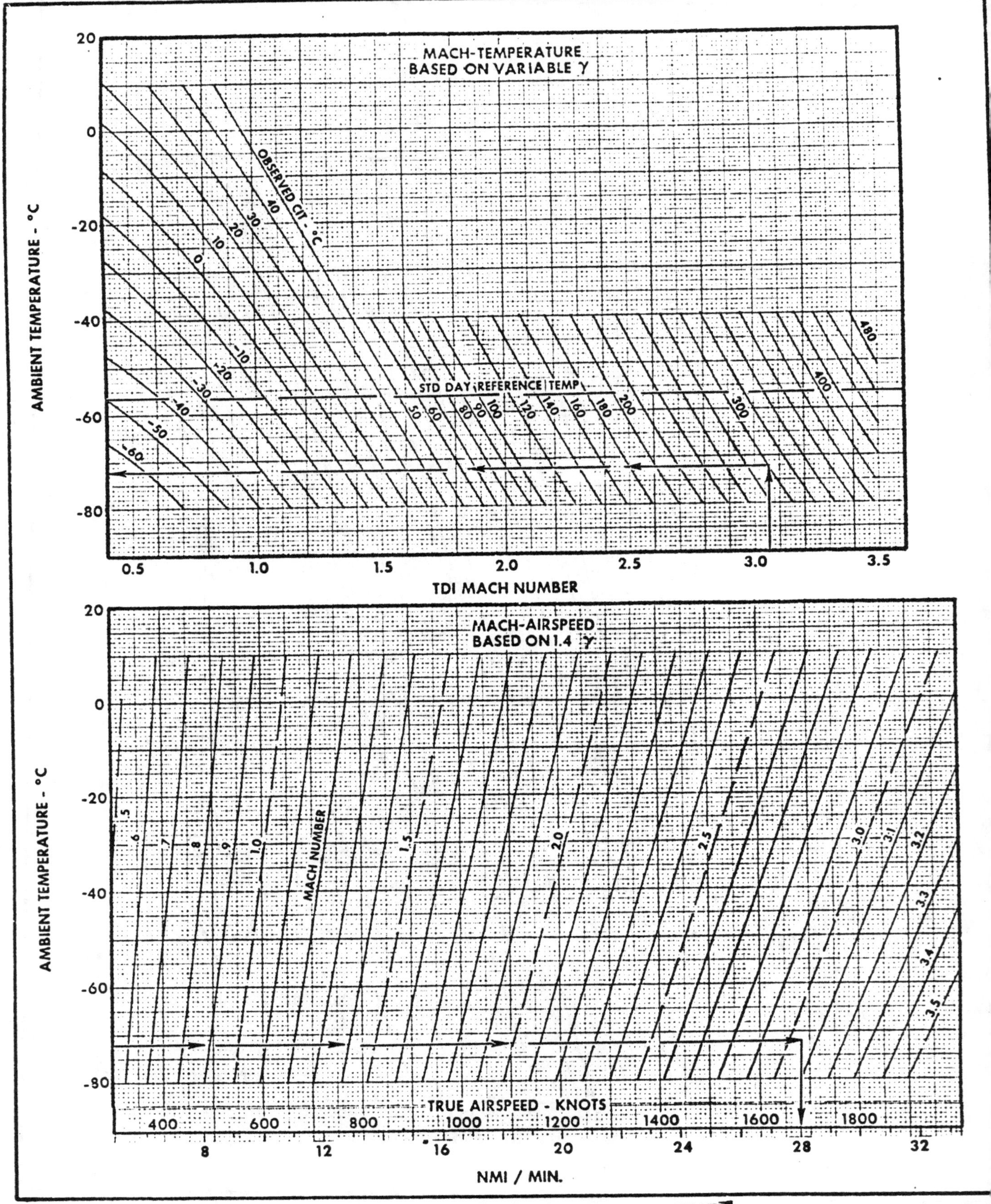

MACH-TEMPERATURE
BASED ON VARIABLE γ

AMBIENT TEMPERATURE - °C

20
0
-20
-40
-60
-80

OBSERVED CIT - °C

40
30
20
10
0
-10
-20
-30
-40
-50
-60

STD DAY REFERENCE TEMP

50
60
80
90
100
120
140
160
180
200
300
400
480

0.5 1.0 1.5 2.0 2.5 3.0 3.5

TDI MACH NUMBER

MACH-AIRSPEED
BASED ON 1.4 γ

AMBIENT TEMPERATURE - °C

20
0
-20
-40
-60
-80

MACH NUMBER

.5 .6 .7 .8 .9 1.0 1.5 2.0 2.5 3.0 3.1 3.2 3.3 3.4 3.5

TRUE AIRSPEED - KNOTS

400 600 800 1000 1200 1400 1600 1800

8 12 16 20 24 28 32

NMI / MIN.

Figure A1-6

STANDARD ATMOSPHERE TABLE

ARDC MODEL ATMOSPHERE – 1956

Alt ft	Temp. t °F	Temp. t °C	Press. P in. Hg	Press. P lb/ft²	$\frac{\rho}{\rho_0}=\sigma$	$\sqrt{\frac{\rho}{\rho_0}}$	$\frac{P}{P_0}=\delta$	$\frac{T}{T_0}=\theta$	$\sqrt{\frac{T}{T_0}}$	$\frac{P}{P_0}\sqrt{\frac{T}{T_0}}$	c ft/sec	Alt ft
0	59.0	15.0	29.92	2116	1.000	1.000	1.000	1.000	1.000	1.000	1117	0
1000	55.4	13.0	28.86	2041	.9711	.9854	.9644	.9931	.9965	.9610	1113	1000
2000	51.9	11.0	27.82	1968	.9428	.9710	.9298	.9862	.9931	.9234	1109	2000
3000	48.3	9.1	26.82	1897	.9151	.9566	.8962	.9794	.9896	.8869	1105	3000
4000	44.7	7.1	25.84	1828	.8881	.9424	.8637	.9725	.9862	.8518	1101	4000
5000	41.2	5.1	24.90	1761	.8617	.9283	.8320	.9656	.9826	.8175	1098	5000
6000	37.6	3.1	23.98	1696	.8359	.9143	.8014	.9587	.9791	.7847	1094	6000
7000	34.0	+1.1	23.09	1633	.8106	.9003	.7716	.9519	.9757	.7529	1090	7000
8000	30.5	−0.8	22.22	1572	.7860	.8866	.7428	.9450	.9721	.7221	1086	8000
9000	26.9	−2.8	21.39	1513	.7620	.8729	.7148	.9381	.9686	.6924	1082	9000
10000	23.3	−4.8	20.58	1455	.7385	.8594	.6877	.9312	.9650	.6636	1078	10000
11000	19.8	−6.8	19.79	1400	.7156	.8459	.6614	.9244	.9615	.6360	1074	11000
12000	16.2	−8.8	19.03	1346	.6932	.8326	.6360	.9175	.9579	.6092	1070	12000
13000	12.6	−10.8	18.29	1294	.6713	.8193	.6113	.9106	.9543	.5834	1066	13000
14000	9.1	−12.7	17.58	1243	.6500	.8062	.5875	.9037	.9506	.5585	1062	14000
15000	5.5	−14.7	16.89	1194	.6292	.7932	.5643	.8969	.9470	.5344	1058	15000
16000	+1.9	−16.7	16.22	1147	.6090	.7804	.5420	.8900	.9434	.5113	1054	16000
17000	−1.6	−18.7	15.57	1101	.5892	.7676	.5203	.8831	.9397	.4889	1050	17000
18000	−5.2	−20.7	14.94	1057	.5699	.7549	.4994	.8762	.9361	.4675	1045	18000
19000	−8.8	−22.6	14.34	1014	.5511	.7424	.4791	.8694	.9324	.4467	1041	19000
20000	−12.3	−24.6	13.75	972.5	.5328	.7299	.4595	.8625	.9287	.4267	1037	20000
21000	−15.9	−26.6	13.18	932.4	.5150	.7176	.4406	.8556	.9250	.4076	1033	21000
22000	−19.5	−28.6	12.64	893.7	.4976	.7054	.4223	.8487	.9212	.3890	1029	22000
23000	−23.0	−30.6	12.11	856.3	.4806	.6933	.4046	.8419	.9176	.3713	1025	23000
24000	−26.6	−32.5	11.60	820.2	.4642	.6813	.3876	.8350	.9138	.3542	1021	24000
25000	−30.2	−34.5	11.10	785.3	.4481	.6694	.3711	.8281	.9100	.3377	1016	25000
26000	−33.7	−36.5	10.63	751.6	.4325	.6576	.3552	.8212	.9062	.3219	1012	26000
27000	−37.3	−38.5	10.17	719.1	.4173	.6460	.3398	.8144	.9024	.3066	1008	27000
28000	−40.9	−40.5	9.725	687.8	.4025	.6344	.3250	.8075	.8986	.2920	1004	28000
29000	−44.4	−42.5	9.297	657.6	.3881	.6230	.3107	.8006	.8948	.2780	999.4	29000
30000	−48.0	−44.4	8.885	628.4	.3741	.6116	.2970	.7937	.8909	.2646	995.1	30000
31000	−51.6	−46.4	8.488	600.3	.3605	.6004	.2837	.7869	.8871	.2517	990.7	31000
32000	−55.1	−48.4	8.106	573.3	.3473	.5893	.2709	.7800	.8832	.2393	986.4	32000
33000	−58.7	−50.4	7.737	547.2	.3345	.5784	.2586	.7731	.8793	.2274	982.0	33000
34000	−62.2	−52.4	7.382	522.1	.3220	.5675	.2467	.7662	.8753	.2159	977.7	34000
35000	−65.8	−54.3	7.041	498.0	.3099	.5568	.2353	.7594	.8714	.2050	973.3	35000
36089	−69.7	−56.5	6.683	472.7	.2971	.5450	.2234	.7519	.8671	.1937	968.5	36089
37000	−69.7	−56.5	6.397	452.4	.2844	.5333	.2138	.7519	.8671	.1854	968.5	37000
38000	−69.7	−56.5	6.097	431.2	.2710	.5206	.2038	.7519	.8671	.1767	968.5	38000
39000	−69.7	−56.5	5.811	411.0	.2583	.5082	.1942	.7519	.8671	.1684	968.5	39000
40000	−69.7	−56.5	5.538	391.7	.2462	.4962	.1851	.7519	.8671	.1605	968.5	40000
41000	−69.7	−56.5	5.278	373.3	.2346	.4844	.1764	.7519	.8671	.1530	968.5	41000
42000	−69.7	−56.5	5.030	355.8	.2236	.4729	.1681	.7519	.8671	.1458	968.5	42000
43000	−69.7	−56.5	4.794	339.1	.2131	.4616	.1602	.7519	.8671	.1389	968.5	43000
44000	−69.7	−56.5	4.569	323.2	.2031	.4507	.1527	.7519	.8671	.1324	968.5	44000
45000	−69.7	−56.5	4.355	308.0	.1936	.4400	.1455	.7519	.8671	.1262	968.5	45000
46000	−69.7	−56.5	4.151	293.6	.1845	.4295	.1387	.7519	.8671	.1203	968.5	46000
47000	−69.7	−56.5	3.956	279.8	.1758	.4193	.1322	.7519	.8671	.1146	968.5	47000
48000	−69.7	−56.5	3.770	265.7	.1676	.4094	.1260	.7519	.8671	.1093	968.5	48000
49000	−69.7	−56.5	3.593	254.1	.1597	.3996	.1201	.7519	.8671	.1041	968.5	49000
50000	−69.7	−56.5	3.425	242.2	.1522	.3901	.1145	.7519	.8671	.09928	968.5	50000
51000	−69.7	−56.5	3.264	230.8	.1451	.3809	.1091	.7519	.8671	.09460	968.5	51000
52000	−69.7	−56.5	3.111	220.0	.1383	.3719	.1040	.7519	.8671	.09018	968.5	52000
53000	−69.7	−56.5	2.965	209.7	.1318	.3630	.09909	.7519	.8671	.08592	968.5	53000
54000	−69.7	−56.5	2.826	199.8	.1256	.3544	.09444	.7519	.8671	.08189	968.5	54000
55000	−69.7	−56.5	2.693	190.5	.1197	.3460	.09001	.7519	.8671	.07805	968.5	55000
56000	−69.7	−56.5	2.567	181.5	.1141	.3378	.08578	.7519	.8671	.07438	968.5	56000
57000	−69.7	−56.5	2.446	173.0	.1087	.3297	.08176	.7519	.8671	.07089	968.5	57000
58000	−69.7	−56.5	2.331	164.9	.1036	.3219	.07792	.7519	.8671	.06756	968.5	58000
59000	−69.7	−56.5	2.222	157.2	.09877	.3143	.07426	.7519	.8671	.06439	968.5	59000
60000	−69.7	−56.5	2.118	149.8	.09414	.3068	.07078	.7519	.8671	.06137	968.5	60000
61000	−69.7	−56.5	2.018	142.8	.08972	.2995	.06746	.7519	.8671	.05849	968.5	61000
62000	−69.7	−56.5	1.924	136.1	.08551	.2924	.06429	.7519	.8671	.05575	968.5	62000
63000	−69.7	−56.5	1.833	129.7	.08150	.2855	.06127	.7519	.8671	.05313	968.5	63000
64000	−69.7	−56.5	1.747	123.6	.07767	.2787	.05840	.7519	.8671	.05064	968.5	64000
65000	−69.7	−56.5	1.665	117.8	.07403	.2721	.05566	.7519	.8671	.04826	968.5	65000
66000	−69.7	−56.5	1.587	112.3	.07055	.2656	.05306	.7519	.8671	.04600	968.5	66000
67000	−69.7	−56.5	1.513	107.0	.06724	.2593	.05056	.7519	.8671	.04384	968.5	67000
68000	−69.7	−56.5	1.442	102.0	.06409	.2532	.04819	.7519	.8671	.04179	968.5	68000
69000	−69.7	−56.5	1.374	97.19	.06108	.2471	.04592	.7519	.8671	.03982	968.5	69000
70000	−69.7	−56.5	1.310	92.63	.05821	.2413	.04377	.7519	.8671	.03795	968.5	70000

Figure A1-7 (Sheet 1 of 2)

STANDARD ATMOSPHERE TABLE

ARDC MODEL ATMOSPHERE — 1956

Alt ft	Temp. t °F	Temp. t °C	Press. P in. Hg	Press. P lb/ft²	$\frac{\rho}{\rho_0} = \sigma$	$\sqrt{\frac{\rho}{\rho_0}}$	$\frac{P}{P_0} = \delta$	$\frac{T}{T_0} = \theta$	$\sqrt{\frac{T}{T_0}}$	$\frac{P}{P_0}\sqrt{\frac{T}{T_0}}$	c ft/sec	Alt ft
71000	-69.7	-56.5	1.248	88.28	.05548	.2355	.04172	.7519	.8671	.03618	968.5	71000
72000	-69.7	-56.5	1.190	84.14	.05288	.2300	.03976	.7519	.8671	.03448	968.5	72000
73000	-69.7	-56.5	1.134	80.19	.05040	.2245	.03789	.7519	.8671	.03285	968.5	73000
74000	-69.7	-56.5	1.081	76.43	.04803	.2192	.03611	.7519	.8671	.03131	968.5	74000
75000	-69.7	-56.5	1.030	72.84	.04578	.2140	.03442	.7519	.8671	.02985	968.5	75000
76000	-69.7	-56.5	.9815	69.42	.04363	.2089	.03280	.7519	.8671	.02844	968.5	76000
77000	-69.7	-56.5	.9355	66.16	.04158	.2039	.03127	.7519	.8671	.02711	968.5	77000
78000	-69.7	-56.5	.8916	63.06	.03963	.1991	.02980	.7519	.8671	.02584	968.5	78000
79000	-69.7	-56.5	.8497	60.10	.03777	.1943	.02840	.7519	.8671	.02463	968.5	79000
80000	-69.7	-56.5	.8099	57.28	.03600	.1897	.02707	.7519	.8571	.02347	968.5	80000
81000	-69.7	-56.5	.7718	54.59	.03431	.1852	.02580	.7519	.8671	.02237	968.5	81000
82021	-69.7	-56.5	.7349	51.98	.03267	.1807	.02456	.7519	.8671	.02130	968.5	82021
83000	-68.1	-55.6	.7012	49.59	.03104	.1762	.02343	.7550	.8689	.02036	970.5	83000
84000	-66.4	-54.7	.6685	47.28	.02947	.1717	.02234	.7582	.8607	.01923	972.5	84000
85000	-64.8	-53.8	.6374	45.08	.02798	.1673	.02130	.7613	.8725	.01858	974.5	85000
86000	-63.2	-52.9	.6079	43.00	.02658	.1630	.02032	.7645	.8744	.01777	976.6	86000
87000	-61.5	-51.9	.5799	41.02	.02525	.1589	.01938	.7677	.8762	.01698	978.6	87000
88000	-59.9	-51.1	.5533	39.13	.02399	.1549	.01849	.7708	.8780	.01623	980.6	88000
89000	-58.2	-50.1	.5280	37.45	.02280	.1510	.01765	.7740	.8798	.01553	982.6	89000
90000	-56.6	-49.2	.5040	35.65	.02167	.1472	.01684	.7772	.8816	.01485	984.6	90000
91000	-54.9	-48.3	.4811	34.03	.02061	.1436	.01608	.7804	.8834	.01421	986.6	91000
92000	-53.3	-47.4	.4594	32.49	.01960	.1400	.01535	.7835	.8852	.01359	988.6	92000
93000	-51.6	-46.4	.4387	31.03	.01864	.1365	.01466	.7867	.8870	.01300	990.6	93000
94000	-50.0	-45.6	.4191	29.64	.01773	.1332	.01401	.7899	.8888	.01245	992.6	94000
95000	-48.3	-44.6	.4004	28.32	.01687	.1299	.01338	.7931	.8906	.01192	994.6	95000
96000	-46.7	-43.7	.3826	27.06	.01606	.1267	.01279	.7962	.8923	.01141	996.6	96000
97000	-45.1	-42.8	.3656	25.86	.01529	.1236	.01222	.7994	.8941	.01093	998.6	97000
98000	-43.4	-41.9	.3495	24.72	.01455	.1206	.01168	.8026	.8959	.01046	1001	98000
99000	-41.8	-41.0	.3341	23.63	.01386	.1177	.01117	.8058	.8977	.01003	1003	99000
100000	-40.1	-40.1	.3195	22.60	.01320	.1149	.01068	.8089	.8994	.009606	1005	100000
110000	-23.7	-30.9	.2062	14.58	.008196	.09053	.006890	.8407	.9170	.006318	1024	110000
120000	-7.19	-21.8	.1352	9.561	.005179	.07197	.004518	.8724	.9340	.004220	1043	120000
130000	+9.27	-12.6	.09000	6.365	.003327	.05768	.003008	.9041	.9508	.002860	1062	130000
140000	25.7	-3.50	.06076	4.297	.002170	.04658	.002031	.9359	.9674	.001965	1081	140000
150000	42.2	+5.67	.04156	2.940	.001436	.03789	.001389	.9676	.9837	.001366	1099	150000
200000	-6.78	-21.5	.00621	0.440	.000238	.01542	.000208	.8732	.9345	.000194	1044	200000

ATMOSPHERIC STANDARDS

	English	Metric
Gravity	32.17405 ft/sec²	9.80665 m/sec²
Absolute zero	–459.688°F	–273.16°C
Standard Values at Sea Level		
Pressure	29.92 in. Hg	760 mm Hg
Pressure	2116 lb/ft²	10332 kg/m²
Temp	59°F	15°C
Abs temp	518.688°R	288.16°K
Specific wt $g\rho_0$	0.076475 lb/ft³	1.2250 kg/m³
Density	0.0023769 lb sec²/ft⁴	0.12492 kg sec²/m⁴
Standard Values at Altitude		
Isothermal alt H	36,089.2 ft	11,000 m
Isothermal temp	–69.7°F	–56.5°C

GENERAL PROPERTIES OF GASES

$Pv = RT$ or $P = \rho gRT$ or $PV = mRT$

Constant Volume $P_1/P_2 = T_1/T_2$

Constant Pressure $V_1/V_2 = T_1/T_2$

Constant Temperature $P_1/P_2 = V_2/V_1$

Reversible Adiabatic $\frac{P_1}{P_2} = \left(\frac{V_2}{V_1}\right)^{\gamma}$

$$\frac{T_1}{T_2} = \left(\frac{V_2}{V_1}\right)^{\gamma-1} = \left(\frac{P_1}{P_2}\right)^{\frac{\gamma-1}{\gamma}}$$

Polytropic $P_1V_1^n = P_2V_2^n$

$$\frac{T_1}{T_2} = \left(\frac{V_1}{V_2}\right)^{1-n} = \left(\frac{P_1}{P_2}\right)^{\frac{n-1}{n}}$$

Figure A1-7 (Sheet 2 of 2)

STANDARD UNITS CONVERSION CHART

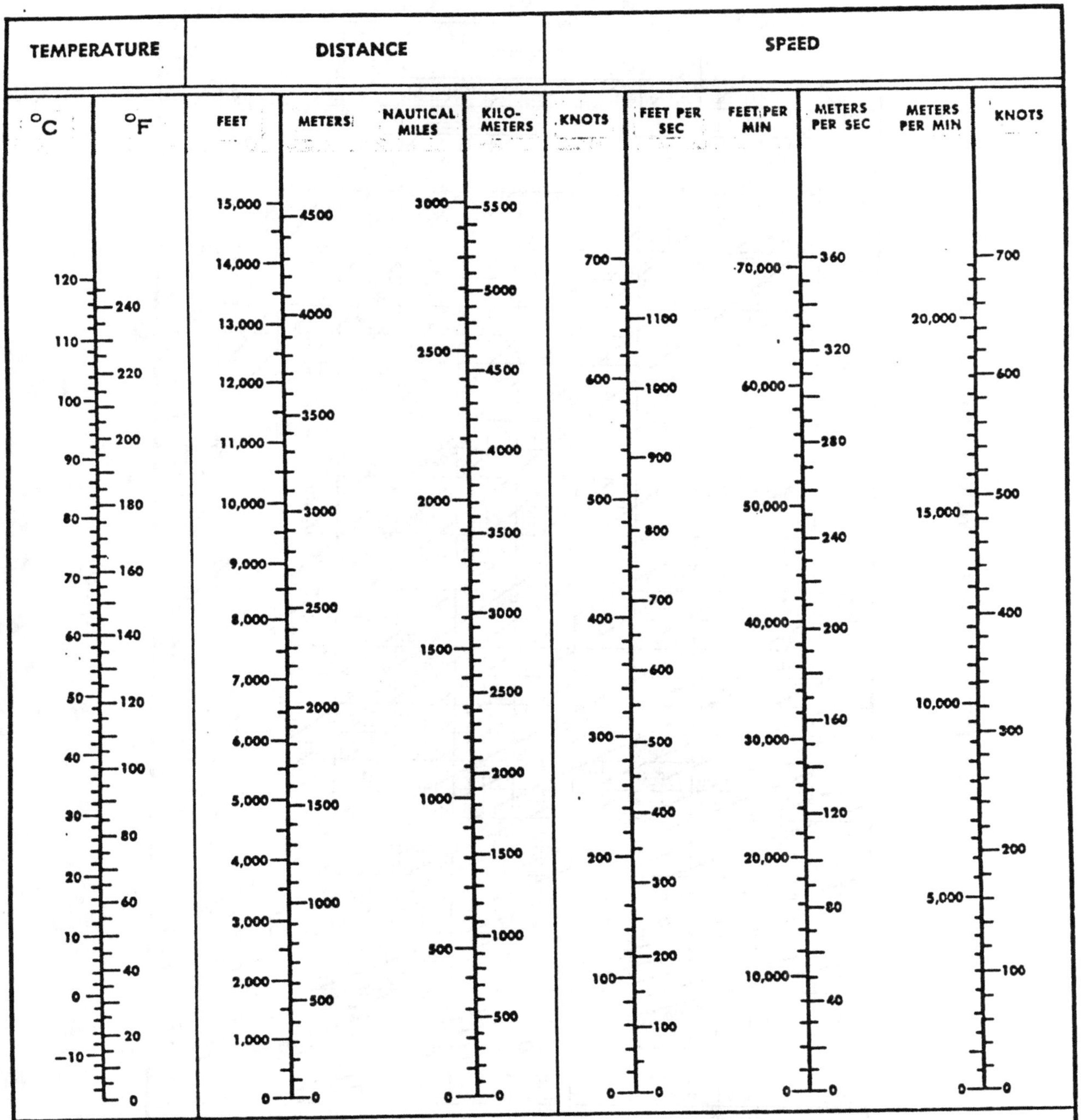

TEMPERATURE		DISTANCE				SPEED					
°C	°F	FEET	METERS	NAUTICAL MILES	KILO-METERS	KNOTS	FEET PER SEC	FEET PER MIN	METERS PER SEC	METERS PER MIN	KNOTS
120	240	15,000	4500	3000	5500	700	1100	70,000	360	20,000	700
110	220	14,000	4000	2500	5000	600	1000	60,000	320	15,000	600
100	200	13,000	3500	2000	4500	500	900	50,000	280	10,000	500
90	180	12,000	3000	1500	4000	400	800	40,000	240	5,000	400
80	160	11,000	2500	1000	3500	300	700	30,000	200	0	300
70	140	10,000	2000	500	3000	200	600	20,000	160		200
60	120	9,000	1500	0	2500	100	500	10,000	120		100
50	100	8,000	1000		2000	0	400	0	80		0
40	80	7,000	500		1500		300		40		
30	60	6,000	0		1000		200		0		
20	40	5,000			500		100				
10	20	4,000			0		0				
0	0	3,000									
−10		2,000									
		1,000									
		0									

NOTE:

TO OBTAIN U S GALLONS MULTIPLY LITERS BY 0.264

TO OBTAIN IMPERIAL GALLONS MULTIPLY LITERS BY 0.220

TO OBTAIN INCHES OF MERCURY MULTIPLY MILLIBARS BY .0255

TO OBTAIN POUNDS MULTIPLY KILOGRAMS BY 2.20

Figure A1-8

W/δ AS A FUNCTION OF GROSS WEIGHT AND ALTITUDE

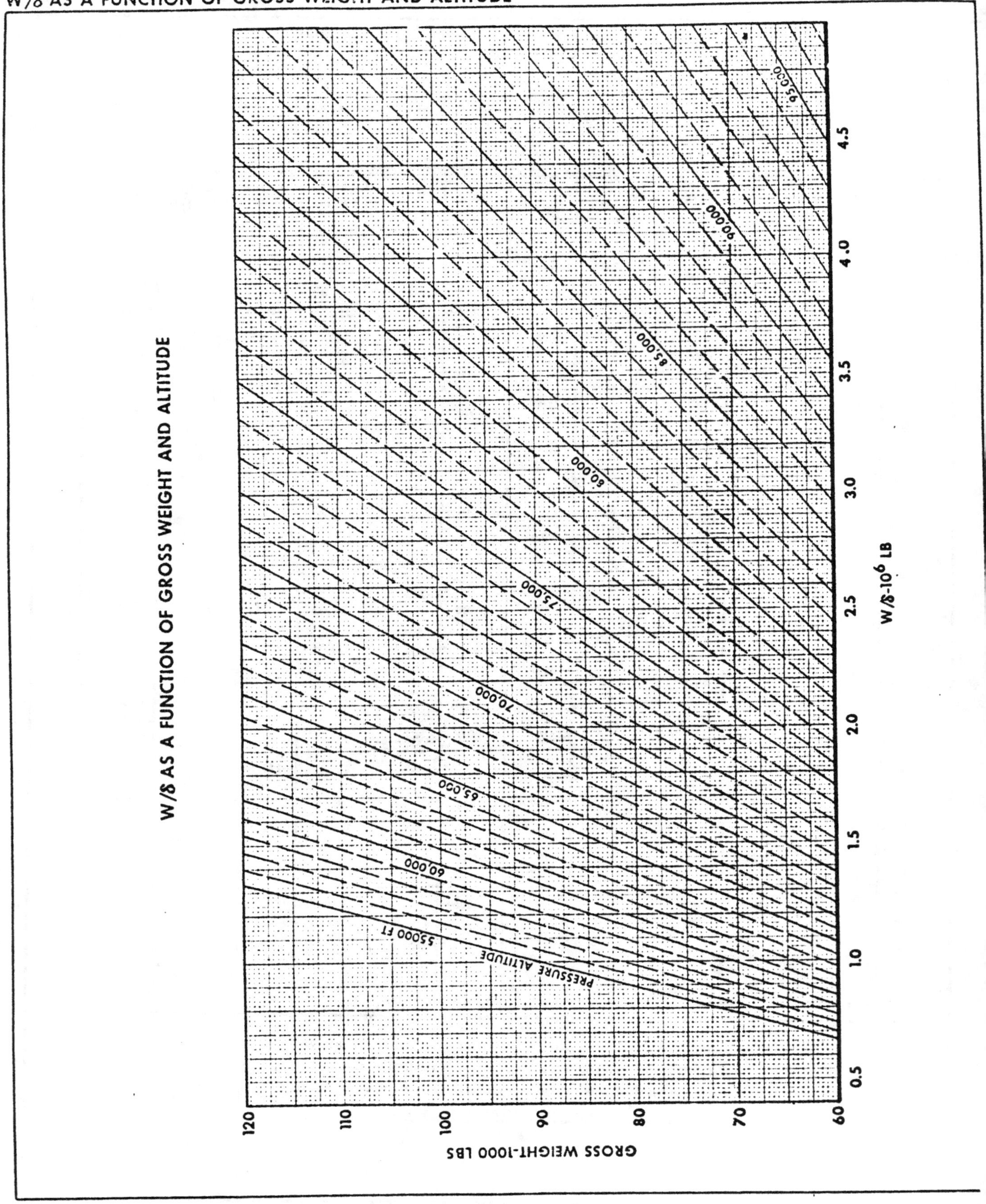

Figure A1-9

TURNING PERFORMANCE

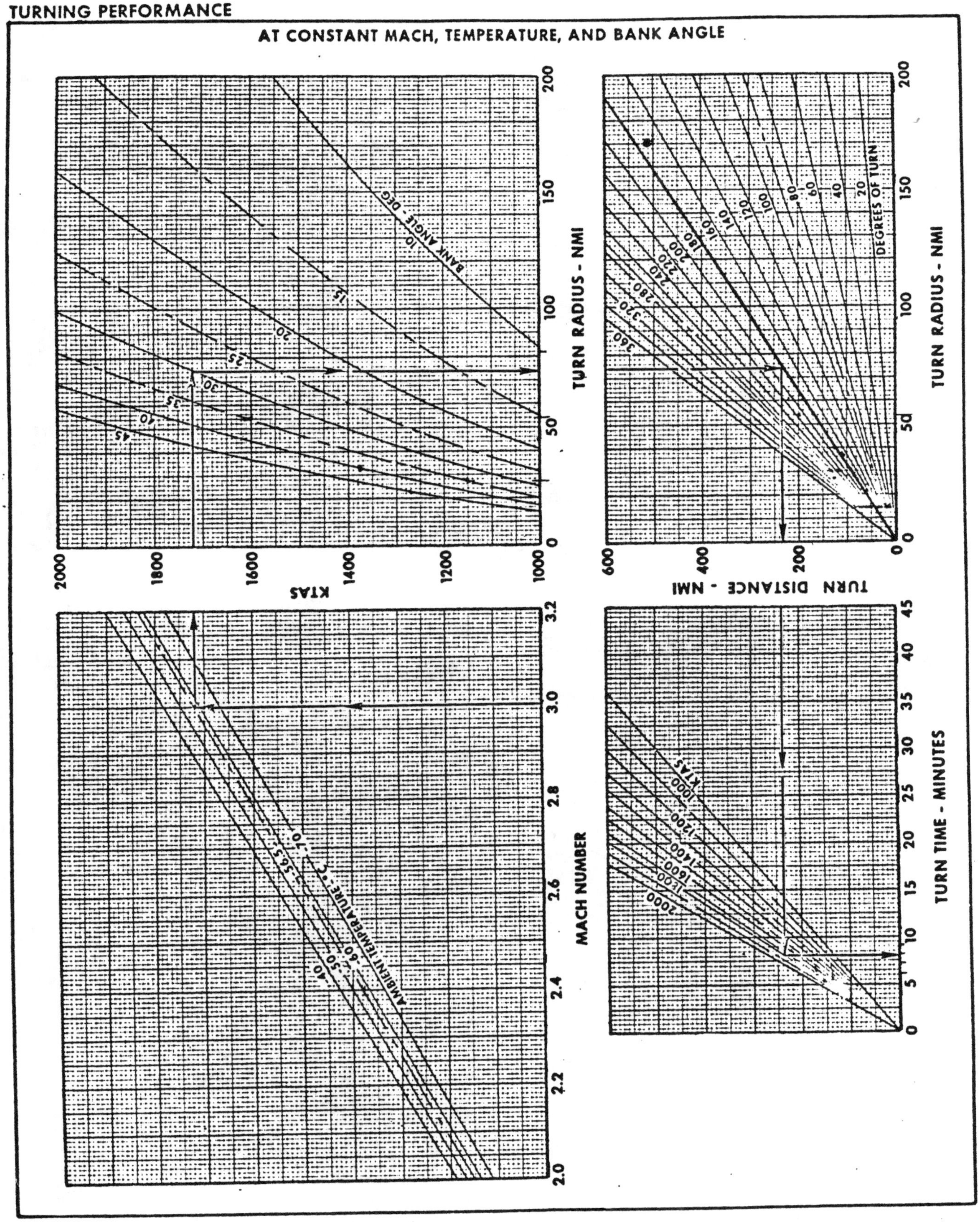

AT CONSTANT MACH, TEMPERATURE, AND BANK ANGLE

Figure A1-10

PART II

FIELD LENGTH REQUIREMENTS

TABLE OF CONTENTS

LIST OF ILLUSTRATIONS

PART II

FIELD LENGTH REQUIREMENTS

INTRODUCTION

This part of the appendix contains data for determining takeoff and landing field length requirements. Charts are also included for computing takeoff acceleration check speed, takeoff refusal speed, single engine climb-out capability, and landing final approach and touchdown speeds. Coverage is provided for a normal operating range of temperature, altitude, gross weight, wind, and runway slope.

Data Basis

All data in this part of the appendix, with the exception of the single-engine climb out information, is based on flight tests. Stopping distances are based on:

Brakes - 7-5-7 rotor type.

Drag chute - 210 knot (-501) drag chute and drag chute failed condition.

Runway surface - Braking performance on dry or wet, grooved or ungrooved runways.

Crosswind Component and Rated Tire Speed Chart

Figure A2-1 provides a means of converting reported wind velocity and direction into runway and crosswind components. The lower portion of the chart may be used to determine whether or not takeoff and landing speeds will exceed rated tire speed.

TAKEOFF PERFORMANCE

Charts are furnished for normal ana maximum performance takeoff procedures. The information supplied by the charts may be abridged in the field and added to the Pilot's and RSO's checklists.

Normal Performance Takeoff

Normal performance takeoff distances are based on rotation at 180 KEAS and liftoff at 210 KEAS. However, rotation at speeds slightly less than 180 KEAS will have no appreciable affect on the takeoff distances shown. At gross weights of less than 110,000 lb., the 210 KEAS takeoff speed may be exceeded if rotation is not begun befc reaching 180 KEAS.

Example:

For a field pressure altitude of 2300 ft., ambient temperature of 86°F (30°C), and a takeoff gross weight of 125,000 lb, determine the Normal Performance takeoff distance with a 10 knot headwind and zero runway slope. Refer to figure A2-2. (Takeoff distance is 6300 feet.)

Intermediate Speed Takeoff

An "intermediate speed" takeoff pr cedure may be desirable when the of the normal takeoff procedure would require excessive runway length, or would approach rated tire speed. Whe

limited by runway length, reduce rotation and takeoff speeds equally by one knot for each percent reduction in takeoff distance desired. When limited by tailwind component, reduce rotation and takeoff speeds equally so as to takeoff within the tire speed rating. Reduce normal performance takeoff distances 1% for each knot reduction in takeoff speed.

Note

Do not reduce takeoff speed to a value less than the maximum performance takeoff speed schedule.

Maximum Performance Takeoff

The maximum performance takeoff procedure will result in minimum takeoff distance. Takeoff speed is scheduled according to gross weight and corresponds to a lift coefficient of 0.60. Since tail clearance at lift-off will be minimum, this procedure is recommended only when required by field length and/or tailwind conditions.

Figure A2-3 presents maximum performance takeoff distances. Takeoff speeds are listed with the corresponding gross weights. Rotation speeds are also shown, based on 4.5 seconds from rotation to lift-off.

Example:

For a field pressure altitude of 2300 ft., ambient temperature of 86°F (30°C) and a takeoff gross weight of 125,000 lb, determine the maximum performance takeoff distance, rotation speed, and takeoff speed with a 10 knot headwind and zero runway slope. Refer to figure A2-3. (Rotation speed is 170 KEAS. Takeoff speed is 196 KEAS. Takeoff distance is 5500 feet.)

TAKEOFF FUEL ALLOWANCE

Fuel load at engine start should be adjusted to allow for fuel consumption prior to takeoff. A value of 110 lb per minute may be used for planning purposes. Fuel consumption during the takeoff run is approximately 1000 pounds.

ACCELERATION CHECK

Figure A2-4 provides a means of checking takeoff acceleration performance during the first part of the takeoff ground run. The check is normally made in terms of indicated airspeed reached after a run of 3000 feet. However, any combination of speed and acceleration check distance can be used.

Takeoff speed and the corresponding takeoff distance are used to establish a guide line on the acceleration check speed chart. The acceleration check speed corresponding to the desired check distance is read at the bottom of the chart. (A 5 knot negative tolerance has been included to provide for less than normal acceleration.)

Example:

For the Normal Performance takeoff example, i.e., 2300 ft. altitude, 86°F temperature, 125,000 lb gross weight, 10 knot headwind, and zero runway slope, find the acceleration check speed corresponding to 3000 feet during the takeoff run. Refer to figure A2-4. (Acceleration check speed is 140 KEAS.)

REFUSAL SPEED

Refusal speed is the maximum speed from which the aircraft can be brought to a stop in a given accelerate-stop distance, usually the available runway length. Maximum refusal speed, when brake energy limited, is the maximum speed from which the aircraft can be brought to a stop without exceeding the energy capacity of the brakes. Either the scheduled rotation speed or the refusal speed with drag chute, whichever is less, is the refusal speed which should be listed on the TOL data card.

Refusal speeds on dry and wet runways, with and without drag chute, may be obtained from figures A2-5, A2-7, and A2-8. The accelerate-stop distance to be used with the charts is the runway length available at the start of the takeoff run. Refusal speeds are given as a function of RCR, ambient temperature, pressure altitude, and gross weight. Brake energy capacity is assumed to be 90% of full rated one-stop capability, thus allowing some normal use prior to a refused takeoff. It is further assumed that brake application is not delayed when airspeed is above the rated brake energy speed (figure 5-8, Section V). Therefore, the maximum refusal speed, brake energy limited charts should always be used in conjunction with the refusal speed charts to ensure that brake energy capacity is not exceeded.

Note

Brake application should not be delayed if it is necessary to abort from a speed greater than the rated brake energy speed. The increased distance traveled at high speed will increase the risk of tire failure, which in turn will decrease the brake energy capacity available for stopping.

Maximum refusal speeds, when brake energy limited, for dry and wet runways, with and without drag chute, may be obtained from figures A2-6 and A2-9. (On a wet runway, with anti-skid "ON" and drag chute deployed, the brakes do not absorb enough energy to exceed the brake energy rating. Therefore no data is presented for this condition.) For abort conditions where brake burn-out will occur before a stop can be made, either the takeoff should be continued, if possible, or a barrier engagement should be anticipated. Various factors may contribute to less than optimum performance, such as blown tires, delayed drag chute deployment, etc.

Refusal Speed With Drag Chute

The refusal speed with drag chute charts are based on operation with the -501, 210 KEAS deploy speed drag chute and 34 K engines.

The assumptions made in calculating refusal speeds with drag chute are:

1. Normal rate of acceleration is continued to the refusal speed. At that point, complete and instantaneous loss of one engine occurs. (If the abort is due to some other cause than engine failure, the right engine should be shutdown.)

2. Maximum afterburning thrust is continued on the operating engine for three seconds, at which time the throttle is retarded to idle, and brake application is initiated. Rotation is not attempted when a takeoff emergency occurs before reaching rotation speed, even though rotation speed may be exceeded during the recognition and action period.

3. Braking torque is obtained one second after brake application. Drag chute deployment is initiated one second after brake application and deployment time is 4.75 seconds.

4. Optimum wheel braking with continuous anti-skid cycling, is used until the aircraft is stopped.

5. The drag chute is jettisoned at 60 knots on a dry runway, and is retained to full stop on a wet runway. (The chute should always be retained until a stop is assured.)

6. Hard surface runway.

7. Zero wind and zero slope.

Example:

For a field pressure altitude of 2300 ft., temperature of 86°F, and gross weight of 125,000 lb. find the refusal speed with drag chute for a 12,000 ft accelerate and stop distance. The runway is dry (RCR = 23). Refer to figures A2-5 and A2-6. (Refusal speed from figure A2-5 is 179 KEAS. Maximum refusal speed if brake energy limited, from figure A2-6, would be 221 KEAS.)

Refusal Speed Without Drag Chute

The assumptions made in calculating refusal speeds without drag chute are the same as above except:

1. Drag chute failure recognition time is normal chute deploy time of 4.75 seconds plus 3.0 seconds.

2. When on a wet runway, up-elevon drag should be used after recognition of chute failure, although not assumed in the calculations. (Nosedown attitude must be maintained.)

Example:

Find the refusal speed without drag chute for the conditions of the previous example. Refer to figures A2-8 and A2-9. (Refusal speed from figure A2-8 is 161 KEAS. Maximum refusal speed if brake energy limited from figure A2-9 would be 177 KEAS.

Note

Use of the lesser of the two speeds determined by figures A2-8 and A2-9 if refusal speed without drag chute is to be listed on the TOL data card.

SINGLE ENGINE CLIMB CAPABILITY

In the event of engine failure after lift-off, marginal control during climb-out will be available if immediate corrective action is taken. Takeoff speeds are higher than minimum single engine control speeds for steady flight. (Refer to Takeoff Emergencies, Section III.)

Figures A2-10 and A2-11 show estimated minimum speeds for level flight with one engine inoperative, with gear down or retracted, in or out of ground effect. (Ground effect extends up to approximately one-half wing span above the ground.) The charts include the effects of pressure altitude, temperature and gross weight. One engine is assumed to be windmilling and the other engine developing maximum afterburning thrust. The drag of the necessary rudder and elevon deflections to maintain a straight flight track is included. The data are based on wind tunnel tests with J engines, and with the c.g. at 24% MAC.

Note

A negative weight increment of 20,000 lb has been applied to original estimates of performance capability, and gross weights shown on the charts include this increment. This adjustment was made on the basis of a preliminary analysis of SR-71 Category II single engine tests.

Gross weights shown on the charts should be increased by 1500 lb for each percent that c.g. is forward of 24% MAC. A nominal increase in rate of climb of 50 feet per minute can be expected for each 1000 lb that the airplane is below the chart weight for a given speed, provided the aircraft is not accelerating or turning.

If operating at less than the maximum weight for single engine flight, the available excess thrust can be used for acceleration instead of climbing. The best single engine climb speed with gear down near the ground is approximately 250 KEAS. Since the gear would ordinarily be retracting during an acceleration to this speed, the value represents a target speed for transition to a shallow climb. The best single engine climb speed with gear up, out of ground effect, is above 400 KEAS. However, 300 KEAS provides an adequate capability for single engine climb to pattern altitude.

Results of additional testing will be published when analysis is accomplished.

Example:

Determine the minimum speed for single engine flight with gear up, out of ground effect, at 2300 ft. altitude, 86°F, and 125,000 lb gross weight. Refer to figure A2-11. (Minimum speed for single engine flight is 249 KEAS.)

TAKEOFF PLANNING

Information obtained from the takeoff and landing distance charts should be summarized on the checklist Takeoff and Landing Data Card prior to flight in addition to any other items which may affect performance.

LANDING FIELD LENGTH REQUIREMENTS

Landing Speed schedules and landing distance information are provided for various Runway Condition Readings (RCR) with and without drag chute, with 7-5-7 rotor brake installations, and for normal and maximum performance landing procedures.

Note

This landing performance data can be used for runways with or without transverse grooving.

APPROACH AND LANDING SPEED SCHEDULES

Figure A2-12 shows the normal final approach and landing speed schedules. Normal final approach and landing speeds are 175 KEAS and 155 KEAS respectively for weights less than 70,000 lb. At 70,000 lb or more, the speeds are a direct function of gross weight. For landings at gross weights of 100,000 lb or more, and for all wet runway landings, use the maximum performance touchdown speed schedule in combination with normal performance final approach speeds. Maximum performance touchdown speeds are 10 knots less than those for normal operation.

Examples:

1. Determine the final approach and landing speeds for normal operation at 75,000 lb gross weight. Refer to figure A2-12. (Final approach speed is 180 KEAS. Touchdown speed is 160 KEAS.)

2. Determine the final approach and landing speeds for a landing at 110,000 pounds gross weight. (Note that this is a heavy weight condition and that the maximum performance landing speed schedule must be used in conjunction with the normal performance final approach speed schedule.) Refer to figure A2-12. (Final approach speed is 215 KEAS. Touchdown speed is 185 KEAS.)

LANDING DISTANCES

Landing distances are provided by figures A2-13 through A2-18 for normal and maximum performance landing procedures. Operating conditions which are considered for each procedure include: 7-5-7 rotor brake configuration and anti-skid "ON" selected, normal or failed drag chute, and dry or wet runway (grooved or ungrooved). The charts show landing distance as a function of ambient temperature, pressure altitude, gross weight, and Runway Condition Reading (RCR). Wind and runway slope correction grids are included on each chart.

Normal Performance Landing Distance

Normal landing distances with drag chute assume that chute deployment is initiated one second after touchdown and completed within six seconds after touchdown. Brakes are applied after the nose gear is on the runway at 120 KEAS. A one-second allowance is included to develop full braking pressure. The chute is jettisoned at 60 knots.

The no-chute distance charts assume that a normal deployment attempt is made. The nose is lowered at or before reaching 120 KEAS and brake applied. Enough up elevon should be used to increase drag without lifting the nose wheel.

Example:

For 86°F air temperature, 2300 ft pressure altitude, and 75,000 lb gross weight, find the normal performance landing ground roll distance with normal chute operation. (RCR = 23). Include corrections for 10 knot headwind component and zero runway slope. Refer to figure A2-13. (Ground roll distance is 3950 feet.)

Maximum Performance Landings

The maximum performance landing procedure should be used for all wet runway landings. Maximum performance landing distances with drag chute assume that chute deployment is initiated and the nose gear lowered as soon as the main gear touches when landing speed is under chute deploy limit speed. If landing at speeds over chute deploy limit speed, initiation of chute deployment and lowering of the nose is delayed until chute deploy limit speed is reached. The chute is assumed to be jettisoned at 60 knots on a dry runway and retained until full stop on a wet runway.

Note

Chute deploy limit speed for "-501" chute is 210 KEAS.

The no-chute distance charts assume that a normal deployment attempt is made and that the nose is lowered and braking initiated as for landings with normal chute operation. (Brakes are applied without regard to brake energy ratings for the no chute condition.) A three second period after the normal 4.75 second deployment allowance is included for recognition of chute failure. One engine is assumed to be shut down at this time.

With wet runway conditions, enough up elevon should be used to increase drag without lifting the nosewheel. Wet runway chart distances are based on anti-skid OFF braking with wheels locked.

Example:

Find the maximum performance landing ground roll distance for 86°F air temperature, 2300 ft pressure altitude, and 75,000 lb gross weight, wet runway (RCR = 14), and chute failed condition. Include corrections for 10 knot headwind and zero runway slope. Refer to figure A2-18. (Ground roll distance is 10,800 ft.)

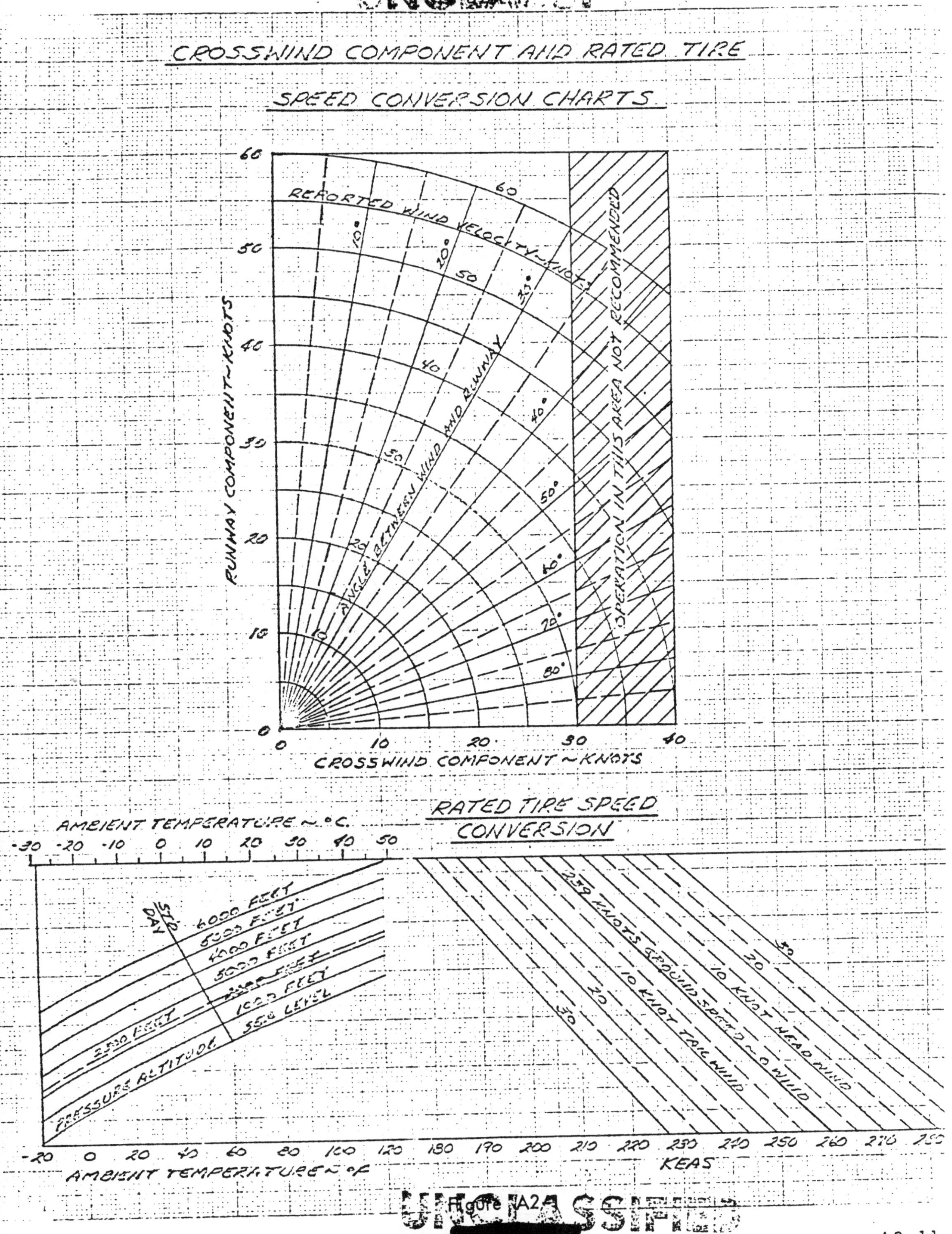
CROSSWIND COMPONENT AND RATED TIRE
SPEED CONVERSION CHARTS
REPORTED WIND VELOCITY~KNOTS
ANGLE BETWEEN WIND AND RUNWAY
OPERATION IN THIS AREA NOT RECOMMENDED
RUNWAY COMPONENT~KNOTS
CROSSWIND COMPONENT~KNOTS
RATED TIRE SPEED
CONVERSION
AMBIENT TEMPERATURE ~ °C.
STD DAY
PRESSURE ALTITUDE
SEA LEVEL
250 KNOTS GROUND SPEED ~ 0 WIND
10 KNOT HEAD WIND
10 KNOT TAIL WIND
KEAS
AMBIENT TEMPERATURE ~ °F

Figure A2-4

NORMAL PERFORMANCE TAKEOFF

180 KEAS ROTATION SPEED   210 KEAS TAKEOFF SPEED

REFER TO FIG. A2-1 FOR RATED TIRE SPEED. IF TAKEOFF SPEED WILL BE WITHIN 5 KNOTS OF RATED TIRE SPEED, DECREASE ROTATION AND TAKEOFF SPEEDS EQUALLY SO AS TO TAKEOFF AT NO MORE THAN 5 KNOTS BELOW RATED TIRE SPEED. DECREASE TAKEOFF DISTANCE 1% PER KNOT DECREASE IN TAKEOFF SPEED. (DO NOT REDUCE TAKE OFF SPEED BELOW MAXIMUM PERFORMANCE TAKEOFF SPEED, FIG. A2-3.)

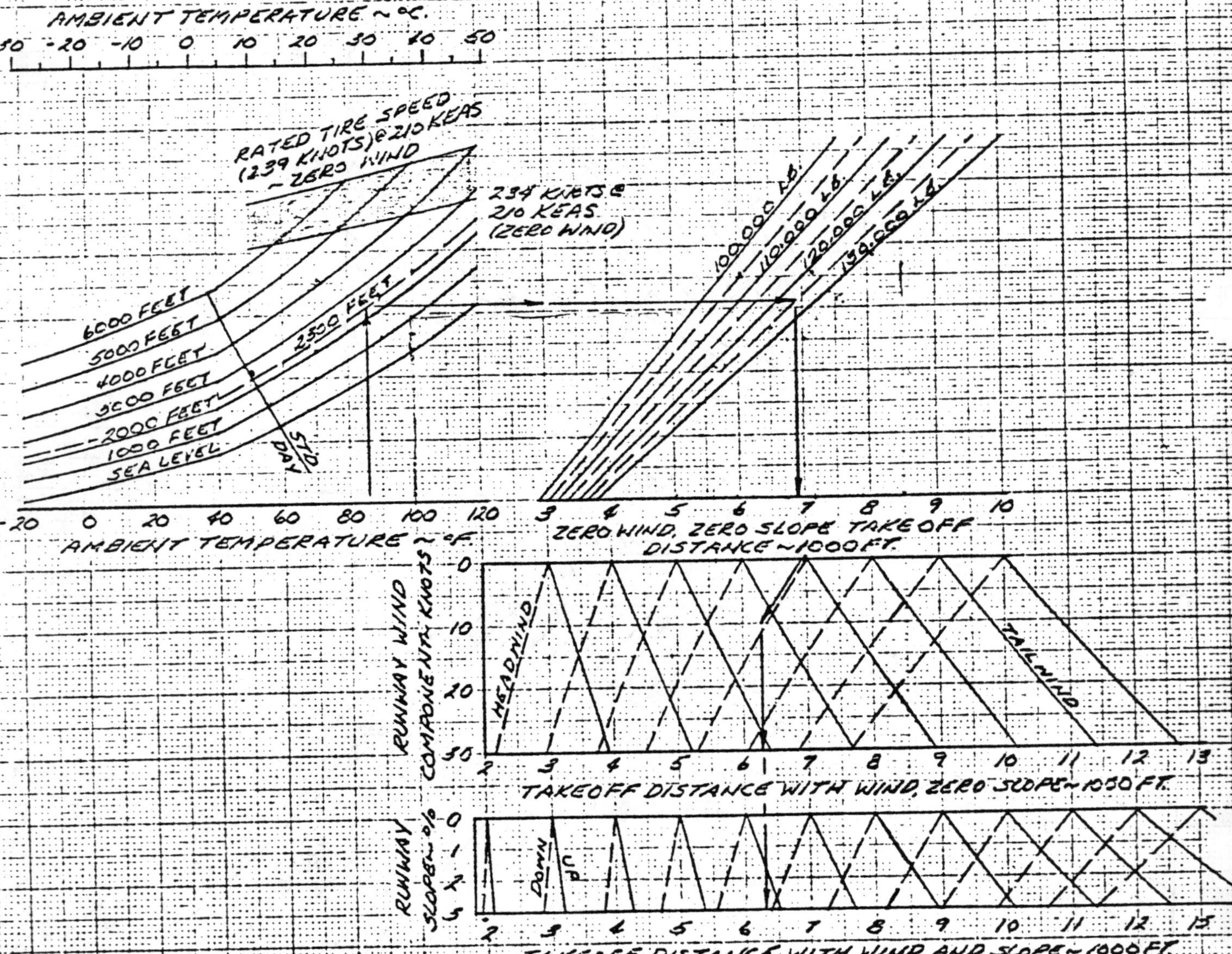

Figure A2-2

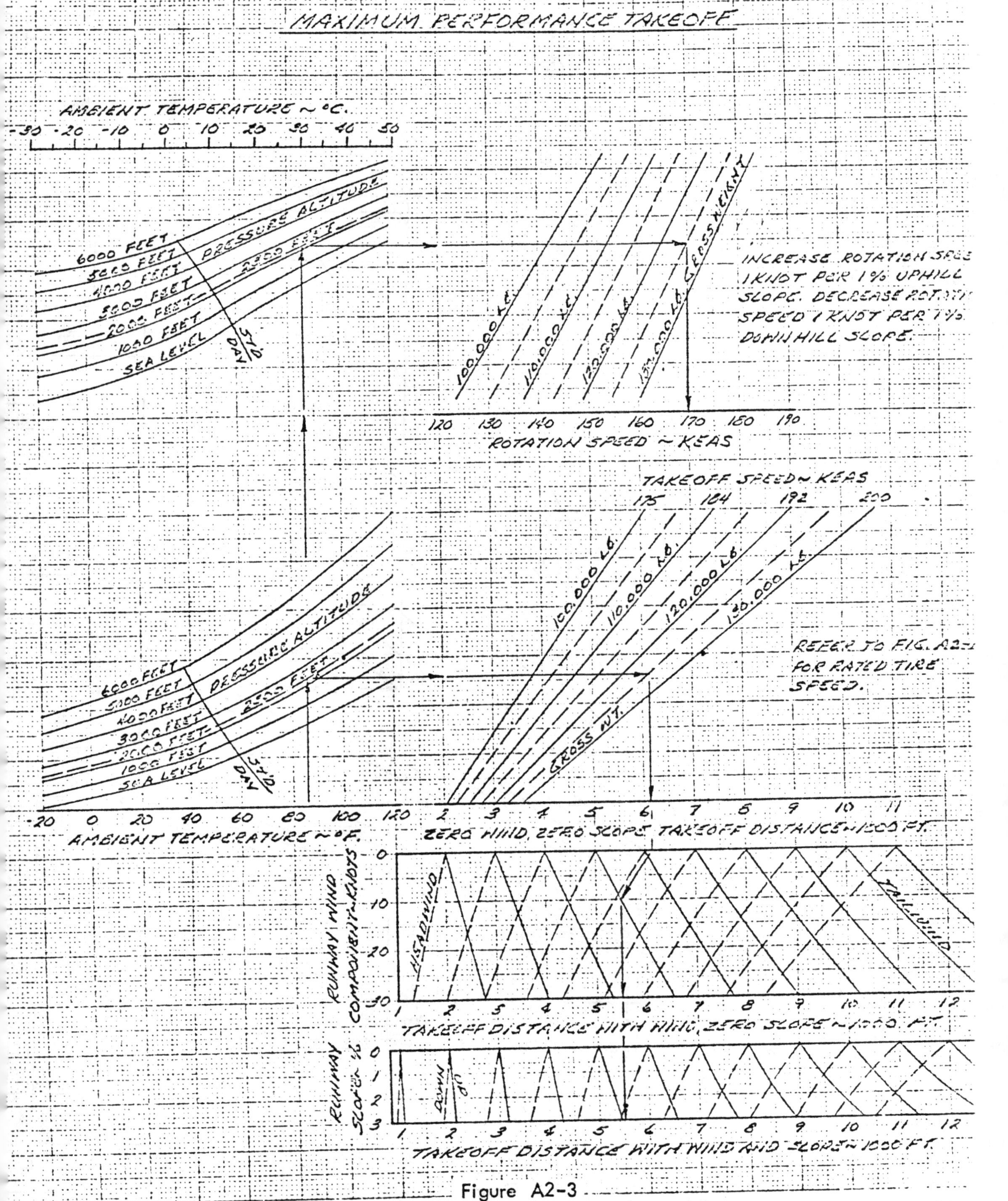

MAXIMUM PERFORMANCE TAKEOFF

Figure A2-3

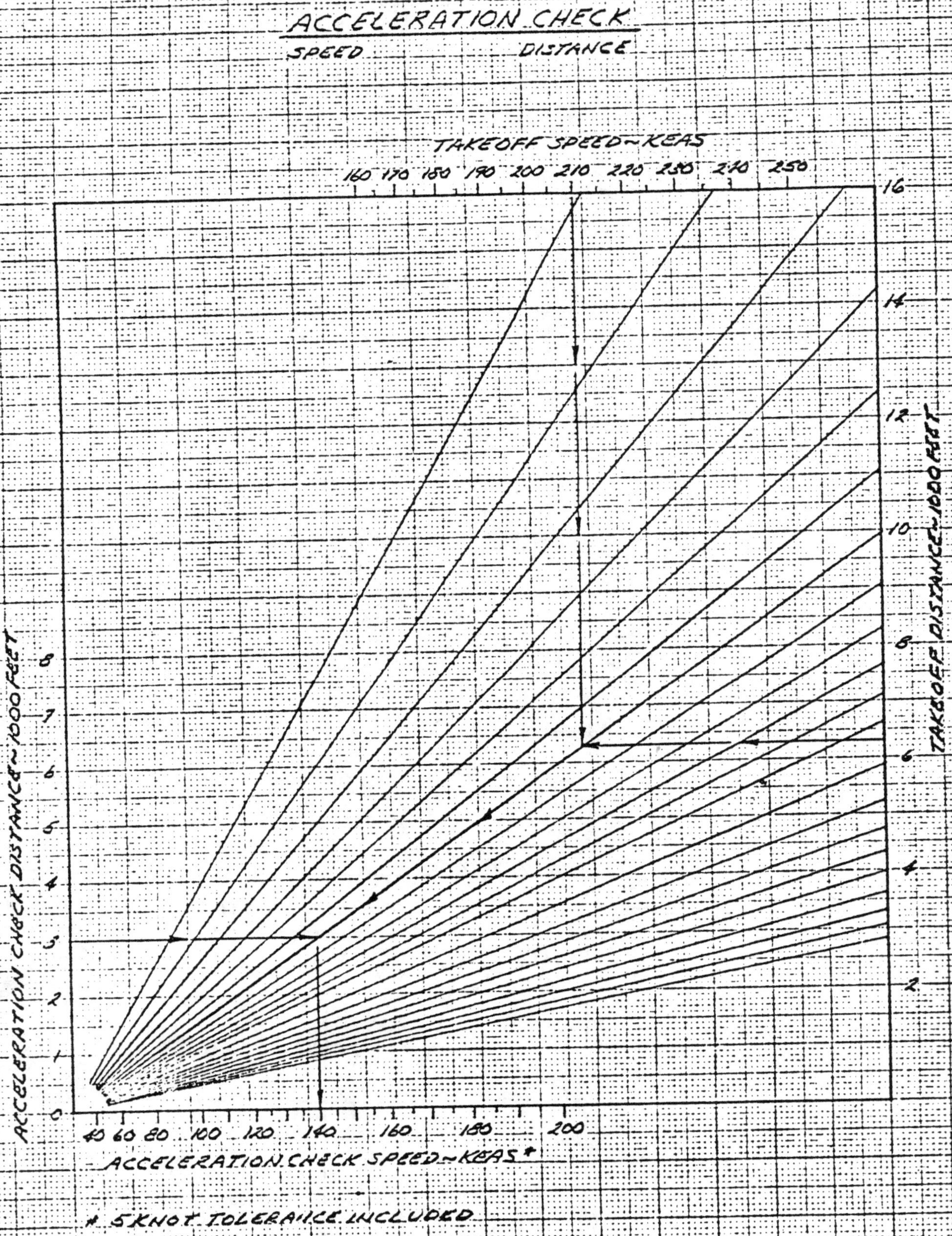

ACCELERATION CHECK

SPEED DISTANCE

TAKEOFF SPEED ~ KEAS

160 170 180 190 200 210 220 230 240 250

16 14 12 10 8 6 4 2

TAKEOFF DISTANCE ~ 1000 FEET

ACCELERATION CHECK DISTANCE ~ 1000 FEET

8 7 6 5 4 3 2 1 0

40 60 80 100 120 140 160 180 200

ACCELERATION CHECK SPEED ~ KEAS*

* 5 KNOT TOLERANCE INCLUDED

Figure A2-4

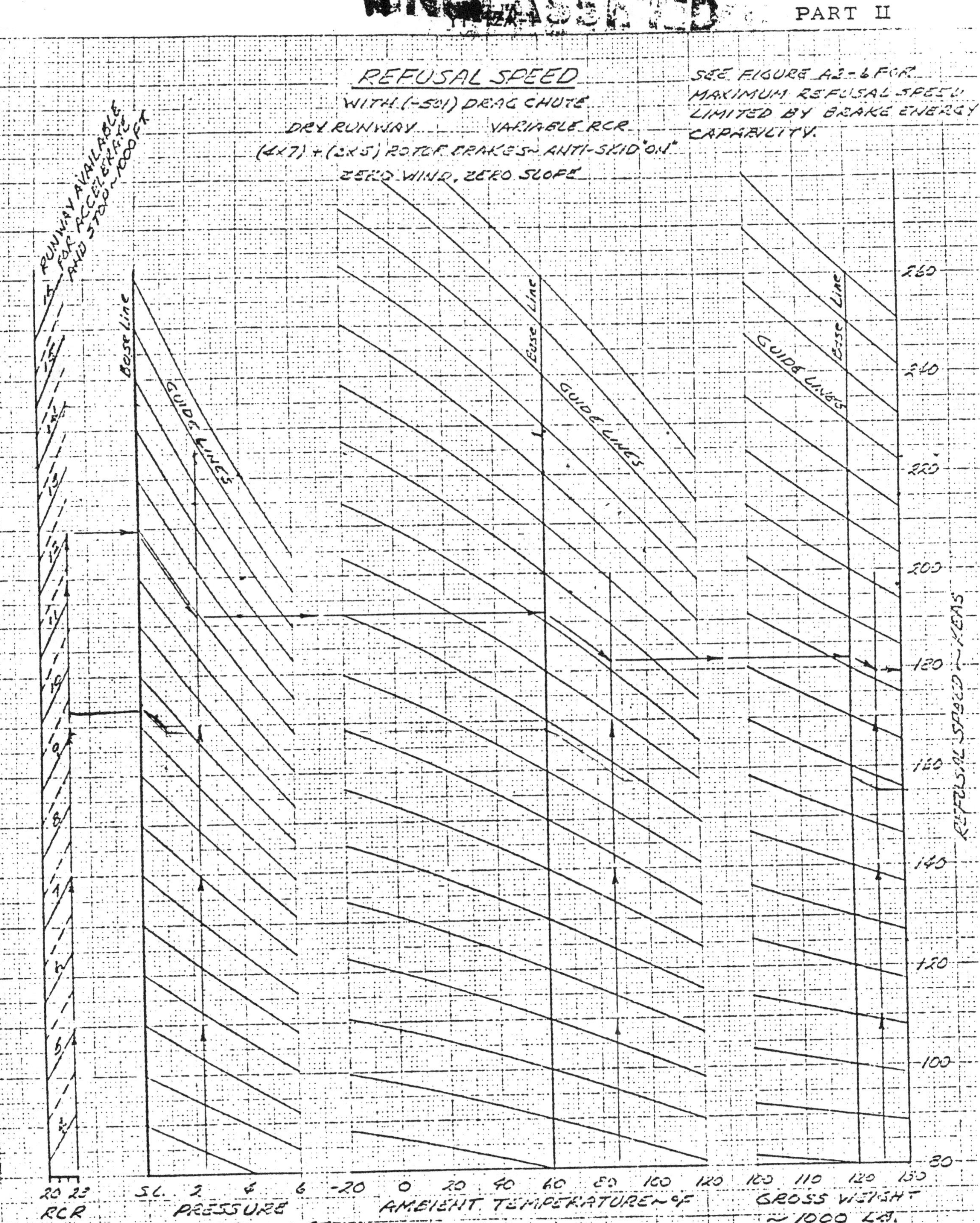
REFUSAL SPEED
WITH (-501) DRAG CHUTE
DRY RUNWAY VARIABLE RCR
(4x7) + (2x5) ROTOR BRAKES ~ ANTI-SKID "ON"
ZERO WIND, ZERO SLOPE
SEE FIGURE A2-6 FOR MAXIMUM REFUSAL SPEED LIMITED BY BRAKE ENERGY CAPABILITY.
RUNWAY AVAILABLE FOR ACCELERATE AND STOP ~ 1000 FT
Base Line
GUIDE LINES
REFUSAL SPEED ~ KEAS
RCR
PRESSURE ALTITUDE ~ 1000 FT.
AMBIENT TEMPERATURE ~ °F
GROSS WEIGHT ~ 1000 LB.

Figure A2-5

MAXIMUM REFUSAL SPEED
BRAKE ENERGY LIMITED*
WITH (-561) DRAG CHUTE
DRY RUNWAY VARIABLE RCR
(4K7) + (2K5) ROTOR BRAKES
ANTI-SKID "ON"
ONE STOP CAPABILITY

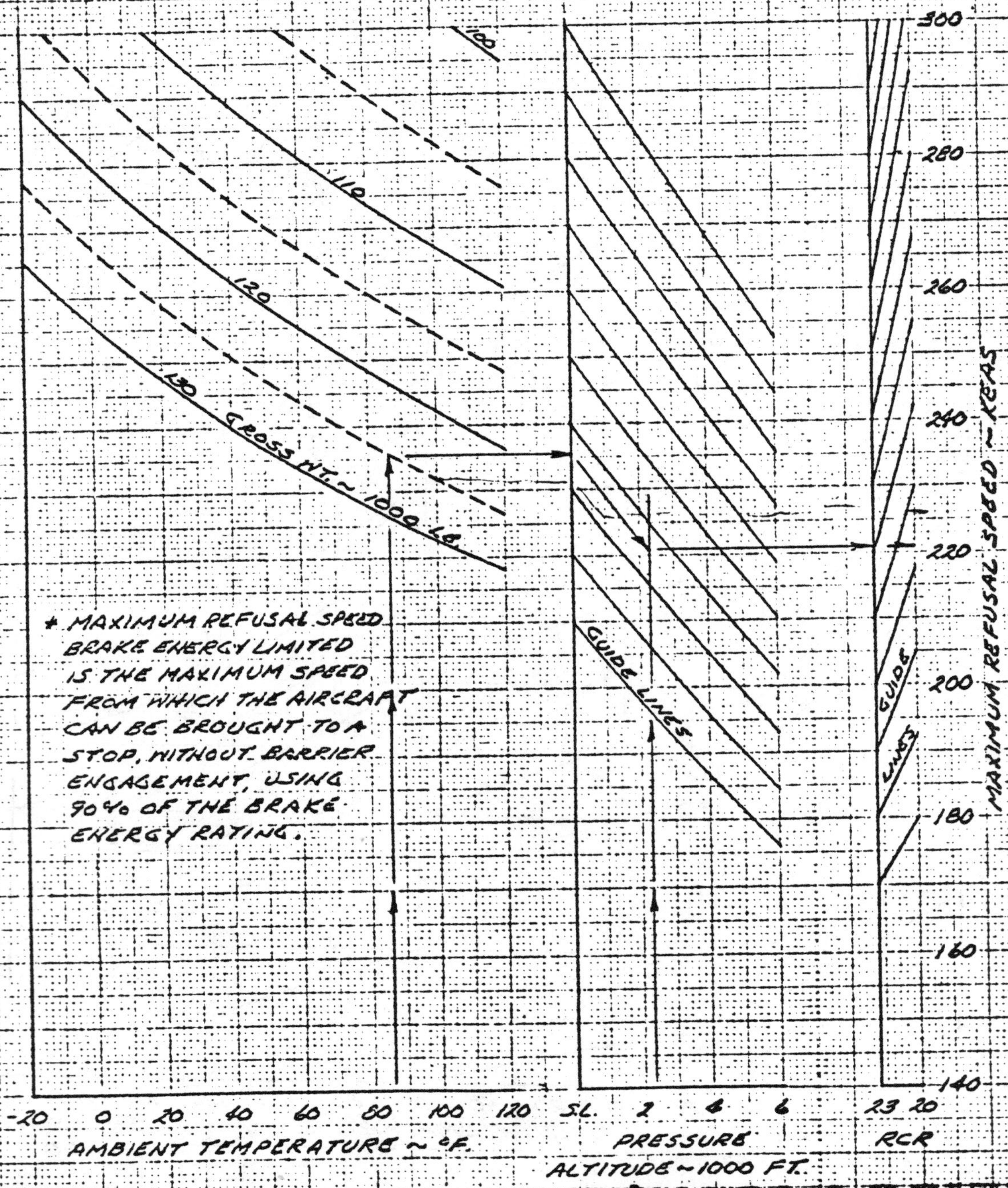

100
110
120
130 GROSS WT. ~ 1000 LB.

GUIDE LINES

GUIDE LINES

300
280
260
240
220
200
180
160
140

MAXIMUM REFUSAL SPEED ~ KEAS

* MAXIMUM REFUSAL SPEED BRAKE ENERGY LIMITED IS THE MAXIMUM SPEED FROM WHICH THE AIRCRAFT CAN BE BROUGHT TO A STOP, WITHOUT BARRIER ENGAGEMENT, USING 90% OF THE BRAKE ENERGY RATING.

-20 0 20 40 60 80 100 120
AMBIENT TEMPERATURE ~ °F.

S.L. 2 4 6
PRESSURE ALTITUDE ~ 1000 FT.

23 20
RCR

Figure A2-6

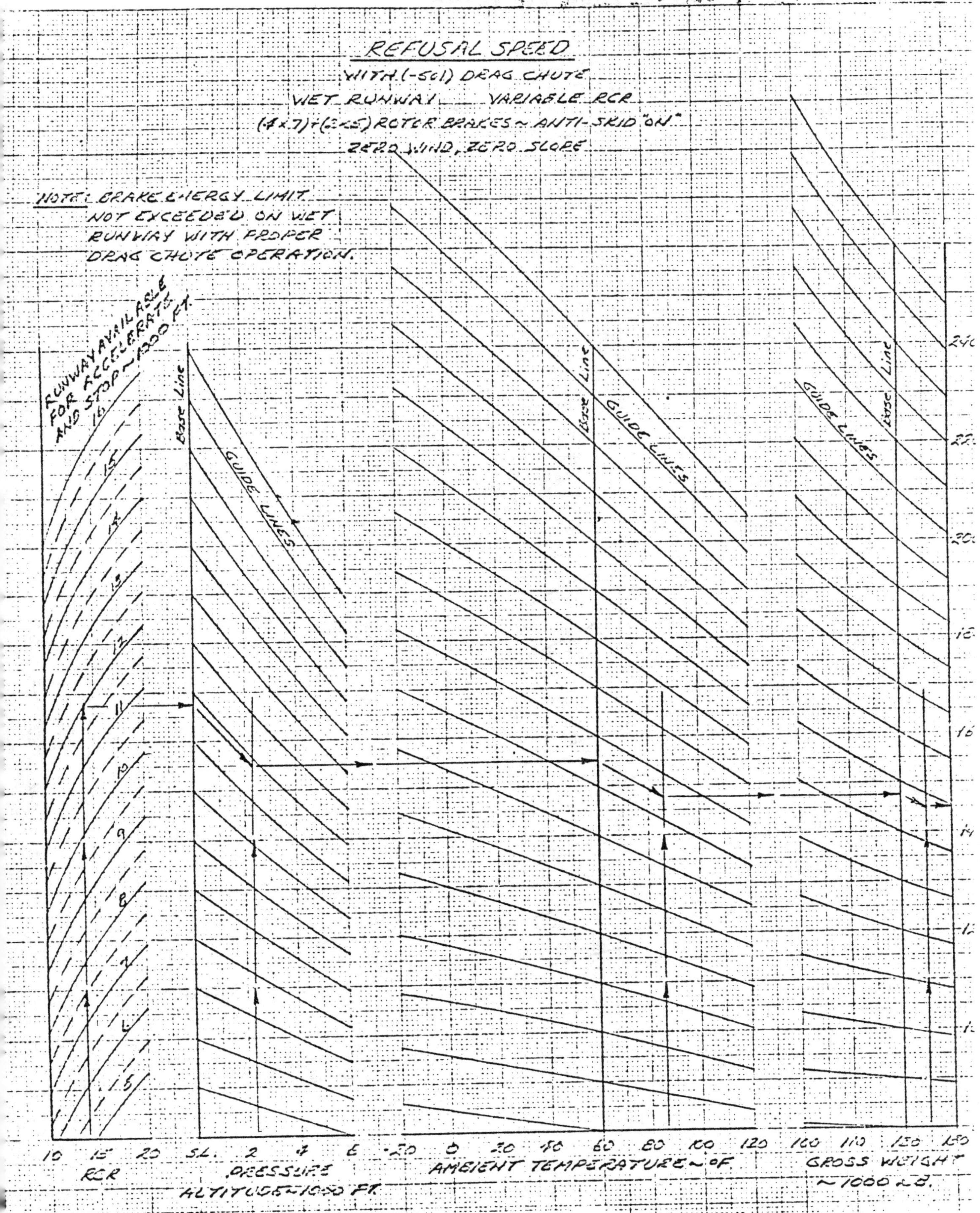

REFUSAL SPEED

WITH (-501) DRAG CHUTE

WET RUNWAY VARIABLE RCR

(4×7)+(2×5) ROTOR BRAKES ~ ANTI-SKID "ON"

ZERO WIND, ZERO SLOPE

NOTE: BRAKE ENERGY LIMIT NOT EXCEEDED ON WET RUNWAY WITH PROPER DRAG CHUTE OPERATION.

RUNWAY AVAILABLE FOR ACCELERATE AND STOP ~1000 FT.

16 15 14 13 12 11 10 9 8 7 6 5

Base Line

GUIDE LINES

Base Line

GUIDE LINES

GUIDE LINES

Base Line

240 220 200 180 160 140 120 100

10 15 20 RCR

S.L. 2 4 6 PRESSURE ALTITUDE~1000 FT.

-20 0 20 40 60 80 100 120 AMBIENT TEMPERATURE ~ °F.

100 110 120 130 GROSS WEIGHT ~1000 LB.

Figure A2-7.

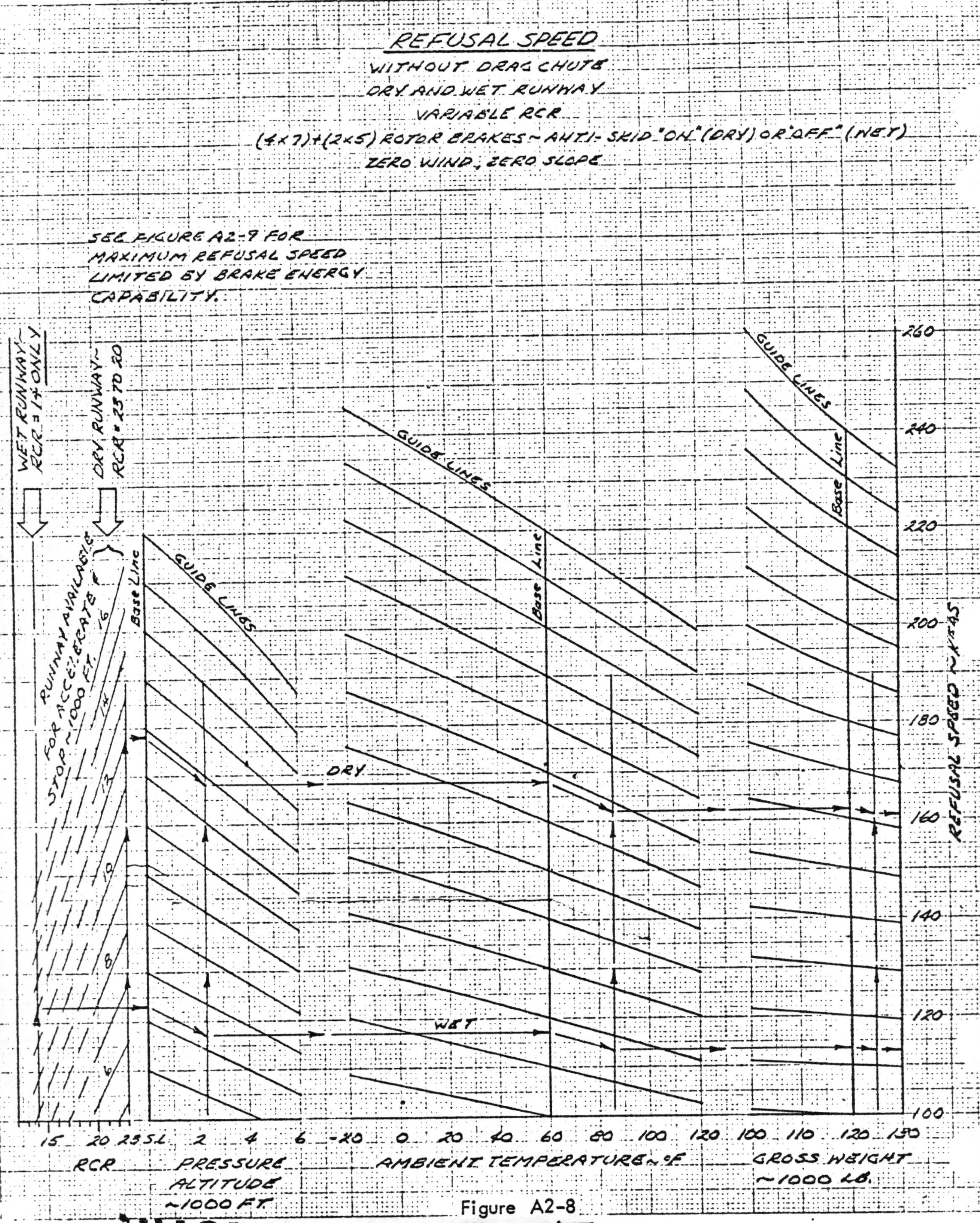
REFUSAL SPEED
WITHOUT DRAG CHUTE
DRY AND WET RUNWAY
VARIABLE RCR
(4×7)+(2×5) ROTOR BRAKES ~ ANTI-SKID "ON" (DRY) OR "OFF" (WET)
ZERO WIND, ZERO SLOPE
SEE FIGURE A2-9 FOR MAXIMUM REFUSAL SPEED LIMITED BY BRAKE ENERGY CAPABILITY.
WET RUNWAY ~ RCR = 14 ONLY
DRY RUNWAY ~ RCR = 23 TO 20
RUNWAY AVAILABLE FOR ACCELERATE & STOP ~ 1000 FT.
6
8
10
12
14
16
Base Line
GUIDE LINES
DRY
WET
REFUSAL SPEED ~ KEAS
260
240
220
200
180
160
140
120
100
15
20
23
S.L.
2
4
6
-20
0
20
40
60
80
100
120
100
110
120
130
RCR
PRESSURE ALTITUDE ~1000 FT.
AMBIENT TEMPERATURE ~ °F
GROSS WEIGHT ~1000 LB.

Figure A2-8

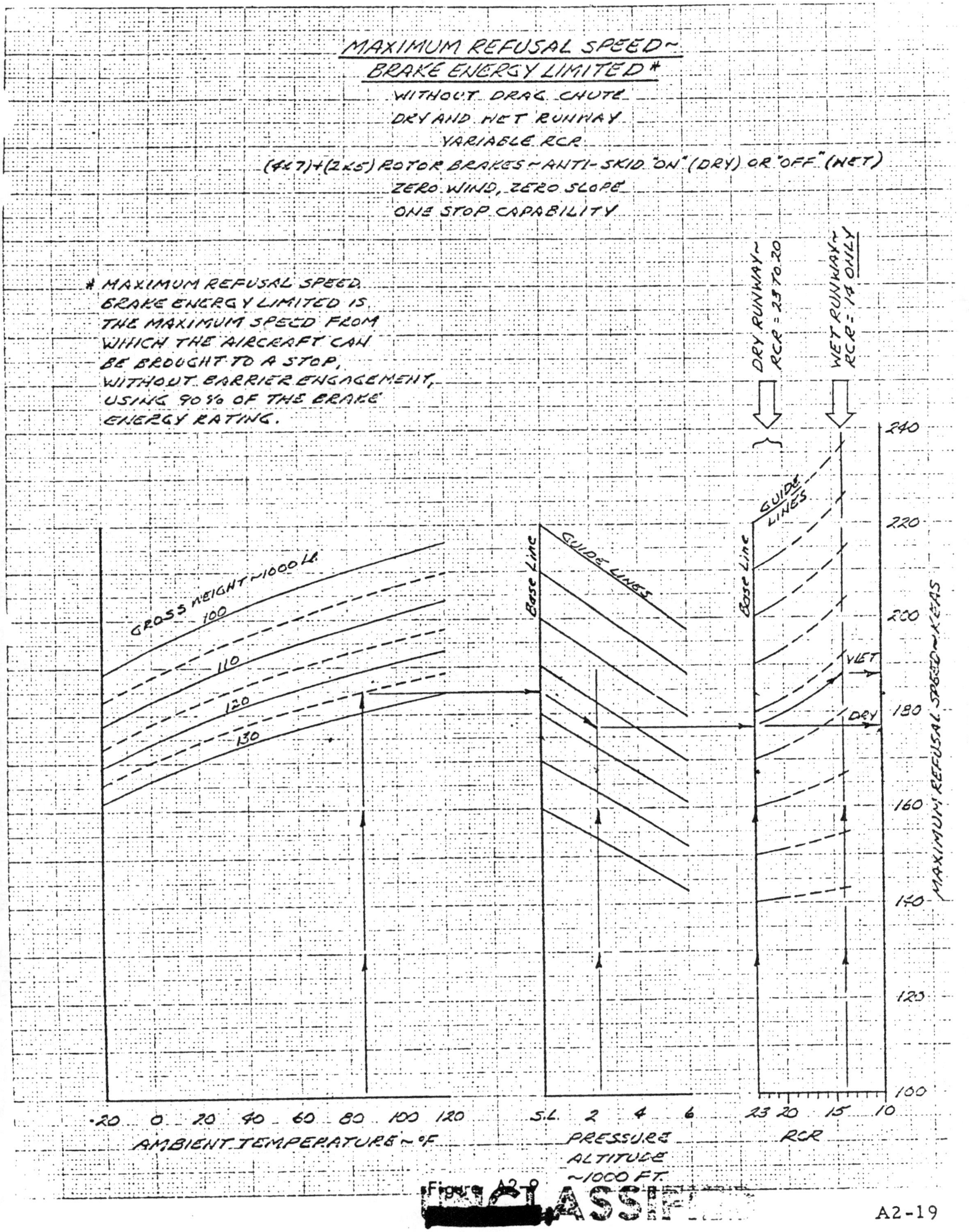

MAXIMUM REFUSAL SPEED~
BRAKE ENERGY LIMITED*

WITHOUT DRAG CHUTE
DRY AND WET RUNWAY
VARIABLE RCR
(4x7)+(2x5) ROTOR BRAKES~ANTI-SKID "ON" (DRY) OR "OFF" (WET)
ZERO WIND, ZERO SLOPE
ONE STOP CAPABILITY

* MAXIMUM REFUSAL SPEED BRAKE ENERGY LIMITED IS THE MAXIMUM SPEED FROM WHICH THE AIRCRAFT CAN BE BROUGHT TO A STOP, WITHOUT BARRIER ENGAGEMENT, USING 90% OF THE BRAKE ENERGY RATING.

Figure A2-9

MINIMUM SPEEDS FOR SINGLE ENGINE LEVEL FLIGHT

C.G. @ 24%
ONE ENGINE AT MAXIMUM THRUST
ONE ENGINE WINDMILLING
RUDDER DEFLECTION AND BANK TO MAINTAIN FLIGHT PATH (SIDESLIP LESS THAN 5°)

IN GROUND EFFECT

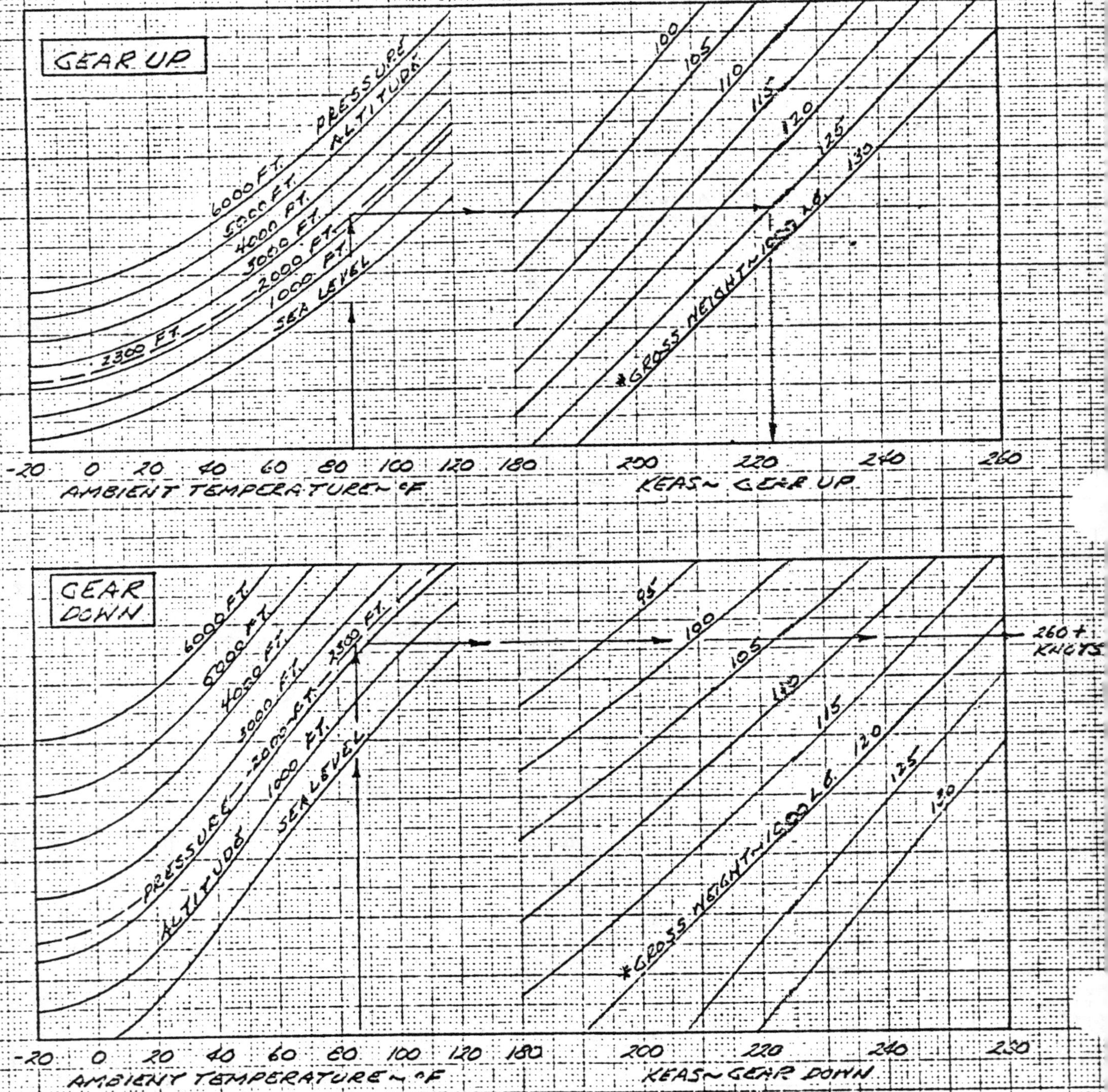

* TEMPORARY DECREMENT OF 20,000 LB. INCLUDED

Figure A2-105

MINIMUM SPEEDS FOR SINGLE ENGINE LEVEL FLIGHT

C.G. @ 24%
ONE ENGINE AT MAXIMUM THRUST
ONE ENGINE WINDMILLING
RUDDER DEFECTION AND BANK TO MAINTAIN FLIGHT PATH (SIDESLIP LESS THAN 5°)

OUT OF GROUND EFFECT

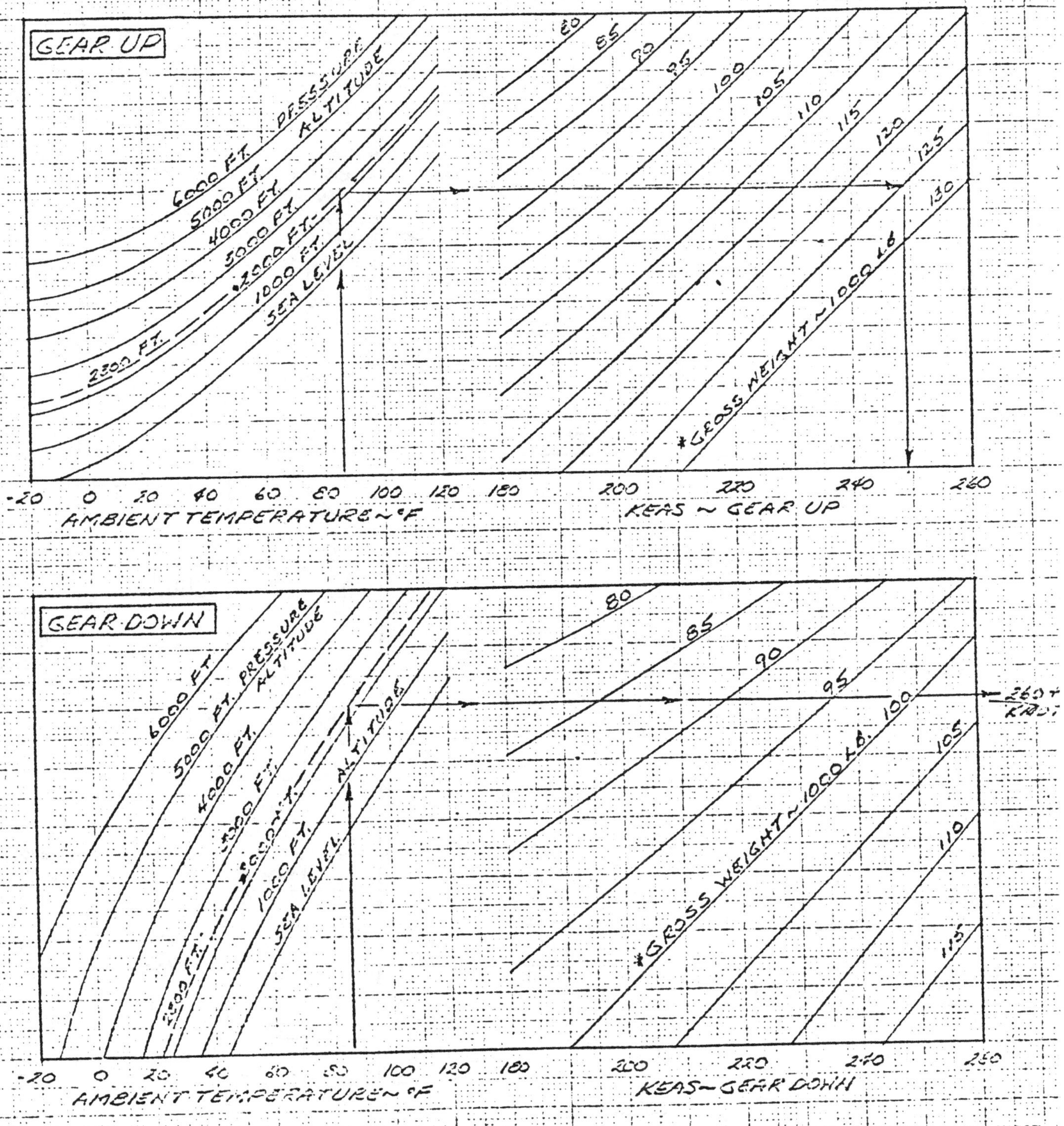

* TEMPORARY DECREMENT OF 20,000 LB. INCLUDED.

Figure A2-11

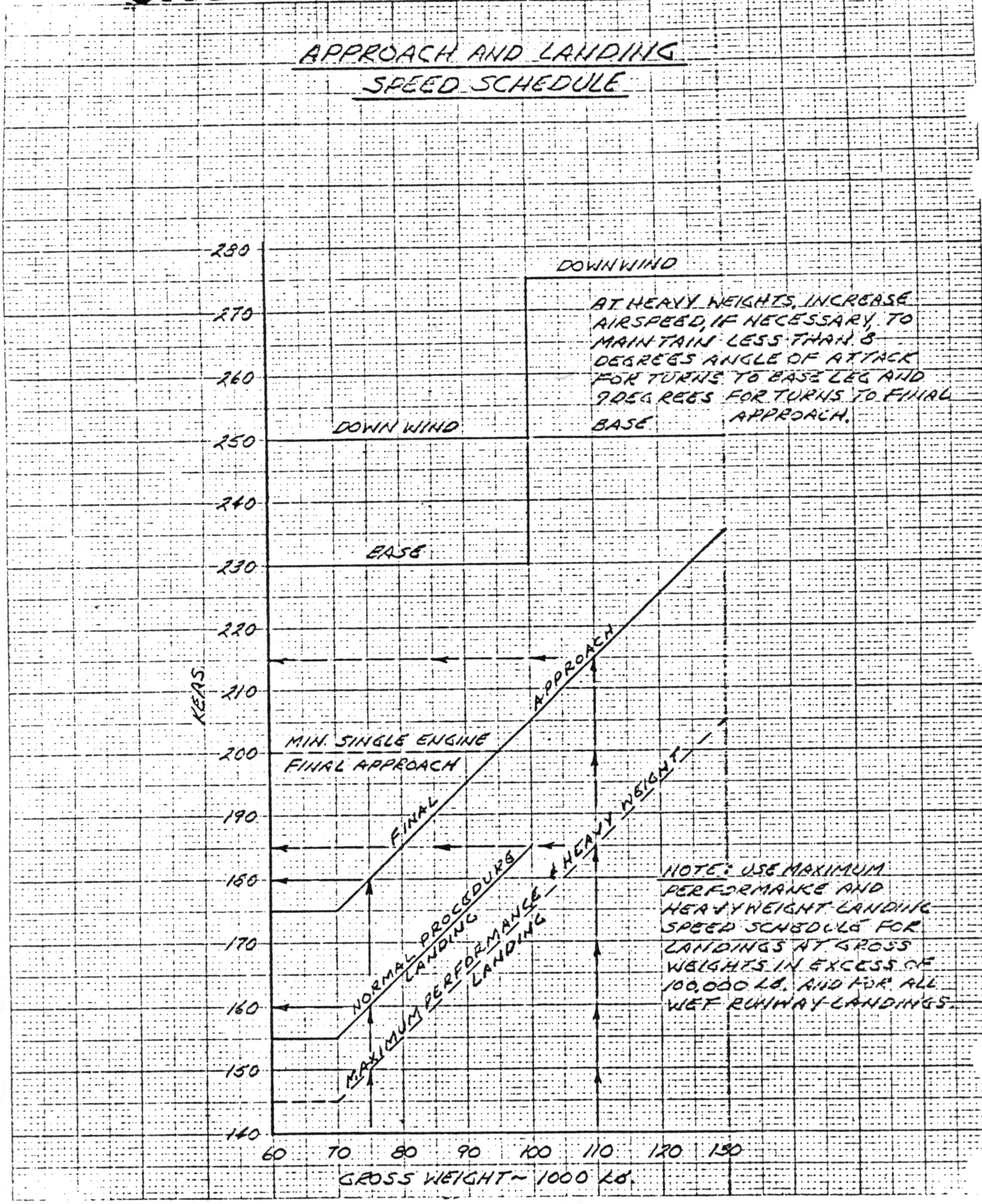
APPROACH AND LANDING
SPEED SCHEDULE
DOWNWIND
AT HEAVY WEIGHTS, INCREASE AIRSPEED, IF NECESSARY, TO MAINTAIN LESS THAN 8 DEGREES ANGLE OF ATTACK FOR TURNS TO BASE LEG AND 9 DEGREES FOR TURNS TO FINAL APPROACH.
DOWN WIND
BASE
BASE
APPROACH
MIN. SINGLE ENGINE
FINAL APPROACH
FINAL
NORMAL PROCEDURE
LANDING
HEAVY WEIGHT
MAXIMUM PERFORMANCE
LANDING
NOTE: USE MAXIMUM PERFORMANCE AND HEAVYWEIGHT LANDING SPEED SCHEDULE FOR LANDINGS AT GROSS WEIGHTS IN EXCESS OF 100,000 LB. AND FOR ALL WET RUNWAY LANDINGS.
KEAS
280
270
260
250
240
230
220
210
200
190
180
170
160
150
140
60
70
80
90
100
110
120
130
GROSS WEIGHT ~ 1000 LB.

Figure A2-12

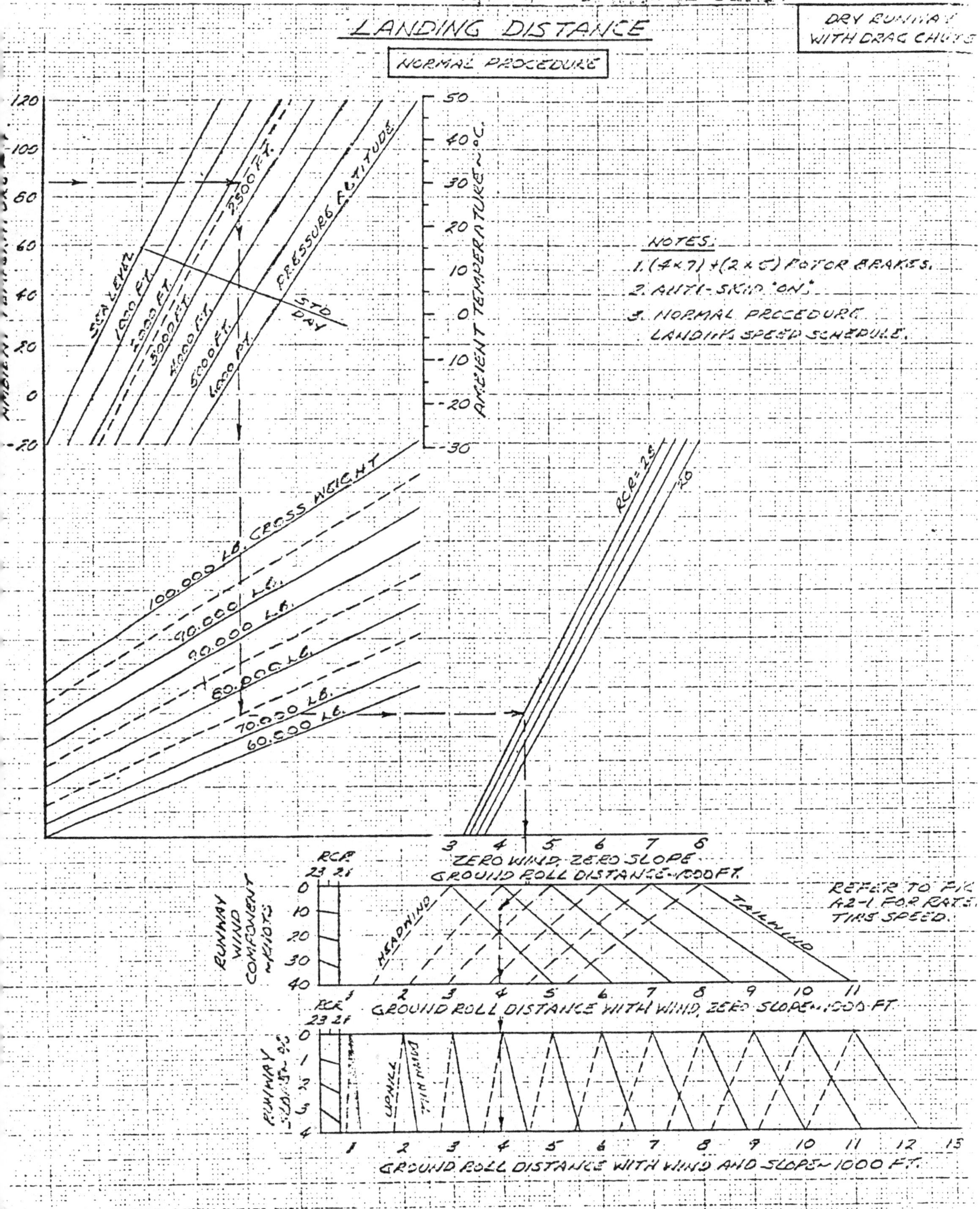

LANDING DISTANCE

NORMAL PROCEDURE

DRY RUNWAY WITH DRAG CHUTE

Figure A2-13

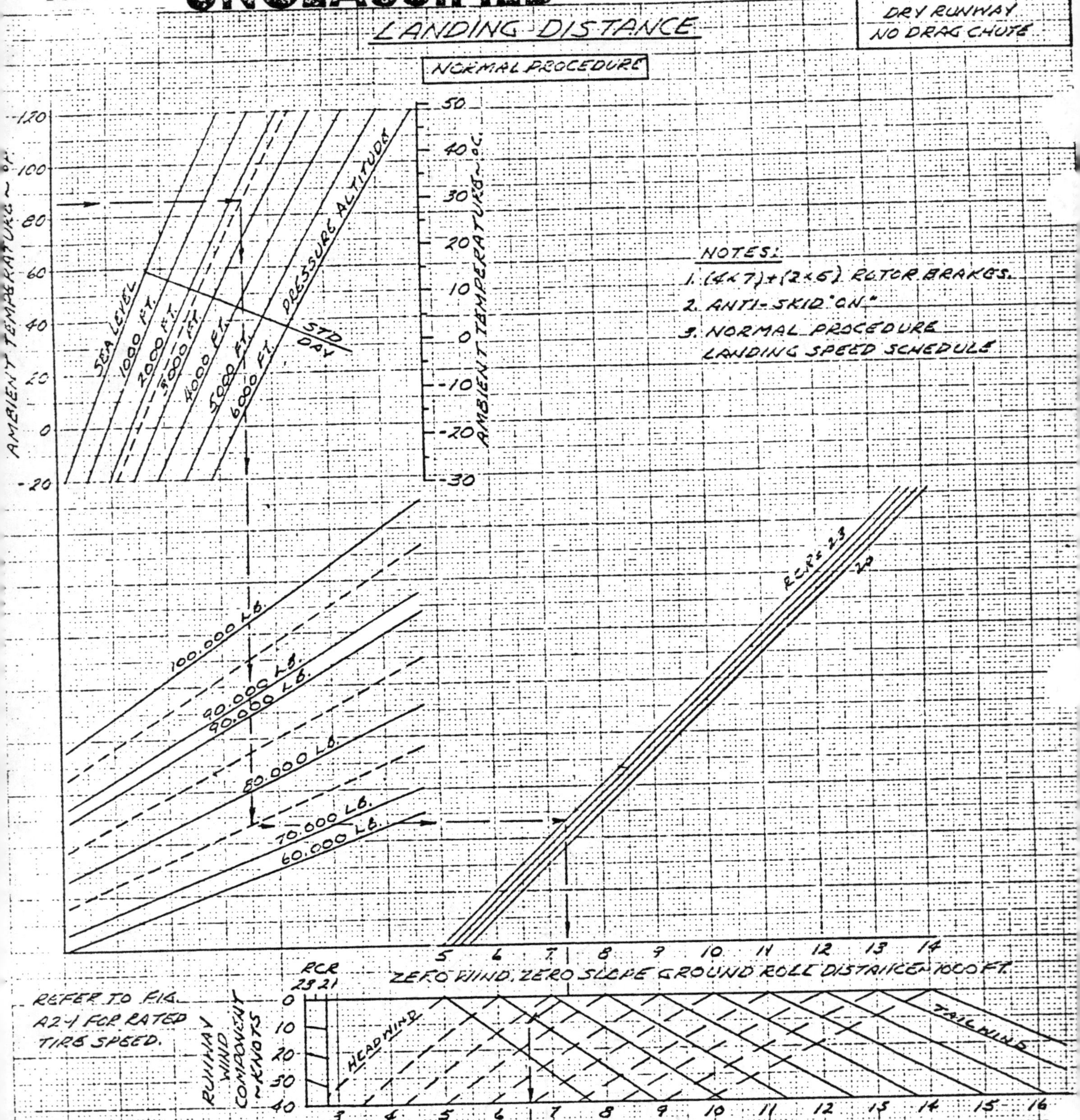

LANDING DISTANCE

DRY RUNWAY
NO DRAG CHUTE

NORMAL PROCEDURE

Figure A2-14

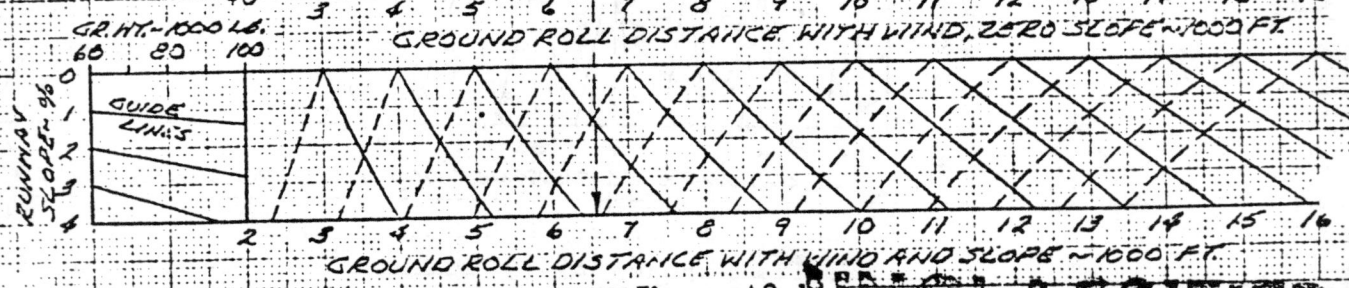

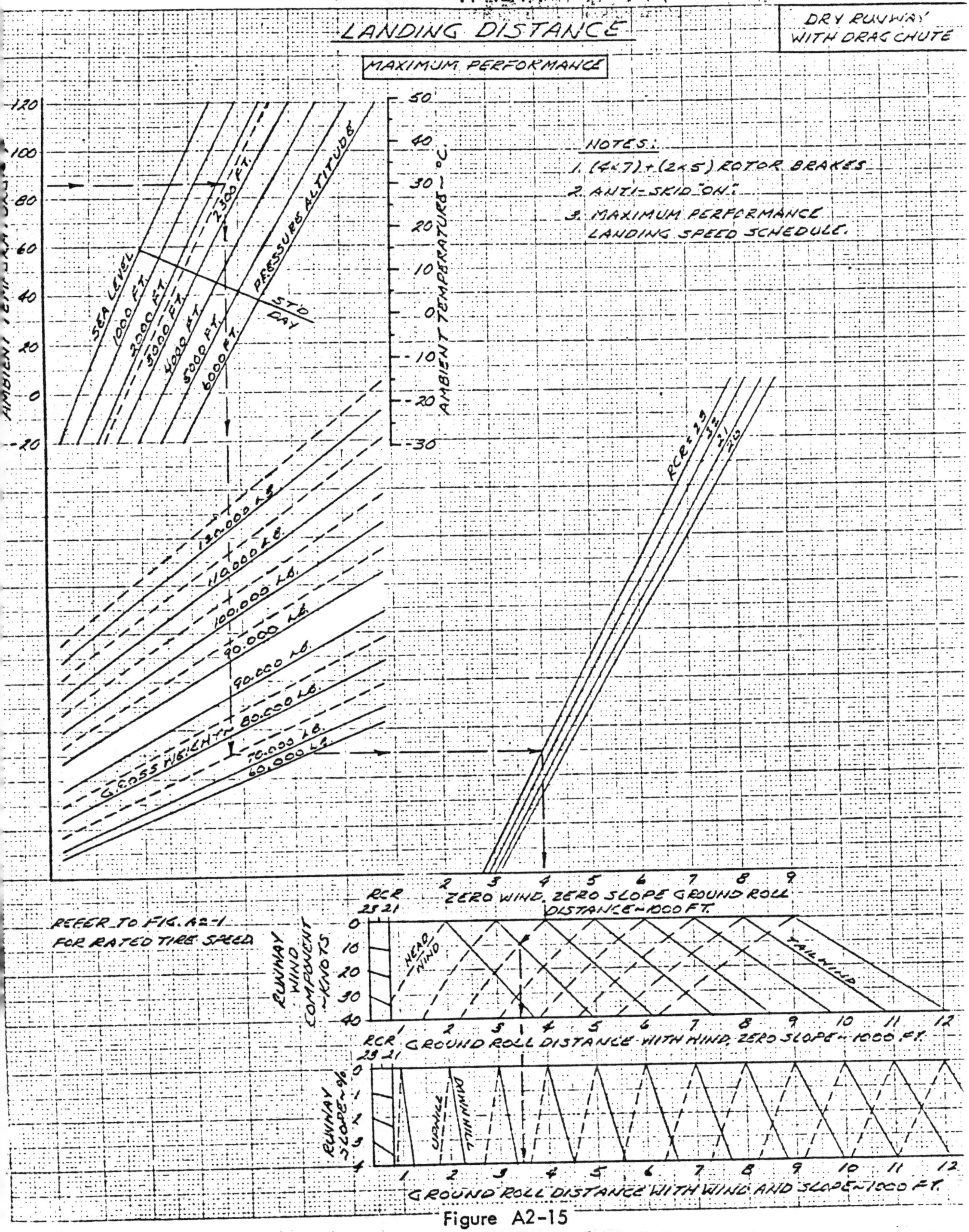

LANDING DISTANCE
DRY RUNWAY WITH DRAG CHUTE
MAXIMUM PERFORMANCE
NOTES:
1. (4×7)+(2×5) ROTOR BRAKES
2. ANTI-SKID "ON"
3. MAXIMUM PERFORMANCE LANDING SPEED SCHEDULE.
AMBIENT TEMPERATURE ~ °F
AMBIENT TEMPERATURE ~ °C
SEA LEVEL
1000 FT.
2000 FT.
3000 FT.
4000 FT.
5000 FT.
6000 FT.
2300 FT.
PRESSURE ALTITUDE
STD DAY
120,000 LB.
110,000 LB.
100,000 LB.
90,000 LB.
80,000 LB.
70,000 LB.
60,000 LB.
GROSS WEIGHT ~
RCR = 23
21
10
ZERO WIND, ZERO SLOPE GROUND ROLL DISTANCE ~ 1000 FT.
REFER TO FIG. A2-1 FOR RATED TIRE SPEED
RUNWAY WIND COMPONENT ~ KNOTS
HEAD WIND
TAIL WIND
GROUND ROLL DISTANCE WITH WIND, ZERO SLOPE ~ 1000 FT.
RUNWAY SLOPE ~ %
UPHILL
DOWNHILL
GROUND ROLL DISTANCE WITH WIND AND SLOPE ~ 1000 FT.

Figure A2-15

YF-12

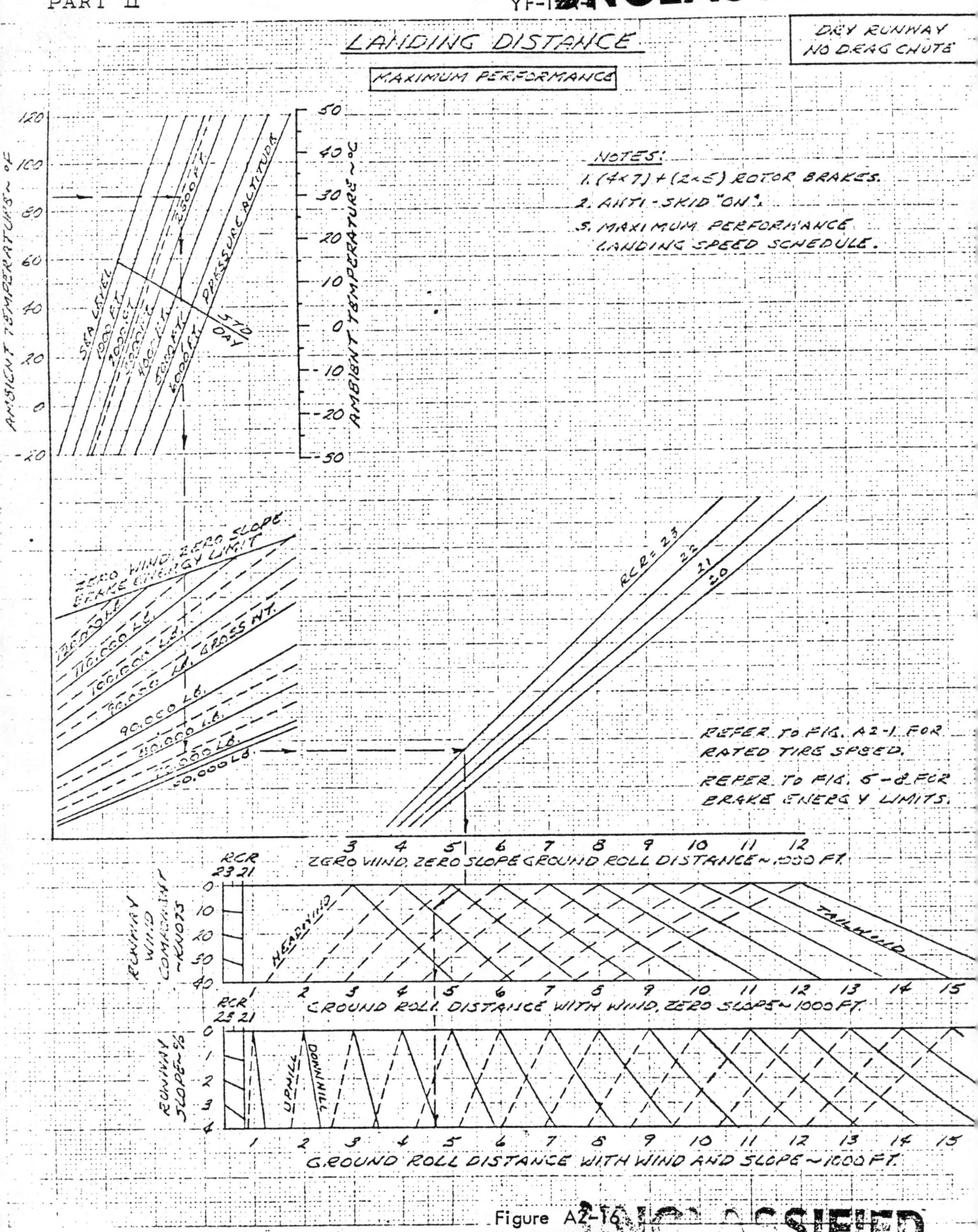
LANDING DISTANCE
MAXIMUM PERFORMANCE
DRY RUNWAY
NO DRAG CHUTE
NOTES:
1. (4×7) + (2×5) ROTOR BRAKES.
2. ANTI-SKID "ON".
3. MAXIMUM PERFORMANCE LANDING SPEED SCHEDULE.
AMBIENT TEMPERATURE ~ °F
AMBIENT TEMPERATURE ~ °C
SEA LEVEL
PRESSURE ALTITUDE
STD DAY
ZERO WIND, ZERO SLOPE BRAKE ENERGY LIMIT
GROSS WT.
RCR = 23
22
21
20
REFER TO FIG. A2-1 FOR RATED TIRE SPEED.
REFER TO FIG. 6-8 FOR BRAKE ENERGY LIMITS.
ZERO WIND, ZERO SLOPE GROUND ROLL DISTANCE ~ 1000 FT.
RUNWAY WIND COMPONENT ~ KNOTS
HEADWIND
TAILWIND
GROUND ROLL DISTANCE WITH WIND, ZERO SLOPE ~ 1000 FT.
RUNWAY SLOPE ~ %
UPHILL
DOWNHILL
GROUND ROLL DISTANCE WITH WIND AND SLOPE ~ 1000 FT.

Figure A2-16

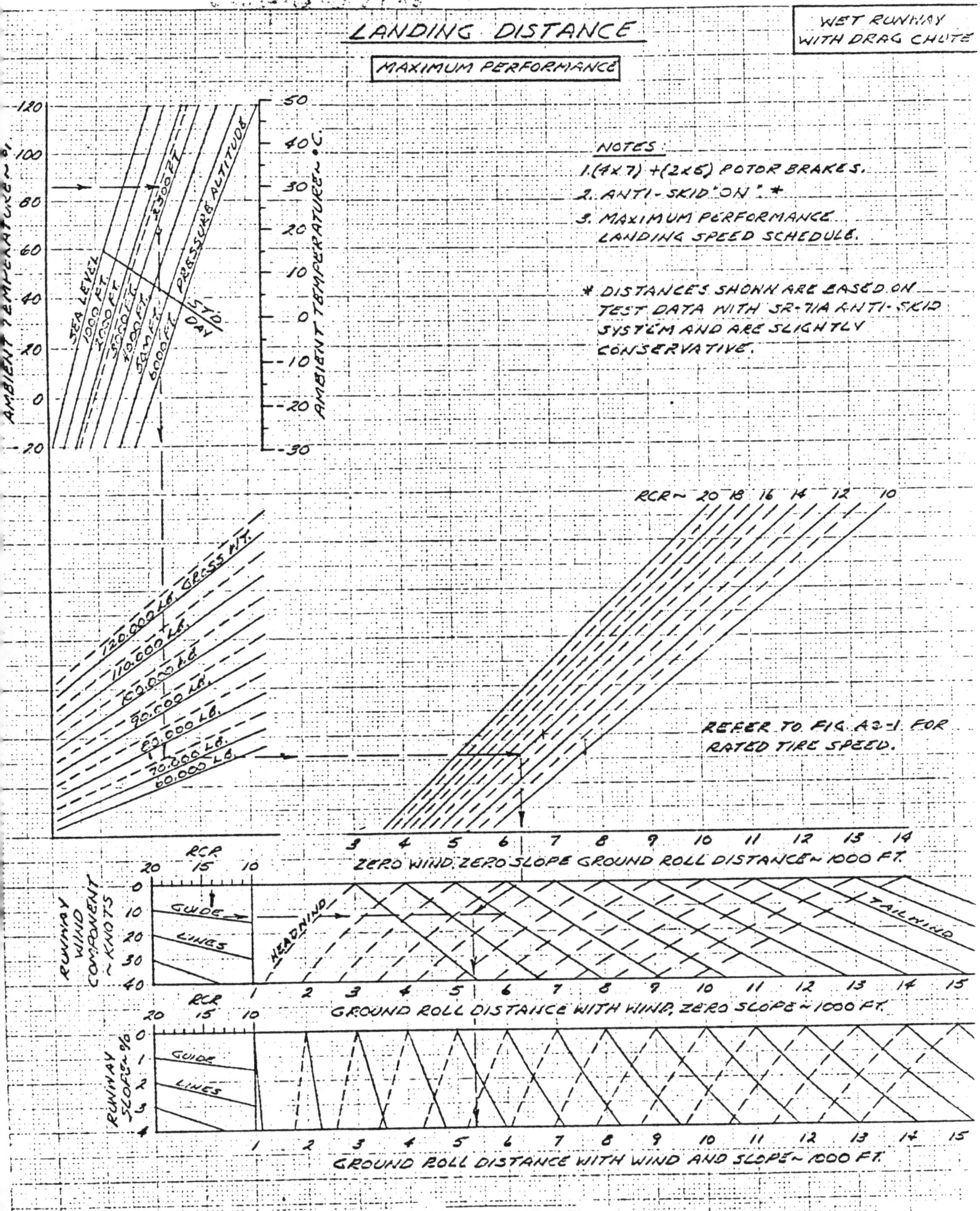

LANDING DISTANCE

WET RUNWAY WITH DRAG CHUTE

MAXIMUM PERFORMANCE

Figure A2-17

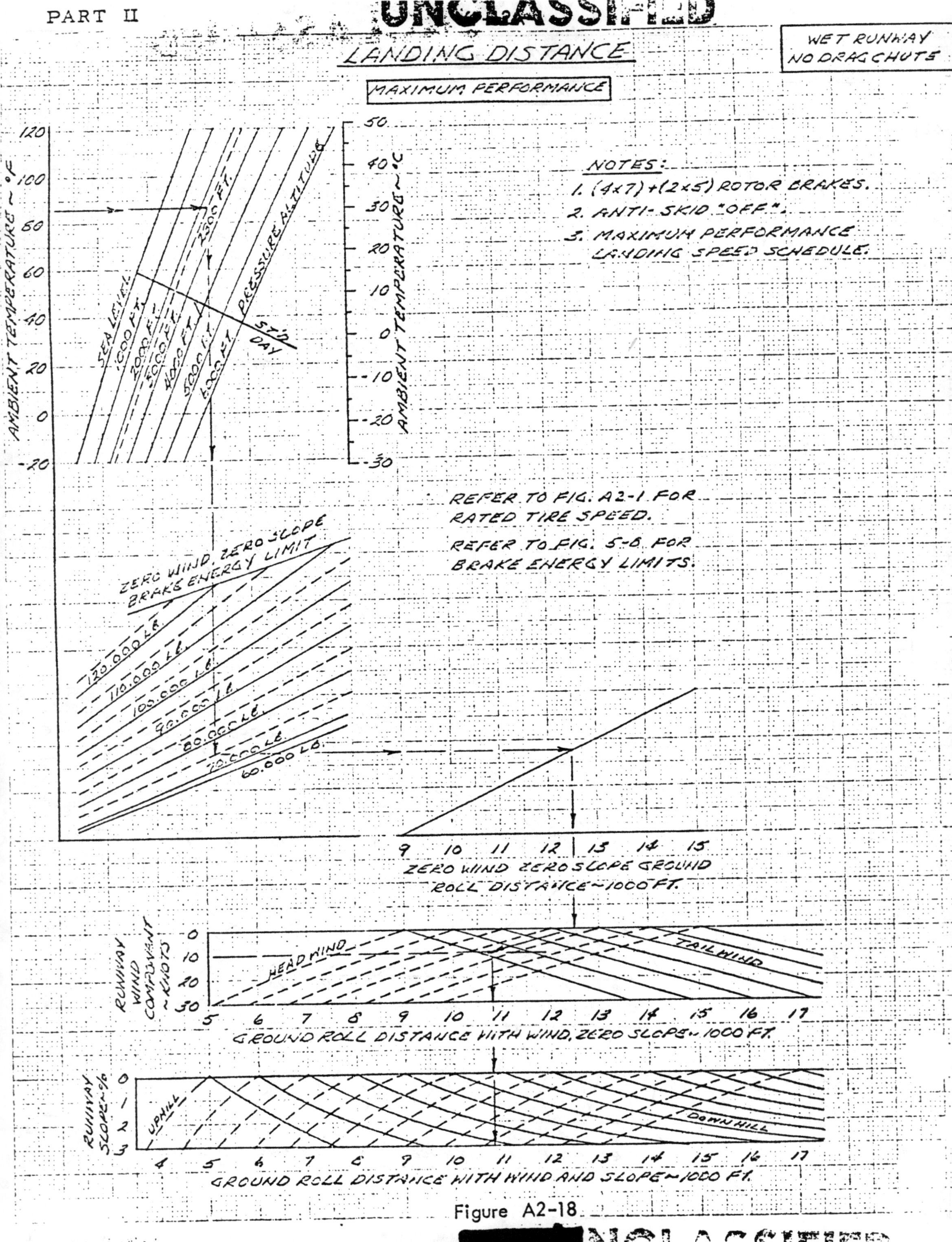
LANDING DISTANCE
WET RUNWAY
NO DRAG CHUTE
MAXIMUM PERFORMANCE
NOTES:
1. (4x7) + (2x5) ROTOR BRAKES.
2. ANTI-SKID "OFF".
3. MAXIMUM PERFORMANCE LANDING SPEED SCHEDULE.
AMBIENT TEMPERATURE ~ °F
AMBIENT TEMPERATURE ~ °C
SEA LEVEL
1000 FT.
2000 FT.
2500 FT.
3000 FT.
4000 FT.
5000 FT.
6000 FT.
PRESSURE ALTITUDE
STD DAY
REFER TO FIG. A2-1 FOR RATED TIRE SPEED.
REFER TO FIG. 5-6 FOR BRAKE ENERGY LIMITS.
ZERO WIND, ZERO SLOPE BRAKE ENERGY LIMIT
120,000 LB.
110,000 LB.
100,000 LB.
90,000 LB.
80,000 LB.
70,000 LB.
60,000 LB.
ZERO WIND ZERO SLOPE GROUND ROLL DISTANCE ~ 1000 FT.
RUNWAY WIND COMPONENT ~ KNOTS
HEADWIND
TAILWIND
GROUND ROLL DISTANCE WITH WIND, ZERO SLOPE ~ 1000 FT.
RUNWAY SLOPE ~ %
UPHILL
DOWNHILL
GROUND ROLL DISTANCE WITH WIND AND SLOPE ~ 1000 FT.

Figure A2-18

PART III

CLIMB AND DESCENT PERFORMANCE

TABLE OF CONTENTS

LIST OF ILLUSTRATIONS

PART III

CLIMB AND DESCENT PERFORMANCE

Introduction

This part of the appendix provides charts for the normal 450 KEAS supersonic climb schedule. An alternate procedure, 400 KEAS supersonic climb chart is also included, as well as normal descent data at 350 KEAS and single-engine descent data at 300, 350, and 400 KEAS.

Data Basis

Performance in this part of the appendix is based on flight tests.

Climb Performance

Time, fuel and distance requirements for the normal procedure, 450 KEAS supersonic climb schedule are shown in figures A3-1, A3-2, and A3-3 for the subsonic, transonic and supersonic climb phases, respectively. Allowances are tabulated on figure A3-1 for ground operation and for takeoff and acceleration to 400 KEAS. Allowances are also shown for the maximum refueled weight condition. (The curves are based on SR-71 flight tests with K engines, but are representative of the level of performance to be expected.)

Figure A3-1 provides climb allowances to reach 30,000 feet and 0.9 Mach no. from sea level and 400 KEAS with minimum afterburning power being used following gear retraction. The data can be used on an incremental basis for climbs between other altitudes in this range. Less time and distance will be required at lower ambient temperatures, and slightly less fuel; however, the differences will not affect overall mission planning.

Figure A3-2 shows the time, fuel and distance requirements for the transonic acceleration maneuver. This segment utilizes a dive technique which is described in Section II. Enter the curve with the end-of-climb weight determined from figure A3-1. (Any necessary heading changes should be accomplished prior to the transonic acceleration maneuver since turns will seriously affect performance during this phase.)

Performance for the supersonic climbing acceleration to the desired cruise altitude is provided by figure A3-3. The data are based on tests with temperatures varying from 10°C above standard at 30,000 feet to 10°C colder than standard at 46,000 feet and then increasing almost linearly with altitude to 7°C above standard at 74,000 feet. Enter the chart with the weight at the end of transonic acceleration determined from figure A3-2.

When constant Mach climb is required to reach cruise altitude, climb at Maximum afterburning power. For planning purposes use 4000 FPM rate of climb and 900 pounds/minute fuel flow.

Sample Use of Charts

Determine the time, fuel and distance required for a climb to M 3.0 at 70,600 feet from sea level, ramp weight of 126,000 pounds.

Segment	Time	Fuel Used	Distance	End Weight
Start engines	0	0	0	126,000
Ground Maneuver (125 lb/min)	(0:24)	3000 3000	0	123,000
Takeoff and accel to 400 KEAS at S. L.	0:02.2 0:02.2	3800 6800	10 10	119,200
Climb to 30,000 feet	0:03.2 0:05.4	3600 10,400	27 37	115,600
Accel to M 1.25	0:02.4 0:07.8	3000 13,400	25 62	112,600
Climb to 70,600 feet (M 3.0, 410 KEAS)	0:09.8 0:17.6	13,900 27,300	206 268	98,700
	Min.	Lb.	N. Mi.	Lb.

Figures A3-4 and A3-5 summarize normal climbing acceleration performance from brake release and from start of acceleration after refuel, respectively. A normal range of operating altitudes and gross weights is presented.

Alternate Climb Performance

Figure A3-6 summarizes time, fuel and distance requirements from takeoff for an alternate, 400 KEAS supersonic climb schedule at a nominal takeoff gross weight of 130,000 lb.

Normal Descent Performance

Time, fuel, and distance for normal 350 KEAS descents are shown in figure A3-7.

Single Engine Descent Performance

Figures A3-8 through A3-12 present single engine descent performance at 300, 350 and 400 KEAS. The charts are based on flight test data with the inlet configuration as listed in the charts. Time, distance, and fuel required are plotted versus altitude. The best range is obtained with the

300 KEAS schedule. Specific range begins to decrease rapidly near 50,000 feet; therefore, the charts are indexed to an altitude of 50,000 feet so that a power reduction technique can be used and the resultant change in performance can be determined. <u>The effect of changing KEAS at the indexed 50,000 feet has not been defined by flight testing and is not included in the data.</u>

Figure A3-8 summarizes the effects of speed on Maximum AB descents for constant values of 300, 350, and 400 KEAS. Figures A3-9 through A3-11 present the effects of decreasing power below the index altitude of 50,000 feet at constant airspeeds of 300, 350, and 400 KEAS respectively. Figure A3-12 presents the effects of a 180° turn at 35° bank angle on a 350 KEAS descent. Approximately 23,000 feet of altitude is required to complete the 180° turn. For convenience in mission planning, a ground track profile is also provided.

Example (1):

Find the time, distance, and fuel required to descend on course from 80,000 feet and Mach 3.10 using the 300 KEAS descent schedule and Minimum AB below 50,000 feet. Enter figure A3-9 at 80,000 feet and read the time, distance, and fuel required to 50,000 feet as 8.6 minutes, 176 nautical miles and 2800 pounds of fuel. Reenter at the final altitude of 31,500 feet on the Minimum AB line and read time, distance, and fuel required as 5 minutes, 52 nautical miles and 1600 lb of fuel. Total the readings and obtain 13.6 minutes, 228 nautical miles, and 4400 pounds of fuel.

Example (2):

Find the track time, distance, and fuel required to descend from 80,000 feet and Mach 3.10 using the 350 KEAS descent schedule. A 90° turn is to be completed above 50,000 feet, and Minimum AB is to be used below 50,000 feet. Enter figure A3-12 and note on the penetration distance curve that 90° of turn is completed at 65,000 feet altitude. Read time at that altitude as 3.0 minutes and fuel used as 1100 pounds. On the ground track profile note that the distance traveled is 80 nautical miles. Enter figure A3-10 at 65,000 feet (end of turn altitude) and read time, distance, and fuel required to 50,000 feet as 4.1 minutes, 75 nautical miles, and 1900 pounds. Reenter at the final altitude of 28,000 feet on the Minimum AB line and read time, distance, and fuel required as 3.4 minutes, 41 nautical miles, and 1100 lb of fuel. Total the readings and obtain 10.5 minutes, 196 nautical miles, and 4100 pounds of fuel.

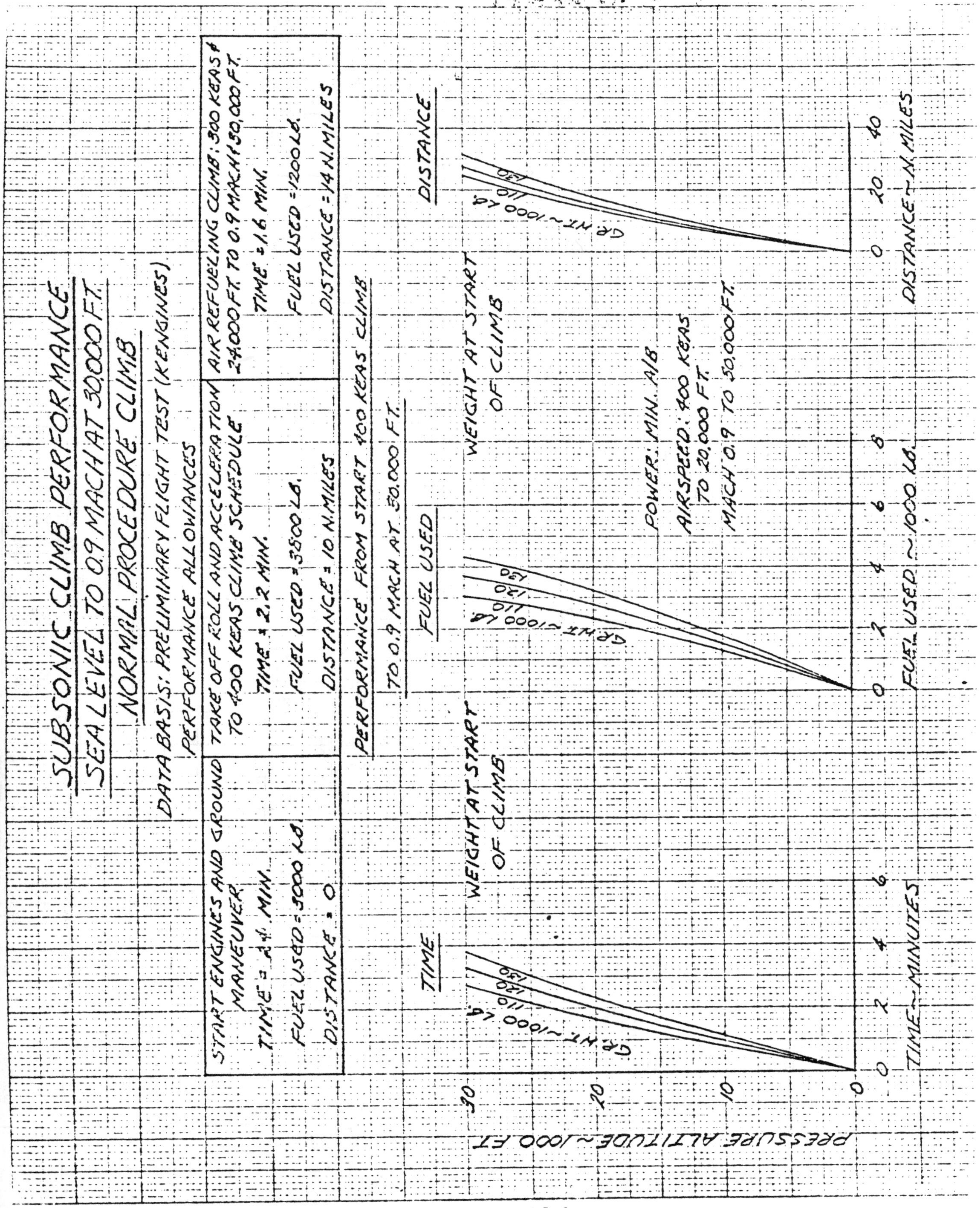

SUBSONIC CLIMB PERFORMANCE

SEA LEVEL TO 0.9 MACH AT 30,000 FT.

NORMAL PROCEDURE CLIMB

DATA BASIS: PRELIMINARY FLIGHT TEST (K ENGINES)

PERFORMANCE ALLOWANCES

START ENGINES AND GROUND MANEUVER	TAKE OFF ROLL AND ACCELERATION TO 400 KEAS CLIMB SCHEDULE	AIR REFUELING CLIMB: 300 KEAS & 24,000 FT. TO 0.9 MACH & 30,000 FT.
TIME = 2.4 MIN.	TIME = 2.2 MIN.	TIME = 1.6 MIN.
FUEL USED = 3000 LB.	FUEL USED = 3800 LB.	FUEL USED = 1200 LB.
DISTANCE = 0	DISTANCE = 10 N. MILES	DISTANCE = 14 N. MILES

PERFORMANCE FROM START 400 KEAS CLIMB TO 0.9 MACH AT 30,000 FT.

TIME

FUEL USED

DISTANCE

WEIGHT AT START OF CLIMB

GR. WT. ~1000 LB.

110

120

130

POWER: MIN. A/B

AIRSPEED: 400 KEAS TO 20,000 FT.

MACH 0.9 TO 30,000 FT.

PRESSURE ALTITUDE ~1000 FT.

0 10 20 30

0 2 4 6

TIME ~ MINUTES

0 2 4 6 8

FUEL USED ~ 1000 LB.

0 20 40

DISTANCE ~ N. MILES

Figure A3-1

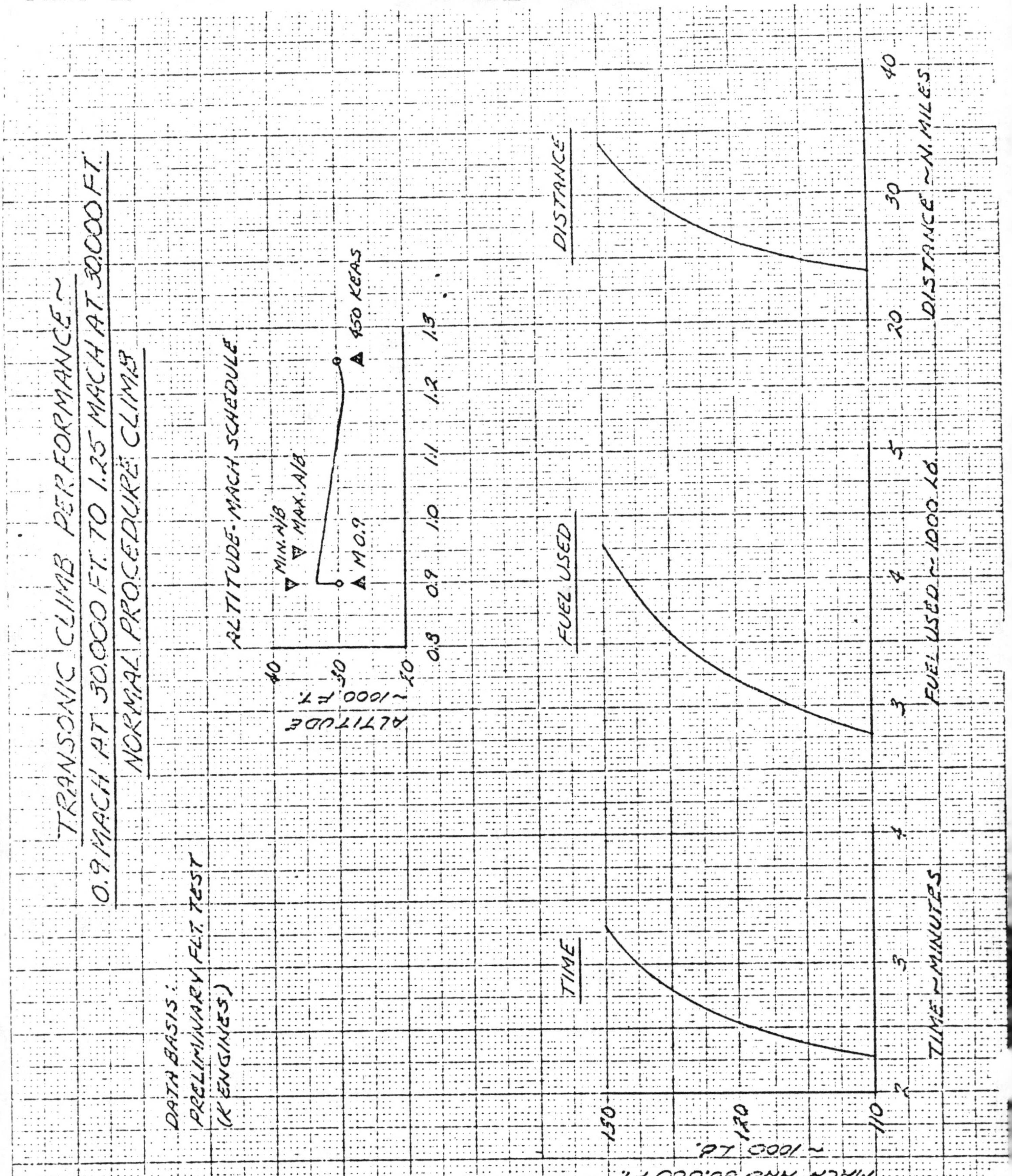

TRANSONIC CLIMB PERFORMANCE ~
0.9 MACH AT 30,000 FT. TO 1.25 MACH AT 30,000 FT.
NORMAL PROCEDURE CLIMB

DATA BASIS:
PRELIMINARY FLT. TEST
(K ENGINES)

ALTITUDE-MACH SCHEDULE
ALTITUDE ~1000 FT.
40
30
20
MIN. A/B
MAX. A/B
M 0.9
450 KEAS
0.8
0.9
1.0
1.1
1.2
1.3

TIME
FUEL USED
DISTANCE
GROSS WEIGHT AT 0.9 MACH AND 30,000 FT. ~1000 LB.
150
120
110
2
3
4
TIME ~ MINUTES
3
4
5
FUEL USED ~ 1000 LB.
20
30
40
DISTANCE ~ N. MILES

Figure A3-2

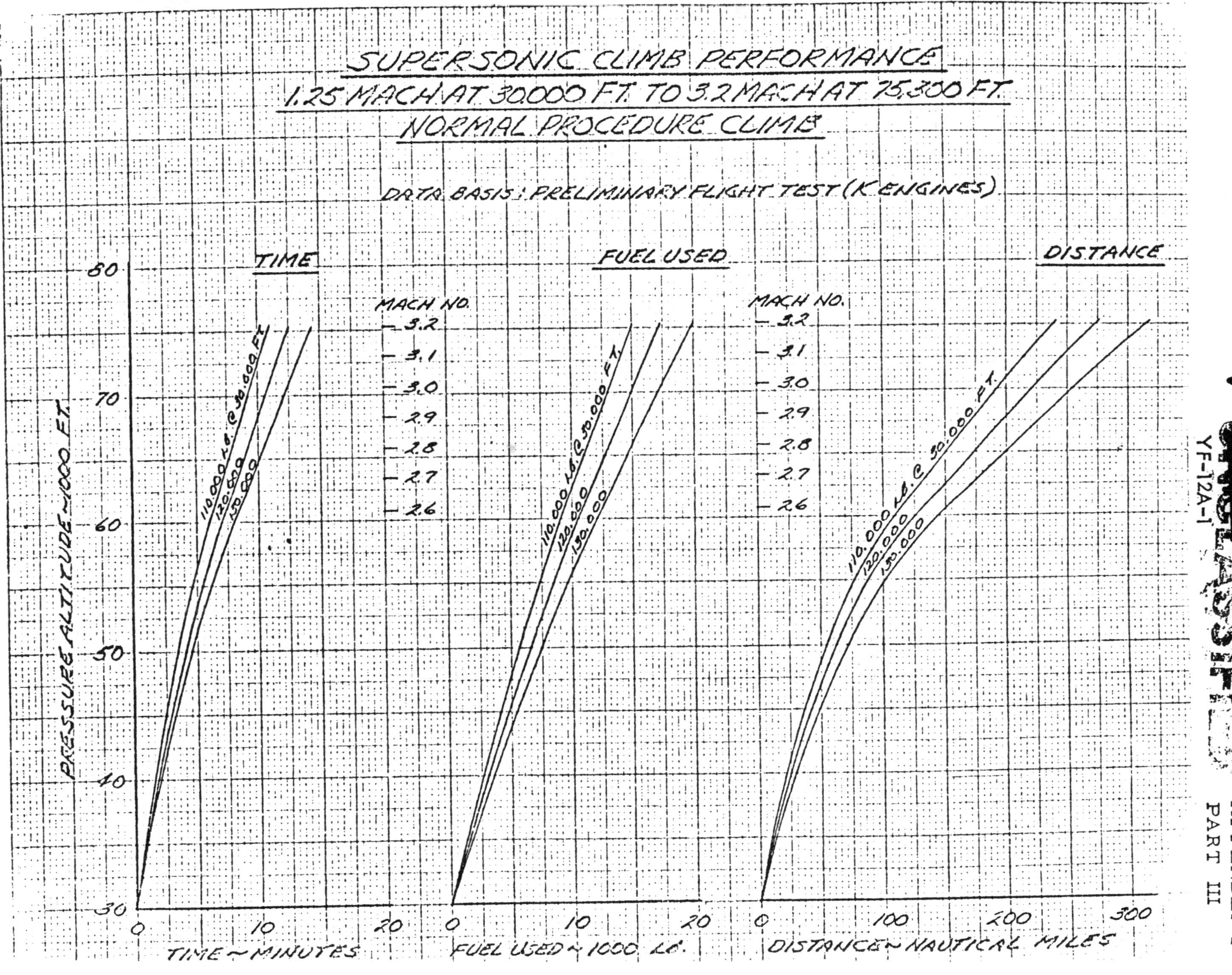
SUPERSONIC CLIMB PERFORMANCE
1.25 MACH AT 30,000 FT. TO 3.2 MACH AT 75,300 FT.
NORMAL PROCEDURE CLIMB
DATA BASIS: PRELIMINARY FLIGHT TEST (K ENGINES)
TIME
FUEL USED
DISTANCE
MACH NO.
3.2
3.1
3.0
2.9
2.8
2.7
2.6
110,000 LB. @ 30,000 FT.
120,000
130,000
PRESSURE ALTITUDE ~ 1000 FT.
30
40
50
60
70
80
TIME ~ MINUTES
0
10
20
FUEL USED ~ 1000 LB.
0
10
20
DISTANCE ~ NAUTICAL MILES
0
100
200
300

Figure A3-3

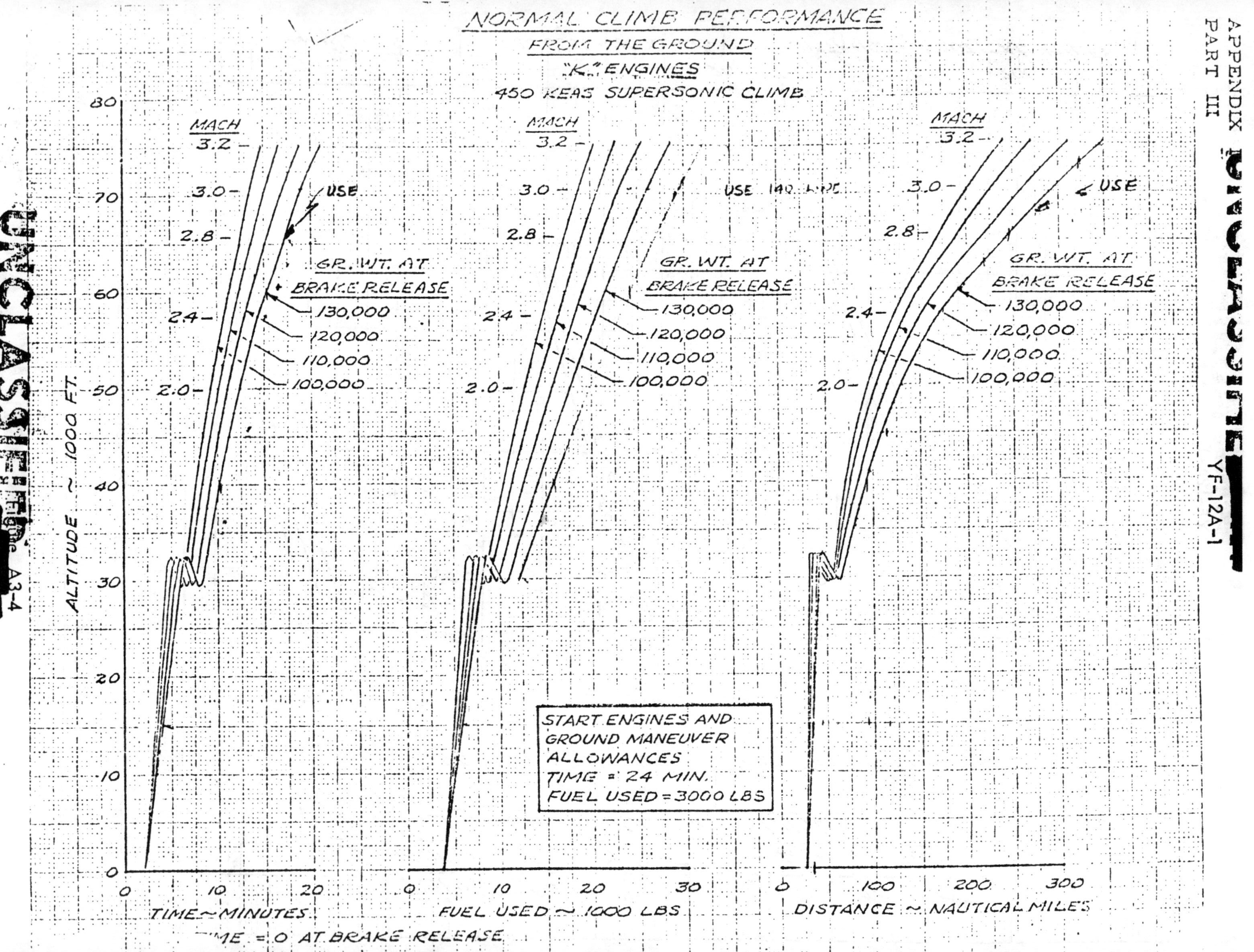
NORMAL CLIMB PERFORMANCE
FROM THE GROUND
"K" ENGINES
450 KEAS SUPERSONIC CLIMB
MACH
3.2
3.0
2.8
2.4
2.0
USE
GR. WT. AT
BRAKE RELEASE
130,000
120,000
110,000
100,000
START ENGINES AND
GROUND MANEUVER
ALLOWANCES
TIME = 24 MIN.
FUEL USED = 3000 LBS
ALTITUDE ~ 1000 FT.
0
10
20
30
40
50
60
70
80
TIME ~ MINUTES
FUEL USED ~ 1000 LBS
DISTANCE ~ NAUTICAL MILES
100
200
300

Figure A3-4

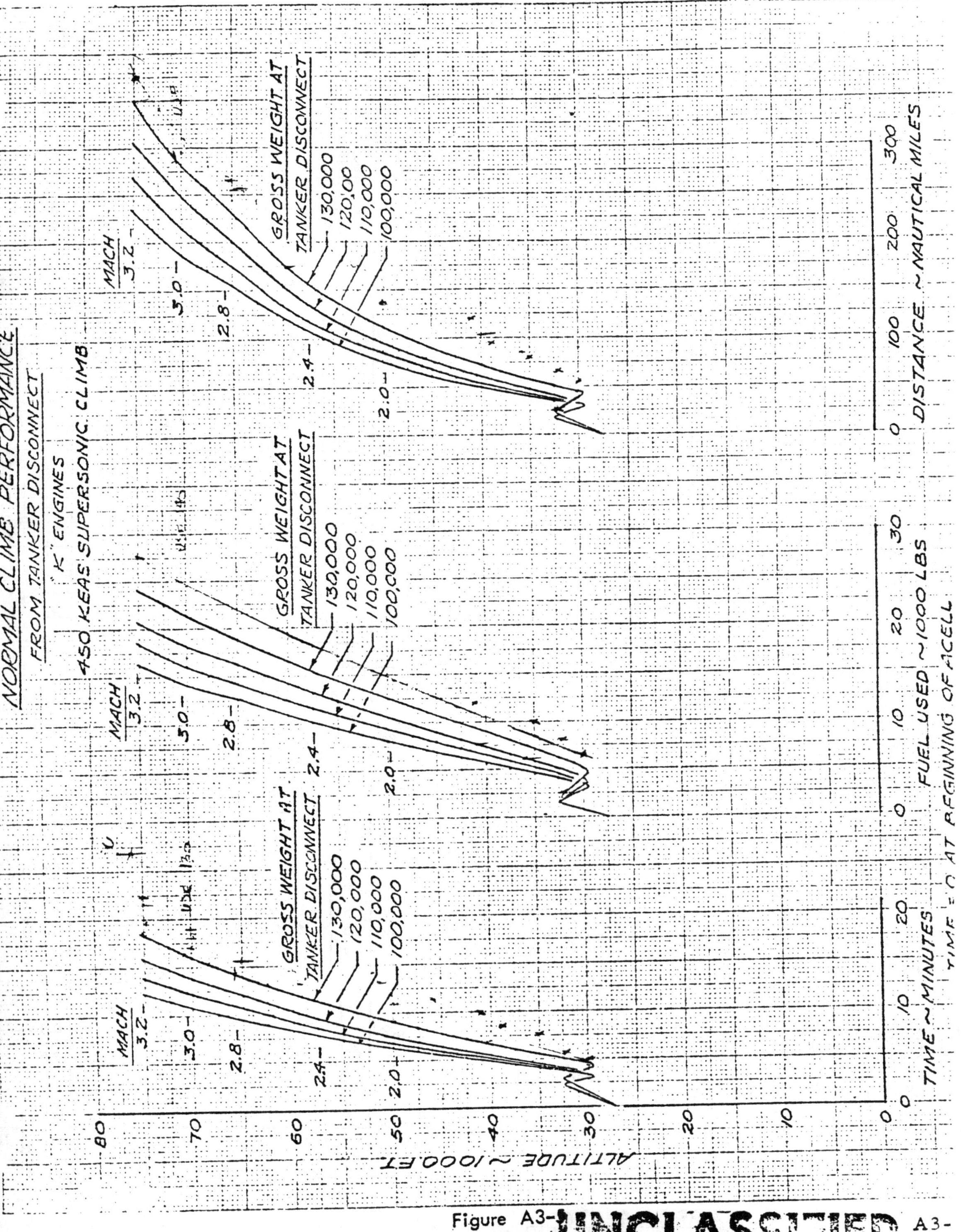

NORMAL CLIMB PERFORMANCE

FROM TANKER DISCONNECT

"K" ENGINES

450 KEAS SUPERSONIC CLIMB

MACH
3.2
3.0
2.8
2.4
2.0

GROSS WEIGHT AT TANKER DISCONNECT
130,000
120,000
110,000
100,000

MACH
3.2
3.0
2.8
2.4
2.0

GROSS WEIGHT AT TANKER DISCONNECT
130,000
120,000
110,000
100,000

MACH
3.2
3.0
2.8
2.4
2.0

GROSS WEIGHT AT TANKER DISCONNECT
130,000
120,00
110,000
100,000

ALTITUDE ~ 1000 FT.
80
70
60
50
40
30
20
10
0

0 10 20
TIME ~ MINUTES

0 10 20 30
FUEL USED ~ 1000 LBS

0 100 200 300
DISTANCE ~ NAUTICAL MILES

TIME = 0 AT BEGINNING OF ACELL

Figure A3-

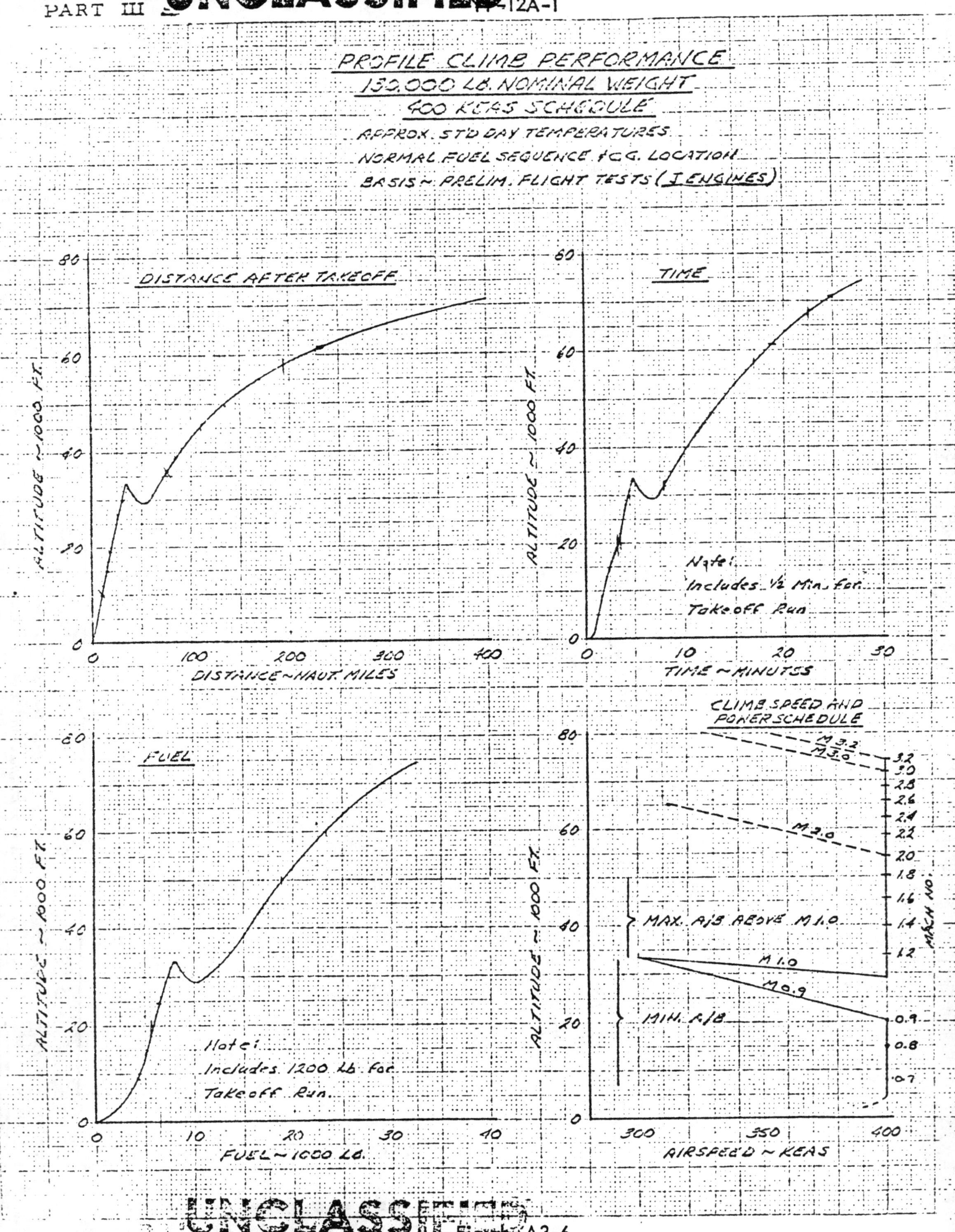
PROFILE CLIMB PERFORMANCE
130,000 LB. NOMINAL WEIGHT
400 KEAS SCHEDULE
APPROX. STD DAY TEMPERATURES
NORMAL FUEL SEQUENCE & C.G. LOCATION
BASIS ~ PRELIM. FLIGHT TESTS (J ENGINES)
DISTANCE AFTER TAKEOFF
ALTITUDE ~ 1000 FT.
DISTANCE ~ NAUT. MILES
TIME
Note:
Includes 1/2 Min. for Takeoff Run
TIME ~ MINUTES
FUEL
Note:
Includes 1200 Lb. for Takeoff Run.
FUEL ~ 1000 LB.
CLIMB SPEED AND POWER SCHEDULE
M 3.2
M 3.0
M 2.0
MAX. A/B ABOVE M 1.0
M 1.0
M 0.9
MIN. A/B
MACH NO.
AIRSPEED ~ KEAS

Figure A3-6

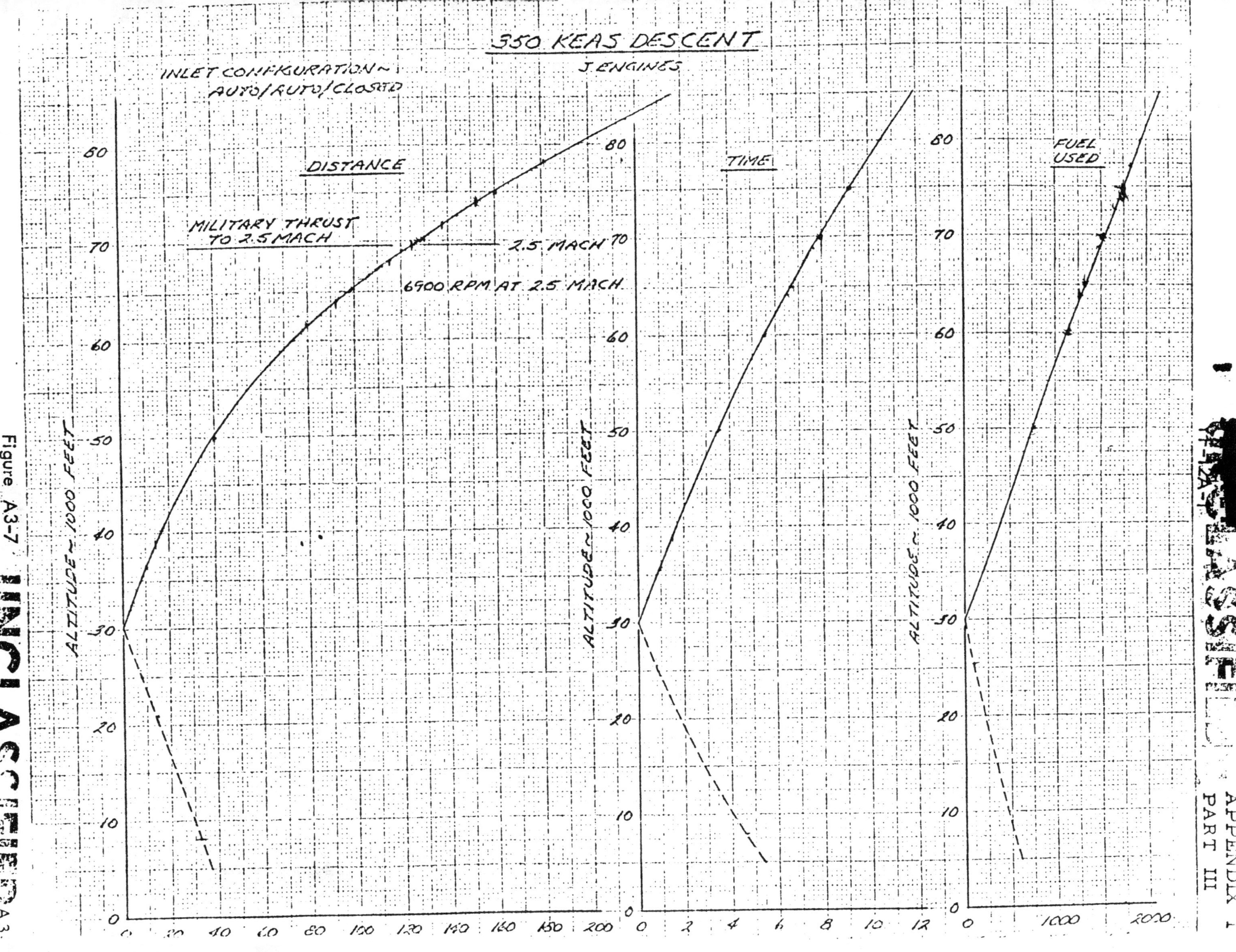
350 KEAS DESCENT
J ENGINES
INLET CONFIGURATION ~
AUTO/AUTO/CLOSED
DISTANCE
MILITARY THRUST
TO 2.5 MACH
2.5 MACH
6900 RPM AT 2.5 MACH
TIME
FUEL USED
ALTITUDE ~ 1000 FEET
ALTITUDE ~ 1000 FEET
ALTITUDE ~ 1000 FEET

Figure A3-7

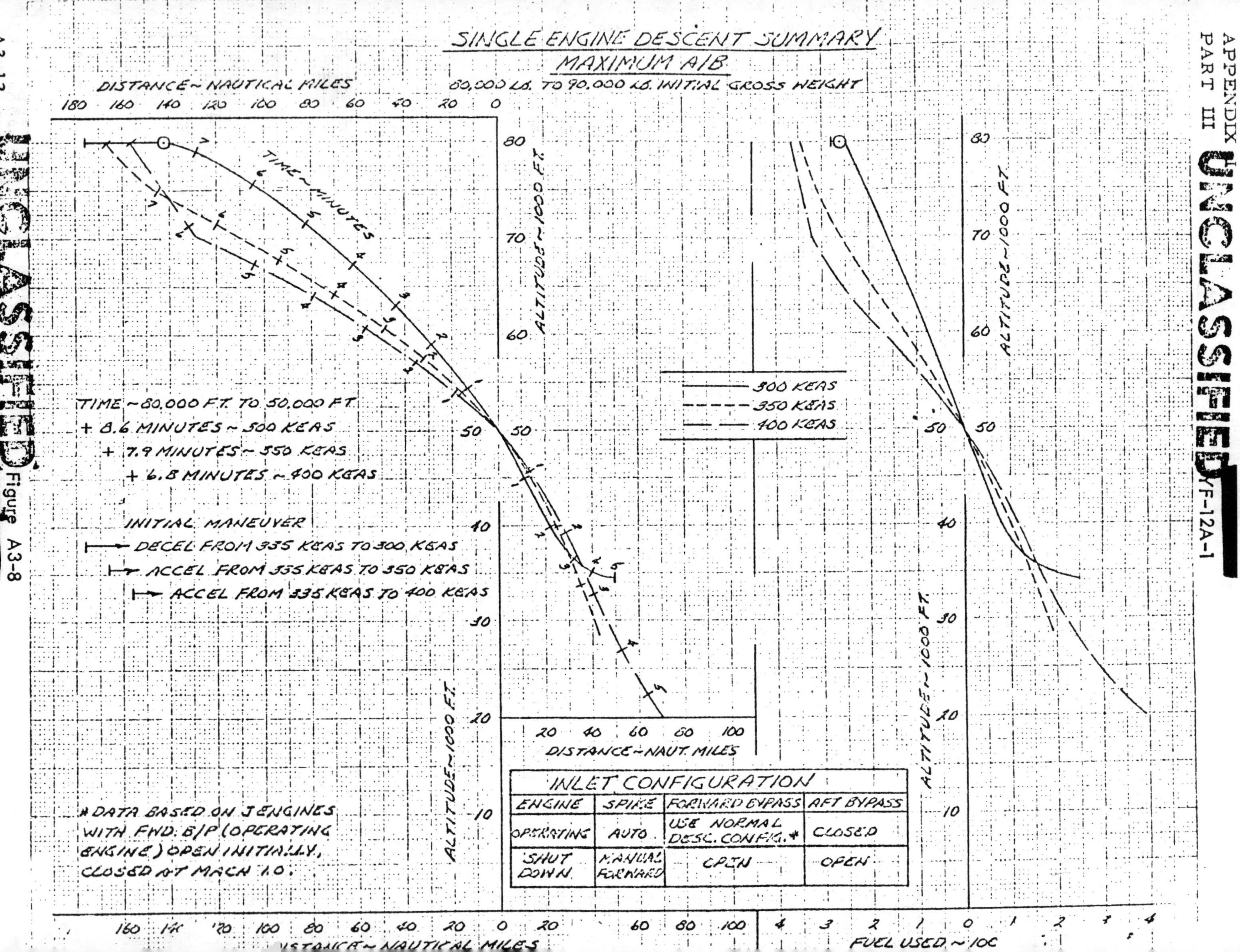
SINGLE ENGINE DESCENT SUMMARY
MAXIMUM A/B
80,000 LB. TO 90,000 LB. INITIAL GROSS WEIGHT
DISTANCE ~ NAUTICAL MILES
TIME ~ MINUTES
ALTITUDE ~ 1000 FT.
300 KEAS
350 KEAS
400 KEAS
TIME ~ 80,000 FT. TO 50,000 FT.
+ 8.6 MINUTES ~ 300 KEAS
+ 7.9 MINUTES ~ 350 KEAS
+ 6.8 MINUTES ~ 400 KEAS
INITIAL MANEUVER
DECEL FROM 335 KEAS TO 300 KEAS
ACCEL FROM 335 KEAS TO 350 KEAS
ACCEL FROM 335 KEAS TO 400 KEAS
DISTANCE ~ NAUT. MILES
INLET CONFIGURATION
ENGINE | SPIKE | FORWARD BYPASS | AFT BYPASS
OPERATING | AUTO | USE NORMAL DESC. CONFIG.* | CLOSED
SHUT DOWN | MANUAL FORWARD | OPEN | OPEN
* DATA BASED ON J ENGINES WITH FWD. B/P (OPERATING ENGINE) OPEN INITIALLY, CLOSED AT MACH 1.0.
DISTANCE ~ NAUTICAL MILES
FUEL USED ~ 10C

UNCLASSIFIED

Figure A3-8

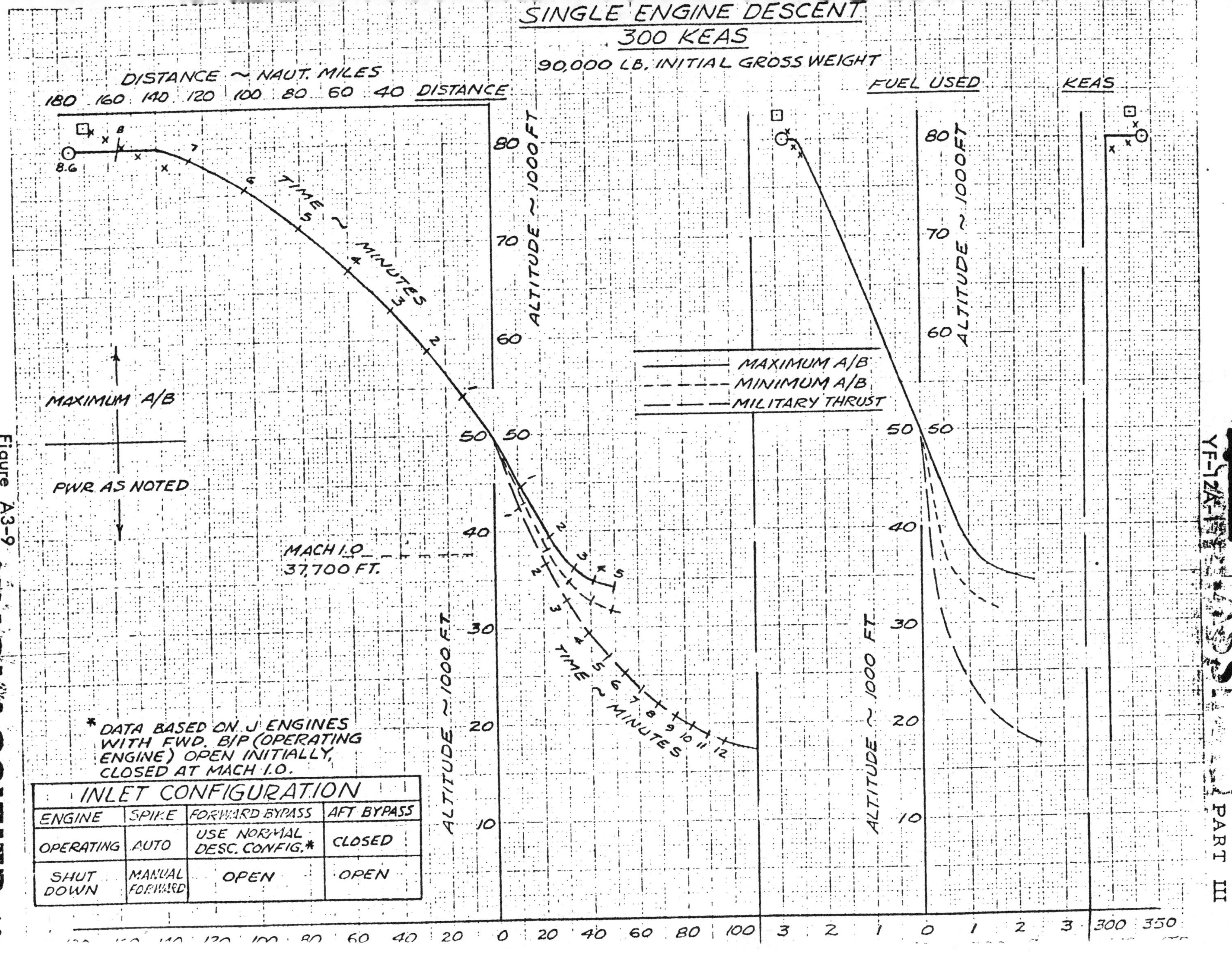
SINGLE ENGINE DESCENT
300 KEAS
90,000 LB. INITIAL GROSS WEIGHT
DISTANCE ~ NAUT. MILES
DISTANCE
FUEL USED
KEAS
ALTITUDE ~ 1000 FT
TIME ~ MINUTES
MAXIMUM A/B
PWR AS NOTED
MACH 1.0
37,700 FT.
MAXIMUM A/B
MINIMUM A/B
MILITARY THRUST
* DATA BASED ON J ENGINES WITH FWD. B/P (OPERATING ENGINE) OPEN INITIALLY, CLOSED AT MACH 1.0.
INLET CONFIGURATION
ENGINE | SPIKE | FORWARD BYPASS | AFT BYPASS
OPERATING | AUTO | USE NORMAL DESC. CONFIG.* | CLOSED
SHUT DOWN | MANUAL FORWARD | OPEN | OPEN

Figure A3-9

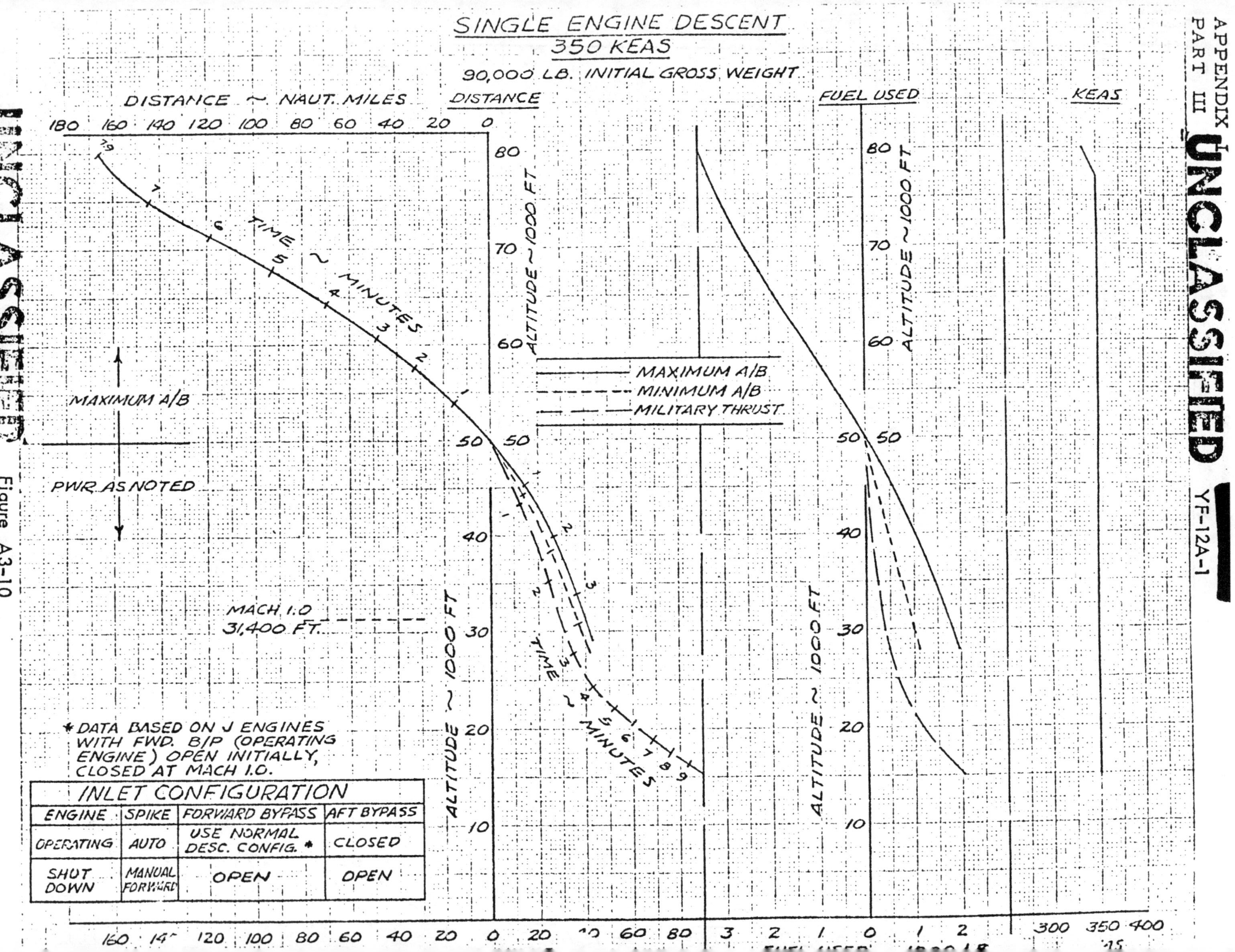

ENGINE	SPIKE	FORWARD BYPASS	AFT BYPASS
OPERATING	AUTO	USE NORMAL DESC. CONFIG. *	CLOSED
SHUT DOWN	MANUAL FORWARD	OPEN	OPEN

UNCLASSIFIED

Figure A3-10

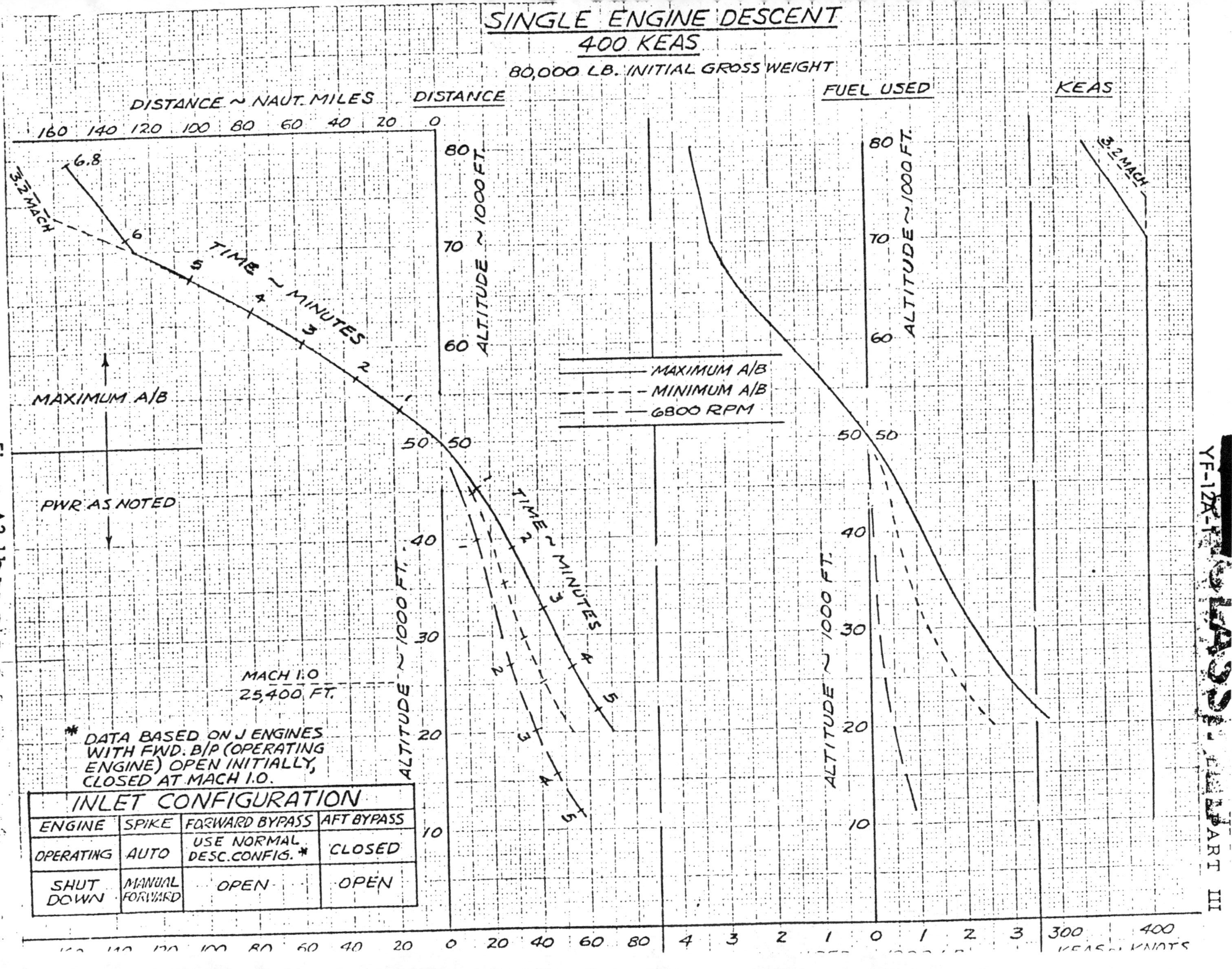

ENGINE	SPIKE	FORWARD BYPASS	AFT BYPASS
OPERATING	AUTO	USE NORMAL DESC. CONFIG. *	CLOSED
SHUT DOWN	MANUAL FORWARD	OPEN	OPEN

Figure A3-11

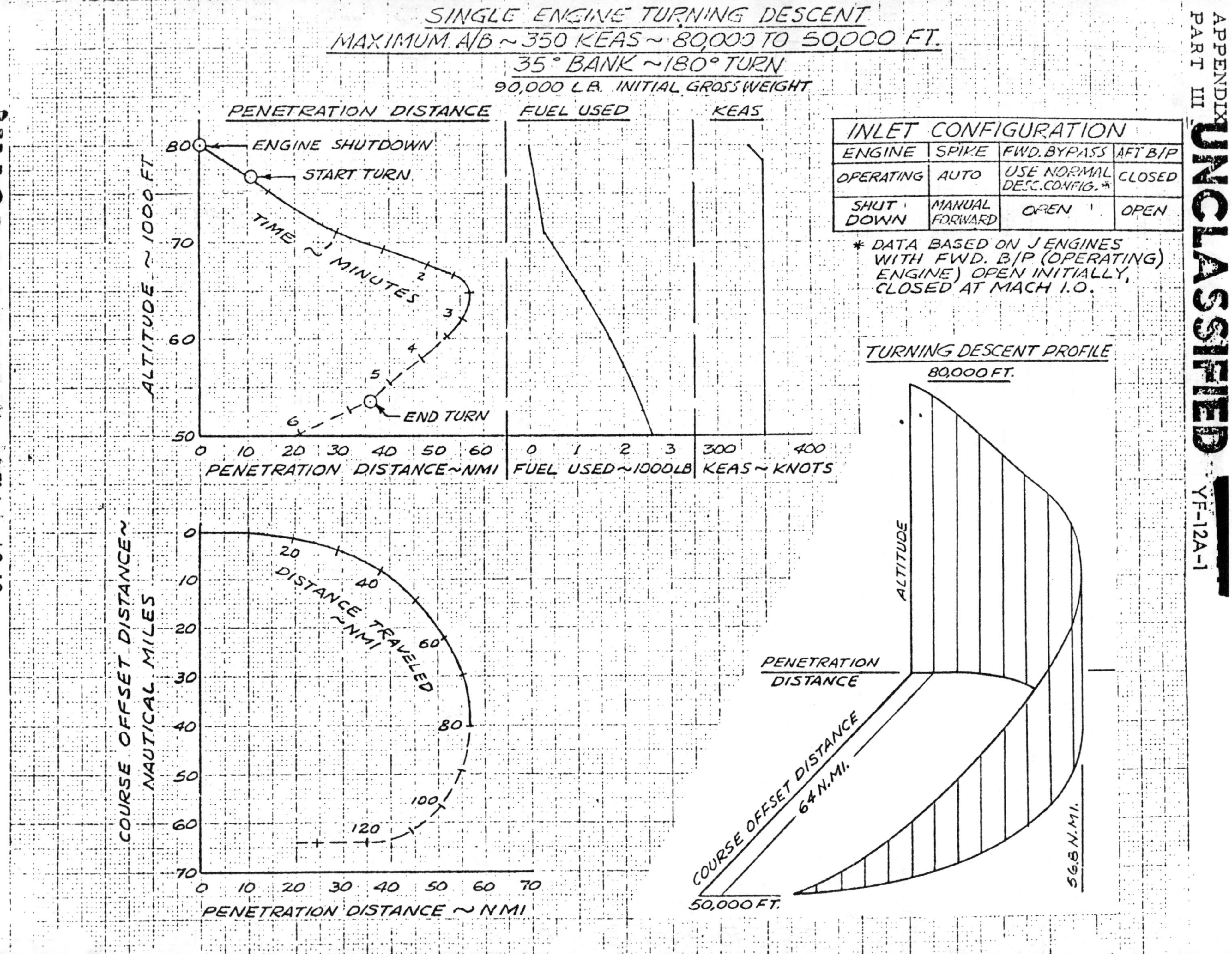
SINGLE ENGINE TURNING DESCENT
MAXIMUM A/B ~ 350 KEAS ~ 80,000 TO 50,000 FT.
35° BANK ~ 180° TURN
90,000 LB. INITIAL GROSS WEIGHT
PENETRATION DISTANCE
FUEL USED
KEAS
ENGINE SHUTDOWN
START TURN
TIME ~ MINUTES
END TURN
ALTITUDE ~ 1000 FT
PENETRATION DISTANCE ~ NMI
FUEL USED ~ 1000 LB
KEAS ~ KNOTS
INLET CONFIGURATION
ENGINE | SPIKE | FWD. BYPASS | AFT B/P
OPERATING | AUTO | USE NORMAL DESC. CONFIG.* | CLOSED
SHUT DOWN | MANUAL FORWARD | OPEN | OPEN
* DATA BASED ON J ENGINES WITH FWD. B/P (OPERATING ENGINE) OPEN INITIALLY, CLOSED AT MACH 1.0.
TURNING DESCENT PROFILE
80,000 FT.
ALTITUDE
PENETRATION DISTANCE
COURSE OFFSET DISTANCE
64 N.MI.
56.8 N.MI.
50,000 FT.
DISTANCE TRAVELED ~NMI
COURSE OFFSET DISTANCE ~ NAUTICAL MILES
PENETRATION DISTANCE ~ NMI

Figure A3-12

PART IV

SUBSONIC CRUISE PERFORMANCE

TABLE OF CONTENTS

PART IV

SUBSONIC CRUISE PERFORMANCE

List of Illustrations

Introduction

This part of the appendix provides the basic miles per pound data for operation at subsonic cruise speeds. Long range profile data for level flight and cruise climb schedules as well as performance summaries for buddy missions and loiter speed schedules are also provided.

Note

This data is applicable only to two-engine operation without afterburning. Refer to Section III, Emergency Procedures, for single-engine capabilities with and without afterburning.

Data Basis

Performance in this section is based on flight tests. All performance information has been corrected to a c.g. of 24% MAC. Operation at c.g. positions forward of 24% reduces range approximately 1% for each percent c.g. shift, as noted on the specific range cruves. The buddy mission schedule, Mach 0.75, is based on performance compatibility with KC-135 tanker aircraft.

Specific Range

Specific Range vs Mach number is shown in figures A4-1 thru A4-13 for altitudes from 10,000 ft to 40,000 ft. The curves show performance for the operational range of gross weights at each altitude and cover the speed range available from maximum endurance to Military thrust. Subscales of equivalent airspeed and standard day true airspeed are provided. The recommended loiter speed schedule is included for each altitude.

Subsonic Maximum Range Cruise–Climb

Figure A4-14 shows the range available by cruise-climbing at 0.88 Mach number and at 382,000 lb W/δ. The curve is indexed to an end-of-cruise gross weight of 70,000 lbs (approximately 10,000 lb of fuel remaining) as a convenience in flight planning. (10,000 lbs is an arbitrary fuel remaining value and is not mandatory for termination of cruise.) The curve can also be used on an incremental basis for any desired start and end cruise condition.

Example:

For a gross weight of 90,000 lbs (approx. 30,000 lbs of fuel remaining) determine the distance remaining to 70,000 lbs gross weight and the distance flown between 90,000 lbs gross weight and 80,000 lbs gross weight. Enter figure A4-14 at 90,000 lbs gross weight and read the distance to 70,000 bls directly as 780 nautical miles. At 80,000 lbs read the distance to 70,000 lbs as 415 nautical miles. The distance flown between 90,000 lbs and 80,000 lbs is (780-415) or 365 nautical miles.

An additional 230 nautical miles is available by ending cruise at 65,000 lbs gross weight (approximately 5000 lbs of fuel remaining).

Range Factor Summary & Max Specific Range

The Range Factor Summary shown in figure A4-15 defines the maximum specific range capability in terms of the gross weight/altitude parameter, W/δ . The Mach number schedule required to obtain this performance is shown on the lower portion of the curve. The curve covers the complete operating weight and non-afterburning power range of the airplane, up to and including Military thrust. Maximum range capability (cruise-climb) is obtained at a W/δ of 382,000 lb and at 0.88 Mach number. Maximum range at Military thrust is obtained by cruise-climbing at 422,000 lb W/δ and at 0.91 Mach number.

Figure A4-16 depicts the information given by the Range Factor Summary in terms of Maximum Specific Range as a function of gross weight and altitude.

Buddy Mission At 0.75 Mach No.

Figure A4-17 shows the specific range available in cruise climb or at constant altitude with constant 0.75 Mach number cruise speed. Greater range capability is obtained by cruise-climbing.

Example:

Determine the range available at 28,000 ft at an initial gross weight of 90,000 lb if 10,000 lb of fuel is to be consumed. Enter figure A4-17 at 90,000 lbs at 28,000 ft and read the specific range available as 32 nautical miles per 1000 lbs of fuel. At 80,000 lb and 28,000 ft read the specific range as 35 nautical miles per 1000 lbs of fuel. The average specific range is 33.5 nautical miles per 1000 lbs and the range available for 10,000 lbs of fuel consumed is 335 nautical miles.

Loiter Performance

Figure A4-18 presents the loiter performance of the aircraft in terms of minutes per 1000 lbs of fuel consumed. The chart is keyed to a loiter speed of 275 KIAS at gross weights less than 80,000 lbs (approximately 20,000 lbs of fuel remaining) and higher speeds as noted for heavier weights.

Example:

Determine the loiter time availabl 20,000 ft at an initial gross weight o 90,000 lbs if 10,000 lbs of fuel is to be consumed. Enter figure A4-18 at 90,000 lbs gross weight and 20,000 and read loiter time available as 4.2 minutes per 1000 lbs of fuel. At 80,000 lbs and 20,000 ft read 4.6 minutes per 1000 lbs of fuel. The average loiter time available is 4.4 minutes per 1000 lbs of fuel or 44 minutes for 10,000 lbs of fuel.

Subsonic Mission Profile

Figure A4-19 presents the constant altitude, cruise climb, and Military thrust range capability of the aircraft in terms of distance to go to 70,000 lbs gross weight (approximately 10,000 lbs of fuel remaining). The curve may also be used on an incremental bas Cruise speeds for constant altitude cruise are tabulated on the chart. Climb performance tables are also provided which apply to takeoff (brake release) at nominal gross weights of 135,200 lb (78,200 lbs of fuel minus 3000 lbs ground maneuver allowance) and 120,000 lb. The climb data are based on Maximum thrust takeoff with power reduction to Minimum afterburning following gear retraction.

Maximum range following refueling is obtained by conducting the refueling operation at or near the altitude corresponding to the desired fuel load. This procedure will permit initiatio of cruise-climb immediately follo disconnect.

Example:

Determine the range available at an initial gross weight of 125,000 lbs if cruise is to be terminated at 10,000 lbs remaining (approximately 70,000 lbs gross weight). Figure A4-19 shows that by cruising at 25,000 feet at an initial gross weight of 125,000 lbs the range will be 1636 nautical miles to 10,000 lbs of fuel remaining. This range increases to 1736 nautical miles by cruising at 30,000 feet. Maximum range is available by cruise climbing at 0.88 Mach number. Under this condition cruise would be initiated at 27,900 feet and ended at 40,200 feet at 10,000 lbs remaining. Distance traveled would be 1795 nautical miles.

SUBSONIC SPECIFIC RANGE
TWO ENGINES
10,000 FT ALTITUDE
STANDARD DAY 24% MAC C.G.

J ENGINES

LOITER: 275 KIAS With 20,000 Lbs Or Less Remaining. Increase KIAS 1.5 Knots Per 1000 Lbs Remaining Above 20,000 Lbs.

Decrease Fuel Economy 1% For 1% Shift Of C.G. Forward Of 24% MAC.

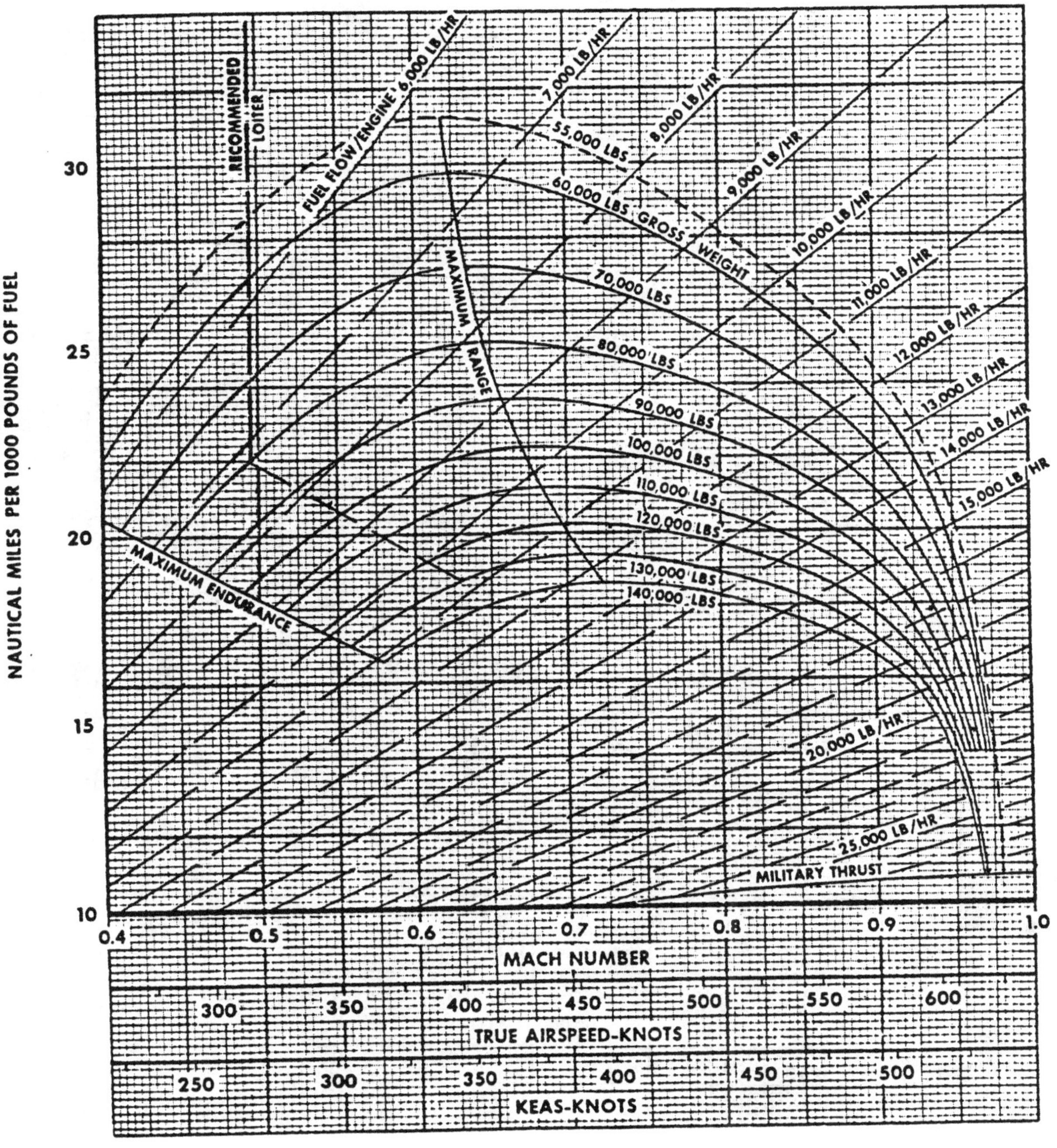

NAUTICAL MILES PER 1000 POUNDS OF FUEL

10
15
20
25
30

RECOMMENDED LOITER

FUEL FLOW/ENGINE 6,000 LB/HR
7,000 LB/HR
8,000 LB/HR
9,000 LB/HR
10,000 LB/HR
11,000 LB/HR
12,000 LB/HR
13,000 LB/HR
14,000 LB/HR
15,000 LB/HR
20,000 LB/HR
25,000 LB/HR
MILITARY THRUST

55,000 LBS
60,000 LBS GROSS WEIGHT
70,000 LBS
80,000 LBS
90,000 LBS
100,000 LBS
110,000 LBS
120,000 LBS
130,000 LBS
140,000 LBS

MAXIMUM RANGE
MAXIMUM ENDURANCE

0.4 0.5 0.6 0.7 0.8 0.9 1.0
MACH NUMBER

300 350 400 450 500 550 600
TRUE AIRSPEED-KNOTS

250 300 350 400 450 500
KEAS-KNOTS

Figure A4-1

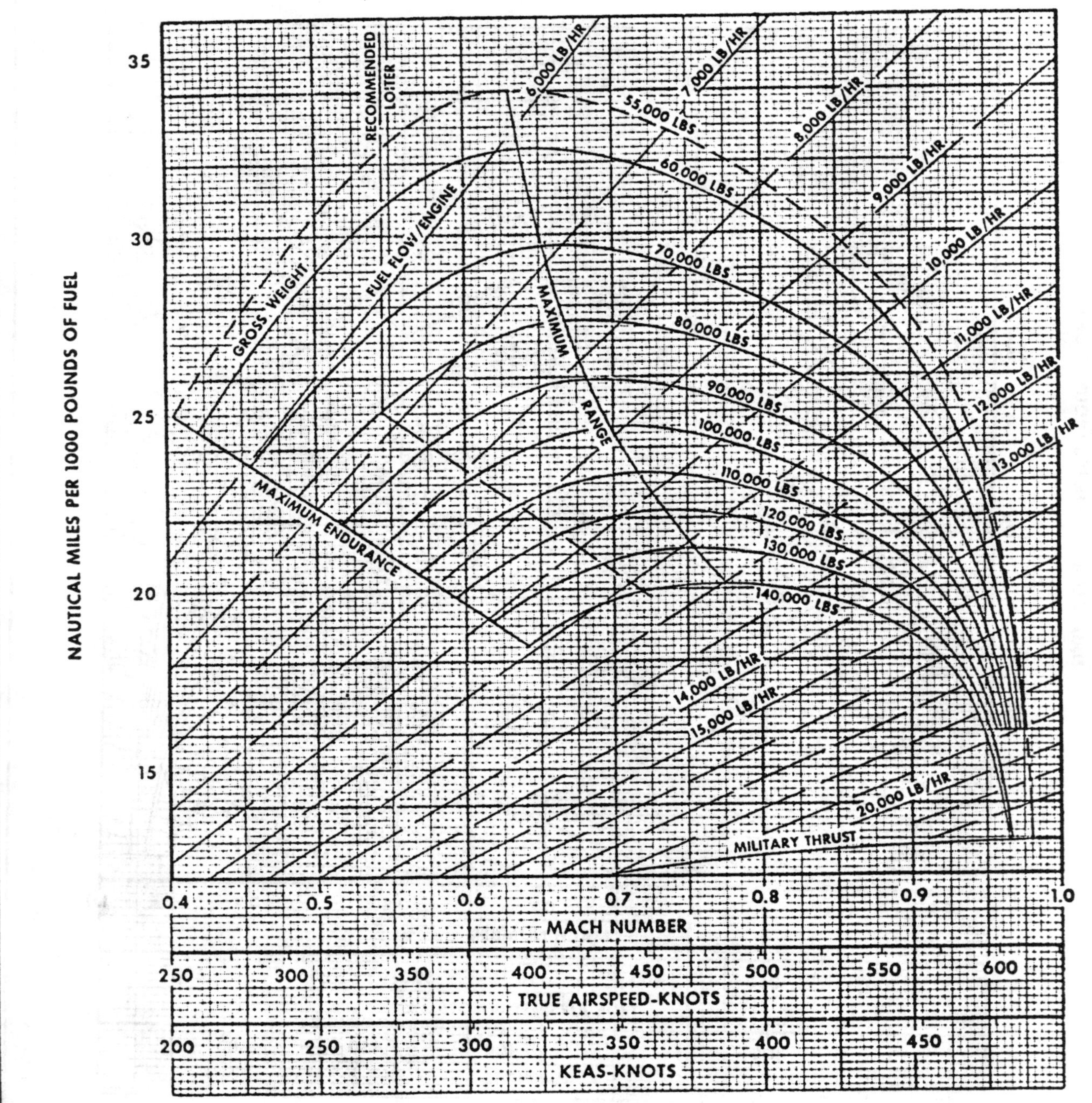
J ENGINES
SUBSONIC SPECIFIC RANGE
TWO ENGINES
15,000 FT ALTITUDE
STANDARD DAY 24% MAC C.G.
LOITER: 275 KIAS With 20,000 Lbs Or Less Remaining. Increase KIAS 1.5 Knots Per 1000 Lbs Remaining Above 20,000 Lbs.
Decrease Fuel Economy 1% For 1% Shift Of C.G. Forward Of 24% MAC.
NAUTICAL MILES PER 1000 POUNDS OF FUEL
35
30
25
20
15
RECOMMENDED LOITER
GROSS WEIGHT
FUEL FLOW/ENGINE
MAXIMUM RANGE
MAXIMUM ENDURANCE
6,000 LB/HR
7,000 LB/HR
8,000 LB/HR
9,000 LB/HR
10,000 LB/HR
11,000 LB/HR
12,000 LB/HR
13,000 LB/HR
14,000 LB/HR
15,000 LB/HR
20,000 LB/HR
55,000 LBS
60,000 LBS
70,000 LBS
80,000 LBS
90,000 LBS
100,000 LBS
110,000 LBS
120,000 LBS
130,000 LBS
140,000 LBS
MILITARY THRUST
0.4
0.5
0.6
0.7
0.8
0.9
1.0
MACH NUMBER
250
300
350
400
450
500
550
600
TRUE AIRSPEED-KNOTS
200
250
300
350
400
450
KEAS-KNOTS

Figure A4-2

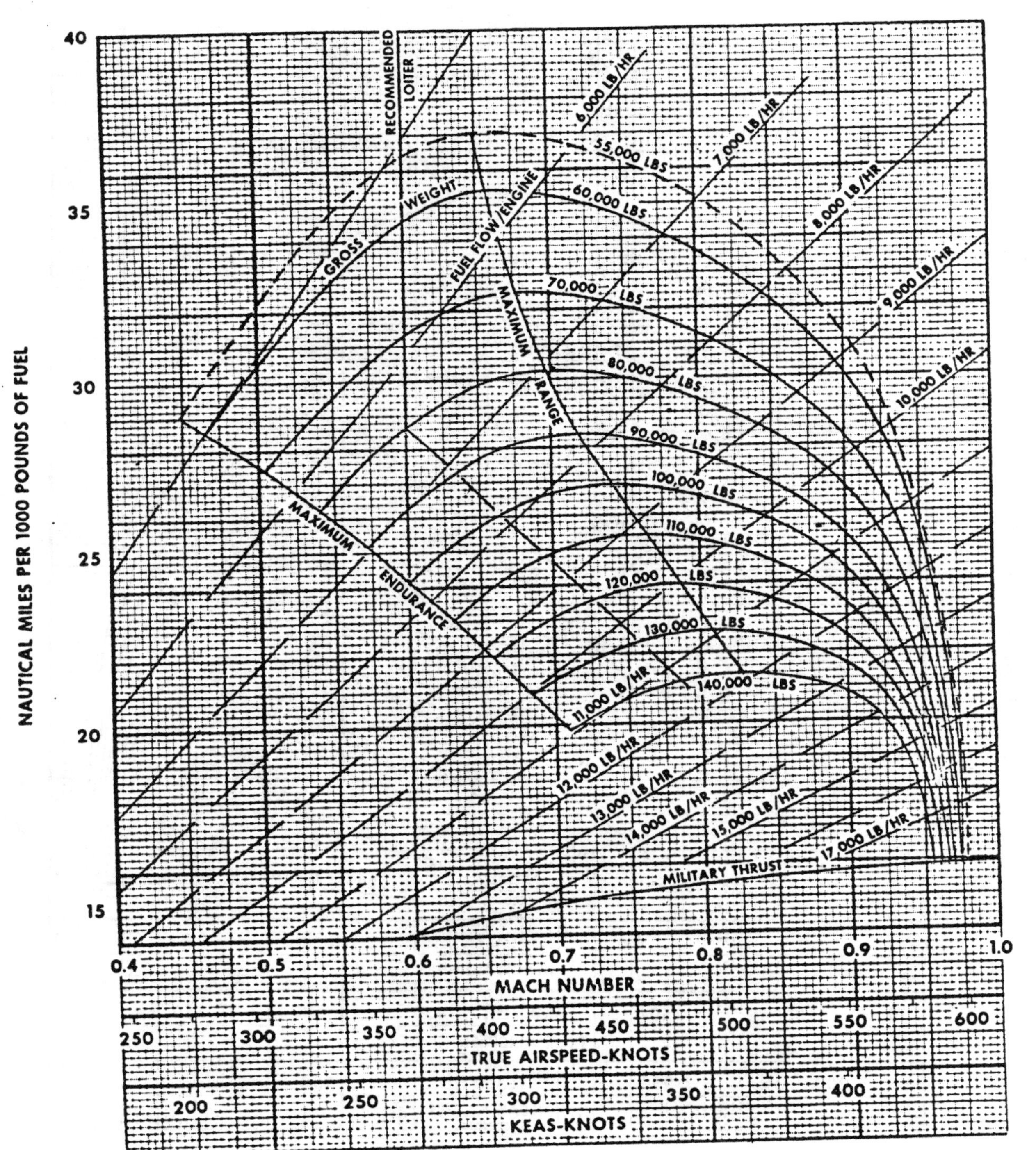
J ENGINES
SUBSONIC SPECIFIC RANGE
TWO ENGINES
20,000 FT ALTITUDE
STANDARD DAY 24% MAC C.G.
OITER: 275 KIAS With 20,000 Lbs Or Less
emaining. Increase KIAS 1.5 Knots Per
00 Lbs Remaining Above 20,000 Lbs.
Decrease Fuel Economy 1% For 1% Shift
Of C.G. Forward Of 24% MAC.
NAUTICAL MILES PER 1000 POUNDS OF FUEL
40
35
30
25
20
15
RECOMMENDED
LOITER
GROSS WEIGHT
FUEL FLOW/ENGINE
MAXIMUM RANGE
MAXIMUM ENDURANCE
6,000 LB/HR
7,000 LB/HR
8,000 LB/HR
9,000 LB/HR
10,000 LB/HR
11,000 LB/HR
12,000 LB/HR
13,000 LB/HR
14,000 LB/HR
15,000 LB/HR
17,000 LB/HR
MILITARY THRUST
55,000 LBS
60,000 LBS
70,000 LBS
80,000 LBS
90,000 LBS
100,000 LBS
110,000 LBS
120,000 LBS
130,000 LBS
140,000 LBS
0.4
0.5
0.6
0.7
0.8
0.9
1.0
MACH NUMBER
250
300
350
400
450
500
550
600
TRUE AIRSPEED-KNOTS
200
250
300
350
400
KEAS-KNOTS

Figure A4-3

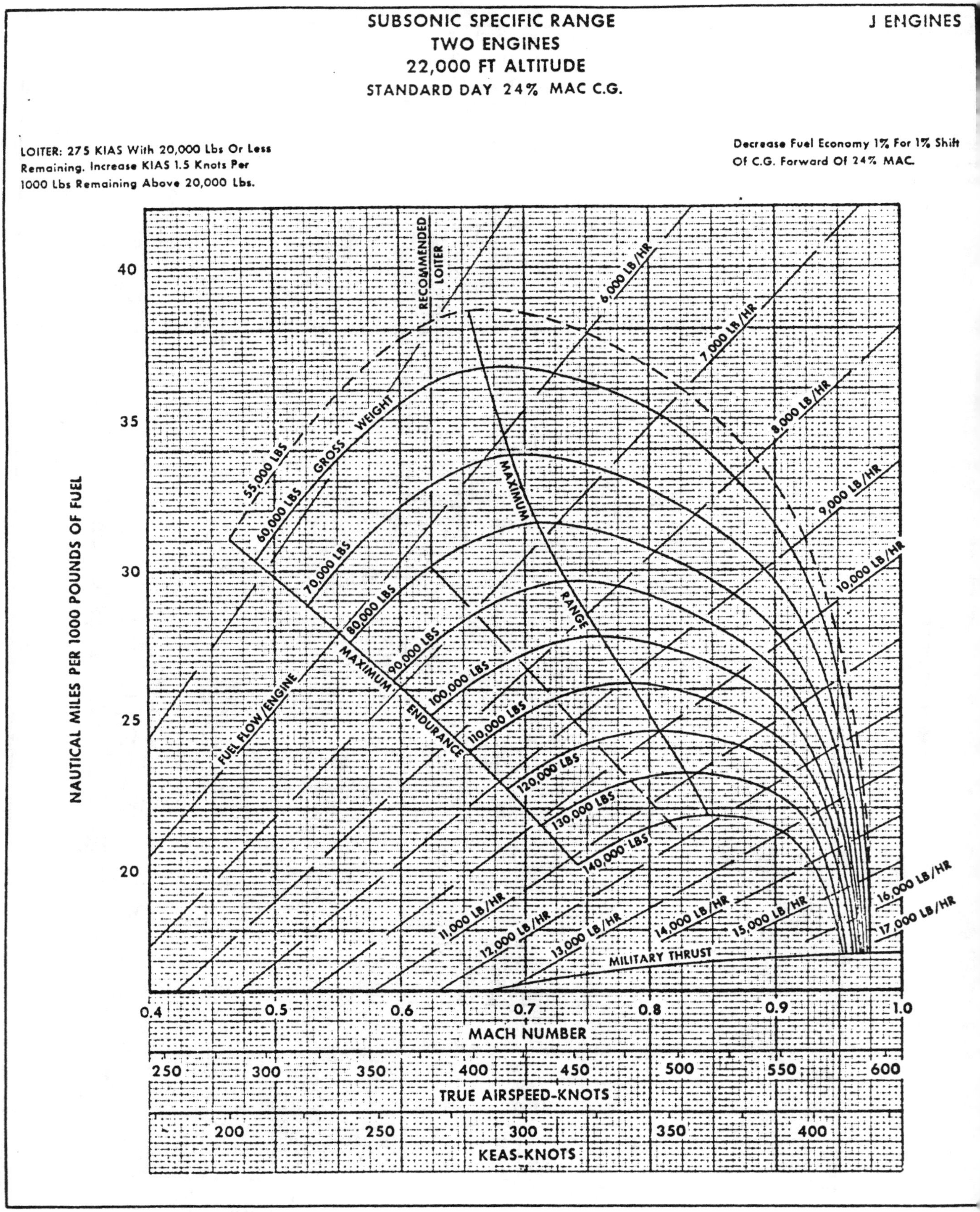

SUBSONIC SPECIFIC RANGE
TWO ENGINES
22,000 FT ALTITUDE
STANDARD DAY 24% MAC C.G.

J ENGINES

LOITER: 275 KIAS With 20,000 Lbs Or Less Remaining. Increase KIAS 1.5 Knots Per 1000 Lbs Remaining Above 20,000 Lbs.

Decrease Fuel Economy 1% For 1% Shift Of C.G. Forward Of 24% MAC.

NAUTICAL MILES PER 1000 POUNDS OF FUEL

40
35
30
25
20

RECOMMENDED LOITER

MAXIMUM RANGE

MAXIMUM ENDURANCE

GROSS WEIGHT

55,000 LBS
60,000 LBS
70,000 LBS
80,000 LBS
90,000 LBS
100,000 LBS
110,000 LBS
120,000 LBS
130,000 LBS
140,000 LBS

FUEL FLOW/ENGINE

6,000 LB/HR
7,000 LB/HR
8,000 LB/HR
9,000 LB/HR
10,000 LB/HR
11,000 LB/HR
12,000 LB/HR
13,000 LB/HR
14,000 LB/HR
15,000 LB/HR
16,000 LB/HR
17,000 LB/HR

MILITARY THRUST

0.4 0.5 0.6 0.7 0.8 0.9 1.0

MACH NUMBER

250 300 350 400 450 500 550 600

TRUE AIRSPEED-KNOTS

200 250 300 350 400

KEAS-KNOTS

Figure A4-4

SUBSONIC SPECIFIC RANGE
TWO ENGINES
24,000 FT ALTITUDE
STANDARD DAY 24% MAC C.G.

J ENGINES

LOITER: 275 KIAS With 20,000 Lbs Or Less Remaining. Increase KIAS 1.5 Knots Per 1000 Lbs Remaining Above 20,000 Lbs.

Decrease Fuel Economy 1% For 1% Shift Of C.G. Forward Of 24% MAC.

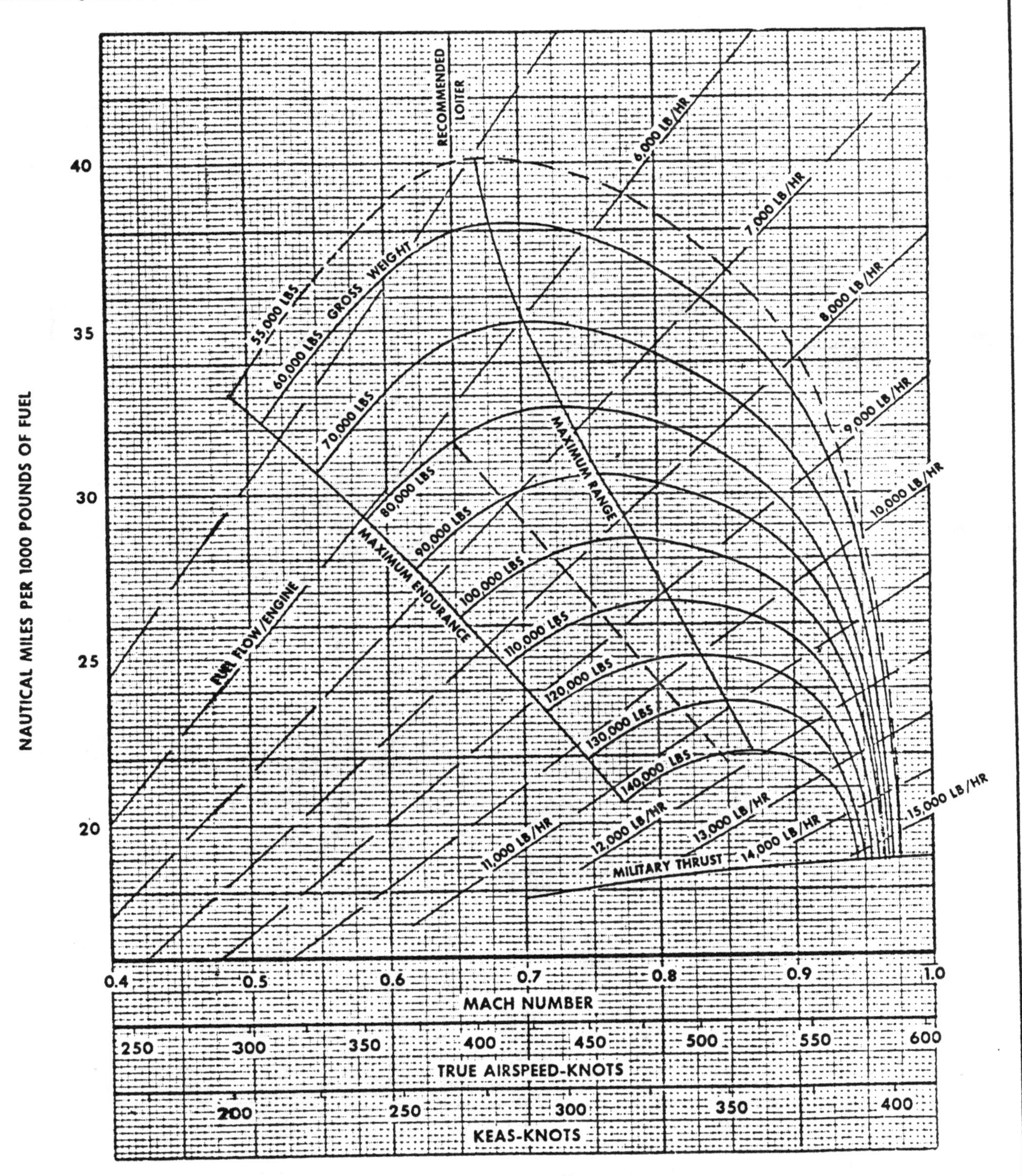

Figure A4-5

SUBSONIC SPECIFIC RANGE
TWO ENGINES
26,000 FT ALTITUDE
STANDARD DAY 24% MAC C.G.

J ENGINES

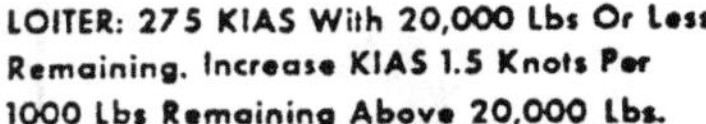

LOITER: 275 KIAS With 20,000 Lbs Or Less Remaining. Increase KIAS 1.5 Knots Per 1000 Lbs Remaining Above 20,000 Lbs.

Decrease Fuel Economy 1% For 1% Shift Of C.G. Forward Of 24% MAC.

⊕ Optimum Cruise Climb Condition.

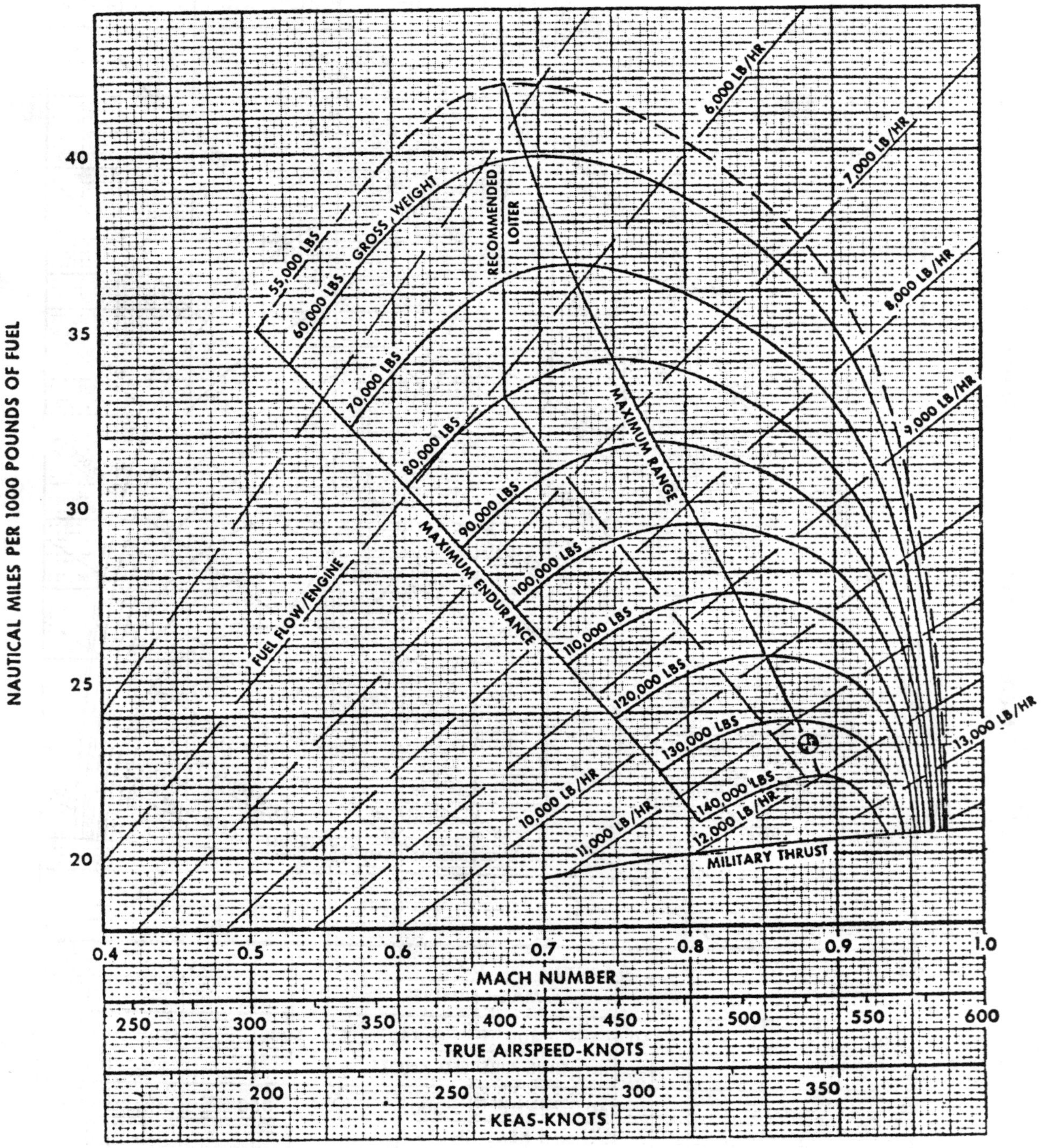

NAUTICAL MILES PER 1000 POUNDS OF FUEL

20
25
30
35
40

55,000 LBS GROSS WEIGHT
60,000 LBS
70,000 LBS
80,000 LBS
90,000 LBS
100,000 LBS
110,000 LBS
120,000 LBS
130,000 LBS
140,000 LBS

RECOMMENDED LOITER
MAXIMUM RANGE
MAXIMUM ENDURANCE
MILITARY THRUST

FUEL FLOW/ENGINE
6,000 LB/HR
7,000 LB/HR
8,000 LB/HR
9,000 LB/HR
10,000 LB/HR
11,000 LB/HR
12,000 LB/HR
13,000 LB/HR

0.4 0.5 0.6 0.7 0.8 0.9 1.0
MACH NUMBER

250 300 350 400 450 500 550 600
TRUE AIRSPEED-KNOTS

200 250 300 350
KEAS-KNOTS

Figure A4-6

J ENGINES

SUBSONIC SPECIFIC RANGE
TWO ENGINES
28,000 FT ALTITUDE
STANDARD DAY 24% MAC C.G.

LOITER: 275 KIAS With 20,000 Lbs Or Less Remaining. Increase KIAS 1.5 Knots Per 1000 Lbs Remaining Above 20,000 Lbs.

Decrease Fuel Economy 1% For 1% Shift Of C.G. Forward Of 24% MAC.

⊕ Optimum Cruise Climb Condition.

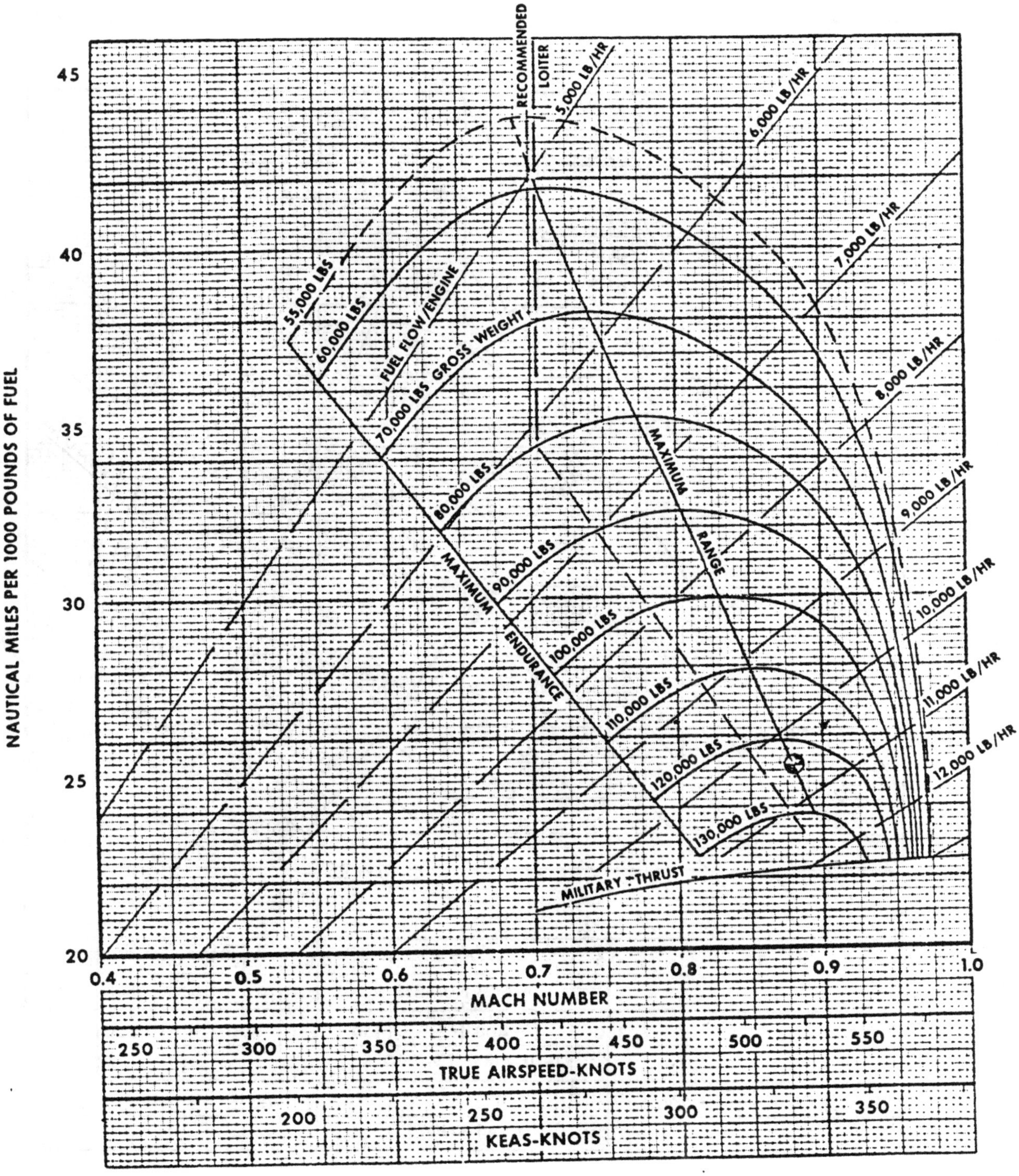

RECOMMENDED LOITER
5,000 LB/HR
6,000 LB/HR
7,000 LB/HR
8,000 LB/HR
9,000 LB/HR
10,000 LB/HR
11,000 LB/HR
12,000 LB/HR
FUEL FLOW/ENGINE
55,000 LBS
60,000 LBS
70,000 LBS GROSS WEIGHT
80,000 LBS
90,000 LBS
100,000 LBS
110,000 LBS
120,000 LBS
130,000 LBS
MAXIMUM RANGE
MAXIMUM ENDURANCE
MILITARY THRUST
NAUTICAL MILES PER 1000 POUNDS OF FUEL
20
25
30
35
40
45
0.4 0.5 0.6 0.7 0.8 0.9 1.0
MACH NUMBER
250 300 350 400 450 500 550
TRUE AIRSPEED-KNOTS
200 250 300 350
KEAS-KNOTS

Figure A4-7

SUBSONIC SPECIFIC RANGE
TWO ENGINES
30,000 FT ALTITUDE

STANDARD DAY 24% MAC C.G.

J ENGINES

LOITER: 275 KIAS With 20,000 Lbs Or Less Remaining. Increase KIAS 1.5 Knots Per 1000 Lbs Remaining Above 20,000 Lbs.

Decrease Fuel Economy 1% For 1% Shift Of C.G. Forward Of 24% MAC.

⊕ Optimum Cruise Climb Condition.

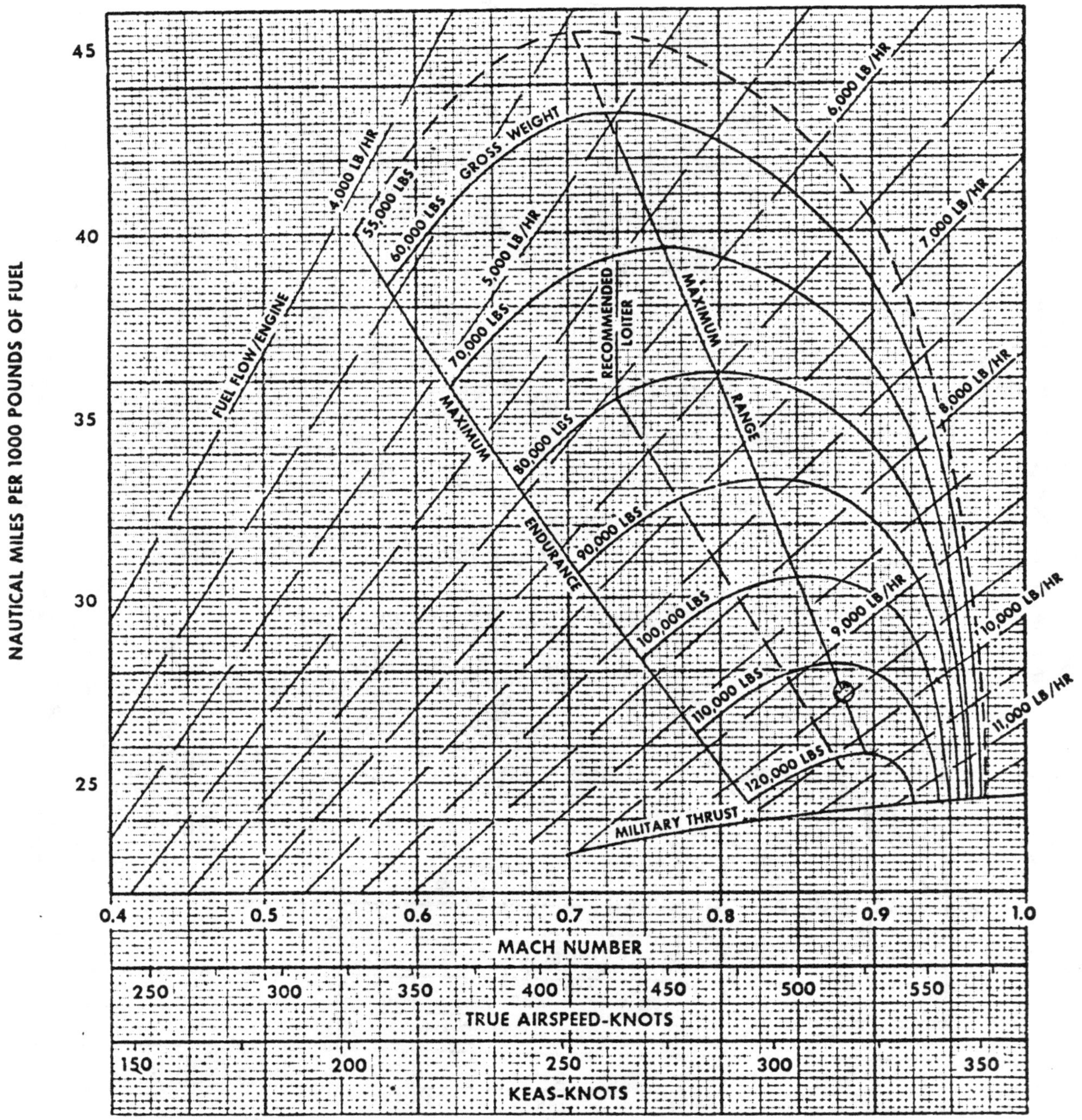

Figure A4-8

SUBSONIC SPECIFIC RANGE
TWO ENGINES
32,000 FT ALTITUDE
STANDARD DAY 24% MAC C.G.

J ENGINES

LOITER: 275 KIAS With 20,000 Lbs Or Less Remaining. Increase KIAS 1.5 Knots Per 1000 Lbs Remaining Above 20,000 Lbs.

Decrease Fuel Economy 1% For 1% Shift Of C.G. Forward Of 24% MAC.

⊕ Optimum Cruise Climb Condition.

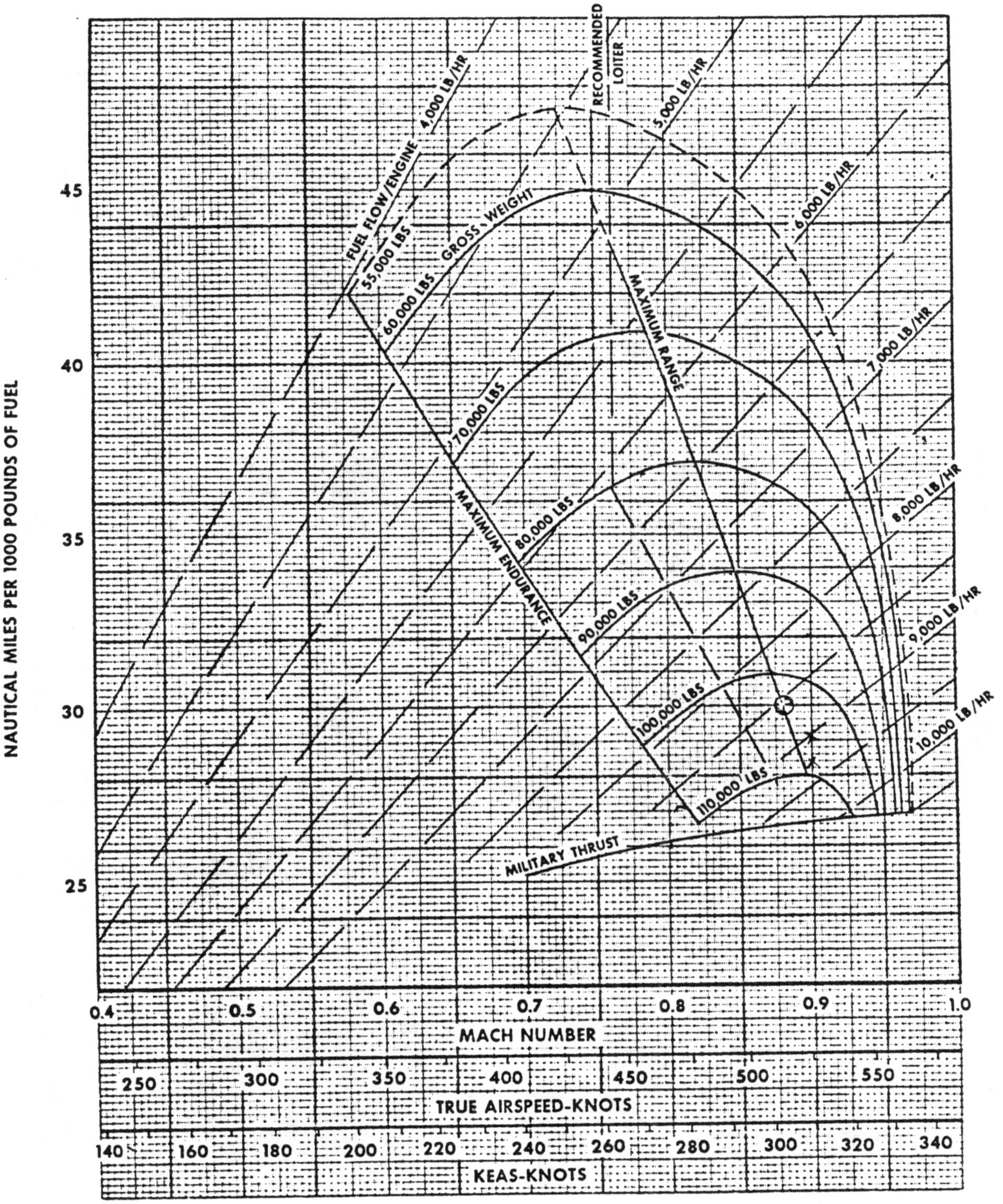

Figure A4-9

J ENGINES

SUBSONIC SPECIFIC RANGE
TWO ENGINES
34,000 FT ALTITUDE
STANDARD DAY 24% MAC C.G.

LOITER: 275 KIAS With 20,000 Lbs Or Less Remaining. Increase KIAS 1.5 Knots Per 1000 Lbs Remaining Above 20,000 Lbs.

Decrease Fuel Economy 1% For 1% Shift Of C.G. Forward Of 24% MAC.

⊕ Optimum Cruise Climb Condition.

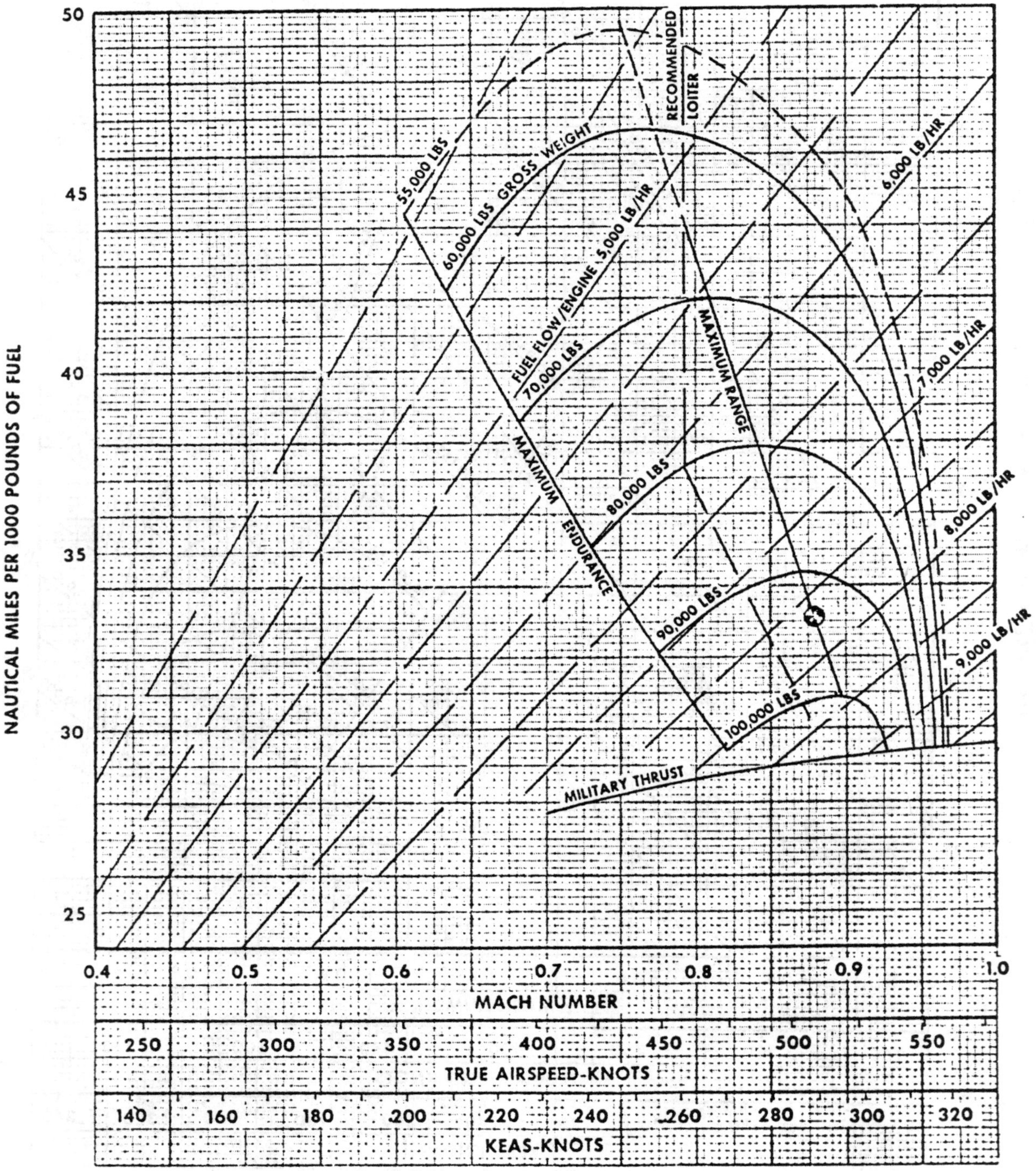

NAUTICAL MILES PER 1000 POUNDS OF FUEL

50
45
40
35
30
25

55,000 LBS
60,000 LBS GROSS WEIGHT
70,000 LBS
80,000 LBS
90,000 LBS
100,000 LBS
FUEL FLOW/ENGINE 5,000 LB/HR
6,000 LB/HR
7,000 LB/HR
8,000 LB/HR
9,000 LB/HR
RECOMMENDED LOITER
MAXIMUM RANGE
MAXIMUM ENDURANCE
MILITARY THRUST

0.4 0.5 0.6 0.7 0.8 0.9 1.0
MACH NUMBER

250 300 350 400 450 500 550
TRUE AIRSPEED-KNOTS

140 160 180 200 220 240 260 280 300 320
KEAS-KNOTS

Figure A4-10

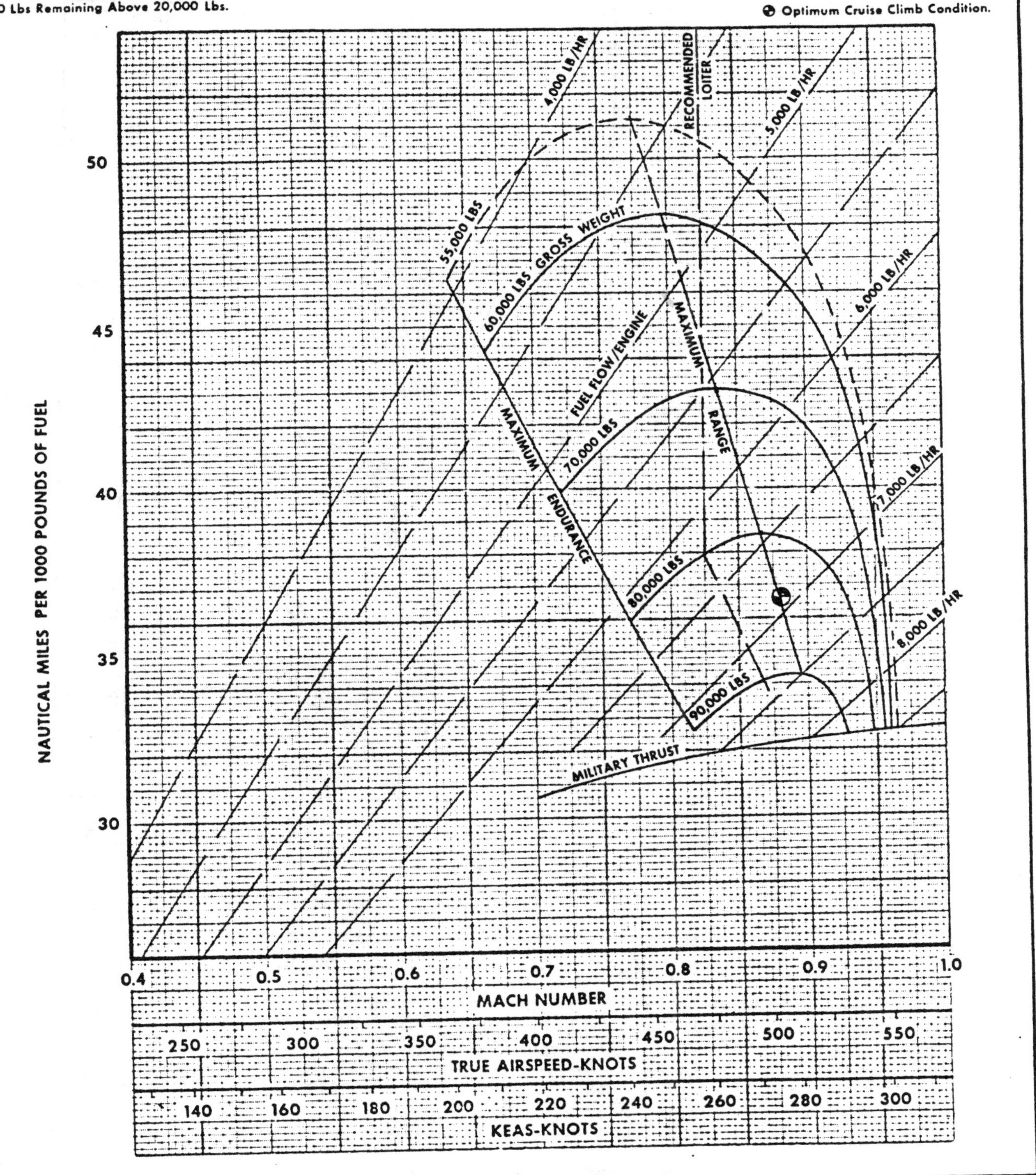
SUBSONIC SPECIFIC RANGE
TWO ENGINES
36,000 FT ALTITUDE
STANDARD DAY 24% MAC C.G.
J ENGINES
LOITER: 275 KIAS With 20,000 Lbs Or Less Remaining. Increase KIAS 1.5 Knots Per 1000 Lbs Remaining Above 20,000 Lbs.
Decrease Fuel Economy 1% For 1% Shift Of C.G. Forward Of 24% MAC.
Optimum Cruise Climb Condition.
NAUTICAL MILES PER 1000 POUNDS OF FUEL
50
45
40
35
30
4,000 LB/HR
5,000 LB/HR
6,000 LB/HR
7,000 LB/HR
8,000 LB/HR
RECOMMENDED LOITER
55,000 LBS
60,000 LBS GROSS WEIGHT
70,000 LBS
80,000 LBS
90,000 LBS
FUEL FLOW/ENGINE
MAXIMUM RANGE
MAXIMUM ENDURANCE
MILITARY THRUST
0.4
0.5
0.6
0.7
0.8
0.9
1.0
MACH NUMBER
250
300
350
400
450
500
550
TRUE AIRSPEED-KNOTS
140
160
180
200
220
240
260
280
300
KEAS-KNOTS

Figure A4-11

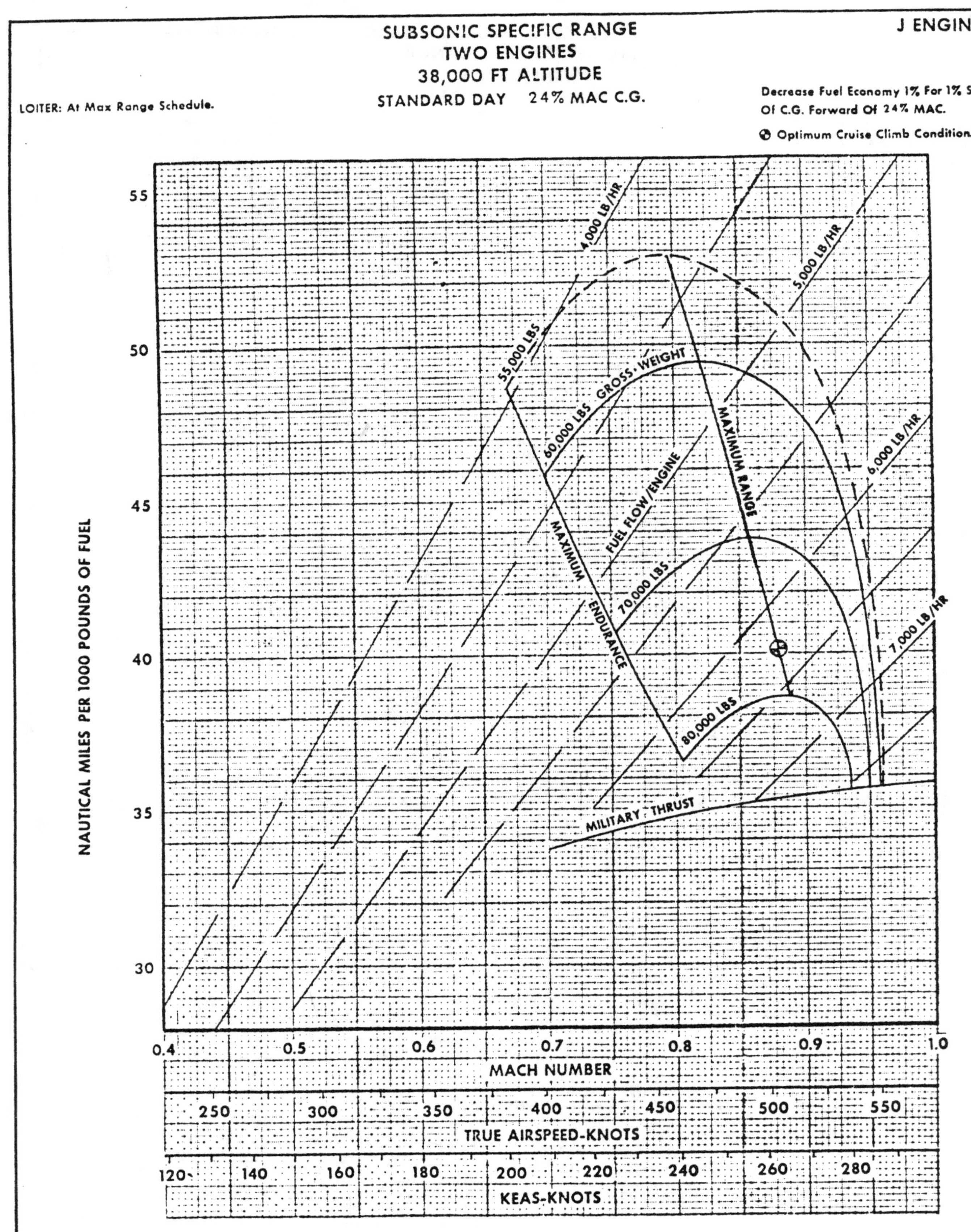
SUBSONIC SPECIFIC RANGE
TWO ENGINES
38,000 FT ALTITUDE
STANDARD DAY 24% MAC C.G.
J ENGINES
LOITER: At Max Range Schedule.
Decrease Fuel Economy 1% For 1% Shift Of C.G. Forward Of 24% MAC.
Optimum Cruise Climb Condition.
NAUTICAL MILES PER 1000 POUNDS OF FUEL
55
50
45
40
35
30
4,000 LB/HR
5,000 LB/HR
6,000 LB/HR
7,000 LB/HR
55,000 LBS
60,000 LBS GROSS WEIGHT
70,000 LBS
80,000 LBS
FUEL FLOW/ENGINE
MAXIMUM RANGE
MAXIMUM ENDURANCE
MILITARY THRUST
0.4
0.5
0.6
0.7
0.8
0.9
1.0
MACH NUMBER
250
300
350
400
450
500
550
TRUE AIRSPEED-KNOTS
120
140
160
180
200
220
240
260
280
KEAS-KNOTS

Figure A4-12

SUBSONIC SPECIFIC RANGE
TWO ENGINES
40,000 FT ALTITUDE
STANDARD DAY 24% MAC C.G.

J ENGINES

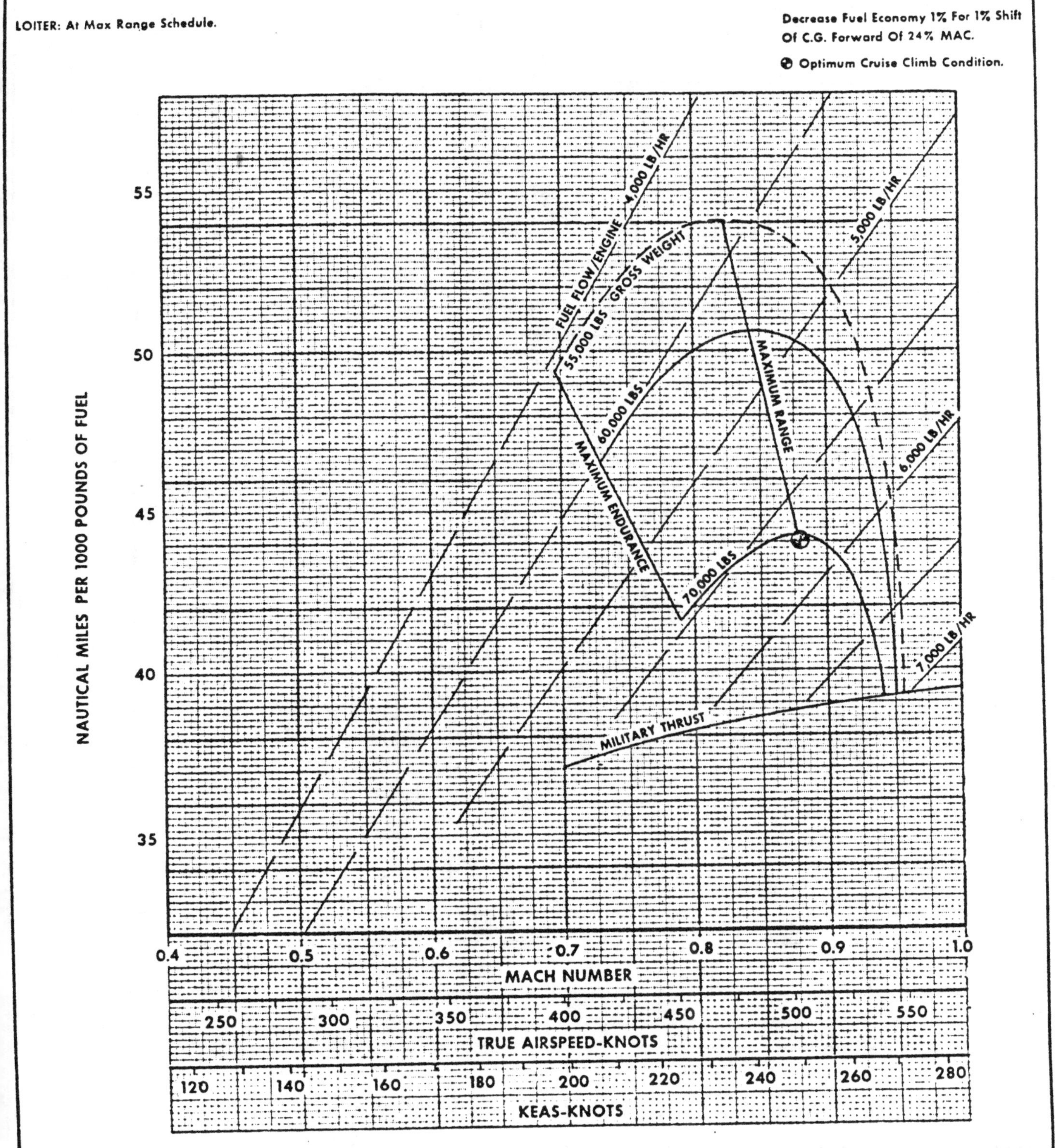

LOITER: At Max Range Schedule.

Decrease Fuel Economy 1% For 1% Shift Of C.G. Forward Of 24% MAC.

⊕ Optimum Cruise Climb Condition.

FUEL FLOW/ENGINE 4,000 LB/HR
5,000 LB/HR
6,000 LB/HR
7,000 LB/HR
55,000 LBS GROSS WEIGHT
60,000 LBS
70,000 LBS
MAXIMUM RANGE
MAXIMUM ENDURANCE
MILITARY THRUST

NAUTICAL MILES PER 1000 POUNDS OF FUEL
55
50
45
40
35

0.4 0.5 0.6 0.7 0.8 0.9 1.0
MACH NUMBER

250 300 350 400 450 500 550
TRUE AIRSPEED-KNOTS

120 140 160 180 200 220 240 260 280
KEAS-KNOTS

Figure A4-13

SUBSONIC MAXIMUM RANGE -CRUISE CLIMB

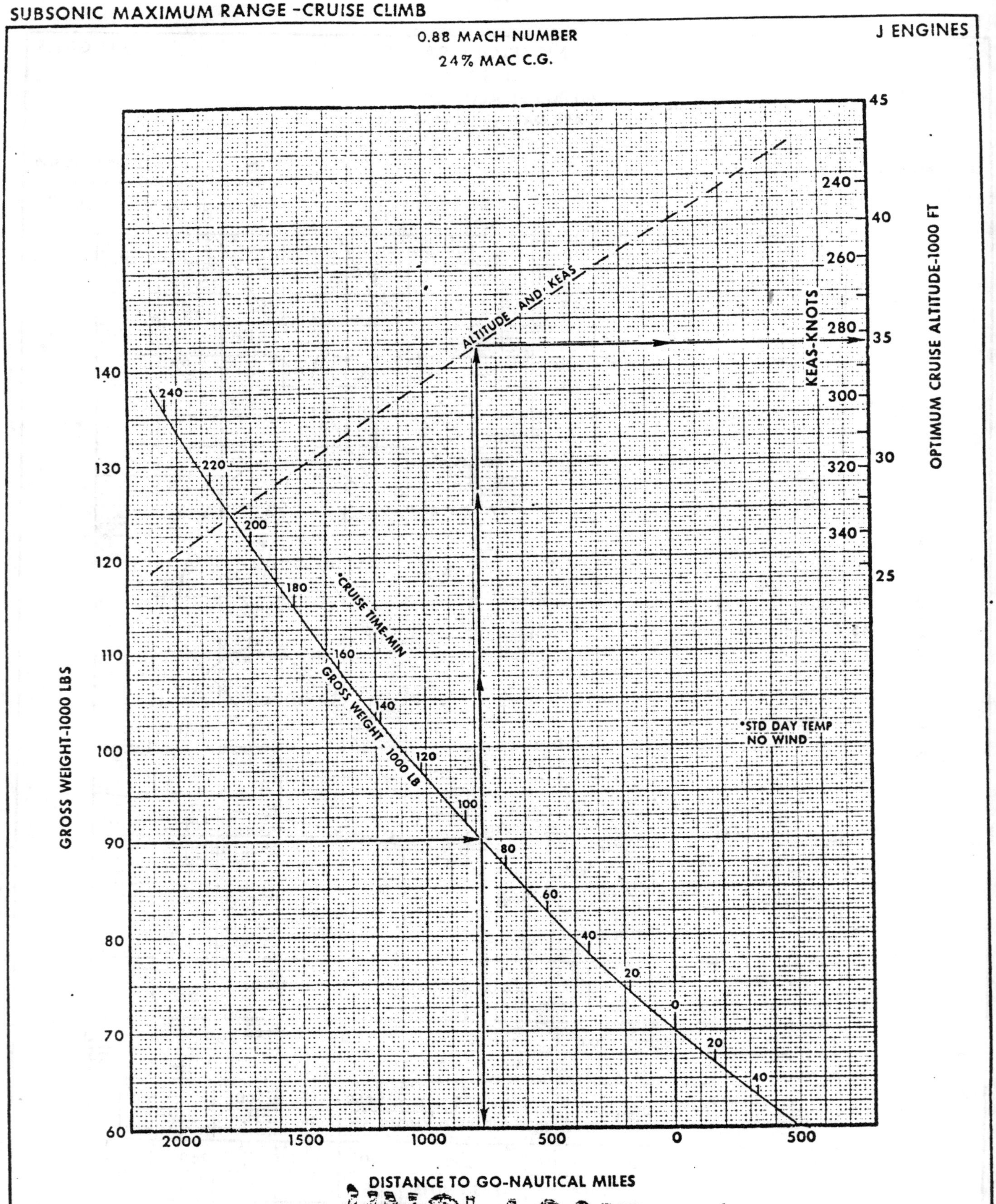

J ENGINES

0.88 MACH NUMBER
24% MAC C.G.

Figure A4-14

SUBSONIC RANGE FACTOR

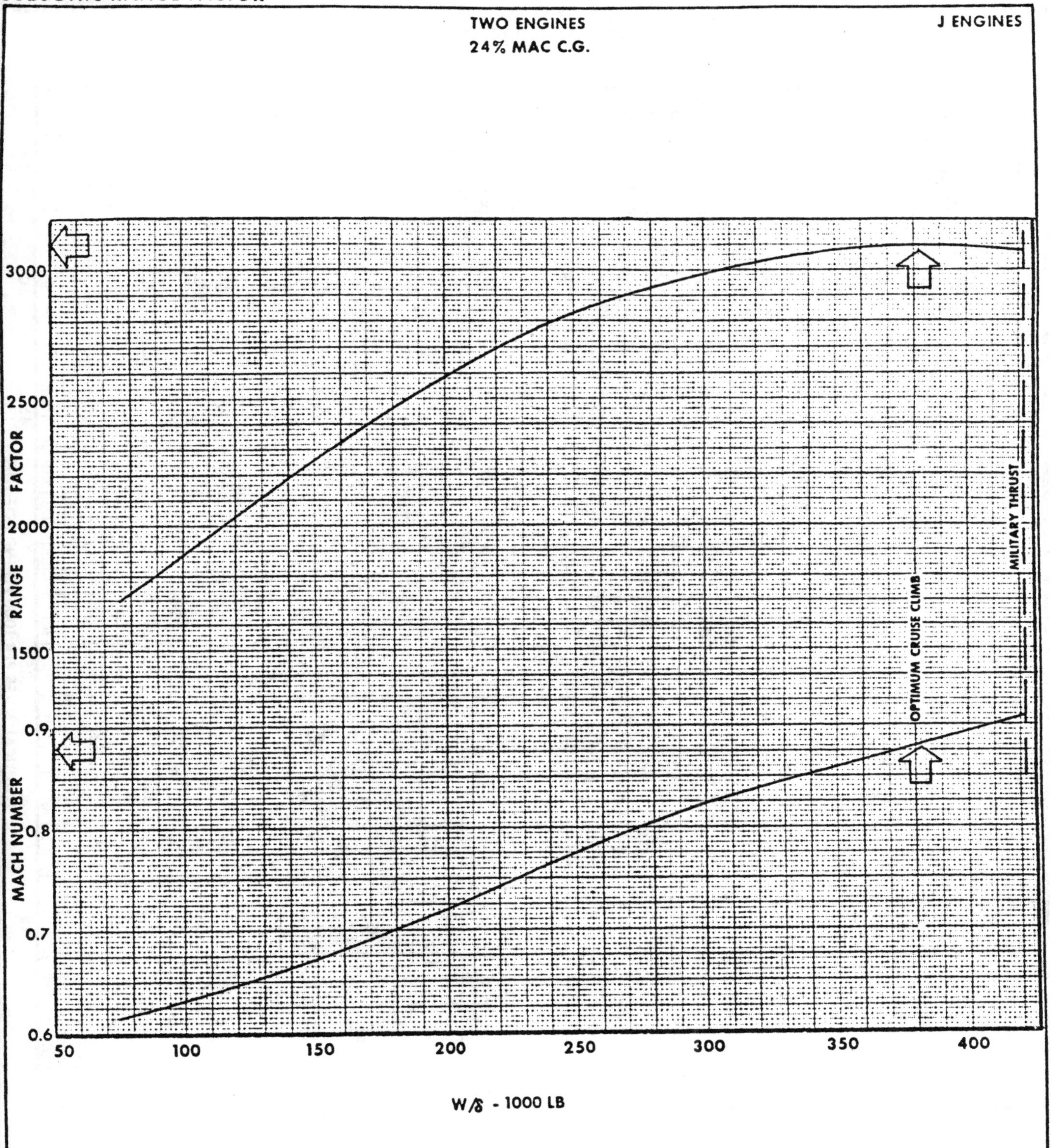

TWO ENGINES
24% MAC C.G.

J ENGINES

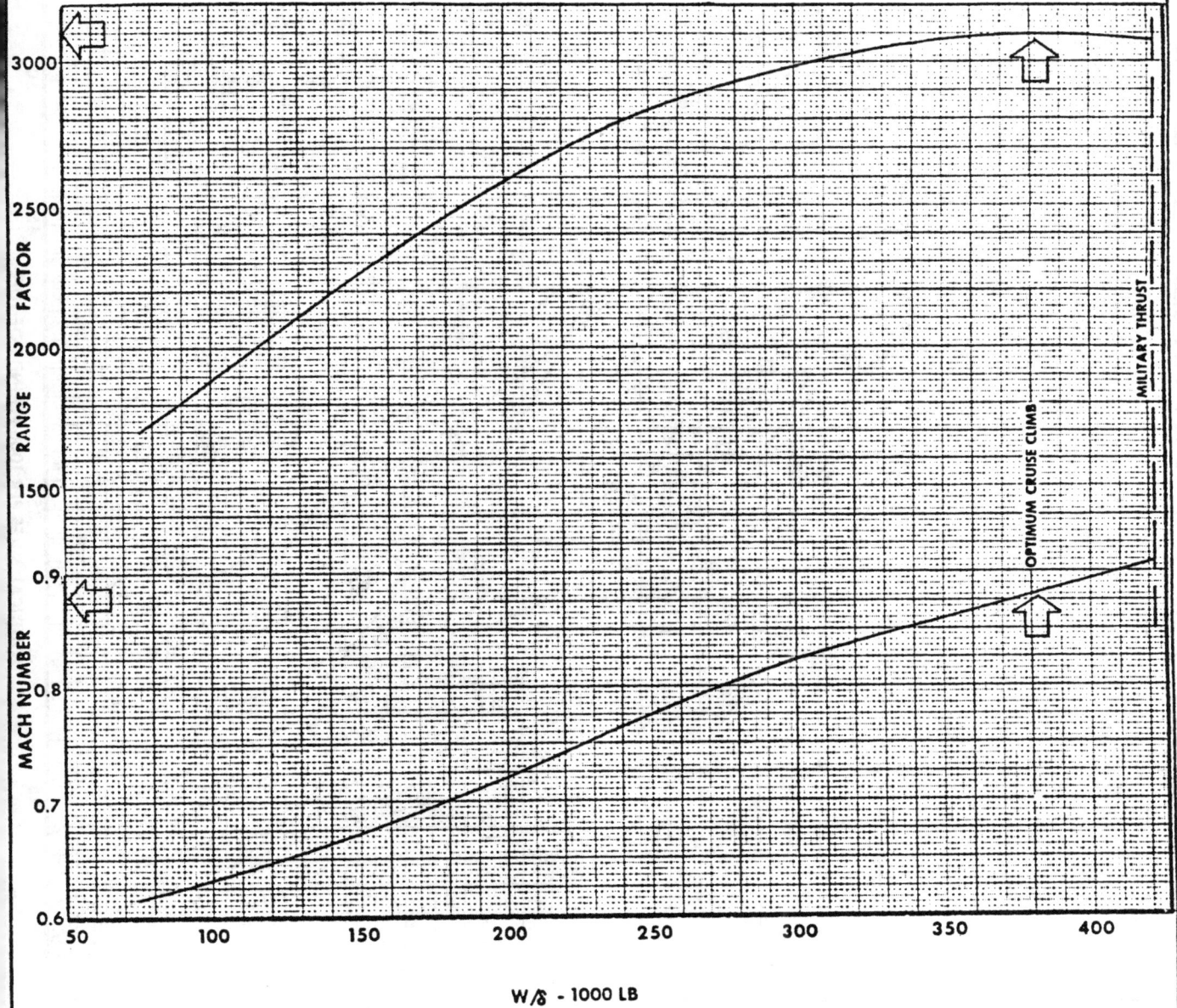

RANGE FACTOR: 1500, 2000, 2500, 3000

MACH NUMBER: 0.6, 0.7, 0.8, 0.9

OPTIMUM CRUISE CLIMB

MILITARY THRUST

W/δ - 1000 LB: 50, 100, 150, 200, 250, 300, 350, 400

Figure A4-15

UNCLASSIFIED

MAXIMUM SPECIFIC RANGE

TWO ENGINES
24% MAC C.G.

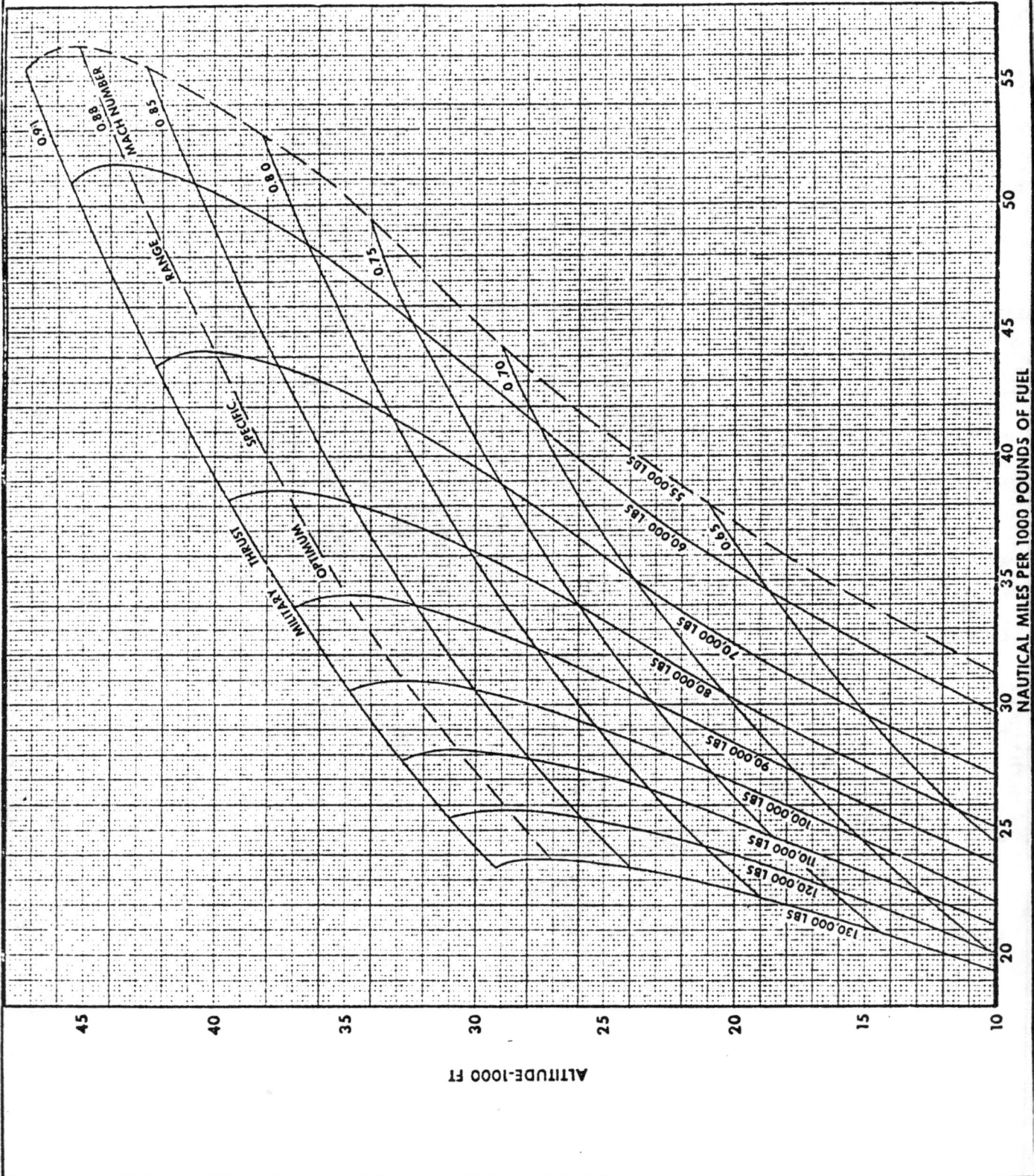

J ENGINES

OPTIMUM SPECIFIC RANGE

MILITARY THRUST

MACH NUMBER 0.91, 0.88, 0.85, 0.80, 0.75, 0.70, 0.65

55,000 LBS
60,000 LBS
70,000 LBS
80,000 LBS
90,000 LBS
100,000 LBS
110,000 LBS
120,000 LBS
130,000 LBS

NAUTICAL MILES PER 1000 POUNDS OF FUEL: 20, 25, 30, 35, 40, 45, 50, 55

ALTITUDE-1000 FT: 45, 40, 35, 30, 25, 20, 15, 10

Figure A4-16

CONSTANT MACH CRUISE

TWO ENGINES
0.75 MACH NUMBER
24% MAC C.G.

BUDDY MISSION

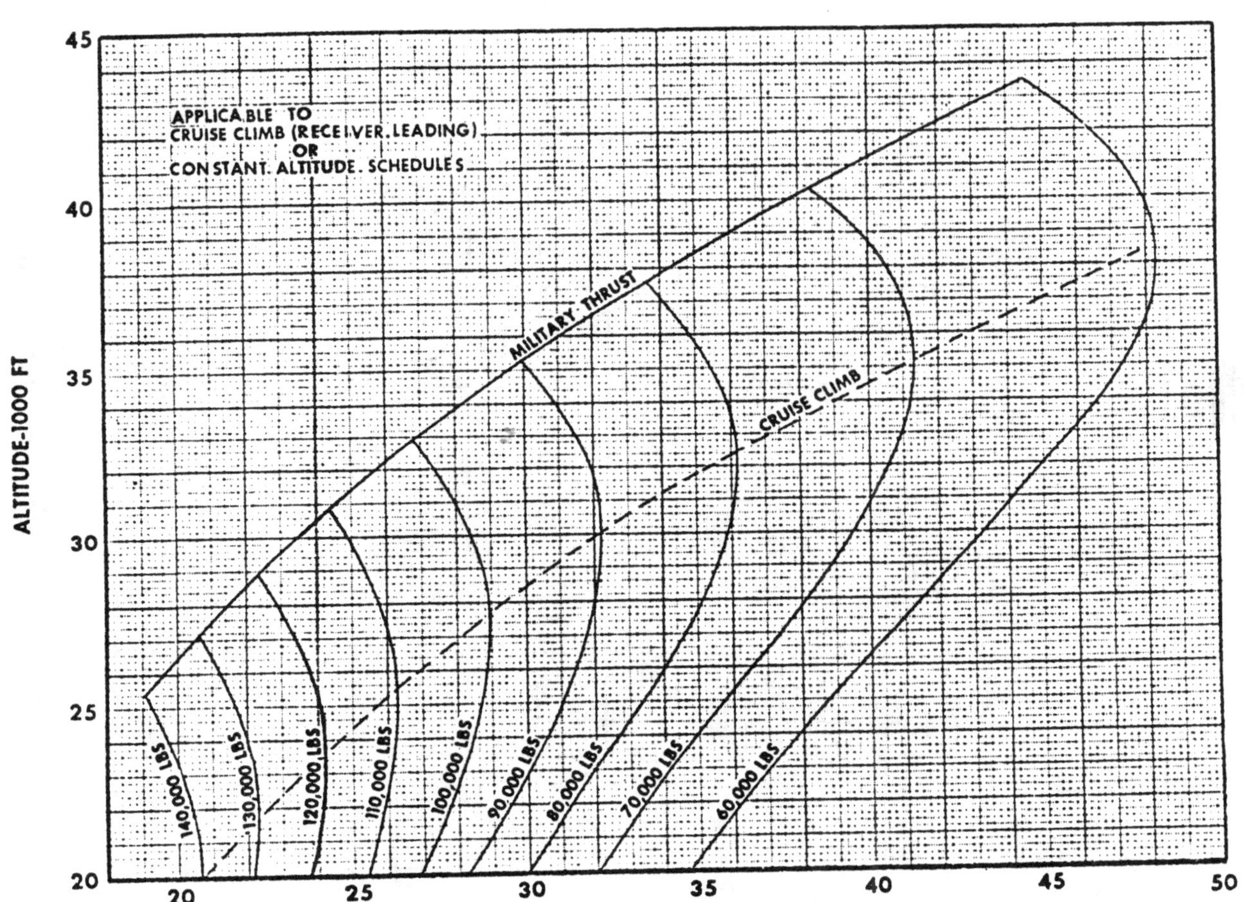

Figure A4-17

LOITER PERFORMANCE

LOITER: 275 KIAS With 20,000 Lbs Or Less Remaining. Increase KIAS 1.5 Knots Per 1000 Lbs Remaining Above 20,000 Lbs.

TWO ENGINES
24% MAC C.G.

J ENGINES

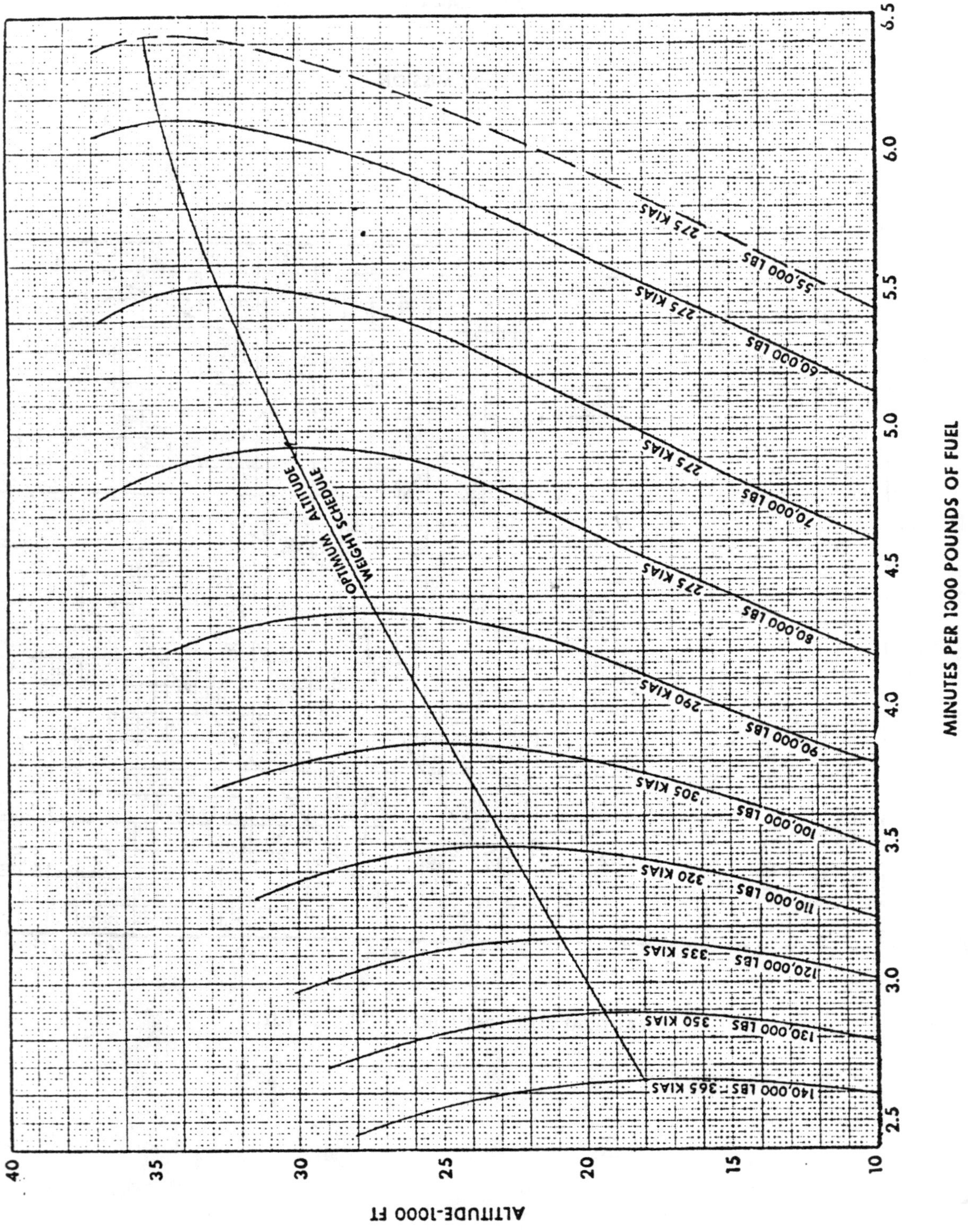

55,000 LBS 275 KIAS
60,000 LBS 275 KIAS
70,000 LBS 275 KIAS
80,000 LBS 275 KIAS
90,000 LBS 290 KIAS
100,000 LBS 305 KIAS
110,000 LBS 320 KIAS
120,000 LBS 335 KIAS
130,000 LBS 350 KIAS
140,000 LBS 365 KIAS

OPTIMUM ALTITUDE WEIGHT SCHEDULE

MINUTES PER 1000 POUNDS OF FUEL

ALTITUDE-1000 FT

Figure A4-18.

Refuel Light Rheostat

A refuel light rheostat installed on the pilot's lower instrument panel may be rotated from OFF to BRT as desired to vary the intensity of a refueling light located in the refueling receptacle.

INTERIOR LIGHTING

The cockpit lighting system includes instrument lights, console panel lights, console floodlights, thunderstorm lights, utility spotlights, circuit breakers and controls. The instruments and consoles are illuminated by integral and edge lighting. Utility spotlights are mounted on the left console in both cockpits. These spotlights are detachable and may be moved about the cockpit. White light may be focused by rotating the lens. Thunderstorm lights are located on the left and right sides of the pilot's canopy and direct light forward onto the instrument panel to avoid blindness caused by lightning.

Cockpit Lighting Switches

A rheostat on the aft end of each spotlight is used to vary light intensity. A pushbutton switch on each light may be used to bypass the rheostat and obtain maximum spotlight brilliance instantaneously. The instrument panel lights, console panel lights and floodlights are controlled by three rheostats located on the pilot's lower instrument panel and the FCO's left console. These rheostats, labeled INSTR, CONSOLE, and FLOOD may be rotated clockwise from OFF to BRT as desired to vary the intensity of the associated lights. The thunderstorm lights are controlled by a two-position switch, on the pilot's lower instrument panel, labeled ON and OFF.

ENVIRONMENTAL CONTROL SYSTEM

The environmental control system consists of air-conditioning and liquid cooling subsystems.

AIR-CONDITIONING SUBSYSTEM

Two complete and independent air-cycle refrigeration systems are provided which use ninth-stage compressor bleed air from each engine to pressurize and cool the pilot and FCO cockpits and ventilated flying suits, electronics bay, and missile bays. (See figure 1-21.) Each system is provided with ram air and fuel-air heat exchangers for primary bleed air cooling. The air temperature of each system is further lowered by fuel-cooled, air-cycle refrigerators before it is used. Cooling fuel is obtained from the engine supply manifolds and is circulated by pumps driven by the accessory drive gearboxes. After the fuel is heated it is returned to temperature sensitive valves and then either used by the engines, or diverted to tank 6. A hot-air bypass manifold, connecting the left and right systems upstream of the refrigerators, provides pressure-regulated air for cockpit heating, windshield defogging, and ASG-18 FCS air-cooled rack temperature control.

ENVIRONMENTAL CONTROL SYSTEM
(Air cycle subsystem)

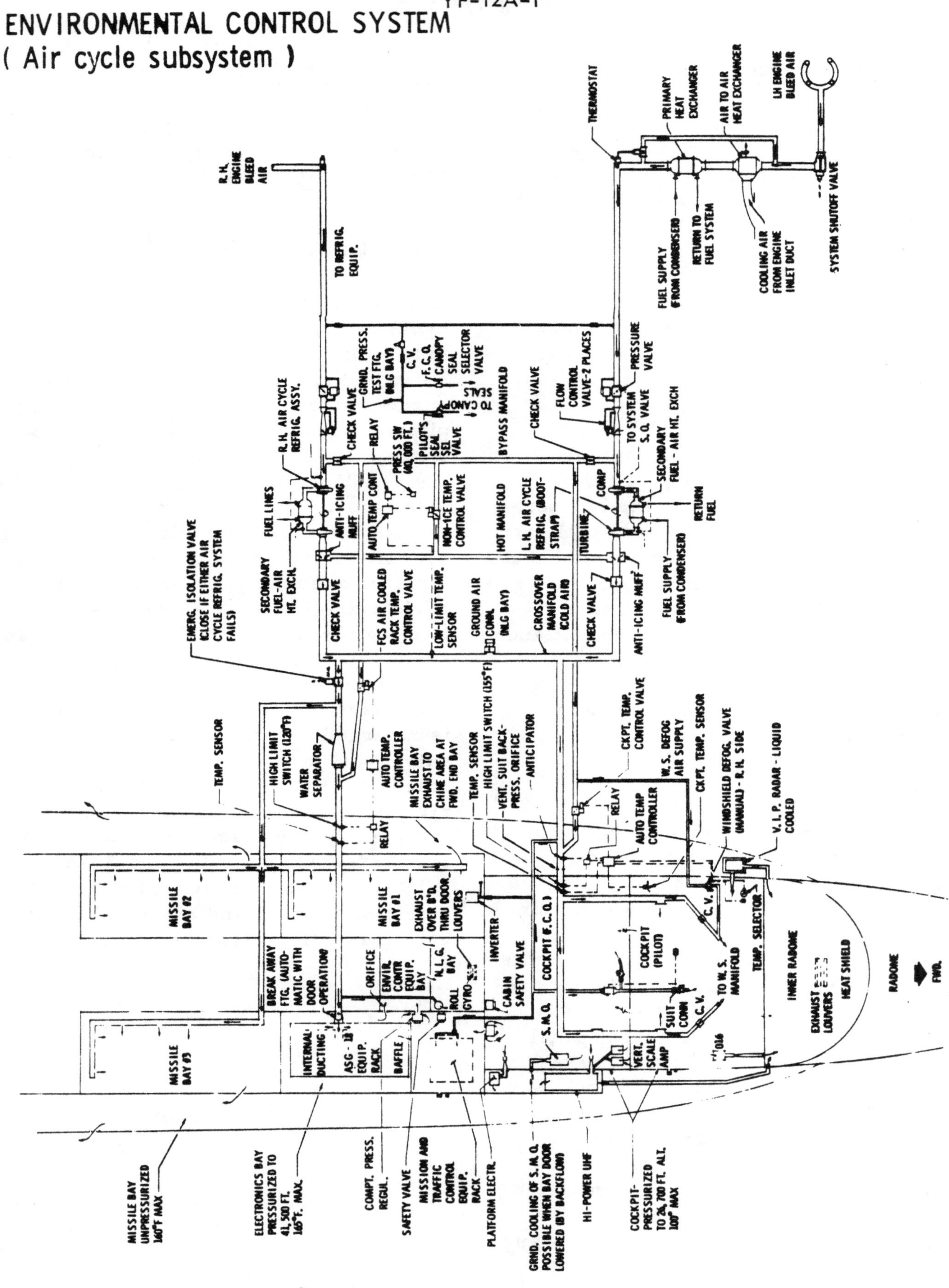

Figure 1-21

www.ingramcontent.com/pod-product-compliance
Lightning Source LLC
Chambersburg PA
CBHW080137220326
41598CB00032B/5093
* 9 7 8 1 9 3 1 6 4 1 6 3 0 *